Nanostructured Catalysts
Selective Oxidations

RSC Nanoscience & Nanotechnology

Series Editors:
Professor Paul O'Brien, *University of Manchester, UK*
Professor Sir Harry Kroto FRS, *University of Sussex, UK*
Professor Harold Craighead, *Cornell University, USA*

Titles in the Series:

1: Nanotubes and Nanowires
2: Fullerenes: Principles and Applications
3: Nanocharacterisation
4: Atom Resolved Surface Reactions: Nanocatalysis
5: Biomimetic Nanoceramics in Clinical Use: From Materials to Applications
6: Nanofluidics: Nanoscience and Nanotechnology
7: Bionanodesign: Following Nature's Touch
8: Nano-Society: Pushing the Boundaries of Technology
9: Polymer-based Nanostructures: Medical Applications
10: Metallic and Molecular Interactions in Nanometer Layers, Pores and Particles: New Findings at the Yoctolitre Level
11: Nanocasting: A Versatile Strategy for Creating Nanostructured Porous Materials
12: Titanate and Titania Nanotubes: Synthesis, Properties and Applications
13: Raman Spectroscopy, Fullerenes and Nanotechnology
14: Nanotechnologies in Food
15: Unravelling Single Cell Genomics: Micro and Nanotools
16: Polymer Nanocomposites by Emulsion and Suspension
17: Phage Nanobiotechnology
18: Nanotubes and Nanowires: 2nd Edition
19: Nanostructured Catalysts: Selective Oxidations

How to obtain future titles on publication:
A standing order plan is available for this series. A standing order will bring delivery of each new volume immediately on publication.

For further information please contact:
Book Sales Department, Royal Society of Chemistry, Thomas Graham House, Science Park, Milton Road, Cambridge, CB4 0WF, UK
Telephone: +44 (0)1223 420066, Fax: +44 (0)1223 420247, Email: books@rsc.org
Visit our website at http://www.rsc.org/Shop/Books/

Nanostructured Catalysts
Selective Oxidations

Edited by

Christian Hess
Technische Universität Darmstadt, Eduard-Zintl-Institut für Anorganische und Physikalische Chemie, Petersenstr. 20, 64287 Darmstadt, Germany

and

Robert Schlögl
Fritz-Haber-Institut der Max-Planck-Gesellschaft, Abteilung Anorganische Chemie, Faradayweg 4–6, 14195 Berlin, Germany

RSC Publishing

RSC Nanoscience & Nanotechnology No. 19

ISBN: 978-0-85404-186-2
ISSN: 1757-7136

A catalogue record for this book is available from the British Library

Published by The Royal Society of Chemistry,
Thomas Graham House, Science Park, Milton Road,
Cambridge CB4 0WF, UK

Registered Charity Number 207890

For further information see our web site at www.rsc.org

Preface

Heterogeneously catalyzed selective oxidation constitutes an important area of chemical industry. A major application is the functionalization of hydrocarbons leading to monomers or modifiers used for synthesis of polymers. Within the area of selective catalytic oxidation the optimization of selectivity is the most important goal. In fact, increasing sustainability by preventing the formation of waste is of enormous relevance in large-scale processes. However, despite extensive research over the last decades heterogeneously catalyzed selective oxidation reactions are still poorly understood due to the complex interplay of processes in different scales of time and length and the multitude of disciplines involved.

This book reviews recent developments in heterogeneously catalyzed selective oxidation reactions with an emphasis on light hydrocarbons. Nanostructuring is an important aspect of these catalysts. In fact, rational catalyst design will heavily depend on the ability to control the catalysts structure and particle size on the nanometer scale, strongly linking heterogeneous catalysis with materials science. While a broad range of catalyst materials is covered, a particular focus is placed on vanadium-containing catalysts because of their importance for selective oxidation catalysis. To this end the effort is made to merge the current state of knowledge into a consistent picture of the working principle of selective oxidation catalysts.

This book gives a comprehensive up-to-date summary of the existing information on heterogeneously catalyzed selective oxidation. Nonetheless, it is not intended for specialists only. Graduate students and researchers both in academia and industry from different areas of catalysis and different background will receive an overview of the field, ranging from historical developments to recent results, which should allow them to quickly catch up with current research issues.

The book is organized as follows. Chapter 1 gives a short introduction to the field of selective oxidation and discusses some basic aspects of the reactions and

RSC Nanoscience & Nanotechnology No. 19
Nanostructured Catalysts: Selective Oxidations
Edited by Christian Hess and Robert Schlögl

Published by the Royal Society of Chemistry, www.rsc.org

materials the book focuses on. Chapter 2 deals with the initial activation step of C–H bonds in selective oxidation reactions on solid oxide catalysts. The first section (Chapters 3–7) is dedicated to oxide-catalyzed selective oxidation reactions of light hydrocarbons with a focus on vanadium and/or molybdenum containing catalysts. The next series of chapters (Chapters 8–11) covers selective oxidation reactions over supported gold, (supported) silver, ruthenium oxide and (potassium-promoted) iron-oxide catalysts. The next section (Chapters 12–14) addresses recent developments regarding supported vanadium-oxide systems as well as mechanistic aspects of ethane/propane oxidative functionalization. In Chapter 15 the characteristics of selective oxidation reactions are discussed focusing on active sites of the V_xO_y type. Chapter 16 addresses engineering aspects of selective oxidation reactions with reference to industrial processes. Chapter 17 gives a short outlook.

We hope that this book, while not exhaustive, will give the reader an idea of the fascinating field of selective oxidation catalysis. We feel that the field is moving towards a profound scientific foundation that will offer plenty of opportunities for better catalyst developments. We thank all contributors to this book and we hope that you enjoy reading it.

Darmstadt and Berlin
Christian Hess
Robert Schlögl

Contents

RSC Nanoscience & Nanotechnology No. 19
Nanostructured Catalysts: Selective Oxidations
Edited by Christian Hess and Robert Schlögl

Published by the Royal Society of Chemistry, www.rsc.org

Section II Other Selective Oxidations (Chapters 8–11)

Section III Supported Vanadium-Oxide Systems and Mechanistic Studies (Chapters 12–14)

CHAPTER 1

Introduction

CHRISTIAN HESS

Technische Universität Darmstadt, Eduard-Zintl-Institut für Anorganische und Physikalische Chemie, Petersenstr. 20, 64287 Darmstadt, Germany

The production of organic chemicals *via* heterogeneously catalyzed *selective oxidations* is one of the most important segments in modern chemical industry. Important products include acrylic acid, acrylonitrile, ethylene oxide, formaldehyde, maleic anhydride, methacrylic acid, and phthalic anhydride.[1] Table 1.1 lists major heterogeneously catalyzed selective oxidation processes discussed within the scope of this book. Due to the high level of empirical development of these processes, further improvements represent a tremendous challenge, which will largely benefit from a mechanistic understanding of selective oxidation catalysts. However, despite extensive research activities over the last decades, still very little is known about the mode of operation of selective oxidation reactions on an atomistic level.[1–3]

A working catalyst requires an interplay of processes over multiple length- and timescales. With respect to time these range from elementary steps such as the breaking of bonds in the substrate and active site (~100 fs) to transport phenomena as well as solid-state transformations of the catalyst (up to years). Simultaneously, in the course of these processes lengthscales from sub-nanometers up to meters are covered.

An important aspect of the rational development of more efficient selective oxidation processes is the ability to control the catalyst structure and particle size on the nanometer scale, strongly linking research in heterogeneous catalysis with material science. Such *nanostructured catalysts* are naturally divided into supported and bulk systems. In general, supported catalysts consist of an oxide

RSC Nanoscience & Nanotechnology No. 19
Nanostructured Catalysts: Selective Oxidations
Edited by Christian Hess and Robert Schlögl

Published by the Royal Society of Chemistry, www.rsc.org

Table 1.1 Heterogeneously catalyzed selective oxidation processes.

Substrate (+air, unless mentioned otherwise)	*Product(s)*	*Catalyst (main components)*
Methane*	Ethylene, methanol, formaldehyde	(non)reducible MO, halogen-containing oxides, SE, TMO
Methanol	Formaldehyde	Ag, FeMoO
Ethylene ($+O_2$)	Ethylene oxide	Ag/Al_2O_3
Propane*	Acrylic acid	MoVTeNbO
Propylene	Acrolein	MCM
Propylene/NH_3	Acrylonitrile	MCM
Propylene ($/H_2$)*	Propylene oxide	Supported Au
Acrolein	Acrylic acid	MoVWO
n-Butane	Maleic anhydride	VPO
Isobutane*	Methacrolein/ Methacrylic acid	HPC
Methacrolein	Methacrylic acid	HPC
o-Xylene/Naphthalene	Phtalic anhydride	V_xO_y/TiO_2
Ethylbenzene (+steam)**	Styrene	KFeO

The marked processes (*) are of interest regarding commercialization. HPC = heteropoly compounds. MCM = multicomponent molybdates. MO = metal oxides. SE = solid electrolytes. TMO = transition-metal oxides. ** Please refer to Chapter 11 for a detailed discussion on the role of steam as oxidant.

support such as Al_2O_3, SiO_2 or TiO_2 onto which either metal nanoparticles are deposited or metal-oxide aggregates are grafted forming monolayer-type systems. If reduced to a small number of atoms, such systems may be designed as "single-site" catalysts, which allow for molecular control of the active site and its surrounding environment. Besides, with recent progress in the development of nanostructured (mesoporous) materials, designed regular-pore systems are now available that can serve as support for the anchoring of active sites. The ultimate goal in rational catalyst synthesis is the preparation of catalysts on the basis of identified active-site structures. The synthesis of bulk systems can then be envisioned as assembly of these sites into nanostructured inorganic solids with high surface area, similar to the synthesis of polymers starting from basic building blocks.[4] A special case of bulk systems are heteropoly compounds, which are built on nanoclusters of a central heteroatom caged by oxygen-linked MO_6 octahedrons.

While methane and ethylbenzene can be considered as limiting cases of low and high reactivity, respectively, the C2–C4 substrates ethane, propane, propene, butane, isobutene and isobutene (see Table 1.1) due to their similar reactivity behavior in oxidative dehydrogenation and oxidative functionalization form a suitable platform for a discussion of general principles. The type of catalysts used for these reactions are in general vanadium and/or molybdenum containing bulk oxide materials including vanadium phosphorus oxides (VPO), heteropoly compounds (HPC) or mixed-metal oxides (MMO) such as MoVTeNb oxide. Many supported systems also constitute efficient catalysts for the above processes. However, with the exception of titania supported vanadium oxide (commercially used for benzene/naphthalene to phthalic anhydride conversion) bulk systems give higher yields as compared to supported systems and are therefore the focus of industrial research. Nevertheless, due to their "simplicity" supported systems can give valuable insights into the operation of selective oxidation reactions, as will be shown in detail below.

There exist various reviews[2–7] and books[1,8–13] covering heterogeneously catalyzed selective oxidation reactions. However, the high level of empirical development of many of the above processes strongly contrasts our current level of scientific understanding. It is probably fair to say that the current development of selective oxidation catalysts is largely based on phenomenological concepts (among which the principle of site isolation and the principle of phase cooperation are fundamental) rather than a profound understanding of their mode of operation. To this end, the purpose of this book is to bring together the current state of knowledge on selective oxidation reactions and, by combination with previous findings, to develop a *consistent picture of the working principle* of selective oxidation catalysts.

Commercially important classes of selective oxidation reactions are the oxidative dehydrogenation of methanol and the epoxidation of ethylene. The epoxidation of propylene has the potential to be commercialized. For these reactions mainly catalysts based on coin metals (Cu, Ag, Au) are used. Ag is a particularly interesting material as it can serve as a catalyst for two completely different processes, methanol oxidation to formaldehyde and epoxidation of ethylene to ethylene oxide. For formaldehyde production besides silver, iron molybdate catalysts are used.

A detailed discussion of all aspects related to selective oxidation catalysts is outside the scope of this book. For example, an important aspect of selective oxidation reactions that has barely been addressed in the literature is the influence of steam on the catalyst structure and dispersion.[14,15] In general, water is a product of selective oxidation reactions. In addition, water vapor is often added to the feed to improve the catalyst performance. It should be mentioned that under the conditions of operation hydrothermal reactions involving oxolation and olation processes may lead to polymerization/depolymerization of an initial M_xO_y condensate, which sets high standards for catalyst stability towards sintering. Another example is the role of carbon deposits on catalytic performance in selective oxidation reactions, which represents a largely unexplored research area.[16]

References

1. G. Centi, F. Cavani and F. Trifiro, *Selective Oxidation by Heterogeneous Catalysis*, Kluwer Academic/Plenum Publishers, New York, 2000.
2. B. Gryzbowska-Swierkosz, *Top. Catal.*, 2000, **11/12**, 23 and references therein.
3. R. Schlögl, A. Knop-Gericke, M. Hävecker, U. Wild, D. Frickel, T. Ressler, R. E. Jentoft, J. Wienold, G. Mestl, A. Blume, O. Timpe and Y. Uchida, *Top. Catal.*, 2001, **15**, 219.

4. R. Schlögl and S. B. A. Hamid, *Angew. Chem.*, 2004, **43**, 1628.
5. R. K. Grasselli, *Top. Catal.*, 2002, **21**, 79.
6. G. Emig and M. A. Liauw, *Top. Catal.*, 2002, **21**, 11.
7. J. M. L. López Nieto, *Top. Catal.*, 2006, **41**, 3.
8. H. H. Kung, *Transition-metal Oxides Surface Chemistry and Catalysis*, Stud. Surf. Sci. Catal., Vol. 45, Elsevier, Amsterdam, 1988.
9. A. Bielanski and J. Haber, *Oxygen in Catalysis*, Marcel Dekker, New York, 1991.
10. B. K. Hodnett, *Heterogeneous Catalytic Oxidation*, John Wiley & Sons Ltd, Chichester, 2000.
11. *Metal oxides*, J. L. G. Fierro (ed.), CRC Press, Taylor & Francis Group, Boca Raton, 2006.
12. *Metal Oxide Catalysis*, S. D. Jackson, J. S. J. Hargreaves (ed.). Wiley-VCH, Weinheim, 2008.
13. *Modern Heterogeneous Oxidation Catalysis*, N. Mizuno (ed.), Wiley-VCH, Weinheim, 2009.
14. G. Landi, E. Lisi and J.-C. Volta, *Catal. Today*, 2004, **91–92**, 275.
15. C. Hess, R. Herbert and G. Tzolova-Müller, *J. Phys. Chem. C*, 2007, **111**, 9471.
16. E. M. Vass, M. Hävecker, S. Zafeiratos, D. Techner, A. Knop-Gericke and R. Schlögl, *J. Phys.: Condens. Matter*, 2008, **20**, 184016.

CHAPTER 2

C–H Activation of Alkanes in Selective Oxidation Reactions on Solid Oxide Catalysts

JOHANNES A. LERCHER AND FREDERIK N. NARASCHEWSKI

Technische Universität München, Department of Chemistry and Catalysis Research Center, Lichtenbergstraße 4, 85747 Garching, Germany

2.1 Introduction

Activation of C–H bonds coupled to the functionalization of the carbon atom is one of the most important and challenging elementary reaction steps in organic synthesis. The challenge does not only lie in the homolytic or heterolytic cleavage of the bond itself, in most cases it has to occur under as mild as possible conditions to allow the subsequent reactions to proceed under very controlled circumstances. As a result, highly active catalysts are needed and the elementary steps in these reactions are dominated by single-electron processes and homolytic C–C bond cleavage, when oxygen is involved, while heterolytic C–H bond breaking is observed only in the minority of cases.

The C–H bond activation and conversion of alkanes in refining processes are somewhat more facile to realize, as the more robust target molecules allow for higher reaction temperatures. Two activation principles dominate, *i.e.*, homolytic cleavage of C–H bonds on metals leading to elimination of hydrogen, a reaction, which is mostly equilibrated under reaction conditions and the

RSC Nanoscience & Nanotechnology No. 19
Nanostructured Catalysts: Selective Oxidations
Edited by Christian Hess and Robert Schlögl

Published by the Royal Society of Chemistry, www.rsc.org

acid-catalyzed addition of a proton to an alkane leading to a carbonium ion, which decomposes spontaneously to smaller fragments and a carbenium ion, as well as the abstraction of a hydride leading directly to the formation of a carbenium ion. Thus, the kinetically dominating C–H activation steps in refining are dominated by processes in which the C–H activation occurs *via* an ionic bond separation.[1,2]

While it is hardly used for the synthesis of energy carriers, selective oxidation is one of the key reactions in chemical industry. Roughly estimated, the worth of chemicals produced by catalytic oxidation processes lies between $20 and $40 billion in 1991.[3] Especially for the synthesis of intermediates and fine chemicals, the pressure to change the feedstock in the chemical industry over the last decade arising from a combination of the limited availability of conventional starting molecules and the pressure to shift to less-expensive ones forced the use of alkanes rather than alkenes in many of the selective oxidation routes to functionalized chemicals. This has led in turn to an intense interest in selective oxidation, but has only materialized in a few heterogeneously catalyzed reactions among which butane to maleic anhydride[4–6] (see Chapter 6) or the oxidation of propane to acrylic acid (see Chapter 4) and the ammoxidation of propane to acrylonitrile[7–9] (see Chapter 5) are the most prominent ones.

Understanding the complexity of the catalysts and the multistep multi-electron processes during selective oxidation at an atomistic and molecular level poses a formidable challenge. More than with any other type of catalytic reaction, it requires that the catalysts are characterized chemically and structurally under realistic reaction conditions in order to be able to draw meaningful conclusions with respect to the surface chemistry (*operando* investigations). This is related to the fact that the sites active for catalysis are only present in small concentrations and that the catalysts change their oxidation state and surface structure in dependence of exogenic influences such as atmosphere, pressure or temperature. In addition, especially alkanes are rather inactive interacting mostly through dispersion forces with oxide surfaces.[10]

The reason for the inertness of alkanes is related to the situation that for carbon and hydrogen the number of valence electrons is equal to the number of valence orbitals and only σ-bonds are present. Reactive modification of these energetically low-lying highest occupied molecular orbitals or energetically high-lying lowest unoccupied molecular orbitals requires very reactive moieties such as radicals or high temperatures. Additionally, the tetrahedral coordination of the carbons in sp^3-hybridization efficiently shields the carbon atoms and complicates possible attacks together with the low polarity of the C–H bond. The primary products after activation contain functional groups or heteroatoms, which are by far more reactive and susceptible for further chemical reactions.

In consequence, the activation of the first C–H bond is the rate-determining step and after addition of the more electronegative substituent to the desired extent, the reaction needs to be kinetically stabilized to prevent total oxidation. Thus, achievable yields in these reactions are strictly depending on the

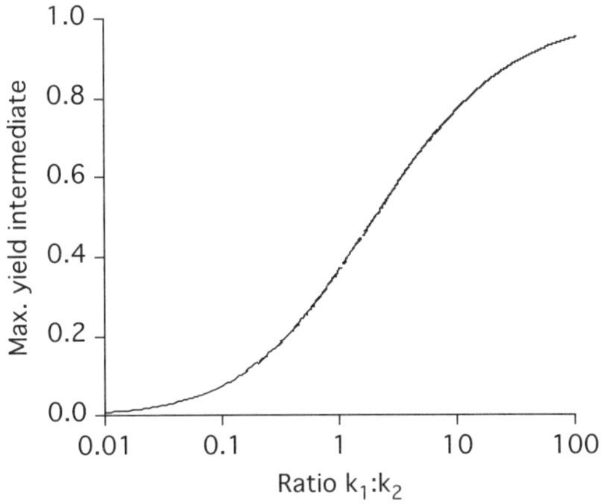

Figure 2.1 Maximum achievable yield *versus* ratio of activation to (unwanted) further oxidation.

ratio of the rate constants for activation to (unwanted) further oxidation (k_1:k_2). Maximum yields in a selective oxidation reaction for a model of two consecutive first-order reactions are shown as an example in Figure 2.1. To reach yields above 70% k_1 needs to be at least one magnitude larger than k_2, posing a major challenge for the design of catalysts and the process environment.

For this chapter we will limit ourselves to the discussion of the oxidative activation of the C–H bond in alkanes. Model experiments, kinetic analysis of complex reactions and theoretical studies will be combined to provide an overview on the current state of insight into the processes as well as their potential and limitations. We will first discuss processes on oxide clusters, followed by the influence of anions and cations on the activation mechanism and finally cover bifunctional catalysts used to achieve oxidative functionalization.

2.2 Models of C–H Bond Activation over Supported and Unsupported Oxide Clusters

We will first discuss the chemistry of small alkanes such as ethane, propane and butane, because of the easier C–H activation in these molecules, and will treat methane activation as a special case. This is done so, as with all these molecules the activated alkanes are able to eliminate hydrogen and may desorb, while this is not possible with methane. It should be emphasized that except for the higher bond strength of the C–H bond of primary carbon atoms, the other principal chemistry will be the same in the initial C–H bond breaking or polarizing step.

2.2.1 Vanadium-Based Clusters

C–H activation on catalysts containing vanadium-oxide species is certainly the most widely studied elementary reaction. The catalytic properties of the vanadium moieties are affiliated with the redox properties of vanadium that changes its oxidation states between +III, +IV and +V (see Chapter 15).[5,6,11–16] The initial step of the C–H activation is so difficult to assess, because the precursor to the homolytic cleavage does not necessarily involve polarization of the C–H or the V–O bonds. Thus, only the final states of the first reaction would be accessible to the spectroscopic characterization.

Most of the experimental studies initially addressed were focused on supported monolayers of vanadia clusters on oxide surfaces (see Chapter 12). Pioneering systematic studies came from the groups of Wachs[17] and Iglesia and Bell.[18] The surface geometry of these clusters was determined with a multitude of indirect spectroscopic characterizations. Overall, it was suggested that monomers, dimers and oligomers are present on the surface (see Figure 2.2) and that the catalytic reactivity and the oxidation state depends critically on the size of the surface cluster.[19] With increasing cluster size of the surface bound vanadia the reactivity towards alkanes increased. The same trend was observed with decreasing electronegativity of the support (according to Sanderson) indicating that a lower charge at the terminal oxygen facilitates C–H bond breaking in the activation of propane.[20] It is interesting to note that the selective oxidation of methanol exhibits the reverse influence of the support oxide, *i.e.*, the catalysts were more active as the charge at the lattice oxygen increased.[21]

One of the central issues in these experimental studies concerned questions with respect to the nature of the active oxygen species or more specifically, the functional group involved in breaking the first C–H bond. In principle, three

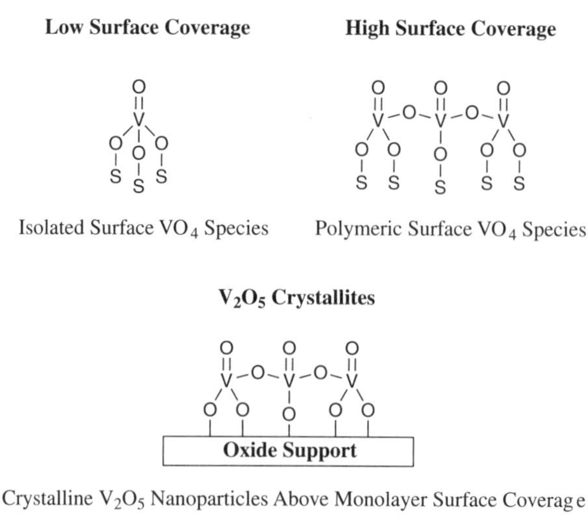

Figure 2.2 Supported vanadium-oxide catalysts.

V–O bonds can be distinguished, the terminal V=O bond, the bridging V–O–V bond and the bridging V–O-support bond. For activation of ethane and propane[22–24] the V–O–V bond was concluded to be catalytically irrelevant as the turnover frequency did not change with the coverage of the support by vanadia species, which should increase the relative concentration of this species. Also, the concentration of the V=O groups did not influence the observed rate of alkane activation leaving the V–O-support bond to be the only remaining oxygen, which was then also concluded to be catalytically active.

However, as this rationale is based on an indirect elimination following more plausibility arguments than rigorous proof, let us turn to model studies using well-defined clusters in the gas phase combined with modeling of the structures and barriers between the intermediates. These clusters may be charged in order to increase their reactivity without changing the relative reactivity of the oxygen atoms in the cluster. The first well-defined case of such an approach was a study of the oxidation of propane on a $[V_3O_7]^+$ cluster by the groups of Schwarz and Sauer.[25,26] The cluster was chosen, as it is the smallest polynuclear entity that formally contains only V in +5 oxidation state. The oxidative dehydrogenation of propane involves the reduction of the metal center, brought about by the addition of two hydrogen atoms.

Detailed DFT calculations show that after assuming a quite stable precursor state ($\Delta G_{298} = -63\,\mathrm{kJ\,mol^{-1}}$) the initial reaction of propane with the cluster is the abstraction of a hydrogen atom with an energetic barrier of $102 \pm 5\,\mathrm{kJ\,mol^{-1}}$.[27,28] Formally, this sequence is initiated as the double bond between vanadium and oxygen cleaves homolytically and forms radicals in the transition state. In the first step, hydrogen is abstracted from C_3H_8 by one of the V=O groups and yields the $C_3H_7^{\bullet}\ V_3O_7H^{+\bullet}$ diradical, which recombines and forms a carbon–vanadium bond through a rebound mechanism.[4,28–33] From this stable adduct a second hydrogen abstraction by a bridging oxygen is possible, which leads directly to propene and $V_3O_7H_2^+$. Also, a pathway forming a propyl cation that rebinds to a hydroxyl group under formation of *i*-propanol can be found (see also Scheme 2.1).

After the initial activation of the alkane to propene, oxygen addition leads to the desired products that are described in more detail in the following chapters of this book (4 and 5 for propane, 6 for *n*-butane).

These experiments demonstrate unequivocally and in agreement with the conclusions by Wachs *et al.*, that V–O–V oxygen does not participate in the initial C–H bond activation. The vanadyl group, however, is indicated in these studies as the decisive reactive oxygen containing group. However, this cluster does not contain a bridging oxygen to another oxide, which has been concluded to be the active site. In order to mimic such bridging oxygen between V and a support cation, a vanadium substituted silsequioxane cluster has been used (Figure 2.3).[30]

The calculations show that also in this case the first hydrogen abstraction at the vanadyl group is the rate-determining step. This practically rules out the involvement of the oxygen link to the surface as the oxygen initiating the

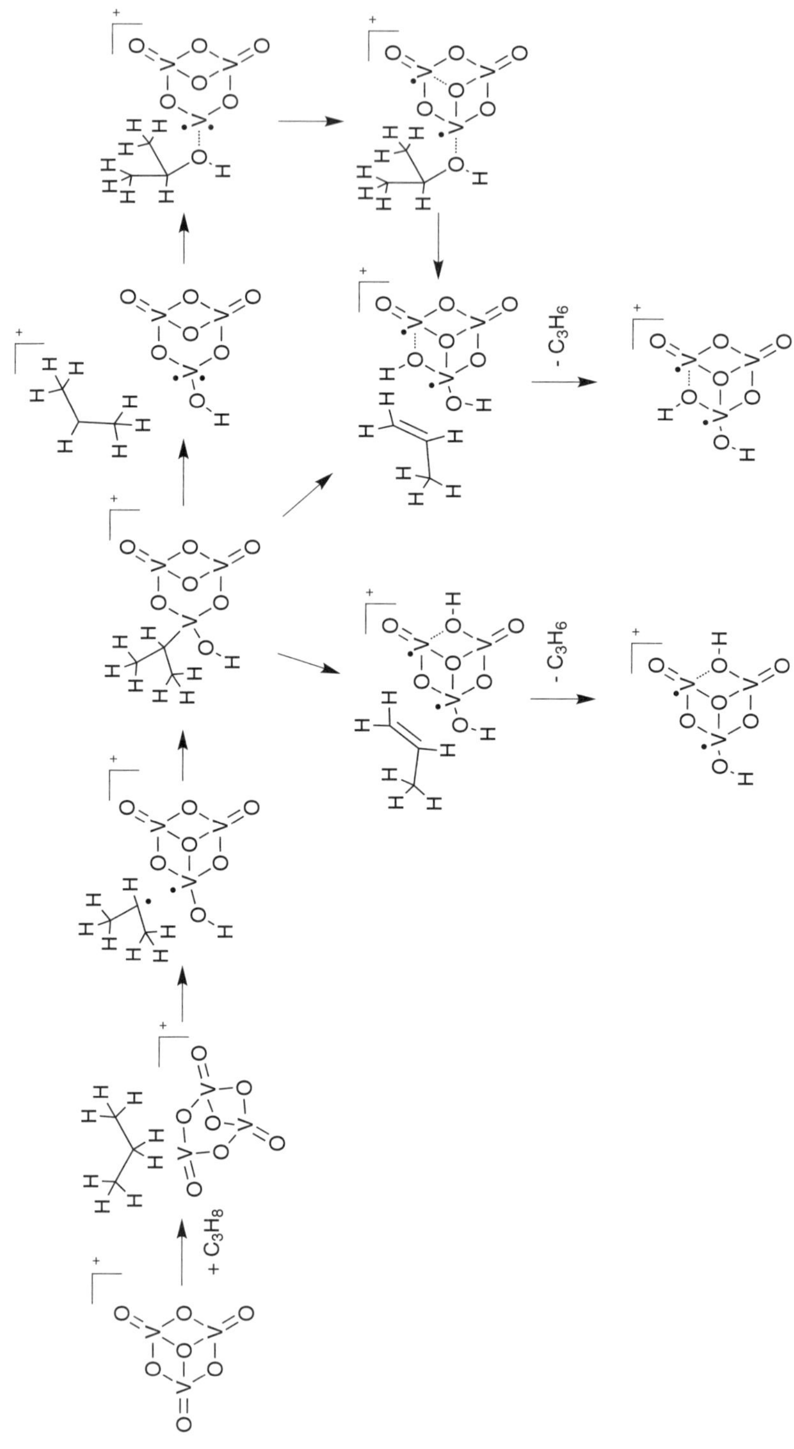

Scheme 2.1 Reaction pathways of propane with $V_3O_7^+$.

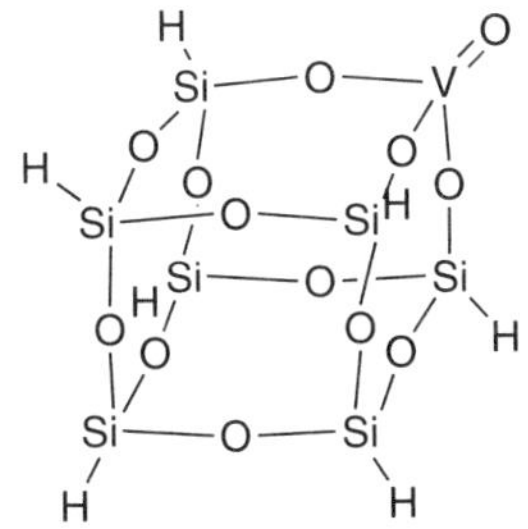

Figure 2.3 Vanadium-substituted silsequioxane.

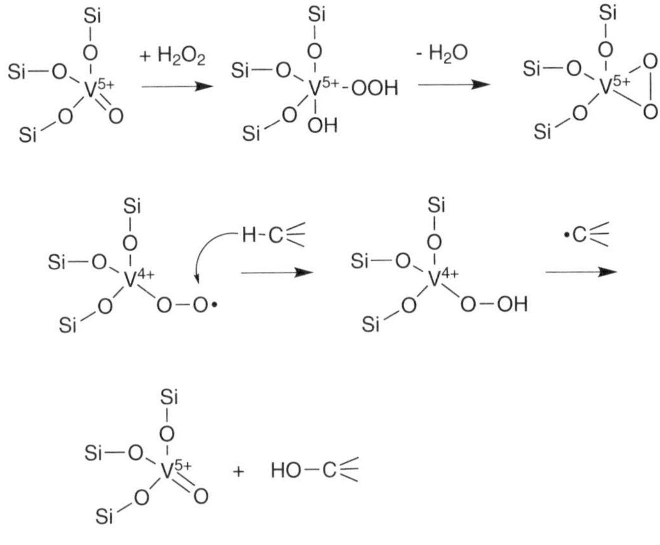

Scheme 2.2 Formation of peroxo-V^{+5} sites.

overall reaction sequence. In passing, it is interesting to note that in a recent contribution the generation of a peroxovanadate has also been proposed that could explain the much higher reactivity of oxygen compared to N_2O (which cannot form a peroxovanadate) in pulse experiments.[34]

This is in line with the observation that vanadium-substituted zeolites are able to activate the C–H bond at the primary carbon atom in light alkanes, while this is not possible with titanium-substituted materials.[35,36] In this case the peroxo-V^{+5} site has been identified by EPR spectroscopy. Alcohols are formed in the presence of both substituents together with H_2O_2 (Scheme 2.2).[37]

It is presently unclear why titanium-substituted zeolites are only able to activate and oxidize (hydroxylate) alkanes at the secondary carbon atom, while vanadium-substituted materials oxidize the alkane at the

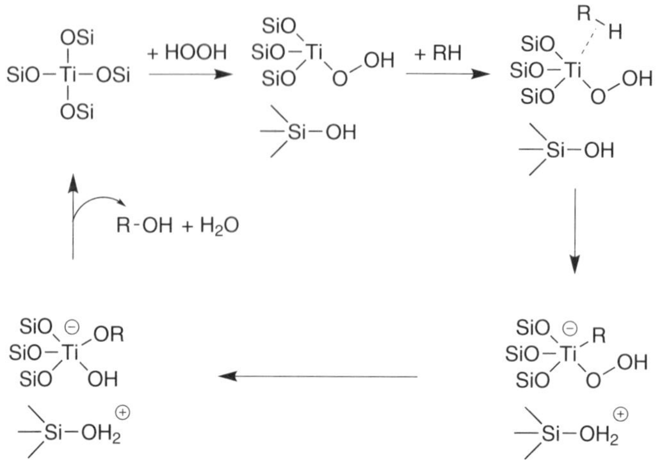

Scheme 2.3 Possible mechanism for alkane activation of TS-1.

primary carbon atom. The catalytic cycle proposed is described in Scheme 2.3.

One possibility to explain the high selectivity for the activation of the C–H bond at the secondary carbon atom with Ti-substituted zeolites could be the fact that the alkane is dehydrogenated in a first step in analogy to the dehydrogenation observed with La^{3+}-exchanged zeolites[38] and that subsequently the olefin is hydroxylated at the TiOOH site, as discussed extensively in the literature. In this case the primary dehydrogenation will occur at a C–H bond of a secondary (or tertiary) carbon atom and will, thus, determine the selectivity of hydroxylation. It is remarkable that in recently reported titanium-rich MFI-type zeolites, the activity towards *n*-hexane was even higher than towards 1-hexene, again only forming products that are oxidized at the secondary carbon atom.[39] Indeed, a detailed study with Ti-MFI shows that activation at the primary carbon atom is about 100 times lower than that at the secondary carbon position.[40] The mechanism speculated by Davis *et al.* would offer a reasonable explanation that has not been contemplated, however, in the paper. It could be that the synergetic interaction between the (basic) oxygen at the SiOH group and the accessible Ti leads to a heterolytic cleavage of the C–H bond (which is significantly easier at the secondary carbon atom) and the formation of a Ti-C bond as postulated. The subsequent β-hydride elimination leads to the formation of an olefin, which in turn is rapidly hydroxylated on the titanium sites.[41]

2.2.2 Molybdenum and Tungsten Oxide Clusters

While vanadium-based transition-metal oxide clusters are the most studied, other transition-metal oxides and in particular those based on molybdenum and tungsten also show remarkable catalytic activity. It is interesting to compare in this case the differences in the particular surface chemistry induced by the metal cation and by the surface structure.

Similar to vanadium, molybdenum-based oxides activate the C–H bond also *via* a homolytic cleavage at a molybdenyl group. This is concluded from experimental and computational studies linking the presence of the Mo=O group to the activity of the catalytic material.[42,43] As much as these results seem to emphasize the similarities to the supported vanadia catalysts, important differences between the two systems exist. The rate normalized to the accessible Mo at the surface increases drastically as the surface concentration of Mo increases (note the contrast to the equal catalytic activity in a similar relation for vanadia clusters). However, what is even more striking is the fact that the ratio of the rate constants between the initial oxidative elimination of hydrogen from propane increases, while that of total oxidation either from propane or from propene decreases. This suggests that the sites for the unselective reactions have to be related to sites at the interface between the support (Al_2O_3 in the study cited[42]) and the molybdenum oxide species on the surface. The molybdenum oxide species show overall a higher reducibility in H_2 with increasing surface concentration. The rate of the supported molybdenum oxide species reducibility (at 673 K) is directly proportional to the rate of oxidative dehydrogenation and, hence, to the rate of C–H activation, which constitutes the rate-determining step in oxidative dehydrogenation. DFT calculations of the C–H activation in propane on trimeric Mo oxide clusters indicate that the lowest barrier exists for the hydrogen abstraction from the secondary carbon atom of propane by one of the terminal oxygen atoms in a one-electron process (Figure 2.4).

Reaction steps involving two electrons and acid–base-type interactions have energetic barriers, which are more than 50 kJ mol^{-1} higher, emphasizing the unlikelihood of such reaction pathways.[43] Also, pathways involving the simultaneous cleavage of two C–H bonds are highly unlikely. Comparing the C–H activation of propane with that in methane one notes that again the one-electron process has the lowest energetic barrier, but that this barrier (according

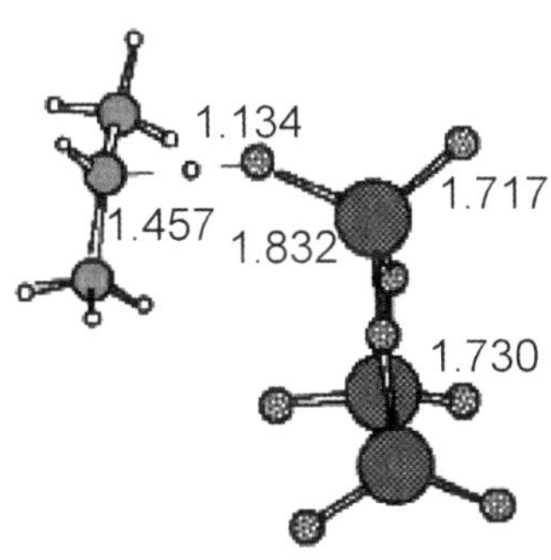

Figure 2.4 Hydrogen abstraction on Mo oxide cluster.

to experiments and DFT calculations) is at least 60 kJ mol^{-1} higher than in the case of propane, emphasizing the difficulty of breaking the C–H bond of methane on these catalysts.[44]

These experiments suggest that the cleavage of the C–H bond is a one-electron process also for Mo oxides and is the more facile the weaker the C–H bond is. Given the weaker C–H bond in propene (compared to propane) one might ask, if propene formed will not be immediately consumed in a sequential activation/oxidation leading to very low selectivities. Let us analyze the situation in this respect by comparing the simplest oxidative activation, the oxidative dehydrogenation of propane with the further reaction of propene, the dominating primary product. For the former reaction, the relative activities between ZrO_2-supported vanadium, molybdenum and tungsten oxides decrease in the order $VO_x/ZrO_2 < MoO_x/ZrO_2 < WO_x/ZrO_2$, while the corresponding apparent energies of activation increase in that order.[45] This indicates that the activities are inversely proportional to the C–H bond-cleavage energies. The activation energies of propane oxidative dehydrogenation are higher than for further total oxidation of propene. The difference in the activation energies (48–61 kJ mol^{-1}) is larger than between bond-dissociation enthalpies for the weakest C–H bond in propane and propene (40 kJ mol^{-1}). It increases in the sequence $VO_x/ZrO_2 < MoO_x/ZrO_2 < WO_x/ZrO_2$, suggesting that relative rates also depend on the differences in the heat of sorption between propane and propene, which will increase as the Lewis acidity ($V^{5+} < Mo^{6+} < W^{6+}$) increases.

2.2.3 Complex Mixed Oxides

So far, the impact of different oxides and supports on the activity for C–H bond activation has been discussed. In this context it has been noted that the primary activity for the one-electron cleavage depends on the ability of the oxide to be reduced and the strength of the C–H bond involved. Any component of a complex oxide that would enhance the strength of the metal–oxygen bond is expected to decrease the activity of the reducible species.

This has been shown for zirconia-supported molybdenum oxide clusters upon addition of alkali metals.[46] The presence of alkali cations (Cs, K, Li) did not affect the structure of MoO_x domains, but influenced their electronic and catalytic properties. Propane C–H activation decreased monotonically with increasing concentration of the alkali cation and for a given concentration with increasing base strength of the corresponding alkali oxide. Also in this case, the subsequent oxidation of propene was slightly suppressed by weakening the acid strength of Mo^{6+} through the formation of a mixed oxide.[47]

The positive influence of the presence of a bridging oxygen between the active vanadium cluster and the support on the selectivity of the oxidative dehydrogenation and the observation that two vanadium atoms form an active site, leads to the question whether it is possible to positively influence the catalytic properties by combining two cations in the active phase.

The simplest conceptual scenario is given by the deposition of a vanadium oxide cluster on a polymolybdate monolayer surface supported on alumina.[48] Without changing the mechanism of the C–H bond activation, the combination enhances the rate of C–H bond activation, by increasing the reducibility of the vanadium oxide clusters. The initial rate of oxidative dehydrogenation (equivalent to the rate of C–H activation) increased proportional nearly up to a monolayer of vanadium, indicating a low sensitivity to the particular structure of the vanadia, but was significantly higher, if vanadia was supported on alumina. Most remarkably, the ratio of the rate constant for the further oxidation and of the primary formation of propene decreased with increasing concentration of vanadia and was for the best catalyst drastically lower than the corresponding ratio in the case of VO_x/Al_2O_3. In line with the arguments used above, the presence of molybdenum is speculated to have reduced the interactions of propene with the accessible cations. The retention of propene is concluded to be critical for the further oxidation.

This effect is not confined to the system vanadia/molybdena though. A positive effect was also observed for chromia-decorated surfaces, although the surface chemistry is more complex.[49] In a similar strategy it was reported recently that the addition of oxides of varying acid–base properties could tune the selectivity and activity of a NiO catalyst for oxidative dehydrogenation. The reasons for these positive effects were explained by a modification of the reducibility of NiO, but neither the site of activity nor the detailed kinetics are currently understood in the same detail as with the vanadia and molybdena systems.[50]

Very complex oxides based on $MoVTeNbO_x$ are currently used for more demanding reactions such as the conversion of propane to acrylic acid (see Chapter 4).[13,51,52] The active phase in these materials is a complex structure consisting of octahedra with Mo and V oxides building up five-, six- and seven-membered channels in which Nb and Te oxides are coordinated. A (001) surface of such a material is displayed in Figure 2.5.

In these oxides the substitution of V for Mo leads to a situation similar to the one described by the layered material described above, *i.e.*, single or paired vanadia units are embedded in a molybdenum matrix. These vanadia octahedra have been identified as the active sites[13,53–55] and the catalytic activity is proportional to the concentration of vanadium. However, the dependence is not straightforward and substantial differences in catalytic activity of the vanadia units seem to exist depending on the location of the substitution. Using the oxidative dehydrogenation of ethane one notes a unique and unparalleled selectivity to ethene with very low rate constants of the oxidation of ethane.[56] Also in this case, the specific catalytic chemistry does not seem to be confined to this particular structure, although a related structure the so-called M2 phase is not active for C–H bond activation (see Figure 2.6).

It is interesting to note that in this structure vanadium is only present in isolated sites, which would be in line with the experimental and theoretical observation that pairs of vanadium oxide units are needed for the initial C–H activation. Indeed several of such materials have been reported to be very active

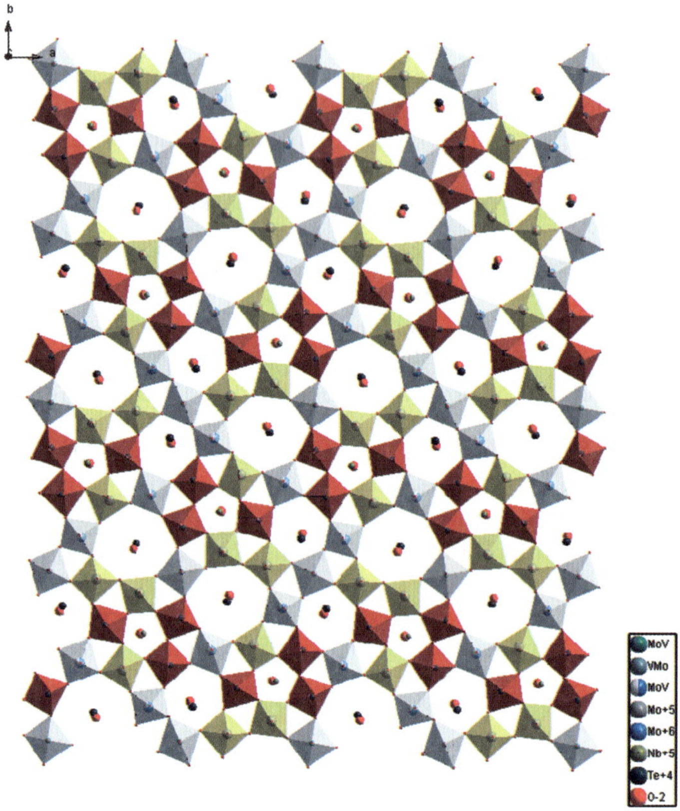

Figure 2.5 Structural model of $MoVTeNbO_x$ M1 phase.

and selective for the oxidative dehydrogenation of propane.[57] Alternatively, the distortion of the V–O octahedron has been claimed to be responsible for its high activity.[54,58] It should be noted that the intense discussion on the site for the multielectron oxidation process of propane to acrylic acid is not decisive for the current contribution, as almost all authors agree that solely the vanadia unit is responsible for the oxidative dehydrogenation of propane, which constitutes the first and in many cases also the rate-determining step.[59]

2.2.4 Transition-Metal Species in Zeolites

The catalytic sites activating the C–H bond in the oxides discussed so far were primarily terminal oxygen atoms (such as vanadyl groups), which are ideally suited for one-electron oxidation steps. Enzymes,[60,61] but also designed catalytic sites such as reducible cation pairs at ion-exchange sites in zeolites, are able to activate alkanes *via* a bridging oxygen groups.[62] Only a few cations, such as Fe, Cu and Co show catalytic activity in such an environment for oxidation reactions. The reason for the existence of only a few cations is related to the steric limitations by the size of the hydrated metal cations (hindering larger

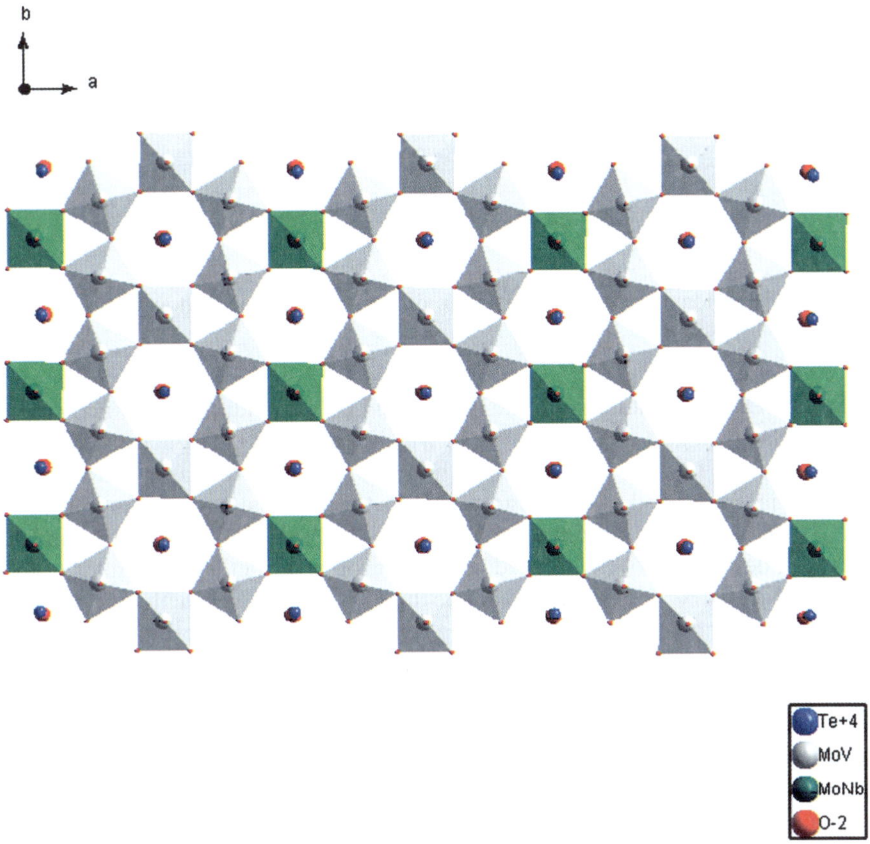

Figure 2.6 Structural model of $MoVTeNbO_x$ M2 phase.

cations to be exchanged) during the ion-exchange procedure and the probability that at least two aluminum cations must be substituted in the zeolites structure at such close distances that the di- or trivalent cation can bridge that distance forming, *e.g.*, a Cu–O–Cu moiety.[63]

There is a quite intense debate on the nature of these complexes involving Fe^{+3} and Cu^{+2} cations that ended quite indecisively as seen in the recent example for Cu-exchanged ZSM-5.[64] The site depicted in Figure 2.7 has been deduced from evidence using primarily spectroscopy. Diffraction results that would allow for a complete structural analysis are not available and will be very difficult to obtain as faults in the zeolite structure impede a precise structure determination. This situation brings attention to the point that the structural information on Cu mono-oxygenase begins only now to reach a level that would permit to start to address the mechanism of the oxygen activation. The current understanding is that a single Cu cation site is located about 2 nm away from a Cu ion pair. The oxidation state of the three cations is not unequivocally settled yet.[60]

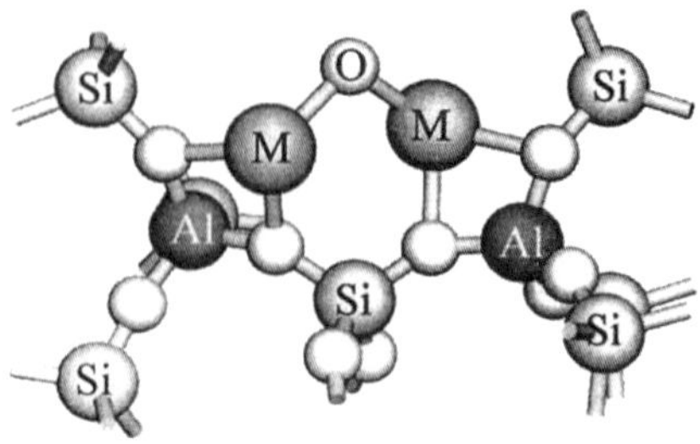

Figure 2.7 Bridging oxygen between two divalent metal cations at exchange sites in zeolites.

Recent DFT calculations on the mechanism of the oxidative C–H activation in Cu-ZSM5 suggest a possible way in which methane and higher alkanes can be activated.[64] *Via* this route even methane can be converted at around 200 °C forming methoxy and hydroxy groups with the simultaneous reduction from Cu^{+2} to Cu^{+}. The constraints in the zeolite and the strong bonding hinder, however, desorption of the formed methoxy groups or methanol. Similar to the surface chemistry described above the hydrogen of, *e.g.*, methane is being cleaved from methane in a one electron oxidation step. The resulting diradical will bind at the newly formed bridging hydroxy group forming methanol. With respect to the inability to desorb this species one needs to emphasize that supported CuO oxide is an excellent steam-reforming catalyst that most likely decomposes the desorbing methanol.[65] Interestingly, Fe-exchanged zeolites need N_2O to generate the active oxygen, while Cu and Co exchanged zeolites can be activated in oxygen.

2.3 Impact of Anions on C–H Activation

2.3.1 Vanadium Pyrophosphate as Catalyst for Butane Activation

Vanadium phosphates are currently the only large-scale catalyst employed for the selective oxidative process of alkanes, specifically for the oxidation of butane to maleic anhydride (see Chapter 6).[4] With vanadium as the main redox-active element in the catalyst, it is critical to ask, whether or not the surface chemistry proceeds analogous to the activation discussed so far for supported vanadia and complex mixed-oxide catalysts.

Two interesting facts have been established early on. The catalysts readily develop the (100) surface in the process of drying and conditioning for the reaction.[66] This specific surface has been claimed to be responsible for the catalytic action. While such a complex reaction involves a wide variety of sites, promoters and variants of catalysts (see, *e.g.*, refs. 67 and 68), the principal chemistry of the C–H activation from *n*-butane seems to be closely related to the vanadyl groups, as discussed above.[69,70] Scheme 2.4 shows the proposed

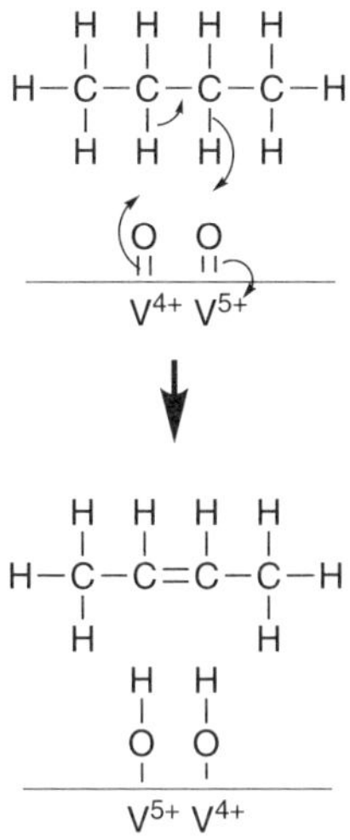

Scheme 2.4 Proposed mechanism for the formation of 2-butene.

mechanism for the formation of 2-butene in this first step. One notes the schematic nature of the proposed transition state along this route (a second pathway involving alkoxy groups has also been proposed in ref 70 above), but formally two vanadyl groups are involved, as also discussed later, as the probably dominating active site. It is striking that a kinetic analysis does not indicate butene as an intermediate with marked concentration in the gas phase. This indicates that at least the interaction between butene and the surface must be very strong.

If one explores the (100) surface of vanadium pyrophosphate using DFT calculations, the termination by P=O and V=O groups, but also the existence of P–O–V bonds is notable.[71] In line with the arguments used above to explain the effect of metal cations on the catalytic activity and selectivity of V=O groups we would like to speculate that the presence of the highly electronegative phosphorus enhances the reducibility of vanadium. Moreover, the top view in Figure 2.8 shows that neighboring V=O groups exist, (well separated by P=O groups) indicating that the surface site that has been identified as required for the alkane activation is also present on this material. As a word of caution though it should be mentioned, however, that the surface structure may be quite disordered and may form over a much longer period during operation, as had been thought originally.[72] This is indeed reflected in the marked influence the relative concentrations of vanadium and phosphorus have on the catalytic activity.[73] A dynamic equilibrium between several phases exists on the surface of the vanadium pyrophosphates, making it very difficult to identify more than the principal features of the atoms involved. There has been, however, a recent elegant *in-situ* study by ^{31}P- and ^{51}V-solid state NMR spectroscopy to probe the phosphorus and vanadium species involved in the oxidation reactions. δ-$VOPO_4$ has been identified to be the phase most

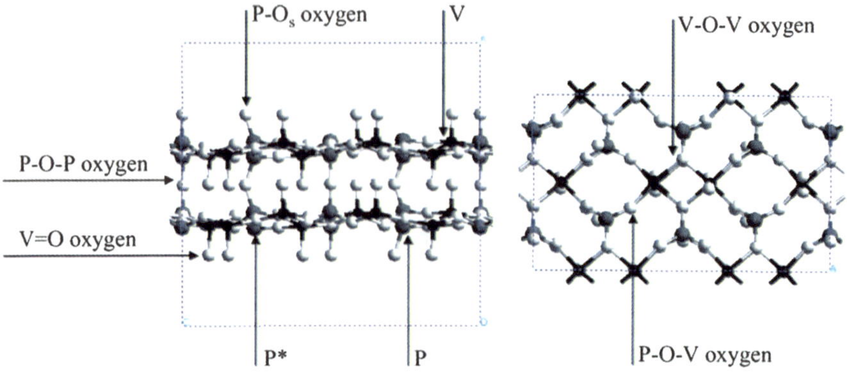

Figure 2.8 Model of $(VO)_2P_2O_7$ (100) surface.

important for the overall conversion of *n*-butane to maleic anhydride. Cycling though oxidation and reduction it could be shown that P/V^{5+}were pairs that were active for the overall reaction. Note that this, unfortunately, does not provide the insight with respect to the initial C–H bond activation.

2.3.2 Supported LiCl as Catalyst for Oxidative Dehydrogenation of Ethane to Ethene

So far anions and cations were used and discussed as promoting species changing the properties of the redox element to activate the C–H bond. In principle, however, the anion itself can undergo the redox step and is able to catalyze the oxidative dehydrogenation of ethene (*e.g.* refs. 74–76).

The most remarkable catalysts in this respect are MgO-supported alkali chlorides, which form at reaction temperatures a melt covering the support surface. Dy_2O_3 has been found to be an excellent promoter that is not interfering with the chemistry, but helps to maintain a reasonably large specific surface area. The active sites are speculated to be hypochlorite anions, which are formed only transiently in the presence of oxygen at the surface of the liquid layer or dissolved in the catalysts. The catalyst is regenerated by eliminating water (see Scheme 2.5).

Mechanistic details are not evaluated, because of the transient nature of the catalytically active sites. The presence of oxygen leads to the formation of positively charged chlorine. It is speculated that the reaction proceeds *via* the electrophilic substitution of hydrogen for chlorine, followed by an immediate elimination of HCl. Concerted elimination of HCl would be very rapid under these conditions, as the reaction temperature is typically above 550 °C. Elimination of water would complete the catalytic reaction cycle and restitute LiCl as catalyst. The catalytic reactivity is not limited to LiCl. Several pure and mixed alkali and alkali-earth chlorides showed promising catalytic properties. It should be noted that the selectivity of the catalysts increased with decreasing

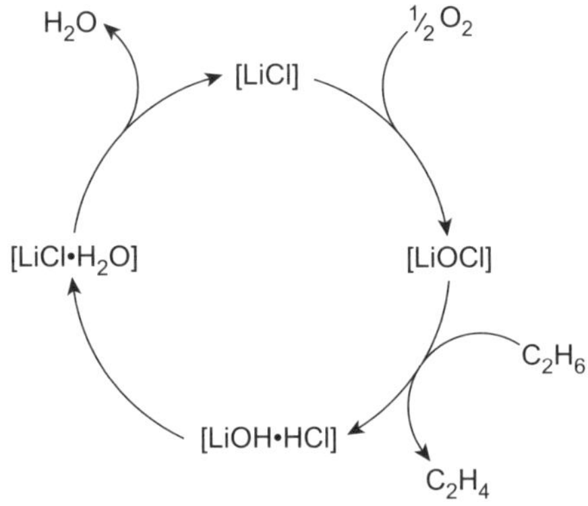

Scheme 2.5 Catalytic cycle of oxidative dehydrogenation over LiCl catalysts.

melting point, indicating that the mobility at the surface leads to a dynamically restructuring material that avoids in this way the formation of under-coordinated metal cations. Such accessible metal cations had been identified as the key sites to adsorb olefins and to provide the required sorption locations for total oxidation.

2.4 Indirect Oxidation *via* Electrophilic Substitution – Chlorination and Bromination

In the surface chemistry described so far, the C–H bond was activated by a homolytic cleavage leading to a diradical species that can rebind. In the case in which the cation may polarize the C–H bond so far that the C–H bond is cleaved, the hydrocarbon moiety will bind to the metal cation and eventually form a carbon–metal bond (*vide infra*). The oxidative elimination of the alkene *via* β-hydride elimination will then lead to the olefin. Here, we will describe an alternative approach using polarization and electrophilic substitution. Once substituted the highly electronegative substituent will rearrange the electron distribution at the carbon atom and so lead to oxidation. This is usually achieved *via* halogenation (chlorination or bromination) of alkanes.

FeO_XCl_Y/Al_2O_3, $TaOF_3$, $NbOF_3/Al_2O_3$, $ZrOF_2/Al_2O_3$, $SbOF_3/Al_2O_3$, SbF_5/graphite and Nafion-H/TaF_5 are known as catalysts for selective chlorination and bromination of methane.[77] The activation of methane is achieved under relatively mild conditions (270 °C). The mechanism is shown in Scheme 2.6. In the first step, a Lewis-acid site of the catalyst polarizes chlorine leading to Cl^- and Cl^+ bound to the catalyst. In the next step, Cl^+ is transferred into methane by an electrophilic insertion. In the transition state carbon is five-coordinated by four methane-hydrogens and one Cl^+ in a cationic complex and finally Cl^+ replaces hydrogen to form methyl chloride together with hydrogen chloride from the negative polarized chloride. This is also the rate-determining step and

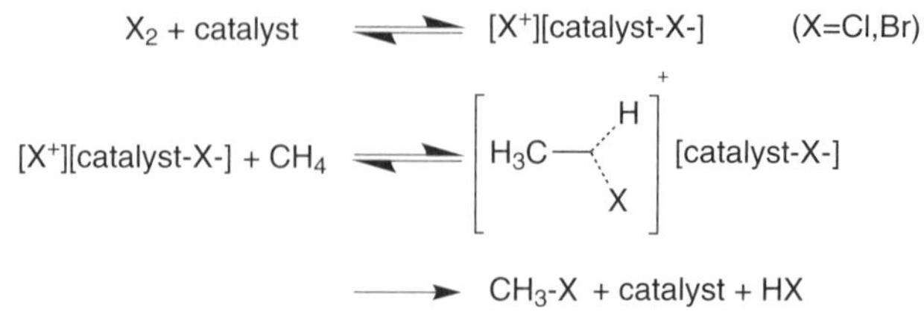

Scheme 2.6 Mechanism of selective methane chlorination over solid acids and Cl_2.

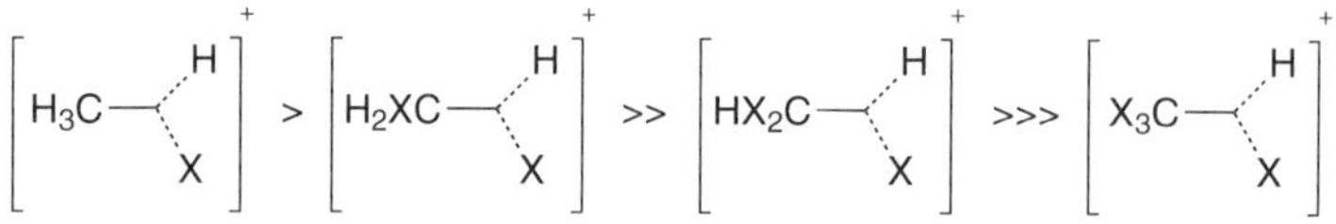

Scheme 2.7 Stability of transition states of selective methane chlorination.

the energy barrier and also the rate of the reaction is dependent on the stability of the five-coordinated transition state.

In the direct radical chlorination the (chlorinated) products react as readily as methane leading to unwanted multiple chlorination steps. In the electrophilic insertion the already chlorinated products also pass over a multichlorinated transition state. In the case of reaction with methyl chlorine two chlorines will be incorporated in the five-coordinated transition state and due to the electron-drawing effect of chlorine, the stability of this cationic complex is lower than the corresponding transition state of the monochlorinated reactant. In consequence, the energy barrier is higher for the second chlorine insertion compared to the first insertion and the rate will be slower. The decreasing stability of multiply chlorinated products (compare Scheme 2.7) leads to an order of reaction rates, in which the rate of formation decreases in parallel to the degree of substitution. If only one equivalent of the chlorine is used, the monochlorination is, in this reaction, strongly preferred. Conversions of 34% and 96% selectivity to monochlorinated product were realized in that way. Thus, the role of the catalyst in the electrophilic substitution is to polarize chlorine and stabilize the formed electrophilic chlorine. Indirectly, the lower reactivity towards electrophilic substitution leads to the decreased rates for higher substituted products.

Catalytic activity for C–H activation reactions has also been reported for lanthanum oxides (La_2O_3) and chlorides (LaOCl, $LaCl_3$), if HCl is used for the chlorination under oxidizing conditions at 400–500 °C. This would be a dramatic commercial benefit, as one mol of HCl is generated for each Cl inserted, which has to be reoxidized to Cl_2 in a separate process.[78–80] It is interesting to note that the catalytic materials, La_2O_3 and LaOCl, have been

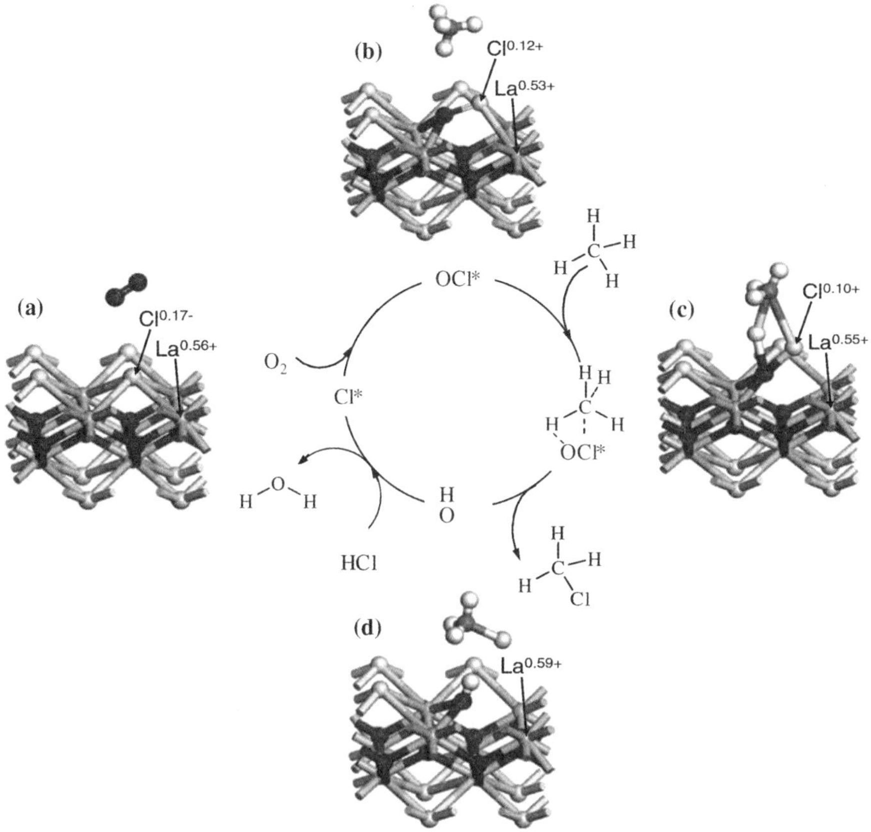

Figure 2.9 Mechanism of selective methane chlorination over LaOCl.

reported as active catalysts for oxidative methane coupling,[81–84] and for ethane oxidative dehydrogenation.[85,86] $LaCl_3$ is known as a promoter in catalysts for the oxydehydrochlorination and ethane conversion to vinyl chloride[87] and is also used in selective chlorination of methane.

The active sites are hypochlorite OCl^- surface species having a formal oxidation state of +1,[78–80] formed from $LaCl_3$ by the reaction with molecular oxygen (see Figure 2.9). In the next step, methane forms a five-coordinated transition state with the positive charged chlorine, which decomposes and leaves a surface hydroxyl group and methyl chloride after decomposition. To close the catalytic cycle the catalyst hydrogen chloride forms water and a surface chloride that is oxidized with molecular oxygen to the original active hypochloride species.[78]

Similarities between the two reaction mechanisms are obvious. Also in the latter case, the rate-determining step proceeds over a five-coordinated carbon and the stability of this transitions state should also decrease with a higher number of chlorines bound to carbon, increasing the selectivity of the monochlorinated product. Catalytic tests with stoichiometric mixtures of $CH_4 : HCl : O_2 = 2:2:1$ at 475 °C show 65% methyl chloride and 25% methylene dichloride (the rest are oxidation products CO and CO_2) at 27% conversion. Thus, this reaction is not as selective as the previously described catalysts and reasons for this are still under discussion. Even changes in temperature between 450–550 °C showed no significant influence on the selectivity of the methyl chloride, but it should also be noted that these temperatures are still much higher than the temperature of the selective FeO_XCl_Y/Al_2O_3 catalysts (270 °C).

2.5 Oxidative C–H Activation *via* Activation with Nonreducible Cations

2.5.1 C–H Activation of Alkanes by Li-Promoted MgO

Li_2O/MgO is one of the most prominent examples for oxidative C–H activation relying on defect sites at surfaces (see Chapter 3), well exemplified by the pioneering work of Lunsford *et al.*[88,89] The active sites have been originally proposed to be $[Li^+O^-]$ sites, which are generated in an MgO host lattice by substitution of divalent magnesium cation by Li^+. This oxygen radical site[74,90,91] catalyzes the abstraction of hydrogen from, *e.g.*, ethane with formation of $[Li^+OH^-]$ and an ethyl radical (eqn (2.1)) in the rate-determining step.

$$C_2H_6 + [Li^+O^-]_s \rightarrow C_2H_5\bullet + [Li^+OH^-]_s \tag{2.1}$$

The initially formed ethyl radical is desorbed into the gas phase, where free-radical reactions proceed. In the gas phase, molecular oxygen abstracts a second hydrogen leading to ethane and a hydroperoxy radical (eqn (2.2))

$$C_2H_5\bullet + O_2 \rightarrow C_2H_4 + HO_2\bullet \tag{2.2}$$

This hydroperoxy radical can again remove a hydrogen atom from ethane under recreation of an ethyl radical, forming a separate catalytic cycle in the gas phase. The formed hydrogen peroxide decomposes to water and oxygen (eqn (2.3)), but formation of two hydroxyl radicals by homolytic cleavage is also possible, so enabling the start of a new chain reaction.

$$HO_2\bullet + C_2H_6 \rightarrow H_2O_2 + C_2H_5\bullet \rightarrow C_2H_5\bullet + H_2O + \frac{1}{2}O_2 \tag{2.3}$$

Overall, ethane is oxidized to ethene by molecular oxygen. The role of the catalysts is to catalytically supply the ethyl radicals needed in the radical cycle of the gas-phase reaction (Scheme 2.8).

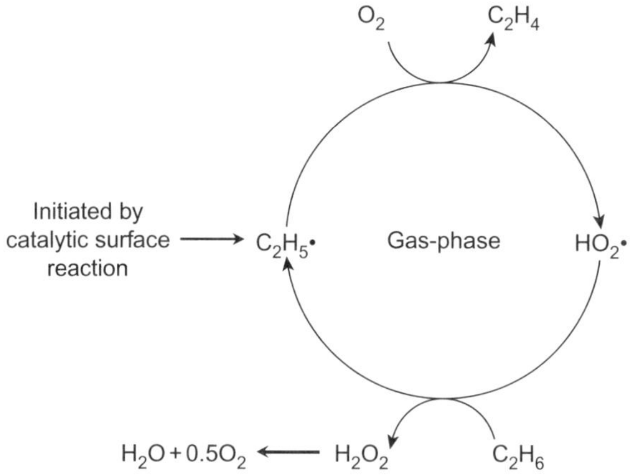

Scheme 2.8 Catalytic cycle in gas phase after activation by $[Li^+O^-]$.

After initiation of the catalytic gas-phase reaction, the active center on the surface of the catalysts needs to be regenerated by reaction with oxygen. At high temperatures (700 °C), regeneration of the surface $[Li^+OH^-]$ occurs under removal of lattice oxygen. At lower temperatures gas-phase oxygen is needed for the regeneration, forming a hydroperoxy radical under recreation of the active center $[Li^+O^-]$. The hydroperoxy radical can react to hydrogen peroxide under regeneration of a second active center and finally after scission to two hydroxyl radicals regenerate two more active centers (eqn (2.4)). Overall, one oxygen molecule is able to regenerate four active centers.

$$\begin{aligned} &[Li^+OH^-]_s + O_2 \rightarrow [Li^+O^-]_s + HO_2\bullet \\ &[Li^+OH^-]_s + HO_2\bullet \rightarrow [Li^+O^-]_s + H_2O_2 \end{aligned} \qquad (2.4)$$

In contrast to the original proposal, surface-science studies by the group of Goodman showed that the concentration of $[Li^+O^-]$ sites does not correlate with the activity for methane activation.[92] Instead, F^0 centers (oxygen vacancies containing two electrons) at or near the surface of MgO have been established to be the sites inducing the C–H bond breaking. One should note that such sites would correspond to such surface sites of low coordination, which are in equilibrium with the near-surface F centers. The presence of Li^+ substituted into the MgO lattice induces a higher concentration of such sites. It is interesting to note also that the $[Li^+O^-]$ sites are mostly found in the bulk and not at the surface. A recent study of F^0 centers on MgO (001) showed for the first time the quantitative measure of the high strength of interaction between the tip of an AFM microscope and these sites, giving reference to their unusual reactivity.[93]

It should be noted at this point that C–H activation can also be achieved on a wide variety of nonreducible oxides (for an excellent review see ref. 94). Most notable, C–H bond cleavage can also be achieved over nonreducible oxides such as alumina on highly defective sites. In contrast to the chemistry discussed above, the C–H bond cleavage involves the heterolytic cleavage into a proton coordinating in the final state to oxygen at the site and a carbanion that interacts with the metal cation.[95] Also, several cases of cations in zeolites, such Ga^{3+}-, La^{3+}- and Zn^{2+}- exchanged zeolites have been found to activate C–H bonds in an analogous way. In these cases a very high strength of interaction of the exposed cations[96] with the C–H bonds leads to the heterolytic dissociation.[38,97] While the final product, *i.e.*, olefins in the simplest case, represent an oxidation step with respect to the starting molecule, we will not discuss this chemistry in detail here, as the C–H activation step is not typical for C–H activation involving oxygen.

2.5.2 Bifunctional Oxidative Catalysis with Main Group Cations at Zeolite Exchange Sites

So far, we have discussed the role of specific catalytic sites involving also the metal–oxygen bonds and anionic species catalyzing the C–H bond separation. However, surprisingly only the coordinating effect of a cation in a large electrostatic field is sufficient to induce C–H bond transfer. This requires the confined space of, *e.g.*, a zeolite that enforces the proximity between an alkane and molecular oxygen. Exchangeable cations provide the high local field apparently necessary to reduce the energy needed to excite the hydrocarbon oxygen charge transfer.[98] It is remarkable that for reactions in the gas phase the energy to perform such charge transfer lies energetically much higher[99] than when both reaction partners are confined to a zeolite cage. The source of the energy to stimulate seems not to be critical, *i.e.*, the reaction can be induced by visible light[100] as well as by thermal energy.[101]

It appears, however, that the local geometry of the arrangement between oxygen the alkane and the cation is critical and strongly determined by the size and equilibration of ion exchange of the cation.

This is best seen in the comparison between Ca^{2+} and Mg^{2+} and their interactions with the alkane. The direct importance and accessibility of the alkaline earth metal cations has been established by a series of ion-exchange experiments, which in essence showed that the catalyst activity drastically increased, once the ion-exchange level led to cations accessible by the reacting molecules.[102] A more subtle local arrangement is important for the type of interaction with the alkane (see Figure 2.10).[103] The small difference in the ionic radius of $Mg^{2+}$and Ca^{2+} leads to a deeper embedding of the former into the plane of the six-ring SII site in zeolite Y. In consequence, the DFT calculations indicate that Mg^{2+} interacts with alkanes mainly in η^2 mode, while with Ca^{2+}interaction with methane and ethane leads preferentially to the η^3 mode. For propane even indications for adsorption in η^4 mode have been found. The larger cation leads to more

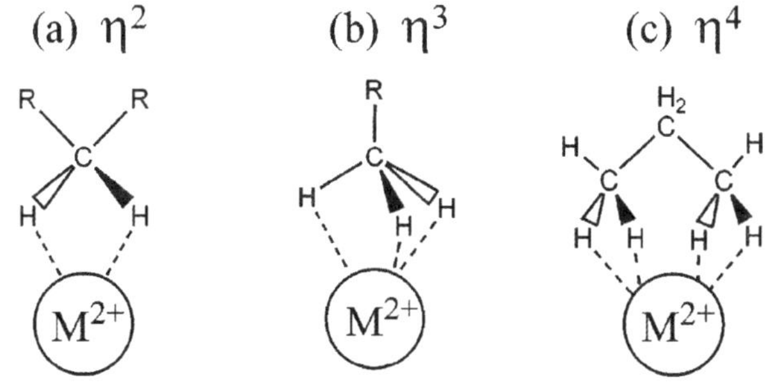

Figure 2.10 Schematic representation of possible conformations of light alkanes to the exchanged cation.

electrostatic contributions of the bonding, while the shielding of Mg^{2+} induces contributions of directed hydrogen bonding to the oxygen of the zeolite lattice.

In-situ IR spectroscopy reveals that indeed not only the alkane interacts with the cations, but that also molecular oxygen forms an adsorption complex, the stretching frequency of which is observable in the adsorbed state, because of the induction of a dipole by these interactions.[104] It should be noted that the interaction with the better-shielded Mg^{2+} is significantly weaker than with Ca^{2+} cations. Molecular oxygen and alkanes compete for the adsorption sites, but indirect evidence exists that both are indeed adsorbed in the proximity of the cation. This geminal adsorption on one cation would lead to a charge-transfer intermediate, which would take the form $[C_3H_8{}^+O_2{}^-]$.[105] Proposed to be aided by protons, it rearranges eventually to hydroperoxide, which decomposes for the case of propane to propan-2-ol or acetone.

2.6 Conclusion and Outlook

The discussed examples show that the catalyzed oxidative C–H activation proceeds in the dominant number of cases *via* single-electron processes induced, however, by a wide variety of catalytically active sites. These span from classic transition-metal terminal or bridging oxygen groups, exemplified best by vanadia, over transiently formed terminal oxygen, such as in the Cu-exchanged zeolite-based catalysts, to the F^0 centers in MgO-type catalysts and the facilitation of the formation of charge-transfer complexes in zeolites by exposed cations. Despite the homolytic cleavage of the C–H bond, the lability of the metal–oxygen bond and the polarizability of the C–H bond facilitate the reaction and appear to lower the energies involved in the reaction pathways. The achievable selectivity remains a great challenge, because most of the materials contain accessible metal cations and labile oxygen, also providing polar pathways to the subsequent oxidation of the intermediate, once the C–H bond has been cleaved in the first (rate-determining) activation step. Certainly, fewer defects and undercoordinated cations would help to improve the selectivity for a particular catalyst.

While impressive progress has been made in the characterization of such catalysts, the sites active in the C–H bond cleavage are frequently minority sites constituting only a very small fraction of the surface. Together with the fact that the surface of the oxides and the active sites in zeolites are very sensitive to the reactive environment, this poses a formidable challenge to characterize the sites. Such insight, together with well-targeted synthesis approaches are, however, the key to the next generation of catalysts for C–H activation and selective functionalization.

References

1. T. F. Narbeshuber, H. Vinek and J. A. Lercher, *J. Catal.*, 1995, **157**, 388.
2. J. A. Lercher, R. A. van Santen and H. Vinek, *Catal. Lett.*, 1994, **27**, 91.
3. G. Centi, F. Cavani, and F. Trifiro in *Selective Oxidation by Heterogeneous Catalysis*, Kluwer Academic/Plenum Publishers, New York, Boston, Dordrecht, London, Moscow, 2001, 1.
4. J. M. M. Millet, *Top. Catal.*, 2006, **38**, 83.
5. F. Cavani and F. Trifiro, *Catal. Today*, 1999, **51**, 561.
6. B. Grzybowska-Swierkosz, *Top. Catal.*, 2000, **11**, 23.
7. T. Ushikubo, K. Oshima, A. Kayou, M. Vaarkamp and M. Hatano, *J. Catal.*, 1997, **169**, 394.
8. J. Holmberg, R. K. Grasselli and A. Andersson, *Top. Catal.*, 2003, **23**, 55.
9. R. K. Grasselli, *Catal. Today*, 1999, **49**, 141.
10. F. Eder and J. A. Lercher, *J. Phys. Chem.*, 1997, **101**, 1273.
11. P. Botella, J. M. L. Nieto, B. Solsona, A. Mifsud and F. Marquez, *J. Catal.*, 2002, **209**, 445.
12. J. M. M. Millet, H. Roussel, A. Pigamo, J. L. Dubois and J. C. Jumas, *Appl. Catal. A*, 2002, **232**, 77.
13. W. Ueda, D. Vitry and T. Katou, *Catal. Today*, 2005, **99**, 43.
14. M. M. Lin, *Appl. Catal. A*, 2003, **250**, 305.
15. M. M. Lin, *Appl. Catal. A*, 2003, **250**, 287.
16. W. A. Goddard, K. Chenoweth, S. Pudar, A. C. T. van Duin and M. J. Cheng, *Top. Catal.*, 2008, **50**, 2.
17. I. E. Wachs, *Catal. Today*, 2005, **100**, 79.
18. M. D. Argyle, K. Chen, E. Iglesia and A. T. Bell, *J. Phys. Chem. B*, 2005, **109**, 2414.
19. M. D. Argyle, K. Chen, C. Resini, C. Krebs, E. Iglesia and A. T. Bell, *J. Phys. Chem. B*, 2004, **108**, 2345.
20. D. E. Keller, S. M. K. Airaksinen, A. O. Krause, B. M. Weckhuysen and D. C. Koningsberger, *J. Am. Chem. Soc.*, 2007, **129**, 3189.
21. G. Deo and I. E. Wachs, *J. Catal.*, 1994, **146**, 323.

22. X. Gao, M. A. Banares and I. E. Wachs, *J. Catal.*, 1999, **188**, 325.
23. X. Gao, J.-M. Jehng and I. E. Wachs, *J. Catal.*, 2002, **209**, 43.
24. Z. Zhao, X. Gao and I. E. Wachs, *J. Phys. Chem. B*, 2003, **107**, 6333.
25. S. Feyel, D. Schröder, X. Rozanska, J. Sauer and H. Schwarz, *Angew. Chem. Int. Ed.*, 2006, **45**, 4677.
26. S. Feyel, D. Schröder and H. J. Schwarz, *J. Phys. Chem. A*, 2006, **110**, 2647.
27. X. Rozanska and J. Sauer, *J. Phys. Chem. A*, 2009, **113**, 11586.
28. X. Rozanska and J. Sauer, *Int. J. Quantum Chem.*, 2008, **108**, 2223.
29. S. Feyel, D. Schröder, X. Rozanska, J. Sauer and H. Schwarz, *Angew. Chem.*, 2006, **118**, 4793.
30. X. Rozanska, R. Fortrie and J. Sauer, *J. Phys. Chem. C*, 2007, **111**, 6041.
31. M. J. Cheng, K. Chenoweth, J. Oxgaard, A. van Duin and W. A. Goddard, *J. Phys. Chem. C*, 2007, **111**, 5115.
32. X. Xu, F. Faglioni and W. A. Goddard, *J. Phys. Chem. A*, 2002, **106**, 7171.
33. H. Fu, Z. P. Liu, Z. H. Li, W. N. Wang and K. N. Fan, *J. Am. Chem. Soc.*, 2006, **128**, 11114.
34. X. Rozanska, E. V. Kondratenko and J. Sauer, *J. Catal.*, 2008, **256**, 84.
35. V. N. Shetti, M. J. Rani, D. Srinivas and P. Ratnasamy, *J. Phys. Chem. B*, 2006, **110**, 677.
36. D. Srinivas and P. Ratnasamy, *Stud. Surf. Sci. Catal.*, 2007, **170**, 1205.
37. A. V. Ramaswamy, S. Sivasanker and P. Ratnasamy, *Microporous Mesoporous Mater.*, 1994, **2**, 451.
38. C. Sievers, A. Onda, A. Guzman, K. S. Otillinger, R. Olindo and J. A. Lercher, *J. Phys. Chem. C*, 2007, **111**, 210.
39. W. Fan, R. G. Duan, T. Yokoi, P. Wu, Y. Kubota and T. Tatsumi, *J. Am. Chem. Soc.*, 2008, **130**, 10150.
40. G. B. Shul'pin, T. Sooknoi, V. B. Romakh, G. Süss-Fink and L. S. Shul'pina, *Tetrahedron Lett.*, 2006, **47**, 3071.
41. C. B. Kouw, C. B. Dartt, J. A. Labinger and M. E. Davis, *J. Catal.*, 1994, **149**, 195.
42. K. Chen, S. Xie, A. T. Bell and E. Iglesia, *J. Catal.*, 2001, **198**, 232.
43. G. Fu, X. Xu, X. Lu and H. Wan, *J. Phys. Chem. B*, 2005, **109**, 6416.
44. G. Fu, X. Xu, X. Lu and H. Wan, *J. Am. Chem. Soc.*, 2005, **127**, 3989.
45. K. D. Chen, A. T. Bell and E. Iglesia, *J. Phys. Chem. B*, 2000, **104**, 1292.
46. K. Chen, S. Xie, A. T. Bell and E. Iglesia, *J. Catal.*, 2000, **195**, 244.
47. J. A. Lercher and H. Noller, *J. Catal.*, 1982, **77**, 152.
48. H. Dai, A. T. Bell and E. Iglesia, *J. Catal.*, 2004, **221**, 491.
49. S. Yang, E. Iglesia and A. T. Bell, *J. Phys. Chem. B*, 2006, **110**, 2732.
50. E. Heracleous and A. A. Lemonidou, *J. Catal.*, 2010, **270**, 67.
51. M. M. Lin, *Appl. Catal. A*, 2001, **207**, 1.
52. P. Beato, A. Blume, F. Girgsdies, R. E. Jentoft, R. Schlögl, O. Timpe, A. Trunschke, G. Weinberg, Q. Basher, F. A. Hamid, S. B. A. Hamid, E. Omar and L. Mohd Salim, *Appl. Catal. A*, 2006, **307**, 137.

53. J. M. Oliver, J. M. López Nieto and P. Botella, *Catal. Today*, 2004, **96**, 241.
54. P. Botella, E. Garcia-Gonzalez, J. M. López Nieto and J. M. González-Calbet, *Solid State Sci.*, 2005, **7**, 507.
55. R. K. Grasselli, D. J. Buttrey, J. D. Burrington, A. Andersson, J. Holmberg, W. Ueda, J. Kubo, C. G. Lugmair and A. F. Volpe Jr, *Top. Catal.*, 2006, **38**, 7.
56. P. Botella, E. García-González, A. Dejoz, J. M. López Nieto, M. I. Vázquez and J. González-Calbet, *J. Catal.*, 2004, **225**, 428.
57. T. Katou, D. Vitry and W. Ueda, *Catal. Today*, 2004, **96**, 235.
58. T. Katou, D. Vitry and W. Ueda, *Catal. Today*, 2004, **91–92**, 237.
59. R. K. Grasselli, D. J. Buttrey, P. DeSanto Jr., J. D. Burrington, C. G. Lugmair, A. F. Volpe Jr. and T. Weingand, *Catal. Today*, 2004, **91–92**, 251.
60. R. Balasubramanian and A. C. Rosenzweig, *Acc. Chem. Res.*, 2007, **40**, 573.
61. S. I. Chan, V. C.-C. Wang, J. C.-H. Lai, S. S.-F. Yu, P. P.-Y. Chen, K. H.-C. Chen, C.-L. Chen and M. K. Chan, *Angew. Chem. Int. Ed.*, 2007, **46**, 1992.
62. G. I. Panov, A. K. Uriarte, M. A. Rodkin and V. I. Sobolev, *Catal. Today*, 1998, **41**, 365.
63. M. J. Rice, A. K. Chakraborty and A. T. Bell, *J. Catal.*, 2000, **194**, 278.
64. J. S. Woertink, P. J. Smeets, M. H. Groothaert, M. A. Vance, B. F. Sels, R. A. Schoonheydt and E. I. Solomon, *Proc. Natl. Acad. Sci. U.S.A.*, 2009, **106**, 18909.
65. M. H. Groothaert, P. J. Smeets, B. F. Sels, P. A. Jacobs and R. A. Schoonheydt, *J. Am. Chem. Soc.*, 2005, **127**, 1394.
66. V. V. Guliants, S. A. Holmes, J. B. Benziger, P. Heaney, D. Yates and I. E. Wachs, *J. Mol. Catal. A: Chem.*, 2001, **172**, 265.
67. B. K. Hodnett, *Catal. Rev. Sci. Eng.*, 1985, **27**, 373.
68. G. Centi, F. Trifiro, J. R. Ebner and V. M. Franchetti, *Chem. Rev.*, 1988, **88**, 55.
69. Y. Zhang-Lin, M. Froissier, R. P. Sneeden, J. C. Vedrine and J. C. Volta, *J. Catal.*, 1994, **145**, 256.
70. Y. Zhang-Lin, M. Froissier, J. C. Vedrine and J. C. Volta, *J. Catal.*, 1994, **145**, 267.
71. D. J. Thompson, I. M. Ciobica, B. K. Hodnett, R. A. van Santen and M. O. Fanning, *Catal. Today*, 2004, **91–92**, 177.
72. M. Hävecker, R. W. Mayer, A. Knop-Gericke, H. Bluhm, E. Kleimenov, A. Liskowski, D. Su, R. Follath, F. G. Requejo, D. F. Ogletree, M. Salmeron, J. A. Lopez-Sanchez, J. K. Bartley, G. J. Hutchings and R. Schlögl, *J. Phys. Chem. B*, 2003, **107**, 4587.
73. F. Cavani, S. Luciani, E. D. Esposti, C. Cortelli and R. Leanza, *Chem. Eur. J.*, 2010, **16**, 1646.

74. S. Gaab, J. Find, T. E. Müller and J. A. Lercher, *Top. Catal.*, 2007, **46**, 101.
75. B. Tope, Y. Zhu and J. A. Lercher, *Catal. Today*, 2007, **123**, 113.
76. C. P. Kumar, S. Gaab, T. E. Müller and J. A. Lercher, *Top. Catal.*, 2008, **50**, 156.
77. G. A. Olah, B. Gupta, M. Farina, J. D. Felberg, W. M. Ip, A. Husain, R. Karpeles, K. Lammertsma, A. K. Melhotra and N. J. Trivedi, *J. Am. Chem. Soc.*, 1985, **107**, 7097.
78. S. G. Podkolzin, E. E. Stangland, M. E. Jones, E. Peringer, A. A. Lemonidou and J. A. Lercher, *J. Am. Chem. Soc.*, 2007, **129**, 2569.
79. E. Peringer, C. Tejuja, M. Salzinger, A. A. Lemonidou and J. A. Lercher, *Appl. Catal. A*, 2008, **350**, 178.
80. E. Peringer, S. G. Podkolzin, M. E. Jones, R. Olindo and J. A. Lercher, *Top. Catal.*, 2006, **38**, 211.
81. J. H. Lunsford, *Angew. Chem., Int. Ed.*, 1995, **34**, 970.
82. A. H. Kaddouri, N. De Blasio and R. Del Rosso, *React. Kinet. Catal. Lett.*, 2001, **72**, 309.
83. K. D. Campbell, H. Zhang and J. H. Lunsford, *J. Phys. Chem.*, 1988, **92**, 750.
84. C. T. Au, H. He, S. Y. Lai and C. F. Ng, *Appl. Catal. A*, 1997, **159**, 133.
85. S. Sugiyama, K. Sogabe, T. Miyamoto, H. Hayashi and J. B. Moffat, *Catal. Lett.*, 1996, **42**, 127.
86. P. Ciambelli, L. Lisi, R. Pirone, G. Ruoppolo and G. Russo, *Catal. Today*, 2000, **61**, 317.
87. A. E. Schweizer, M. E. Jones and D. A. Hickman, WO Patent 6 452 058, 2002.
88. S. J. Conway and J. H. Lunsford, *J. Catal.*, 1991, **131**, 513.
89. S. J. Conway, D. J. Wang and J. H. Lunsford, *Appl. Catal. A*, 1991, **79**, L1.
90. S. Fuchs, L. Leveles, K. Seshan, L. Lefferts, A. Lemonidou and J. A. Lercher, *Top. Catal.*, 2001, **15**, 169.
91. S. Gaab, M. Machli, J. Find, R. K. Grasselli and J. A. Lercher, *Top. Catal.*, 2003, **23**, 151.
92. M. C. Wu, C. M. Truong, K. Coulter and D. W. Goodman, *J. Catal.*, 1993, **140**, 344.
93. T. König, G. H. Simon, U. Martinez, L. Giordano, G. Pacchioni, M. Heyde and H.-J. Freund, *ACS Nano*, 2010, **4**, 2150.
94. C. Copéret, *Chem. Rev.*, 2010, **110**, 656.
95. J. Joubert, A. Salameh, V. Krakoviack, F. Delbecq, P. Sautet, C. Copéret and J. M. Basset, *J. Phys. Chem. B*, 2006, **110**, 23944.
96. E. A. Pidko, R. A. van Santen and E. J. M. Hensen, *Phys. Chem. Chem. Phys.*, 2009, **11**, 2893.
97. E. A. Pidko, E. J. M. Hensen and R. A. van Santen, *J. Phys. Chem. C*, 2007, **111**, 13068.
98. H. Frei, *Science*, 2006, **313**, 309.

99. H. Tsubomura and R. S. Mulliken, *J. Am. Chem. Soc.*, 1960, **82**, 5966.
100. F. Blatter and H. Frei, *J. Am. Chem. Soc.*, 1993, **115**, 7501.
101. J. Xu, B. L. Mojet, J. G. van Ommen and L. Lefferts, *J. Phys. Chem. B*, 2004, **108**, 15728.
102. J. Xu, B. L. Mojet, J. G. van Ommen and L. Lefferts, *J. Catal.*, 2005, **232**, 411.
103. E. A. Pidko, J. Xu, B. L. Mojet, L. Lefferts, I. R. Subbotina, V. B. Kazansky and R. A. van Santen, *J. Phys. Chem. B*, 2006, **110**, 22618.
104. J. Xu, B. L. Mojet, J. G. van Ommen and L. Lefferts, *J. Phys. Chem. B*, 2005, **109**, 18361.
105. P. Li, Y. Xiang, V. H. Grassian and S. C. J. Larsen, *J. Phys. Chem. B*, 1999, **103**, 5058.

Section I

Oxide-Catalyzed Selective Oxidations of Light Hydrocarbons

CHAPTER 3

Catalysis of Oxidative Methane Conversions

EVGENII V. KONDRATENKO[1] AND MANFRED BAERNS[2]

[1] Leibniz Institute for Catalysis at the University of Rostock, Albert-Einstein-Str. 29a, D-18059 Rostock, Germany; [2] Fritz-Haber-Institut der Max-Planck-Gesellschaft, Dept. of Inorganic Chemistry, Faradayweg 4-6, 14195 Berlin, Germany

3.1 Introduction

This chapter deals with the catalytic conversions of methane to C_2 hydrocarbons, *i.e.* ethane and its consecutive product ethylene, as well as to methanol and formaldehyde. The heterogeneous catalysts and the mechanisms of these reactions are described and discussed. For an advanced understanding of catalytic methane conversions, mechanistic aspects of enzymatic conversion of methane to methanol are also dealt with, assuming that this might be a contribution for an improved design of respective heterogeneous catalysts.

3.2 Methane as a Feedstock for Oxidative Conversions

In the future, the reserves of methane being the main constituent of natural gas will exceed those of crude oil. Currently, combustion of natural gas is mainly used for heating and generating electrical power. However, such utilization of

RSC Nanoscience & Nanotechnology No. 19
Nanostructured Catalysts: Selective Oxidations
Edited by Christian Hess and Robert Schlögl

Published by the Royal Society of Chemistry, www.rsc.org

methane does not meet the future needs for efficient use of raw materials. Therefore, a strong economical interest exists in developing processes enabling methane conversion to higher-valued products. Selected research programs underline the present interest in this field: for example, since 2007 Dow Chemicals[1] has been funding a research program "Methane Challenge". This focuses on methods for methane conversion to ethylene and propylene. At Cardiff University, Graham Hutchings' group is designing supported gold and bimetallic gold alloy catalytic materials for methane oxidation below 473 K. At Northwestern University K. Tobin Marks' team is aiming on achieving selective methane oxidation to olefins or other useful feedstocks. According to ref. 1 this group approaches the task by (i) controlling oxidant concentrations by nano/subnanoscale catalyst environment engineering, (ii) application of μ-oxo complexes creating centers tuned for methane partial oxidation to methanol, and (iii) use of soft oxidants (presumably N_2O and SO_3) to avoid overoxidation. From an applied point of view, strategies for the utilization of methane depend on its price, location, the demand for products, and the required effort for separating reaction products and unconverted feed within such a process. From a chemical point of view, the methane molecule is very stable; the strength of the C–H bond amounts to 439 kJ/mol (the molecule does not have any functional groups that might facilitate chemical attacks). The situation is changed in the presence of oxidizing agents. The most attractive present technologies for utilization of methane are shown on the right-hand side of Figure 3.1.

Steam reforming of methane is a mature technology, which is widely used for producing syngas, *i.e.* carbon monoxide and hydrogen.[2] An important industrial utilization of methane is the production of hydrocyanic acid *via* the Andrussow[3] or Degussa processes.[3] Hüls[4] and BASF[4] have developed processes for acetylene production from methane providing the required heat of this endothermic process either by an electric arc or by partial oxidation of natural gas above 1573 K in a special reactor at very short contact times for preventing cracking of acetylene formed. The most attractive alternative technologies for oxidative utilization of methane are shown on the left-hand side of Figure 3.1. These process routes employing novel technologies and

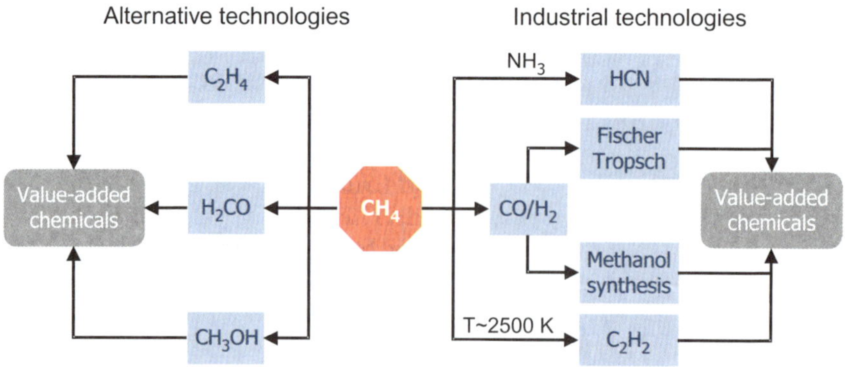

Figure 3.1 Conversions of methane into value-added chemicals.

catalysts could result in significant cost advantages for raw materials compared to current technologies.

3.3 Oxidative Coupling of Methane (OCM)

3.3.1 Catalytic Materials and Modes of Process Operation

Catalytic materials for the oxidative coupling of methane (OCM) can be divided into four groups according to the systematization of Lee and Oyama:[5]

i) reducible metal oxides, *e.g.* $NaMnO_4/MgO \cdot SiO_2$;
ii) nonreducible metal oxides, *e.g.* Li/MgO, SrO/La_2O_3;
iii) halogen-containing oxide materials, *e.g.* BaF_2/Y_2O_3;
iv) solid electrolytes, *e.g.* $SrCe_{0.9}Yb_{0.1}O_{3-x}$.

Best OCM performances and catalytic materials reported in open and patent literature are summarized in Figure 3.2. The black and gray lines illustrate upper bonds of C_2 ($C_2H_4 + C_2H_6$) yields predicted for O_2 and CH_4 cofeeding[6] and alternating feeding[7] of the reactants, respectively; the simulations have been performed on the basis of gas-phase and surface kinetics. Various engineering means have been suggested to improve product separation and overall C_2 yields. These are the countercurrent moving-bed chromatographic reactor,[8] staged catalytic fixed-bed reactors with distributed and cofeed oxygen addition,[9] and product removal over several stages,[10] oxygen-conducting membrane reactors,[11–13] recycling reactors with continuous removal of ethylene over 5A zeolite[14] or as an aqueous silver ion complex *via* a membrane contactor.[15]

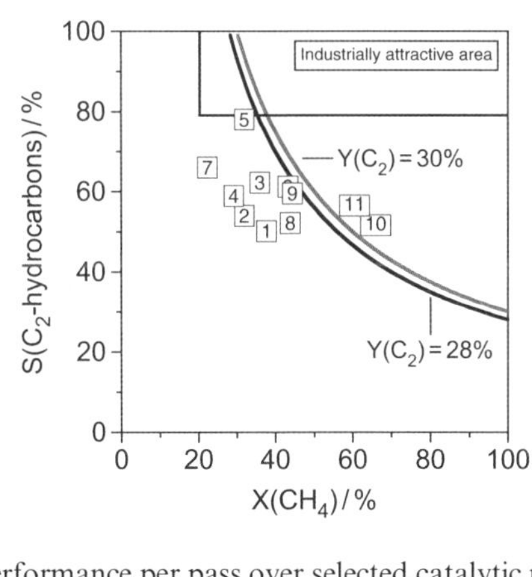

Figure 3.2 OCM performance per pass over selected catalytic materials: 1 – Li/MgO,[16] 2 – 30%BaO/Ga_2O_3,[17] 3 – 95 mol%BaF_2/Y_2O_3,[18] 4 – La/MgO,[19] 5 – Rb_2WO_4/SiO_2,[20] 6 – $Bi_{1.5}Y_{0.3}Sm_{0.2}O_{3-\delta}$,[21] 7 – La_2O_3–CeO_2,[22] 8 – Na_2WO_4/SiO_2,[20] 9 – 10%Y_2O_3–89%ZrO_2–1%TiO_2 membrane with $Li_2O/MgO/ZrO_2$,[23] 10 – $Ba(1{-}0.5x)TiO_3SnCl_2(0.05x)$ (x is 1 mol),[24] 11 – $Bi_{1.5}Y_{0.3}Sm_{0.2}O_{3-\delta}$.[25]

None of these approaches appear presently economically viable due to the recycle of huge amounts of unconverted methane and expensive product separation.

3.3.2 Interaction of Methane on Solid Catalytic Surfaces

From a fundamental point of view the oxidative coupling of methane is an interesting reaction; it occurs *via* a mechanism including heterogeneous and homogeneous steps coupled with each other. Since a methane molecule does not possess any functional groups but four equivalently strong C–H bonds, as pointed out above, the OCM reaction is initiated by breaking one of these bonds by oxygen species on the surface of the catalyst. For the mechanism of C–H bond activation, two concepts exist:

(1) The assumption of heterolytic breaking of the C–H bond is based on the well-established experimental observation that selective catalytic materials are of basic nature.[26,27] According to this concept a proton is abstracted by the basic surface site of O^{2-}, while the remaining methyl anion CH_3^- is linked to the surface cation M^{n+} (eqn (3.1)). Thus, it may be concluded that there is no change of the valence state of the metal center and that methane is reversibly activated as suggested by the experimental measurement of the extent of CH_4/CD_4 isotopic equilibration.[28] Moro-oka, however, concluded that no such strong acid–base pairs exist on most metal-oxide catalysts and that heterolytic cleavage would be less preferable on most metal-oxide surfaces.[29] In fact, there was no direct experimental evidence for the heterolytic cleavage of the C–H bond in the rate-determining step on metal-oxide catalysts.

$$CH_4 + M^{n+}O^{2-} \rightarrow M^{n+}CH_3^- + O^{2-}H^+ \tag{3.1}$$

(2) In the homolytic cleavage mechanism (eqn (3.2)), there is still an open question whether the initial oxidation step involves a two-electron (2e) transfer or single-electron (1e) transfer. Haber postulated,[30,31] that in the activation of a C–H bond on oxide surfaces, the proton would attach to the surface oxygen to form a hydroxyl group (OH), while the alkyl cation would attach to the surface oxygen forming an alkoxy group (OR). Two electrons originally localized in the C–H bond would be injected into the conductive band of the solid.[30] The presence of hydroxyl and alkoxy groups detected by IR spectroscopy on oxide surfaces was interpreted as experimental support of Haber's mechanism.[32] In support of the homolytic cleavage, Sinev, who has applied kinetic and thermochemical analysis of the OCM reaction, suggested that a single-electron transfer process involving the free radical formation was likely to be the most favorable process.[33]

$$CH_4 + M^{n+}O^{2-} \rightarrow CH_3^{\bullet} + M^{(n-1)+} + OH^- \tag{3.2}$$

In contrast to the mechanism of breaking C–H bond in CH_4, there is no doubt that methyl radicals are primary gas-phase products. Lunsford and associates[34,35] observed methyl radicals by means of matrix-isolation electron-spin resonance over various catalytic materials. The formation of gas-phase methyl radicals *via* heterogeneous reaction of methane with adsorbed oxygen species on MgO and Sm_2O_3 was also proven by Buyevskaya *et al.*[36] using the temporal analysis of products (TAP) reactor operated under Knudsen-diffusion conditions. Later, methyl radicals were observed *in situ* upon methane oxidation over platinum at temperatures between 1273 and 1573 K.[37]

3.3.3 Mechanistic Insights into Ethane, Ethylene and CO_x Formation

Methyl radicals generated on the catalyst surface recombine in the gas phase to ethane. In the presence of gas-phase O_2, these radicals can react to $CH_3O_2^{\bullet}$ and/or $CH_3O^{\bullet}$ radicals, which are precursors of CO_x. Alternatively, methyl radicals can undergo heterogeneous transformations to CO_x. Ethylene is formed either *via* homogeneous or heterogeneous ethane dehydrogenation. As reported in ref. 38, methane inhibits homogeneous oxidation of ethane. Similar to methane oxidation, the initial step of heterogeneously catalyzed ethane oxidation to ethylene is breaking a C–H bond of the molecule by a surface-oxygen species. Thus, this reaction competes with methane activation. Due to the lower bonding energy of the C–H bond in C_2H_6 compared to CH_4, the rate of ethane consumption is 5–20 times higher than that of methane consumption.[39,40] Ethylene is even more reactive than ethane. Due to the strong difference in the reactivities of C_1 and C_2 hydrocarbons, consecutive oxidation of ethane and ethylene is the main reaction pathway leading to CO_x at high methane conversions. These nonselective reactions occur *via* homogeneous and heterogeneous pathways, the latter playing the most significant role. Therefore, the main challenge of the OCM reaction is minimizing CO_x formation. Since the nonselective gas-phase reactions are difficult to control, significant improvements in the OCM performance may presumably only be achieved by controlling heterogeneous reaction pathways, which are influenced by various physicochemical properties of the catalysts applied. Therefore, possible approaches for developing novel selective catalytic materials comprise knowledge about fundamental relationships between selectivity and activity on the one hand and physicochemical properties on the other hand, as well as knowledge about reaction mechanism and microkinetic parameters.

3.3.4 Oxygen Species in the OCM Reaction

The most essential reaction step of the OCM reaction is the generation of active oxygen species from gas-phase oxygen on the catalyst surface. The type of oxygen species formed from O_2 on the catalyst surface and the physicochemical properties of solids are crucial selectivity-governing factors. Therefore, establishing relationships between these properties and individual selective and

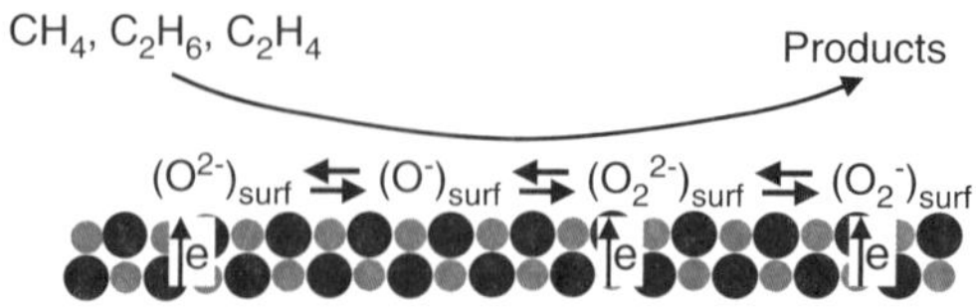

Figure 3.3 Oxygen species stabilized on the surface of oxide catalysts upon activation of gas-phase O_2.[42–44]

nonselective reaction pathways of the OCM reaction is a highly important requirement for designing novel catalytic materials. The diversity of oxygen species, which can exist on the catalyst surface and participate in the formation of selective and nonselective reaction products, is illustrated in Figure 3.3. However, the nature of these species in the OCM reaction is still debated. This is due to the variety of OCM catalysts and due to the experimental limitations for an unambiguous identification of active sites at high temperatures (>873 K), which are required for methane activation. According to ref. 41, the following physicochemical properties of solid materials are required for generation of active oxygen species:

i) the presence of oxygen vacancies or impurity metal ions or interstitial sites;
ii) the catalysts should be p-type semiconductors with a bandgap of 5–6 eV.

The latter requirement is due to the fact that the bandgap determines the concentration of charge carriers. If the bandgap is too low, a high concentration of adsorbed oxygen species is expected, which favors total oxidation over the selective formation of C_2-products. The poor performance of n-type conductors is ascribed to their low ability for generating methyl radicals but their high activity for total oxidation of the radicals.

Irrespective of the nature of the OCM catalysts, oxygen adsorption takes place on defects in the structural lattice, if the oxide structure enables oxygen to enter these sites.[41] Lattice oxygen (O^{2-}) is directly formed from gas-phase O_2 upon its interaction with reducible metal oxides: the metal cation changes its oxidation state with simultaneous electron transfer to oxygen. Since the bandgap in oxides of nonreducible metals is high compared to the reducible ones, differently charged oxygen species (O^-, O_2^-, O_2^{2-}, and O^{2-}) can exist on the catalyst surface. Intrinsic factors (anion vacancies, interstitial sites) and impurity defects in these oxides play a significant role in electronic transfers, since the metal cations of the host oxide matrix cannot change their oxidation state.

O^- is the most discussed oxygen species in the OCM reaction. Lunsford and associates[34,45] identified $M(Li\ or\ Na)^+O^-$ centers in Li/MgO, Li/ZnO and Na/CaO catalysts by means of ESR (electron spin resonance) spectroscopy. Since M^+O^- centers were identified in the bulk of the catalysts, these authors suggested that there exists equilibrium between bulk and surface species. The high reactivity of O^- oxygen species in the OCM reaction was supported by periodic DFT (density-functional theory) calculations over La_2O_3.[46,47] However, the authors suggested a mechanism considering the participation of O_2^{2-} in the breaking of a C–H bond in the methane molecule. The role of O^- was not completely excluded.

The ability of O_2^{2-} species for methane activation was supported by the results of methane oxidation over Na_2O_2 and BaO_2 in the absence of gas-phase oxygen.[48,49] By means of *in-situ* Raman analysis, O_2^{2-} species were identified over Na/La_2O_3 and Sr/La_2O_3[50] at 973 K and over BaO/MgO[51] at 973 and 1023 K upon methane oxidation with oxygen. In addition to O_2^{2-}, O_2^- species were also observed over catalytic materials based on rare-earth oxides in a recent *in-situ* Raman study.[52] It was suggested that O_2^- and O_2^{2-} originate from O_3^{2-}. The latter species is formed *via* reversible coupling of O_2 with a neighboring O^{2-}. A correlation between the rate of the disappearance of the O_2^- species and C_2H_4 formation was observed over a SrF_2/La_2O_3 catalyst.[53] This result suggests that O_2^- may be an active oxygen species responsible for the oxidative coupling of methane.

In previous studies,[54–56] it was demonstrated that the presence of Na in Na/CaO catalysts influenced the nature of surface oxygen species originating from gas-phase O_2 as well as catalytic performance in the OCM reaction and in the oxidative dehydrogenation of ethane (ODE) to ethylene. The latter reaction occurs also under conditions of the OCM reaction. For all the catalysts, C_2 (C_2H_4 and C_2H_6) and C_2H_4 selectivity in the OCM and ODE reactions, respectively, are affected by the alkane-to-oxygen ratio and sodium concentration. Some of the dependencies are illustrated in Figure 3.4.

An increase in oxygen partial pressure results in a decrease in selectivity towards desired products; this was accompanied by an increase in CO_x selectivity. CO_2 formation prevails over CO production in the whole range of oxygen partial pressures. This indicates that C_2H_6 and C_2H_4 are oxidized to CO_2. Besides the oxygen effect, selectivity towards C_2 products increases with an increase in sodium loading at a constant oxygen partial pressure as shown in Figure 3.4. In contrast to C_2 selectivity, the selectivity towards CO and CO_2 decreases with an increase in Na content. Therefore, it can be concluded that nonselective consecutive oxidation of C_2 hydrocarbons to CO_x is inhibited upon doping of CaO with Na. Based on the results of measurements of the electrical conductivity and work function of electrons,[56] it was concluded that the incorporation of Na^+ into the cationic sublattice of CaO results in the creation of anion vacancies in the anionic sublattice. The latter are active sites for O_2 adsorption. The concentration of surface anion vacancies in Na–CaO materials was determined from multipulse O_2 experiments performed in the TAP reactor;[56] the higher the Na loading, the higher the concentration of anion vacancies. In order to derive additional insights into the effect of Na in Na–CaO materials on the mechanism of O_2 activation, a detailed microkinetic analysis of O_2 activation on the basis of elementary reaction steps including surface intermediates was performed.[55] From this microkinetic analysis it was concluded that the alkali dopant does not change the mechanism of oxygen activation and the state of surface oxygen species. Gas-phase O_2 adsorbs over anion vacancies reversibly and dissociatively *via* a moleculary adsorbed

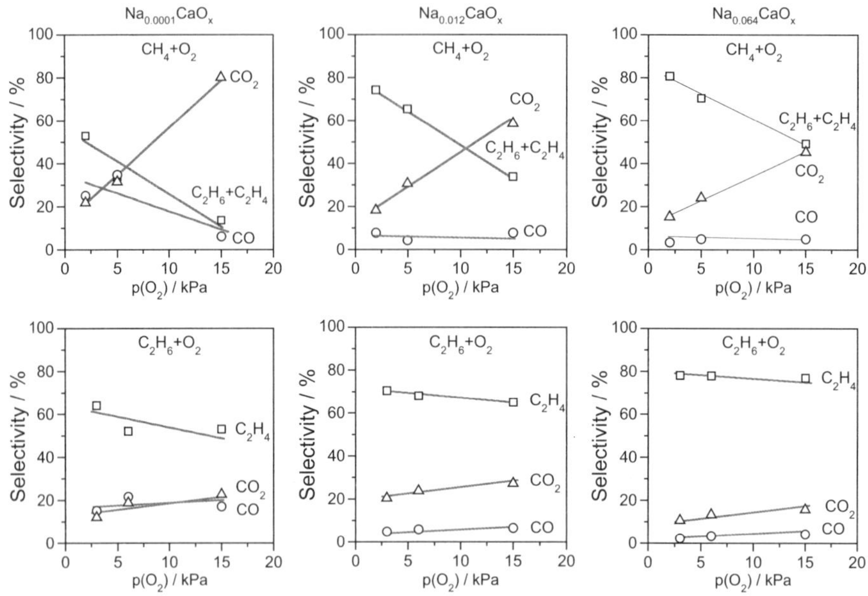

Figure 3.4 Effect of oxygen partial pressure on selectivity of the OCM and ODE reactions over an Na–CaO catalyst at 1023 K and 923 K, respectively ($X(CH_4) \sim 10\%$, $p(CH_4) = 30$ kPa, $X(C_2H_6) \sim 10\%$, $p(C_2H_6) = 30$ kPa). Data are from refs. 54–56.

precursor. Only the catalyst ability for dissociation of biatomic oxygen species seems to distinguish differently doped Na–CaO catalysts; the higher the sodium loading, the faster the dissociation. This is related to the creation of additional anion vacancies in the lattice of CaO upon incorporation of Na^+ into cationic lattice positions. By taking into account the microkinetic knowledge of O_2 adsorption, the ability of the catalyst for fast transformation of biatomic oxygen species into atomic ones was identified as a key catalyst property for achieving high C_2 selectivity. It was suggested that monoatomic oxygen species are responsible for selective oxidation (ethane and ethylene formation), while biatomic oxygen species take part in consecutive oxidation of C_2 hydrocarbons to CO_x. The dissociation of biatomic oxygen species can be accelerated by creation of anion vacancies in the oxide lattice upon adding low-valence additives into the host oxide matrix. This statement was experimentally proven by doping of CaO with Na_2O,[56] CeO_2 with CaO[57] and Nd_2O_3 with SrO.[58]

Comparing the results of the OCM reaction using O_2 and N_2O as oxidizing agents indirectly supports the importance of the nature of oxygen species. In refs. 59–63, higher C_2 selectivity was found over Na/CaO and Li/CaO catalytic materials with low (<2 at.%) Li and Na loading with N_2O than with O_2. The difference was related to the different oxygen species generated from O_2 and N_2O. Adsorption of gas-phase O_2 was suggested to occur *via* a molecular precursor followed by its dissociation to O^{2-}_{cus} (cus: coordinatively unsaturated sites) and noncharged atomic oxygen (O). The latter is selective and also generated upon N_2O activation. The O^{2-}_{cus} species was assumed to participate in CO formation, since the rate of CO formation strongly decreased upon replacing O_2 by N_2O.

3.3.5 Selectivity- and Activity-Governing Factors

The importance of structural defects of catalytic materials for generating selective and nonselective oxygen species as well as for achieving high C_2 selectivity has been repeatedly reported for the OCM reaction.[41,55,56,58,64–66] The authors of ref. 64 divided all known OCM catalysts into two groups: (i) multiphase catalysts and (ii) single-phase catalysts. For the former group of catalysts, particular phases appear to be required for good performing catalysts. Oxygen vacancies and ions of transitional metals as impurities are the main defects of single-phase catalysts. Structural defects influence the type of electrical conductivity of solid materials. The electrical properties play an essential role in the OCM reaction.[66] However, the defect structure is changed under OCM conditions as outlined in refs. 60,61,64,67. These reaction-induced defects influence the catalytic performance. For M/CaO (M = Li or Na < 2 at.%), $[MCO_3]^-$, CO_2^{2-} and O_3^{2-} defects were identified in the volume of the CaO lattice after the OCM reaction took place at 1023 K. An increase in the fraction of the $[M(Li\ or\ Na)CO_3]^-$ centers in doped CaO resulted in a rise of C_2 selectivity using an O_2-containing reaction feed.[60,61] CO_2^{2-} species were considered as precursors for CO, while O_3^{2-} was assumed to be a nonselective oxygen species for consecutive oxidation of C_2 products. However, when N_2O was used as oxidant, no structural changes in the catalyst volume were observed. The differences in catalytic performance were related to the nature of oxygen species generated from O_2 and N_2O.

Nagy *et al.*[67] demonstrated the importance of reaction-induced changes in the silver-catalyzed OCM reaction. Ag faceting promotes the formation of O_γ species (nucleophilic oxygen embedded in the uppermost layer of silver atoms) as determined by the authors. These species were suggested to be responsible for improved catalytic performance.

Based on past and more recent characterization studies, it has been suggested that the catalytic performance of various solid materials for selective oxidative transformation of light hydrocarbons was negatively affected by surface acidity[68,69] but positively by basicity.[70] However, Choudhary and coworkers[71] did not observe a direct relationship between surface basicity and catalytic activity/selectivity over CaO doped with alkali-metal oxides. For example, a Na/CaO catalyst (Na:Ca = 0.1 before calcination) showed better OCM performance (C_2 selectivity of 68.8% at 24.7% methane conversion) than Rb/CaO materials. In a recent OCM study with CO_2,[72] it was suggested that the strength and distribution of surface basic sites are important factors determining the OCM performance. According to this study, selective OCM catalysts should possess low concentrations of very strong basic sites (CO_2 desorption above 923 K).

Based on the above background it is clear that the OCM reaction is influenced by various physicochemical properties, which, however, have not been previously considered together for deriving a final conclusion on their relevance for catalyst design. Therefore, a better understanding of the various physicochemical properties and their effect on reaction mechanism, microkinetics and, as results thereof, on catalytic performance would certainly allow controlling the overall process of methane conversion and, hence providing useful guidelines for a rational catalyst design and development. It needs to be seen whether there are synergistic or possibly even adverse effects of different solid-state properties on the desired catalytic performance. This can be achieved by systematic catalyst characterization using modern surface and bulk techniques as well as microkinetic analysis.

3.4 Methane Oxidation to Methanol and Formaldehyde

3.4.1 Interaction of Oxygen and Methane on Transition-Metal Oxides

The direct oxidation of methane into methanol and/or formaldehyde is even more challenging than the OCM reaction. In contrast to the heterogeneously catalyzed methane oxidation, the natural enzyme catalyst mono-oxygenase in methanotrophic bacteria demonstrates that highly selective methane oxidation to methanol is possible even at room temperature.[73] This is an enticing example of what could be achieved with other catalytic materials. Also, as proven by theoretical analysis and experimental data, methyl radicals and methanol were formed after the interaction of methane with gaseous $[NiO]^+$,[74] $[FeO]^+$,[74] $[MoO_3]^+$,[75] $[OsO_n]^+(n=0\text{–}4)$,[76] $[V_4O_{10}]^+$,[77] and $[MgO]^+$[78] clusters. This leads to a recent review by Fu *et al.*[79] who analyzed different C–H activation pathways over terminal [=O] and bridged [-O-] oxygen species. On this basis, Mo_3O_9 was used as a model system to support previous findings. The results provide some additional insights into the fundamental issues with respect to C–H activation by metal oxo-species involved in many heterogeneous, homogeneous and enzymatic catalytic processes.

In the direct conversion of methane and oxygen to methanol and formaldehyde, activation of the reactants is of fundamental and practical importance. Two issues need to be solved:

- First, the chemical inertness of methane as demonstrated by the high C–H bond strength (104.8 kcal/mol), high ionization potential (12.5 eV), low proton affinity (4.4 eV), and low acidity ($pK_a = 48$) of the molecule has to be overcome.
- Secondly, the propensity for full oxidation in the presence of oxygen is to be surmounted.

The authors of ref. 79 refer to various studies on the mechanisms of alkane oxidation by transition-metal oxides. A detailed and final mechanism of C–H bond activation on these catalysts remains, however, elusive.[80–85]

Some interesting but partly contradictory results were obtained in recent years on the mechanism of the oxidation of methane. Lunsford *et al.*[86] reported that MoO_3 supported on silica was a catalyst for the selective oxidation of methane to methanol and formaldehyde when N_2O is the oxidant. The O^- ions coordinated to Mo^{VI} at the surfaces were claimed to be the active species, capable of abstracting a hydrogen atom from methane. The methyl radicals formed reacted rapidly with the oxidic surface to form methoxide complexes, which then decompose to HCHO, or react with water to form CH_3OH. Barbaux *et al.*[87] investigated the properties of surface oxygen species generated by different oxidants on the MoO_x systems. They found that O^- generated by O_2 led to the formation of CO_x, while O^{2-} generated by N_2O led to higher selective oxidation to HCHO. Smith and Ozkan examined the partial oxidation of methane to formaldehyde over MoO_3/SiO_2.[88] Emphasis was on the surface species formed on the supports and their effects on reactivity and selectivity: low MoO_3 loadings led to the formation of silicomolybdic species with terminal Mo=O sites, while higher loadings gave rise to the formation of polymolybdate species with Mo–O–Mo bridges, the former having a better selectivity for HCHO than the latter.[88,89] Interestingly, other experiments seem to support the opposite view: it was concluded that bridge oxygen [-O-] was responsible for selective oxidation, while terminal oxygen [=O] was related to total oxidation.[90]

3.4.2 Methane to Methanol

3.4.2.1 Homogeneous and Heterogeneous Methanol Formation

Methanol can be directly formed from methane *via* homogeneous gas-phase reactions of radical nature at high pressures (50–100 bar) in the temperature range from 573 to 973 K.[82] The methanol yields achieved are rather low and cannot be easily tuned, since radical reactions are difficult to control. Therefore, heterogeneously catalyzed methane oxidation has been intensively investigated besides liquid-phase oxidation. Methanol selectivities up to 80% at methane conversions below 4% have been reported over Mo-, Fe-, and V-based catalytic materials.[80,91–94] From a mechanistic point of view, methanol formation occurs either *via* direct insertion of active oxygen species into a C–H bond in CH_4 or *via* oxidative abstraction of an H atom yielding a methyl radical and an OH group. The methyl radical formed undergoes further surface or gas-phase oxidations to methoxide species. The latter can react with water to form methanol or can decompose to formaldehyde. Both these oxygenates are easily oxidized to carbon oxides. This gives rise to a fundamental problem of selective CH_4 conversion to CH_3OH (similar to OCM) without its further consecutive oxidation to CO_x *via* H_2CO. This situation results in the need to isolate the selective reaction products before side products are formed. The extent of such isolation depends on the relative rates of formation and further oxidation of

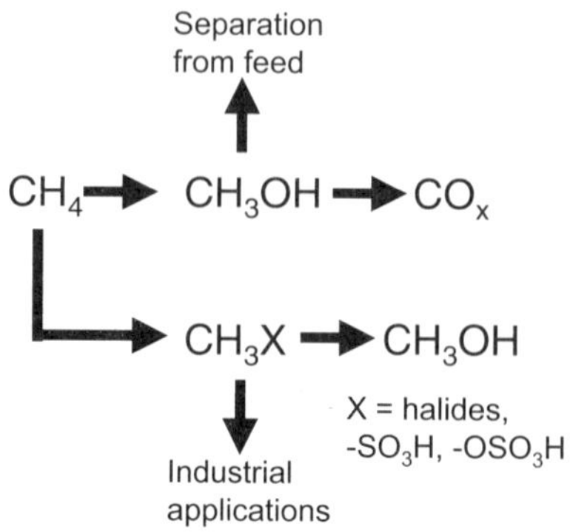

Figure 3.5 Possible alternatives for preventing methanol overoxidation to CO_x.

CH_3OH. As shown schematically in Figure 3.5, methanol overoxidation can be avoided by continuous removal of methanol from the reactive zone or generating methyl derivatives, which are chemically protected from overoxidation to carbon oxides. Hereafter, these derivatives might be transformed to methanol or directly used for industrial purposes.

Chemical protection for liquid-phase conversion of methane to methanol was originally demonstrated by Periana *et al.*[95,96] *via* the formation of methyl bisulfate in oleum at 373 K. A bipyrimidyl platinum (II) complex has been identified as the most efficient and highly selective catalyst resulting in the formation of methyl bisulfate with a one-pass yield of 72% (81% selectivity based on methane). Although the achieved catalytic performance appears to be attractive, this process scheme suffers from the production of large amounts of diluted sulfuric acid, and from hydrolyzing the methyl ester to methanol. Muehlhofer *et al.*[97] have alternatively reported the formation of trifluoroacetic acid methyl ester upon methane oxidation in a mixture of trifluoroacetic acid and trifluoroacetic acid anhydride at a methane pressure of 20–30 bar and 353–373 K. In this case, palladium (II) complexes with N-heterolytic carbine ligands were used as catalysts. The authors mentioned that trifluoroacetic acid methyl ester can be easily hydrolyzed and the acid recovered can be used for the reaction, although this was not demonstrated.

Inspired by the studies of liquid-phase methane oxidation to methanol derivatives, Li and Yuan[98] suggested a process scheme for gas-phase methane oxidation to methanol using various solid materials as catalysts: $Pt(SO_4)_2$, $HgSO_4$, $Ce(SO_4)_2$ and $Pd(SO_4)_2$ supported over $BaSO_4$ nanotubes. Methanol yields up to 28% were reported in the presence of gas-phase oxygen as oxidant in the temperature range from 443 to 503 K. The catalysts were pretreated in concentrated sulfuric acid at 393 K for 3 h before they were applied for methane oxidation. Among the catalysts investigated, $Pt(SO_4)_2/BaSO_4$ showed the best performance; methanol selectivities of about 70% at methane conversions of around 40% were achieved. The authors suggested that sulfuric acid deposited on the catalysts is used for methane activation, yielding methyl bisulfate followed by its hydroxylation to methanol and sulfuric acid. Gas-phase O_2 oxidizes SO_2 formed during the methane activation by SO_3. Although this latter approach might be interesting for possible industrial application, the authors did not provide any data on the time-on-stream performance, which may be influenced by possible removal of the acid during the reaction course. Moreover, corrosion problems have to be taken into account.

CH_3OH selectivity above 55% and CO_2 selectivity below 10% at 14% CH_4 conversion were reported by Li and Yuan[99] upon gas-phase oxidation of CH_4 by O_2 over $MoO_2Br_2(H_2O)$/Zn-MCM-48 at 493 K and 10 bar. Other reaction products were dimethyl ether and methane bromides. This high catalytic performance was, however, not stable with time-on-stream and depended on the content of ZnO in the catalyst; the higher the ZnO loading, the more stable the performance and the higher the production of dimethyl ether. Catalyst deactivation with time-on-stream was related to bromine leaching as bromine fumes and methane bromides. This leaching is a significant drawback. In order to overcome this drawback, the catalyst ability for converting methane bromides to methanol and ZnO to $ZnBr_2$ as well as MoO_3 to MoO_2Br_2 should be significantly improved.

The abstraction of hydrogen from methane can be facilitated upon feeding gaseous radical initiators like NO_x ($x = 1, 2$) into CH_4–O_2 feeds. Previous literature dealing with the effect of NO_x on methanol and formaldehyde production from methane has been thoroughly reviewed in 2002.[80] In general, NO increases the catalyst activity and selectivity towards oxygenates. Up to now, the highest overall yield of methanol and formaldehyde of 16% was achieved over a low surface area V_2O_5/SiO_2 catalyst.[100] Methane was oxidized with O_2 in the presence of NO_x. The positive effect of NO_x is related to acceleration of gas-phase reactions leading to methanol and formaldehyde.

3.4.2.2 Enzymatic Conversion of Methane to Methanol

As opposed to the various types of chemical conversions of methane to methanol as mentioned above, its enzyme-catalyzed reaction to methanol may also be of high interest due to the studies on a molecular level. The knowledge derived from such investigations may contribute to possible new routes in designing novel and improved inorganic catalysts for the reaction of methane to methanol.

Shan and Que[101] outlined in a review article that the initial step in the enzymatic (biomimetic) oxidation of alkanes is the activation of oxygen. High-valent iron-oxo species ($Fe^{IV}=O$ or $Fe^{V}=O$) are often invoked as the key oxidizing agents in the catalytic cycles of oxygen-activating nonheme iron enzymes. A principal mechanistic scheme of oxygen activation at biological iron sites is shown in Figure 3.6.

Arguments in support of a metal-based oxidant instead of oxygen radical are ascribed to the high degree of conversion of the feed molecule and the nature of the reaction products. For example, it has been assumed that among the mononuclear nonheme iron enzymes one class, *i.e.* the Rieske

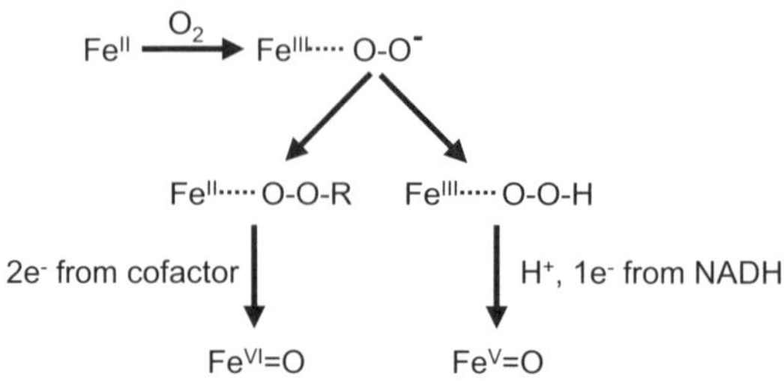

Figure 3.6 Mechanistic scheme for oxygen activation at biological iron sites [ ]; $Fe^{IV}=O(porph^*)$ for heme enzymes, $Fe^{IV}{}_2O_2$ for MMO, $Fe^{V}=O(OH)$ for Rieske dioxygenases.[102]

dioxygenases, involves a formal $Fe^{V}=O$ oxidant in the oxygen-activation mechanism.[102]

Recently Rinaldo *et al.*[103] dealt in a theoretical study with the intermediates in dioxygen activation by soluble methane mono-oxygenase (MMO) for oxidation of methane to methanol. They combined QM/MM and broken-symmetry DFT methods. The work revealed that two iron sites (see above) need a distinct environment. A protein environment helps driving the reactants toward products along the reaction path. In summary, various reaction steps are required for converting methane to methanol on $Fe=O$ sites as illustrated in Figure 3.7.

Despite their common oxidation state of the two iron atoms the electronic structure reveals that they are not equivalent. They are in different environments, with one iron having an octahedral geometry and the other one being in a square pyramidal environment; hereby, their interaction with oxygen, and hence, with methane *via* the attached oxygen is supposed to be different.

From the above results as obtained by Shan and Que[101] as well as by Rinaldo *et al.*[103] indications may be derived for developing novel or modifying known catalytic solid materials for the gas-phase oxidation of methane to methanol.

3.4.3 Methane to Formaldehyde

Various catalytic materials have been tested for methane oxidation to formaldehyde. It is generally suggested that the active and selective catalysts should fulfil a double role: (i) dehydrogenation of methane and (ii) oxygen addition. Most of the catalysts studied are supported ones; most often silica or silica-alumina are used as support, while oxides of vanadium, molybdenum, iron, tungsten, and nickel are the active components. It is suggested that the effectiveness of silica is related to its low ability for consecutive formaldehyde conversion to CO_x. Figure 3.8 compares formaldehyde selectivities obtained over different materials as a function of methane conversion;

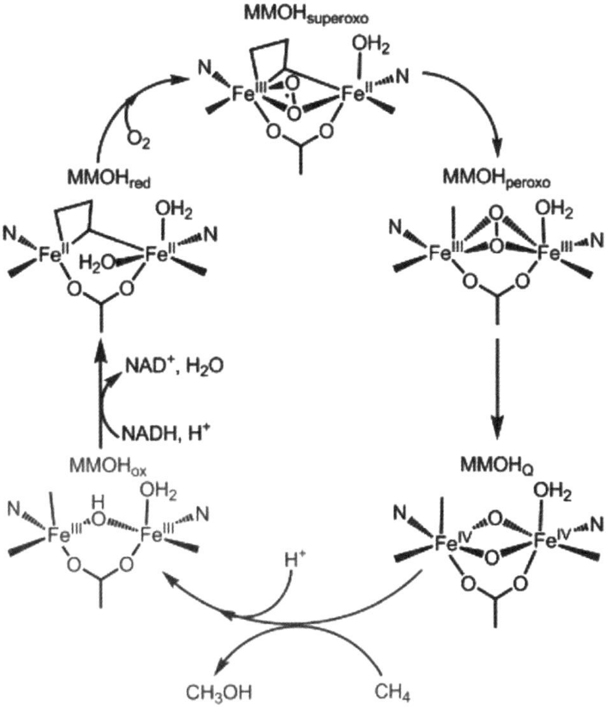

Figure 3.7 The catalytic cycle of MMO (in its various forms). The part of the cycle corresponding to activation of dioxygen is shown in black, while the oxidation of the substrate (hydroxylation of methane to methanol *in vivo*) is in light gray. The data are from ref. 103.

catalysts yielding less than 2% of CH_4 conversion and formaldehyde selectivities below 20% are not included in this figure. It should be stressed that these data were obtained under various reaction conditions using both O_2 and N_2O as oxidizing agents and NO_x as gas-phase dopants. None of the catalysts investigated up to now led to formaldehyde yields of at least 20%. The low formaldehyde selectivity is due to heterogeneous reaction steps, which occur on both active components and supports and by homogeneous formaldehyde combustion.

Aiming at fundamental knowledge for developing novel well-performing catalytic materials, the effect of physicochemical properties, but also reaction mechanisms and kinetics, have been widely investigated. A key parameter for good formaldehyde selectivity is the dispersion of MoO_x, VO_x and FeO_x species on the support surface.[111,117–123] A deeper kinetic analysis of direct methane oxidation to formaldehyde has been performed by Bell and coworkers[124] who used a MoO_x/SiO_2 catalyst with a nominal surface density of molybdenum of 0.44 Mo nm^{-1}. The gas-phase reactions did not contribute significantly to the product distribution between 848 and 923 K. Formaldehyde formation from

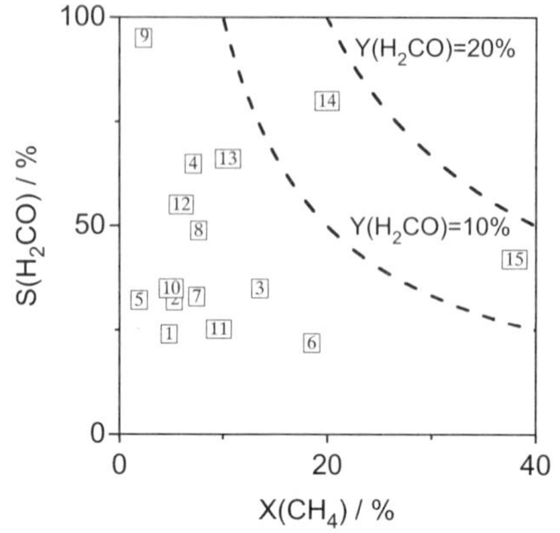

Figure 3.8 Formaldehyde selectivity *versus* methane conversion over selected catalytic materials: 1 – SiO_2,[104] 2 – MoO_3/SiO_2,[105] 3 – V_2O_5/SiO_2,[106] 4 – MoSnP/SiO_2,[107] 5 – Zr-P-O_x,[108] 6 – Zn-$Fe_2(MoO_4)_3$,[109] 7 – $TiO_2 + CH_2Cl_2$,[110] 8 – V_2O_5/SiO_2,[111] 9 – VO_x(3 wt.%)/SBA-15,[111] 10 – SiO_2 4-5P,[112] 11 – $Sn_{0.8}Ge_{0.2}O_2$,[113] 12 – MoO_3/SiO_2 with N_2O,[114] 13 – 40% $FePO_4$/MCM-41 with N_2O,[115] 14 – $MoCl_5/R_4Sn/SiO_2$ R is an alkyl group,[116] 15 – total selectivity of H_2CO and CH_3OH over V_2O_5/SiO_2 with NO addition.[100]

methane occurs predominantly over MoO_x species, while formaldehyde combustion takes place additionally over bare silica. Since methane and formaldehyde compete for the same active sites (lattice oxygen of MoO_x), the highest single-pass yield of formaldehyde can be achieved when the rate of methane conversion is increased compared to that of consecutive formaldehyde consumption. The kinetics predicts that the rate of methane consumption depends more strongly on the coverage by oxidized sites than the rate of the consecutive formaldehyde reaction. Moreover, the activation energy of the reaction of methane conversion to formaldehyde is higher than that of formaldehyde consumption. Therefore, the maximal formaldehyde yield is predicted when the methane oxidation is performed at high CH_4/O_2 ratios and reaction temperatures as high as possible where gas-phase reaction can be still avoided.

3.5 Outlook

Although oxidative methane conversion towards the coupling products ethane and ethylene as well as to methanol and formaldehyde are of significant industrial interest the fundamentals of its heterogeneous catalysis are still not satisfactory for designing highly selective catalysts. Thorough studies including experimental as well as theoretical work are still needed; the results on the biocatalytic conversion of methane to methanol as mentioned above might be supportive.

References

1. *http://www.dow.com/innovation/news/2008/&10080124a.htm.*
2. P. F. van den Oosterkamp, *Encyclopedia of Catalysis*, ed. I. Horvath, Wiley-VCH Verlag GmbH & Co. KGaA, Weinheim, 2003, **6**, 770.
3. Ullmann's Encyclopedia of Industrial Chemistry, *Cyano Compounds, Inorganic*, Wiley-VCH, Weinheim, 2003, **10**, 774.
4. Ullmann's Encyclopedia of Industrial Chemistry, *Acetylene*, Wiley-VCH, Weinheim, 2003, **1**, 785.
5. J. S. Lee and S. T. Oyama, *Catal. Rev. Sci. Eng.*, 1988, **30**, 249.
6. Y. S. Su, J. Y. Ying and W. H. Green, *J. Catal.*, 2003, **218**, 321.
7. J. A. Labinger, *Catal. Lett.*, 1988, **1**, 371.
8. A. V. Kruglov, M. C. Bjorklund and R. W. Carr, *Chem. Eng. Sci.*, 1996, **51**, 2945.
9. M. Baerns and W. Hinsen, US 4608449, 1996.
10. I. P. Androulakis and S. C. Reyes, *AIChem. J.*, 1999, **45**, 860.
11. D. Eng and M. Stoukides, *Catal. Rev. Sci. Eng.*, 1991, **33**, 375.
12. Y. Zeng and Y. S. Lin, *J. Catal.*, 2000, **193**, 58.
13. Y. Zeng, Y. S. Lin and S. L. Swartz, *J. Membrane Sci.*, 1998, **150**, 87.
14. J. Jiang, I. V. Yentekakis and C. G. Vayenas, *Science*, 1994, **264**, 1563.
15. E. M. Cordi, S. Pak, M. P. Rosynek and J. H. Lunsford, *Appl. Catal. A*, 1997, **155**, L1.
16. T. Ito, J. Wang, C.-H. Lin and J. H. Lunsford, *J. Am. Chem. Soc.*, 1985, **107**, 5062.
17. C. T. Au, K. D. Chen and C. F. Ng, *Appl. Catal. A*, 1998, **170**, 81.
18. C. T. Au, X. P. Zhou, Y. W. Liu, W. J. Ji and C. F. Ng, *J. Catal.*, 1998, **174**, 153.
19. V. R. Choudhary, V. H. Rane and S. T. Chaudhari, *Fuel*, 2000, **79**, 1487.
20. A. Palermo, J. P. H. Vazquez and R. M. Lambert, *Catal. Lett.*, 2000, **68**, 191.
21. Y. Zeng, F. T. Akin and Y. S. Lin, *Appl. Catal. A*, 2001, **213**, 33.
22. A. G. Dedov, A. S. Loktev, I. I. Moiseev, A. Aboukais, J. F. Lamonier and I. N. Filimonov, *Appl. Catal. A*, 2003, **245**, 209.
23. E. A. Hazbun, US 4791079, 1989.
24. H. Abbas, H. Azzis and E. Bagherzadeh, WO 2005005042, 2005.
25. F. T. Akin and Y. S. Lin, *Catal. Lett.*, 2002, **78**, 239.
26. V. D. Sokolovskii and E. A. Mamedov, *Catal. Today*, 1992, **14**, 415.
27. O. V. Krylov, *Catal. Today*, 1993, **18**, 209.
28. G. A. Martin and C. Mirodatos, *Fuel Proc. Technol.*, 1995, **42**, 179.
29. Y. Moro-oka, *Appl. Catal. A*, 1999, **181**, 323.
30. J. Haber, *Stud. Surf. Sci. Catal.*, 1997, **110**, 1.
31. J. Haber and M. Witko, *J. Catal.*, 2003, **216**, 416.
32. G. Busca, E. Finocchio, V. Lorenzelli, G. Ramis and M. Baldi, *Catal. Today*, 1999, **49**, 453.
33. M. Y. Sinev, *J. Catal.*, 2003, **216**, 468.
34. D. J. Driscoll, W. Martir, J.-X. Wang and J. H. Lunsford, *J. Am. Chem. Soc.*, 1985, **107**, 58.

35. J. H. Lunsford, P. G. Hinson, M. P. Rosynek, C. Shi, M. Xu and X. Yang, *J. Catal.*, 1994, **146**, 301.
36. O. V. Buyevskaya, M. Rothaemel, H. W. Zanthoff and M. Baerns, *J. Catal.*, 1994, **146**, 346.
37. M. Geske, K. Pelzer, R. Horn, F. C. Jentoft and R. Schlögl, *Catal. Today*, 2009, **142**, 61.
38. S. Pak, P. Qiu and J. H. Lunsford, *J. Catal.*, 1998, **179**, 222.
39. M. Y. Sinev, V. N. Korchak and O. V. Krylov, *Kinet. Katal.*, 1989, **30**, 860.
40. E. M. Kennedy and N. W. Cant, *Appl. Catal.*, 1991, **75**, 32.
41. E. N. Voskresenskaya, V. G. Roguleva and A. G. Anshits, *Catal. Rev. Sci. Eng.*, 1995, **37**, 101.
42. A. Bielanski, J. Haber, *Oxygen in Catalysis*, Marcel Dekker, New York, 1991.
43. Y. Moro-oka, *Catal. Today*, 1988, **45**, 3.
44. G. I. Panov, A. K. Uriarte, M. K. Rodkin and V. I. Sobolev, *Catal. Today*, 1998, **41**, 365.
45. Y. Tong, M. P. Rosynek and J. H. Lunsford, *J. Phys. Chem.*, 1989, **93**, 2386.
46. M. S. Palmer, M. Neurock and M. M. Olken, *J. Am. Chem. Soc.*, 2002, **124**, 8452.
47. M. S. Palmer, M. Neurock and M. M. Olken, *J. Phys. Chem. B*, 2002, **106**, 6543.
48. K. Otsuka, Y. Murakami, Y. Wada, A. A. Said and A. Morikawa, *J. Catal.*, 1990, **121**, 122.
49. M. Y. Sinev, N. V. Korchak and O. V. Krylov, *Kinet. Katal.*, 1986, **27**, 1274.
50. G. Mestl, H. Knözinger and J. H. Lunsford, *Ber. Buns. Phys. Chem.*, 1993, **97**, 319.
51. J. H. Lunsford, X. Yung, K. Haller, J. Laane, G. Mestl and H. Knözinger, *J. Phys. Chem.*, 1993, **97**, 13810.
52. H. B. Zhang, G. D. Lin, H. L. Wan, Y. D. Liu, W. Z. Weng, J. X. Cai, Y. F. Shen and K. R. Tsai, *Catal. Lett.*, 2001, **73**, 141.
53. L. Wang, X. Yi, W. Weng, C. Zhang, X. Xu and H. Wan, *Catal. Lett.*, 2007, **118**, 238.
54. E. V. Kondratenko, O. V. Buyevskaya and M. Baerns, *J. Mol. Catal. A*, 2000, **158**, 199.
55. E. V. Kondratenko, O. V. Buyevskaya, M. Soick and M. Baerns, *Catal. Lett.*, 1999, **63**, 153.
56. E. V. Kondratenko, D. Wolf and M. Baerns, *Catal. Lett.*, 1999, **58**, 217.
57. D. Wolf, M. Heber, W. Grünert and M. Muhler, *J. Catal.*, 2001, **199**, 92.
58. G. Gayko, D. Wolf, E. V. Kondratenko and M. Baerns, *J. Catal.*, 1998, **178**, 441.
59. A. G. Anshits, V. G. Roguleva and E. V. Kondratenko, *Stud. Surf. Sci. Catal.*, 1994, **82**, 337.

60. E. V. Kondratenko, N. G. Maksimov and A. G. Anshits, *Kinet. Katal.*, 1995, **658**.
61. E. V. Kondratenko, N. G. Maksimov, G. E. Selyutin and A. G. Anshits, *Catal. Today*, 1995, **24**, 273.
62. V. G. Roguleva, E. V. Kondratenko, N. G. Maksimov, G. E. Selyutin and A. G. Anshits, *Catal. Lett.*, 1992, **16**, 165.
63. V. G. Roguleva, M. A. Nikiphorova, N. G. Maksimov and A. G. Anshits, *Catal. Today*, 1992, **13**, 219.
64. A. G. Anshits, E. N. Voskresenskaya, E. V. Kondratenko and N. G. Maksimov, *Catal. Today*, 1995, **24**, 217.
65. P. J. Gellings and H. J. M. Bouwmeester, *Catal. Today*, 2000, **58**, 1.
66. Z. Zhang, X. E. Verikios and M. Baerns, *Catal. Rev. Sci. Eng.*, 1994, **36**, 507.
67. A. Nagy and G. Mestl, *Appl. Catal. A*, 1999, **188**, 337.
68. O. V. Buyevskaya, A. I. Suleimanov, S. M. Aliev and V. D. Sokolovskii, *React. Kinet. Catal. Lett.*, 1988, **22**, 223.
69. V. D. Sokolovskii, G. M. Aliev, O. V. Buyevskaya and A. A. Davydov, *Catal. Today*, 1988, **4**, 293.
70. M. Baerns, in *Methane Conversion by Oxidative Process. Fundamentals and Engineering Aspects*, ed. E. E. Wolf, Van Nostrand Reinhold, New York, 1992, p. 123.
71. V. H. Rane, S. T. Chaudhari and V. R. Choudhary, *J. Natur. Gas Chem.*, 2008, **17**, 313.
72. N. A. S. A. Istadi, *J. Mol. Catal.*, 2006, **259**, 61.
73. J. Colby, D. I. Stirling and H. Dalton, *Biochem. J.*, 1977, **165**, 395.
74. D. Schröder and H. Schwarz, *Angew. Chem. Int. Ed.*, 1995, **34**, 1973.
75. I. Kretzschmar, A. Fiedler, J. N. Harvey, D. Schröder and H. Schwarz, *J. Phys. Chem. A*, 1997, **101**, 6252.
76. K. K. Irikura and J. L. Beauchamp, *J. Am. Chem. Soc.*, 1989, **111**, 75.
77. S. Feyel, J. Döbler, D. Schröder, J. Sauer and H. Schwarz, *Angew. Chem. Int. Ed.*, 2006, **45**, 4681.
78. D. Schröder and J. Roithová, *Angew. Chem. Int. Ed.*, 2006, **45**, 5705.
79. G. Fu, X. Xu, X. Lu and H. Wan, *J. Am. Chem. Soc.*, 2005, **127**, 3989.
80. K. Tabata, Y. Teng, T. Takemoto, E. Suzuki, M. A. Banares, M. A. Pena and J. L. G. Fierro, *Catal. Rev.*, 2002, **44**, 1.
81. N. R. Foster, *Appl. Catal.*, 1985, **19**, 1.
82. H. D. Gesser, N. R. Hunter and C. B. Prakash, *Chem. Rev.*, 1985, **85**, 235.
83. M. A. Banares, *Catal. Today*, 1999, **51**, 319.
84. M. M. Bettahar, G. Costentin, L. Savar and J. C. Lavalley, *Appl. Catal. A*, 1996, **145**, 1.
85. M. M. Lin, *Appl. Catal. A*, 2001, **207**, 1.
86. H. F. Liu, R. S. Liu, D. Y. Liew, R. E. Johnson and J. H. Lunsford, *J. Am. Chem. Soc.*, 1994, **106**, 4117.
87. Y. Barbaux, A. Elamrani and J. P. Bonnelle, *Catal. Today*, 1987, **1**, 147.
88. M. R. Smith and U. S. Ozkan, *J. Catal.*, 1993, **141**, 124.

89. A. de Lucas, J. L. Valverde, P. Canizares and L. Rodriguez, *Appl. Catal. A*, 1999, **184**, 143.
90. B. Grzybowska-Swierkosz, *Top. Catal.*, 2000, **11/12**, 23.
91. M. Baerns and O. V. Buyevskaya, *Erdöl Erdgas Kohle*, 2000, **116**, 25.
92. J. H. Lunsford, *Catal. Today*, 2000, **63**, 165.
93. K. Otsuka and Y. Wang, *Appl. Catal. A*, 2001, **222**, 145.
94. M. Baerns, G. Grubert, E. V. Kondratenko, D. Linke and U. Rodemerck, *Oil Gas Eur. Mag.*, 2003, **1**, 36.
95. R. A. Periana, D. J. Taube, E. R. Evitt, D. G. Loffler, P. R. Wentrcek and T. Masuda, *Science*, 1993, **259**, 340.
96. R. A. Periana, D. J. Taube, S. Gamble, H. Taube, T. Satoh and H. Fujii, *Science*, 1998, **280**, 560.
97. M. Muehlhofer, T. Strassner and W. A. Herrmann, *Angew. Chem. Int. Ed.*, 2002, **41**, 1745.
98. F. Li and G. Yuan, *Chem. Commun.*, 2005, **2238**.
99. F. Li and G. Yuan, *Angew. Chem. Int. Ed.*, 2006, **45**, 6541.
100. J. A. Barbero, M. C. Alvarez, M. A. Banares, M. A. Pena and J. L. G. Fierro, *Chem. Commun.*, 2002, **1184**.
101. X. Shan and L. Que, *J. Inorg. Biochem.*, 2006, **100**, 421.
102. J. L. Que, *J. Biol. Inorg. Chem.*, 2004, **9**, 684.
103. D. Rinado, D. M. Philipp, S. J. Lippard and R. A. Friesner, *J. Am. Chem. Soc.*, 2007, **129**, 3155.
104. G. N. Kastanas, G. A. Tsigdios and J. Schwank, *Appl. Catal.*, 1998, **44**, 33.
105. N. D. Spencer, *J. Catal.*, 1988, **109**, 187.
106. A. Parmaliana, F. Frusteri, A. Mezzapica, M. S. Scurrel and N. Giordano, *J. Chem. Soc., Chem. Commun.*, 1993, **751**.
107. T. Weng and E. E. Wolf, *Appl. Catal.*, 1993, **96**, 383.
108. M. Y. Sinev, S. Setaidi and K. Otsuka, *Mendeleev Commun.*, 1993, **10**.
109. K. Otsuka, Y. Wang, I. Yamanaka, A. Morikawa and M. Y. Sinev, *Stud. Surf. Sci. Catal.*, 1994, **81**, 503.
110. T. R. Baldwin, R. Burch, G. D. Squire and S. C. Tsang, *Appl. Catal.*, 1991, **75**, 153.
111. L. D. Nguyen, S. Loridant, H. Launay, A. Pigamo, J. L. Dubois and J. M. M. Millet, *J. Catal.*, 2006, **237**, 38.
112. F. Arena, G. Gatti, G. Martra, S. Coluccia, L. Stievano, L. Spadaro, P. Famulari and A. Parmaliana, *J. Catal.*, 2005, **231**, 365.
113. K. Tabata, T. Kawabe, Y. Yamaguchi, E. Suzuki and T. Yashima, *J. Catal.*, 2005, **231**, 430.
114. H. Iwamoto, S. Kagawa and J. H. Lunsford, *Shokubai (Catalysis)*, 1985, **23**, 278.
115. X. Wang, Y. Wang, Q. Tang, Q. Guo, Q. Zhang and H. Wan, *J. Catal.*, 2003, **217**, 457.
116. V. Amir-Ebrahimi and J. J. Rooney, *J. Molec. Catal.*, 1989, **50**, L17.
117. A. Parmaliana, V. Sokolovskii, D. Miceli, F. Arena and N. Giordano, *J. Catal.*, 1994, **148**, 514.

118. A. Parmaliana and F. Arena, *J. Catal.*, 1997, **167**, 57.
119. F. Arena, N. Giordano and A. Parmaliana, *J. Catal.*, 1997, **167**, 66.
120. H. Berndt, A. Martin, A. Brückner, E. Schreier, D. Müller, H. Kosslick, G. U. Wolf and B. Lücke, *J. Catal.*, 2000, **191**, 384.
121. V. Fornés, C. López, H. H. López and A. Martínez, *Appl. Catal. A*, 2003, **249**, 345.
122. H. Launay, S. Loridant, D. L. Nguyen, A. M. Volodin, J. L. Dubois and J. M. M. Millet, *Catal. Today*, 2007, **128**, 176.
123. H. Launay, S. Loridant, A. Pigamo, J. L. Dubois and J. M. M. Millet, *J. Catal.*, 2007, **246**, 390.
124. N. Ohler and A. T. Bell, *J. Catal.*, 2005, **231**, 115.

CHAPTER 4

Propane-Selective Oxidation to Acrylic Acid

ANNETTE TRUNSCHKE

Fritz-Haber-Institut der Max-Planck-Gesellschaft, Faradayweg 4-6, 14195 Berlin, Germany

4.1 Introduction

Current manufacture of acrylic acid is based on propylene generated mainly by high-temperature processes, such as steam reforming or fluid catalytic cracking using petroleum feedstock. Alternative technologies directed to the use of natural gas or biotransformation concepts are motivated by the requirement to reduce CO_2 emissions and to optimize utilization of resources and energy. From an industrial point of view there are clear economic and environmental incentives to develop catalysts for direct oxidation of propane to acrylic acid. Selective transformation of propane into acrylic acid corresponds to an eight-electron oxidation and involves propylene and acrolein as formal reaction intermediates (Scheme 4.1).

Considering the multitude of steps, selectivity is an essential issue particularly because allylic C–H bonds in propylene-derived intermediates are more easily activated compared to the nonpolar C–H bonds in propane. In addition, the abundance and spatial arrangement of active sites that simultaneously bring about H-abstraction and O-insertion is critical. The high selectivity to acrylic acid that has been achieved over chemically and structurally complex multi-component oxides, like quinternary Mo–V–Te–Nb mixed oxides, may imply

RSC Nanoscience & Nanotechnology No. 19
Nanostructured Catalysts: Selective Oxidations
Edited by Christian Hess and Robert Schlögl

Published by the Royal Society of Chemistry, www.rsc.org

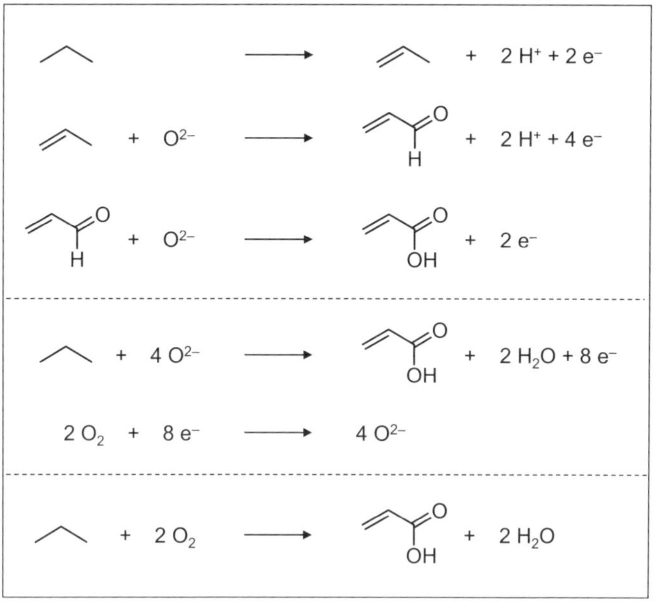

Scheme 4.1 Formal reaction steps in selective oxidation of propane to acrylic acid.

that a balance of isolation and cooperation of active sites or ensembles on the catalyst surface is needed.

Direct oxidation of propane to acrylic acid has been touched on in reviews previously.[1–5] In the present chapter, an attempt is undertaken to examine critically relations between catalyst synthesis, bulk structure, surface termination, and dynamics of the catalyst surface in the feed of the reactants bringing high-performance multimetal oxides into the focus of analysis and compiling thereby mainly results from the open literature. Multimetal-oxide catalysts are characterized by high crystallinity, well-defined crystal structures and high selectivity at comparatively high conversion. These properties make them attractive also with regard to fundamental studies aiming at a deeper understanding of the essential aspects that determine selectivity in oxidation catalysis.

4.2 Technological Background

Acrylic acid is an important intermediate in the chemical industry for manufacture of acrylates and polyacrylic acid. The global production capacity lies in the range of 4 million tons per year and is growing at approximately 3–4% per year. Acrylates are used for the production of, *e.g.*, coatings, sealants, adhesives and textile fibers, whereas glacial acrylic acid is mainly converted to polyacrylates for superabsorbent applications. Historically, acrylic acid was manufactured mainly by C–C coupling reactions in processes based on Reppe chemistry (eqn (4.1)). Acetylene was converted in the presence of carbon monoxide and water over catalysts containing carbonyl-forming metals, such as $NiBr_2/CuI$ in the BASF process.[6]

$$HC{\equiv}CH + CO + H_2O \rightarrow CH_2{=}CHCOOH \tag{4.1}$$

The process was operated until the mid-1990s. Acid-catalyzed hydrolysis of acrylamide was another technology that is no longer used for production of acrylic acid.

Today, acrylic acid is exclusively manufactured by oxidation of acrolein derived from propylene in two-stage processes.[7]

$$H_2C{=}CHCH_3 + O_2 \rightarrow H_2C{=}CHCHO + H_2O \quad \Delta_R H = -340.8\,kJ/mol \tag{4.2}$$

$$2\,H_2C{=}CHCHO + O_2 \rightarrow 2H_2C{=}CHCOOH \quad \Delta_R H = -254.1\,kJ/mol \tag{4.3}$$

The oxidation of propylene is performed at a reaction temperature of 623 ± 50 K over molybdenum and bismuth oxides. In commercial catalysts, a number of promoter elements such as Fe, Co, Ni, W, and K have been added to optimize the catalyst performance approaching propylene conversion and acrolein selectivity close to 100%. The consecutive oxidation of acrolein to acrylic acid requires lower temperatures. Typically, a reaction temperature of 543 K is applied. Molybdenum–vanadium oxides promoted for instance with W, and Cu are used as catalysts. The whole process works very efficiently giving total acrylic acid yields of almost 90%. Commercial technologies include multitubular fixed-bed reactors with 25 mm inner diameter and a length of 5–7 m, placed within a molten salt as the heat-transfer fluid. 30–40 thousand tubes are bundled in the respective reactors for the two steps, which are connected either in series or one reactor is divided into two temperature zones. The fundamental studies undertaken to improve the understanding of selective oxidation of propylene and acrolein have been summarized in a number of review articles, *e.g.* refs. 1,7–11.

Economic reasons have initiated research activities aimed at exploration of alternative routes to acrylic acid. In this respect, the propylene price that dominates the crude acrylic acid production costs is the main driving force. For substitution of propylene, the following three options have been considered:

- direct oxidation of propane to acrylic acid;
- integration of nonoxidative or oxidative dehydrogenation of propane to propylene (three-step process);
- pathways based on renewable raw materials such as lactic acid[12] or glycerol.[13]

Propane-based processes are very attractive routes for utilization of abundant natural-gas resources, however, industrial implementation of such technologies has failed so far. Despite improved energy efficiency of the exothermal oxidative dehydrogenation of alkanes, olefin selectivities achieved till now do not justify a replacement of established olefin production processes based on steam reforming or fluid catalytic cracking (FCC), because fuel-grade feedstocks are already used in these technologies and, therefore, the raw material price is not a crucial factor.[14] This is different for the direct conversion of alkanes to oxygenates, such as propane to acrylic acid.[15]

$$C_3H_8 + 2O_2 \rightarrow H_2C{=}CHCOOH + 2H_2O \quad \Delta_R H^0 = -852\,\text{kJ/mol} \qquad (4.4)$$

Current propylene prices require a yield of acrylic acid of about 60–70% for a process to become competitive. However, in spite of intense research activities in academia and industry over the past 30 years, an efficient and stable catalyst that matches this requirement is still missing. Limitations in selectivity and the higher heat of reaction have an impact on process design and reactor technology for the prospective direct oxidation of propane. Preferences of fluidized-bed technologies have been predicted, because in fixed-bed reactors the heat-transfer surface is limited and more reactors than in propylene oxidation may be needed due to the restricted yield of acrylic acid.[15,16] Selection of the process in turn is determined by the physical and mechanical properties and the chemical stability of the catalyst under operation conditions.[17,18] In order to identify new catalysts, rational approaches of catalyst synthesis as well as high-throughput techniques have been applied and an extensive library of catalysts has been generated. In the following, a brief summary of the various types of catalysts that catalyze the direct oxidation of propane to acrylic acid is given.

4.3 Catalysts and Product Distribution in Selective Oxidation of Propane

The catalysts claimed to be active in direct oxidation of propane to acrylic acid are generally bulk catalysts that contain in most of the cases vanadium as an essential component. However, vanadium is not necessarily required to activate the alkane.[14,19] With respect to the chemical composition, the numerous materials studied have been classified into (i) molybdenum-based mixed oxides, (ii) heteropoly acids and their salts, and (iii) phosphates. However, such a strict classification is questionable, because the catalyst precursors undergo structural changes during catalysis and especially heteropoly compounds are generally not stable under reaction conditions, but are transformed into mixed oxides or phosphates and, consequently, similarities exist between the catalysts derived from different types of precursors.

In Table 4.1, catalyst examples are summarized. Despite the chemical variety, some common aspects of the most efficient catalysts can be identified, such as

(i) high crystallinity of the catalyst bulk;
(ii) phase purity;
(iii) balanced redox properties not restricted to vanadium;

Table 4.1 Performance of various catalysts in selective oxidation of propane to acrylic acid.

Catalyst	*Feed C3/O_2/ H_2O/N_2(He)/ i.s.(Ne or He) (vol%)*	*Flow rate*	m_{cat} [g]/ V_{cat} [ml]	T [K]	X_{C3} [%]	S_{AA} [%]	Y_{AA} [%]	*Formation rate [$mmol_{AA}$/ $g_{cat}h$]*	*Ref.*
$V_1P_{1.15}Te_{0.1-0.15}O$	2/76/22/0/0	$C3 + O_2 = 274$ mL/min $H_2O = 0.195$ mol/h	40/–	663	30	30	10.5	0.0418	23
$V_1P_{1.1}O$	1/35/64/0/0	6.9 s	10/10	693	46.8	32	14.54	0.036	44
$(PyH)_3PMo_{12}O_{40}$	20/10/20/50/0	50 mL/min	3/–	613	7.5	29	2	0.178	57
$H_{1.26}Cs_{2.5}Fe_{0.08}PVMo_{11}O_{40}$	30/40/0/30/0	15 cm^3/min	1/–	653	47	28	13	1.5	55
$NbPMo_{11}V_{pyr}$	27/13/33/27/0	120 mL/min/36 000 h^{-1}	–	653	20.9	50	10.5	37.38	61
$MoV_{0.3}Te_{0.23}Nb_{0.12}O_n$	3/11/39/47/0	1.9 s	0.61/–	653	80	61	49	1.08	81
$MoV_{0.44}Te_{0.10}O_n$	8/10/45/37/0	20 mL/min	0.5/–	653	36.2	46.6	17	1.4	136
$MoV_{0.25}Te_{0.11}Nb_{0.12}O_n$	8/10/45/37/0	20 mL/min	0.5/–	653	33.4	62.4	21	1.7	136
Te–Ni–Mo–O	12/8/40/40/0	1500 mL/g_{cat} h	1/–	693	30	55	17	1.4	79
$MoV_{0.45}Te_{0.17}Nb_{0.12}O_n$	4/8/30/58/0	205 $g_{cat}h/mol_{C3}$	0.3–3/–	653	35.2	54.3	19.1	0.93	134
$MoV_{0.26}Te_{0.11}Nb_{0.12}O_n$	5/10/45/35/5	30 mL/min	0.5/–	653	33	55	18.15	0.43	108
$MoV_{0.28}Sb_{0.13}Nb_{0.15}O_n$	5/10/45/35/5	30 mL/min	0.5/–	653	35	49	17.15	0.41	108
$MoV_{0.36}Te_{0.17}Nb_{0.12}O_n$	4/8/30/58/0	100 $g_{cat}h/mol_{C3}$	–/–	673	21.5	58.5	12.6	1.2	109
$MoV_{0.3}Te_{0.23}Nb_{0.12}O_n$	2/6/50/42/0	20–80 Ncm^3/s	215–340/–	633	38	65	24.7	0.908	158
$MoV_{0.17}Te_{0.22}Nb_{0.15}O_n$	4/8/30/58/0	50 $cm^3 min^{-1}$	0.5–2.5/–	653	21.1	33.2	7	0.14	132
$MoV_{0.25}Te_{0.23}Nb_{0.10}O_n$	6/10/43/36/5	50 $cm^3 min^{-1}$	0.5–2.5/–	653	58	52	30	0.27	132
$Mo_1V_{0.26}Te_{0.17}Nb_{0.29}O_n$	3/6/40/51/0	3 g_{cat}s/mL	0.5/–	673	56	79	44	0.66	166
$Mo_1V_{0.26}Te_{0.17}Nb_{0.29}O_n$	3/6/40/51/0	0.75 g_{cat}s/mL	0.5/–	673	49	73	36	2.15	166

(iv) low surface acidity; and
(v) the admixture of phosphorous and especially tellurium.

The various materials require different reaction conditions and feed compositions for optimum productivity. In the following, the different types of catalysts will be discussed in more detail.

4.3.1 Phosphates

Vanadium phosphates have been reported to show activity in propane oxidation to acrylic acid in oxygen-rich feed at temperatures between 543 and 673 K.[20–46] Acrylic acid, acetic acid and carbon oxides were observed as products. Approximately 1 mol% propane was converted into acrylic acid over pure vanadium pentoxide.[23] The yield of acrylic acid increases with the addition of phosphorous oxide, but does not exceed 10 mol%. Feed composition and operation conditions strongly influence activity and product distribution over V–P–O catalysts, whereas high oxygen content and the presence of steam in the feed are essential requirements for formation of acrylic acid.[21,26] The best catalysts are exclusively composed of vanadyl pyrophosphate as the single phase with high crystallinity and low surface acidity.[32] These properties depict analogies to critical characteristics of molybdenum-based mixed-oxide catalysts as will be discussed below. Addition of tellurium oxide to V–P–O has a positive effect on the yield of acrylic acid.[23] Performance and stability of V–P–O were also improved by addition of zirconium.[27] Other phosphorous containing mixed oxides, such as Fe–P–O,[20] Mn–P–O, and Cu–P–O[47] have been reported to activate propane, but the selectivity to partial oxidation products was very low, except for boron–phosphorous oxides. B–P–O converts 22% propane at 823 K in an equimolar mixture of propane and oxygen in dry He as carrier gas with 50% selectivity to propylene and 13% selectivity to C3 oxygenates.[47]

4.3.2 Catalysts Derived from Heteropoly Compounds

Heteropoly compounds or catalysts derived from heteropoly compounds constitute another type of catalyst for selective oxidation of propane.[19,48–66] 12-Molybdophosphoric acid has been reported to show some activity in the formation of acrylic acid.[54–56,59,60] Exchange of the protons by other counterions, like Cs^{+}, Fe^{3+}, or Ni^{2+}, or substitution of the addenda atom Mo^{6+} by V^{5+}, enhanced the catalytic activity. Infrared spectroscopy and XRD results suggest stability of the Keggin structure during the reaction in the range 673–773 K, however, some indication of segregation of vanadium ions has also been found. In agreement with this observation, enhanced catalytic activity has been observed after introduction of vanadium species into the secondary structure of a cesium salt of 12-molybdophosphoric acid.[65] Ga^{3+} seems to be another effective counterion that promotes the formation of acrylic acid over Keggin-type heteropoly oxometalates.[48] The investigation of reduced heteropoly molybdates by Ueda has shown that vanadium is not necessarily required to selectively activate propane.[66] Acrylic acid and acetic acid are the main oxygenated products in propane oxidation over molybdophosphoric acid reduced with pyridine. The initial heteropoly compound undergoes substantial structural changes during the activation procedure that involves thermal treatment in nitrogen. The active catalyst was composed of a single orthorhombic phase and the formation of acrylic acid was associated with the reduction of molybdenum. The nonreduced acidic form of $H_3PMo_{12}O_{40}$ gave propylene as the main reaction product, illustrating the relevance of redox properties of the catalyst for selective oxidation of alkanes to oxygenates. The addition of niobium increased the productivity of catalysts derived from heteropoly compounds substantially.[61] $P(Mo,V)_{12}$, exchanged sequentially with niobium oxalate and pyridine giving $NbPMo_{11}V_{pyr}$ and pretreatment of the salt in He at 693 K displayed space-time yields of $0.6\,mmol_{aa}\,min^{-1}\,g^{-1}$, which were two orders of magnitude higher compared to the reference catalyst without niobium (Table 4.1). The thermal pretreatment that removes all the organic species present in the precursor was required to generate an active catalyst consisting of a mostly amorphous molybdenum oxide phase and highlighting close relations of catalysts derived from heteropoly compounds with mixed multimetal oxides that are often prepared *via* aqueous routes involving heteropoly compounds as synthesis intermediates.[52,62,67–69]

4.3.3 Multimetal Oxides

The best performances in direct oxidation of propane to acrylic acid have been achieved with mixed oxides composed of approximately 50–70 wt% molybdenum oxide and oxides of V, Te, and Nb. Catalysts for selective oxidation based on molybdenum oxide originally trace back to the work of Thorsteinson *et al.*[70] They reported that molybdenum–vanadium oxides composed of a crystalline phase with *d*-spacing at 4.0 Å were active for oxidative dehydrogenation of ethane to ethylene and selective oxidation of ethane to acetic acid at temperatures as low as 473 K. Many ternary and quaternary transition-metal oxides with good catalytic performance in selective oxidation are characterized by shear structures with disorder in the *ab* plane and long-range order in the *c* direction with *d*-spacing at 4.0 Å.[71–74] Figure 4.1 shows typical XRD patterns of a nanocrystalline Mo–V–W oxide catalyst for oxidation of acrolein or propylene to acrylic acid.

In the *ab* plane, crosslinking of corner-shared metal-oxide polyhedra is observed that results in the formation of channels with different sizes typical of oxidic bronze structures. The needle-like particles grow perpendicular to the *ab* plane along the crystallographic *c* direction. In ethane oxidation, a composition $Mo_{0.61}V_{0.31}Nb_{0.08}$ appeared to show the optimum performance.[70] The high stability of the catalyst bulk structure against oxidation and reduction was allocated to niobium that seems to keep the catalyst in an optimal oxidation

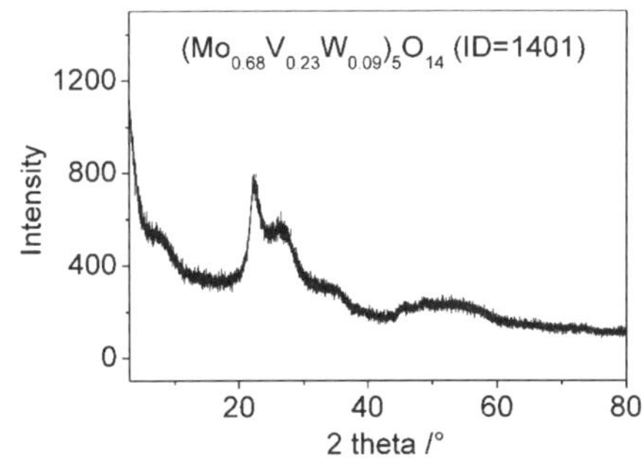

Figure 4.1 Typical XRD patterns (measured using $CuK_{\alpha 1}$ radiation) of a semi-crystalline Mo–V–W oxide catalyst for oxidation of propylene or acrolein to acrylic acid.

state. Addition of palladium further accelerated the formation of acetic acid. Palladium is well known as an olefin oxidation catalyst from homogeneous Wacker chemistry. Concepts that combine active materials in oxidative dehydrogenation of alkanes with components that carry functionalities for selective oxidation of alkenes have been applied frequently in the design of catalysts for oxo-functionalization of alkanes. Similarly, nickel molybdates, which were reported to be active in the oxidative dehydrogenation of propane to propylene,[75–77] have been modified with tellurium and phosphorous. The resultant Ni–Mo–Te–P oxide catalysts showed improved selectivity to acrolein and acrylic acid.[78,79] The success of this strategy, however, remained limited, resulting in 20% maximum yield of acrolein and acrylic acid.[79] Sb-modified nickel molybdates have also been investigated, but the selectivity to acrylic acid was significantly lower.

Tellurium in general seems to have a beneficial effect, when acrylic acid is the desired reaction product. Doping vanadium phosphates as well as mixed oxides with tellurium usually resulted in improved selectivity to acrylic acid.[23,78–80] Incorporation of Te into the bulk structure of the catalyst turned out to be the most successful strategy. The synthesis of bronze-like Mo–V–Te–Nb–O_x phases that accommodate Te on lattice sites of the crystal structure provided the highest yield of acrylic acid so far. As a result of empirical optimization, Mitsubishi Kasai Cooperation achieved acrylic acid yields close to 50% with a catalyst composed of Mo, V, Te and Nb in the molar ratio 1 : 0.30 : 0.23 : 0.125.[81] Similar catalysts have been covered by patents before claiming high performance in ammoxidation of propane to acrylonitrile (see Chapter 5).[82] Various phases have been identified to constitute Mo–V–Te–Nb mixed oxides, including $(Mo_{0.93}V_{0.07})_5O_{14}$, MoO_3, MoO_2, elemental Te, and $TeMo_5O_{16}$.[83,84] However, the major focus of attention lies on two crystal structures, originally called M1 and M2, respectively, by Ushikubo *et al.*[85,86] In the patents by Mitsubishi, specific diffraction patterns corresponding to M1/M2 phase mixtures have been claimed to ensure high productivity.[81] Interestingly, the two phases are also characterized by having diffraction of *d* spacing at 4 Å.

4.4 Crystal Structure of the Key Phases in Mo-Based Multimetal-Oxide Catalysts

The crystal structure of the M1 phase has been analyzed by several groups.[67,87–91] Refinement of the M1 and M2 crystal structures based on a combination of synchrotron X-ray powder diffraction, neutron powder diffraction, and electron microscopy has been reported by De Santo *et al.*[88,89] The structural model of M1 adopted from De Santo *et al.*[88] is illustrated in Scheme 4.2 that shows a two-dimensional projection of the polyhedral network and a stick-and-ball model in the crystallographic (001) plane.

Corner-linked MO_6 (M=Mo, V) octahedra are two-dimensionally assembled forming 6-, and 7-membered rings. Niobium has been reported to be mainly located in the center of a MO_7 pentagonal bipyramidal configuration sharing edges with the surrounding octahedra.[88,92] An isostructural variant of the Mo–V–Te–Nb–O_x M1 phase with tantalum substituting niobium has been synthesized aimed at understanding the location of Nb in M1 better.[90,93] Tantalum resembles niobium chemically, but, unlike Nb, Ta is clearly distinguishable from molybdenum from scattering experiments. It was shown that Ta preferentially occupies pentagonal bipyramidal sites. High-angle annular dark-field (HAADF) imaging provided evidence of the coexistence of Ta-rich and Ta-poor domains. Such inhomogeneities might also exist in the Nb compound. In the structural model postulated by DeSanto *et al.*, complete occupation of all pentagonal bipyramidal sites in the unit cell with Nb was assumed *a priori*. In agreement with this model, niobium may be located in octahedral positions as

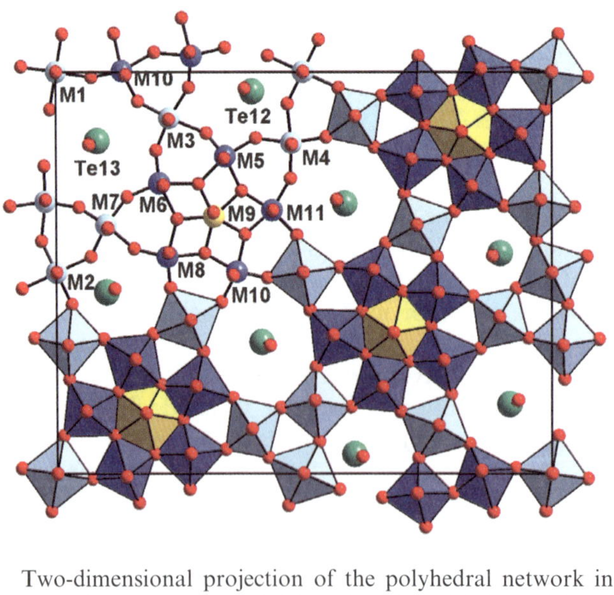

Scheme 4.2 Two-dimensional projection of the polyhedral network in the crystallographic (001) plane (right) and stick-and-ball model (left) of the M1 structure.

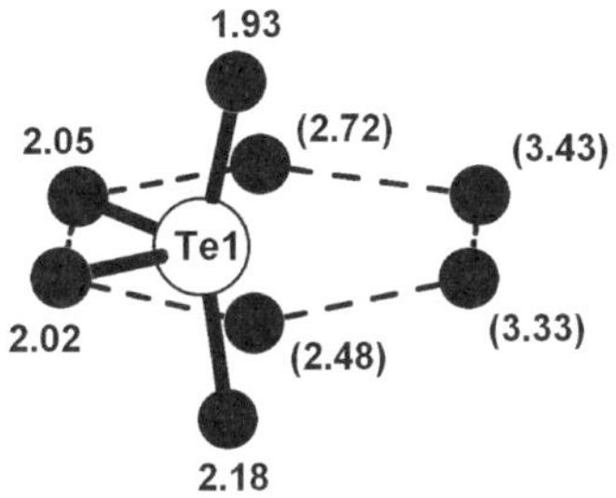

Scheme 4.3 Schematic representation of the coordination of tellurium in the hexagonal channel of M1.

well in phase-pure and highly crystalline M1 with Nb content higher as the specimen analyzed by DeSanto *et al.*, as shown by EXAFS.[92]

In the M1 structure, the {001} planes formed by networking the polyhedra are congruently stacked along the crystallographic [001] direction by sharing corner oxygen atoms, thereby giving rise to a prismatic particle morphology and a channel-like structure. The {001} planes are arranged perpendicular to the length axis of the crystals representing the basal planes of the prisms. The heptagonal and hexagonal channels are partially occupied by finite $(TeO)_n$ zigzag chains.[88,91,94,95] The occupancy of the channels depends on the Te content and the preparation of M1.[88,94,95] De Santo *et al.* reported 80% and 20% occupancy for the hexagonal and heptagonal channels, respectively.[88] The heptagonal channels in the hydrothermally prepared M1 studied by Murayama *et al.* were empty.[91] The Te atoms in the hexagonal channel are bound to four oxygen atoms in distorted trigonal bipyramidal coordination assuming that the active lone pair occupies the fifth coordinate position in the equatorial plane (Scheme 4.3).

Such a coordination and the Te–O distances ranging from 1.93 Å to 2.18 Å are typical for Te(IV) oxides.[96] In the heptagonal channel, $(TeO)_n$ oligomers have been refined to be arranged near the center with distances >3 Å between the Te atom in the chain and the oxygen atoms in the polyhedrons forming the channel wall.[88]

The Te-rich M2 phase is less complex, being composed exclusively of corner-linked octahedra that are connected in the (001) plane forming 6-membered rings occupied by $(TeO)_n$ chains (Scheme 4.4). Formulas of the refined unit cells have been reported to be $Mo_{7.8}V_{1.2}NbTe_{0.937}O_{28.9}$ for M1, and $Mo_{4.31}V_{1.36}Te_{1.81}Nb_{0.33}O_{19.81}$ for M2, respectively.[88]

4.5 Phase Cooperation and Crystallinity

Symbiosis between the two phases M1 and M2 has been suggested to be responsible for the maximum yield of acrylic acid or acrylonitrile in the oxidation or ammoxidation of propane, respectively.[97–100] The phenomenon has been discussed in terms of phase cooperation concepts.[9,99,101–103] The M1 phase is crucial for selective oxidation of propane to acrylic acid. M2 on its own does not[98] or poorly[104] activate propane, but selectively oxidizes the propylene intermediate.[98,105] Correlations between phase composition and catalytic

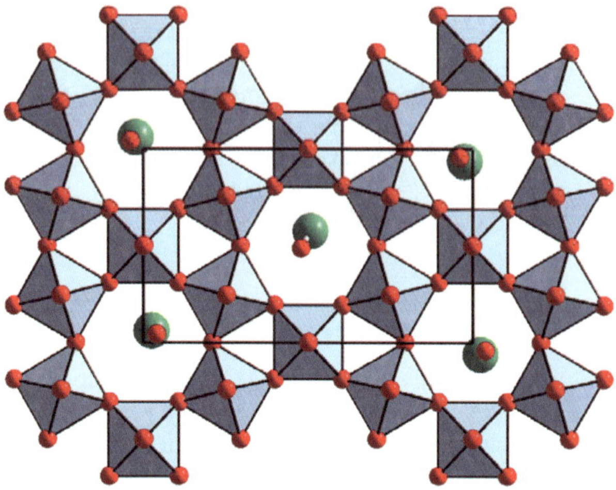

Scheme 4.4 Two-dimensional projection of the polyhedral network in the crystallographic (001) plane of the M2 structure.

properties are still actively debated.[68,69,97,99,100,104,106–108] However, it has been generally accepted that the high selectivity towards acrylic acid is associated with the M1 phase.[83,107,109–111] An even higher yield of acrylic acid has been found after removal of M2 from Mo–V–Sb–Nb oxide catalysts prepared by hydrothermal synthesis.[111]

Similar to V–P–O,[32] high crystallinity of the catalyst seems to be essential for good performance. In a study concerning the M2 phase in (amm)oxidation of propylene, crystalline M2 catalysts without amorphous overlayers are vastly superior to those encumbered with an amorphous overlayers of tellurium molybdate. The former give a maximum acrylonitrile yield of approximately 45%, the latter approximately 24%. In the selective oxidation of propylene to acrylic acid, the difference in performance between the two types of catalysts becomes even more obvious. The yield comprises 34% for the crystalline M2 catalyst, but 3% for the catalyst with the amorphous overlayer.[112]

In Figure 4.2, the selectivity to acrylic acid is plotted as a function of propane conversion exemplary for several catalysts including phosphates and materials derived from heteropoly compounds irrespective of the different reaction conditions that have been applied in the catalytic studies. Reaction conditions and space-time yields achieved are summarized in Table 4.1. The trajectory in Figure 4.2 indicates the desired minimum yield of acrylic acid of 50% beyond which a catalyst may become attractive from commercial point of view. The distribution of catalyst performance is quite broad, revealing generally a lower selectivity of the catalysts derived from heteropoly compounds. It becomes apparent that phase-pure M1 catalysts are characterized by high selectivity even at high propane conversions. However, the space-time yield of acrylic acid of on average 100 $g_{aa}\,kg_{cat}^{-1}\,h^{-1}$ is comparatively low. For a phase pure M1 catalysts that shows a space-time yield of 0.58 $\mu mol_{aa}\,g_{cat}^{-1}\,s^{-1}$, a monolayer

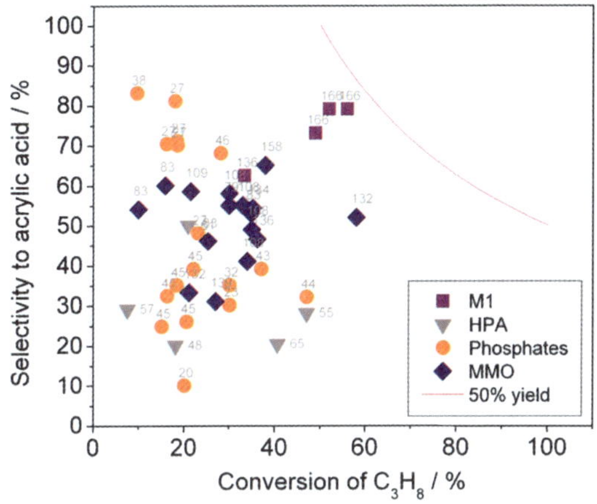

Figure 4.2 Selectivity to acrylic acid as a function of propane conversion of phase-pure M1 Mo–V–Te–Nb oxide catalysts (squares), materials derived from heteropoly acids (triangles), phosphates (circles), and multimetal oxides (MMO) (diamonds) tested in selective oxidation of propane to acrylic acid under different reaction conditions (the numbers indicate the reference). The red line corresponds to 50% yield acrylic acid.

coverage of 35 $\mu mol_{propane}\ g_{cat}^{-1}$ has been determined. Assuming that all propane adsorption sites would be active sites, a quite low turnover number of 0.016 s^{-1} would result that is still low if only 1% of the propane adsorption sites are considered as active sites (TON = 1.6 s^{-1}). Nevertheless, the exceptional selectivity makes the M1 structure attractive from an industrial point of view. Beyond that, the M1 phase is particularly suited for investigations targeted at improved understanding of structural and chemical requirements for selectivity in alkane activation. The well-defined and stable bulk structure provides a basis for studying surface dynamics necessarily associated with catalysis of redox reactions. In this respect, controlled synthesis of phase-pure and highly crystalline materials is vital, since the presence of metal-oxide admixtures in the bulk of a catalyst makes it difficult to elucidate the specific catalytic functionality of individual phases. Although, in many cases, the final catalyst has been the object of comprehensive investigation, the preparation of Mo–V–Te–Nb oxides has been studied to a lesser extent. However, the preparation parameters are reported to have a significant effect on the catalytic activity.[67,113] In the following, the synthesis parameters that direct phase composition and properties of Mo–V–Te–Nb oxides are discussed.

4.6 Synthesis of Phase-Pure Mo-Based Multimetal Oxides

In general, Mo–V–Te–Nb oxide catalysts are prepared by coprecipitation with subsequent rapid evaporation of the solvent ("slurry" or "dry-up" method) or by hydrothermal synthesis. Subsequent annealing in an inert atmosphere is necessary to crystallize the phases and to derive balanced oxidation states of the metals. Although sophisticated preparation techniques are known, it is still a nontrivial task to synthesize phase-pure M1 catalysts in a controlled manner. In particular, it is difficult to avoid formation of biphases during material synthesis, such as M2 phase, Te, $(Mo_{0.93}V_{0.07})_5O_{14}$, $V_{0.95}Mo_{0.97}O_5$, $TeMo_5O_{16}$, MoO_3 and MoO_2.

4.6.1 Preparation of Mo–V–Te–Nb Oxides by Precipitation/ Rapid Solvent Evaporation

The so-called "slurry method"[114] is a rather complex preparation procedure consisting of several steps, which are in part highly sensitive to the reaction conditions. Consecutive mixing of aqueous solutions of ammonium heptamolybdate $(NH_4)_6Mo_7O_{24} \cdot 4H_2O$, ammonium metavanadate $NH_4VO_3 \cdot xH_2O$, hexaoxotelluric acid $Te(OH)_6$, and ammonium niobium oxalate $NH_4[NbO(C_2O_4)_2] \cdot xH_2O$ gives an orange, jelly-like precipitate. By fast evaporation of water from the precipitate suspended in the mother liquor, orange-colored precursors are obtained. It is well known that the nature of isopoly and heteropoly anions formed in the initial solutions and in the mother liquor and the degree of condensation and/or protonation of these ions strongly depend on temperature, concentration and pH.[69,115] Raman spectroscopic investigations indicate that parallel formation of various phases seems to be predetermined by the chemical equilibria in the initial ternary Mo–V–Te solution that contains the metals in a nonuniform distribution.[69] At pH=5, vanadium mainly coexists with the Anderson-type heteropoly anion $[TeM_6O_{24}]^{n-}$, (M=Mo, V, $n \geq 6$)[116–119] in the form of protonated decavanadate species $[H_xV_{10}O_{28}]^{(6-x)-}$. NMR spectroscopy indicated the additional presence of Te- and Mo-modified decavanadates $TeV_5Mo_3O_{27}$.[5–115] The addition of niobium oxalate triggers precipitation, but does not change the structure of the Mo–V–Te polyoxometalate species in the supernatant mother liquor and also drying preserves the preassembled Anderson-type heteropoly anions, as observed in the Raman spectrum of the dried material.[69] Separate processing the jelly-like precipitate results in formation of a crystalline solid enriched in M1, whereas the mother liquor gives M2 (Figure 4.3). The chemical composition of the two phases in the suspension is different. The isolated and dried X-ray amorphous precipitate contains the metals in a stoichiometry of $Mo_1V_{0.26}Te_{0.23}Nb_{0.26}$, indicating an enrichment of Nb compared to the metal stoichiometry $Mo_1V_{0.30}Te_{0.23}Nb_{0.12}$ of the final catalyst. In contrast, the dried mother liquor $Mo_1V_{0.25}Te_{0.19}Nb_{0.04}$ contains only little Nb, but mainly Mo, V, and Te, which is in agreement with the spectroscopic observations described above. Crystallization of the dried precipitate gives a solid that is rich in M1 phase and that contains M2 as a secondary phase, whereas the dried mother liquor results in M2, Mo_5O_{14}-like phase, and MoO_3, but no M1. Impregnation of a dried $MoV_{0.3}Te_{0.23}$ precursor with a solution of $Nb(OEt)_5$ in pentane and subsequent crystallization of the amorphous solid results in a M1-free phase mixture containing M2, Mo_5O_{14}-like phase, $V_{0.95}Mo_{0.97}O_5$, and MoO_2. This experiment

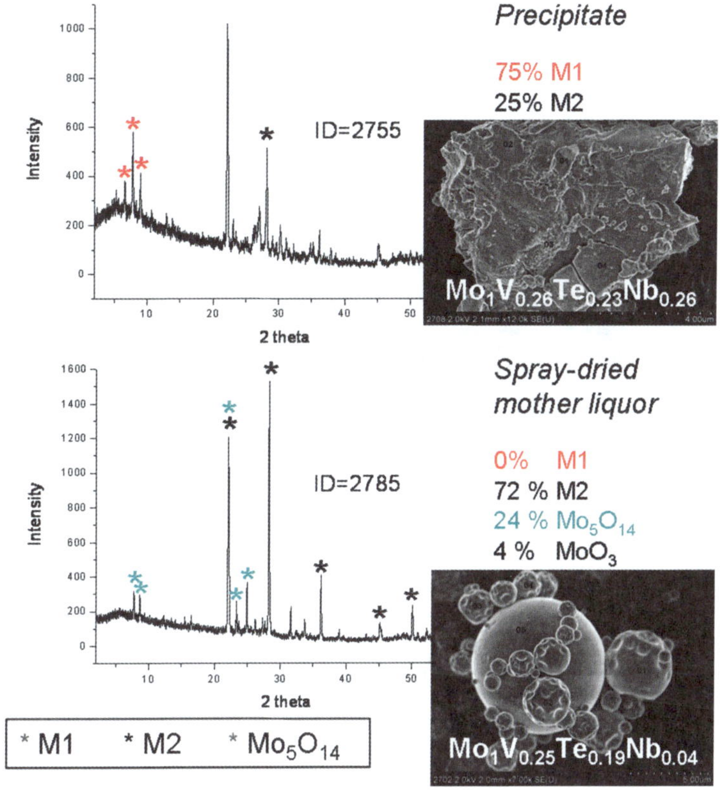

Figure 4.3 Phase composition, morphology and chemical composition of the materials worked up separately from the precipitate (top) and mother liquor (bottom) of orange Mo–V–Te–Nb precursor suspension.

demonstrates that Nb presumably acts already in the solution as a linker that connects the $[TeM_6O_{24}]^{n-}$, (M=Mo,V) units. Oliver *et al.* have reported a change in crystallinity of Mo–V–Te–Nb–O_x catalysts depending on the pH of the slurry.[120] Lowering the pH was assumed to have two effects, first, to decrease the crystallinity of the molybdotellurates formed, and, secondly, facilitating the partial introduction of heteroatoms.

The drying process is critical as well. Lin *et al.* observed high catalytic activity only after spray-drying.[113] Phase separation tends to occur when other drying techniques were applied, such as freeze-drying or drying in a rotary evaporator, indicating that quenching the situation in the slurry is beneficial.

Figure 4.4 shows the SEM image of spherical particles typically obtained by spray drying of the slurry. From the coexistence of a variety of different species in solution and in the spray-dried catalyst precursor and from the crystallization experiments with precipitate and dried mother liquor one can expect that the thermal treatment will not result in a phase-pure material. While the dried precursor is not active in propane oxidation to acrylic acid, stepwise

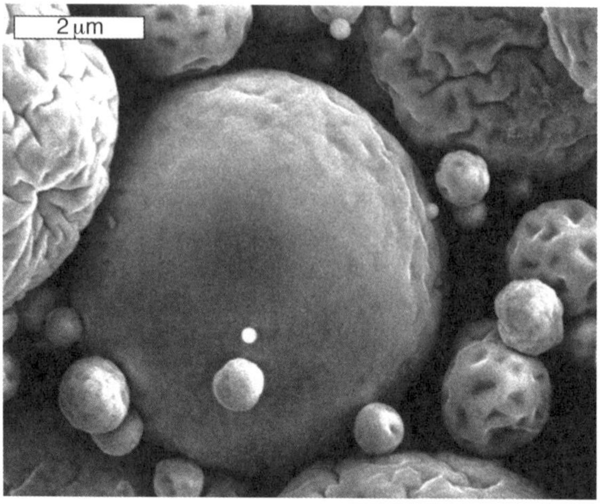

Figure 4.4 SEM image of spray-dried Mo–V–Te–Nb oxide precursor.

thermal treatment at first in air and then in an inert atmosphere activates the catalyst prepared by the slurry method. Crucial seems to be the calcination in air at 548 K that results in dehydration and decomposition of the ammonium salt of the Anderson-type anion $[TeM_6O_{24}]^{n-}$ establishing nanocrystalline precursors of the finally formed phases.[69,121,122] The final thermal treatment in an inert atmosphere at 873 K obviously consolidates these phases and, concurrently, may cause the formation of minor phases. *In-situ* XRD (Figure 4.5) shows that the formation of M2 starts at significantly lower temperatures than that of M1. Furthermore, the parallel genesis of M1, M_5O_{14} and MoO_3 indicates that these phases may be formed competitively, *i.e.* during decomposition of a common precursor stock.

The crystallization behavior of Mo–V–Te–Nb oxides bears a resemblance to the genesis of a $Mo_{0.68}V_{0.23}W_{0.09}$ oxide with Mo_5O_{14}-like structure.[74] In the synthesis of $Mo_{0.68}V_{0.23}W_{0.09}$ oxide, a precursor prepared by spray drying of mixed aqueous metal-salt solutions is typically calcined in air and subsequently treated in helium at different temperatures. X-ray diffraction, HRTEM, ^{51}V magic-angle spinning (MAS) NMR, ESR, UV/Vis DR spectroscopy and oxygen and hydrogen-adsorption measurements have been applied to monitor the preparation procedure. It was found that a structure closely related to that of Mo_5O_{14} already appears at the nanoscale level after calcination of the spray-dried precursor in air at 623 K. At this stage, the material comprises of crystalline particles less than 3 nm in size stabilized by an amorphous matrix. Further heating causes nanostructural rearrangements that lead to the formation of the final $Mo_{0.68}V_{0.23}W_{0.09}$ oxide with phase-pure polycrystalline structure.[74] The formation of minor phases during crystallization of Mo–V–Te–Nb oxides may be related to partial vapor-phase transport of tellurium and molybdenum oxide at high temperatures in the inert atmosphere. Under these conditions, MoO_3 and Te could be involved in solid-state redox reactions as described by Bart *et al.*[123] and may also affect the surface arrangement.

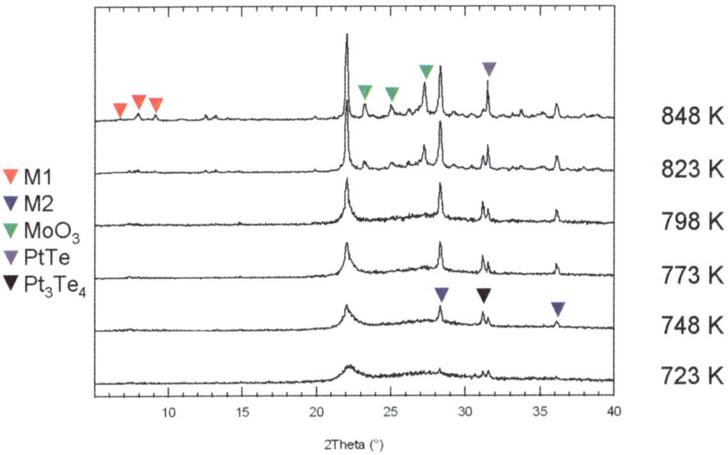

Figure 4.5 *In-situ* XRD during activation of a calcined Mo–V–Te–Nb oxide precursor in 100 mL/min He and a heating rate of 2 K/min with isothermal sections for the measurement. The Pt_xTe_y peaks arise from the reaction of liberated Te with the Pt sample holder.

Accordingly, the nature of the Te source[124,125] and organic additives such as oxalic acid[126] have an impact on the redox chemistry during thermal processing and, consequently, on the phase composition. Strongly reducing conditions, *e.g.* due to the presence of organic material in the precursor, may lead to reduction of Te. Elemental tellurium melts at 722.51 K and is lost by sublimation at the high temperatures that are required for crystallization of the mixed oxide, being no longer available as an inorganic template that may direct the self-assembly of metal-oxide polyhedra towards the formation of the 6- and 7-membered rings in the desired M1 phase. Heating rates, gas composition and flow rates during thermal activation also affect the characteristics of the final product.[127] Therefore, it is not surprising that attempts to reproduce the catalytic benchmarks reported in the patent literature might fail occasionally.

The "dry-up" or "slurry" method described above gives access to M1/M2 phase mixtures, which show reasonable catalytic productivities. In order to gain a deeper understanding of the specific catalytic function of the two phases, the synthesis of phase-pure materials is desirable. Phase-pure M2 is easily available by "slurry synthesis" applying appropriate stoichiometries, *e.g.*, according to the formula $Mo_{49}V_{26}Te_{25}O_{250}$.[128] Purification of M1/M2 phase mixtures with nitric acid[68] or H_2O_2,[129,130] respectively, gives access to phase-pure M1. However, the chemical composition of M1 might be affected by leaching.[106]

4.6.2 Preparation of Mo–V–Te–Nb Oxides by Hydrothermal Synthesis

Hydrothermal synthesis represents an alternative synthesis route to Mo–V–Te–Nb oxide catalysts, in particular for the synthesis of phase-pure materials.[131]

Usually, the synthesis is performed at $T = 448$ K in a nitrogen atmosphere applying a synthesis time of 48 h.[84,110,132–137] The hydrothermal synthesis may be accelerated by applying microwave irradiation.[138] López Nieto and coworkers conducted the synthesis under these conditions in a Teflon-lined stainless-steel autoclave using different metal-containing compounds and various nominal metal stoichiometries.[84,132–135,137,139] Generally, phase mixtures including Mo_5TeO_{16}, MoO_3, M_5O_{14} (M = Mo, V, Nb), $Te_2M_{20}O_{57}$ (M = Mo, V, Nb), and $Te_{0.33}MO_{3.33}$ (M = Mo, V, Nb) were obtained after heat treatment at 873 K in nitrogen of the hydrothermally prepared precursor. The latter two phases correspond to M1 and M2, respectively. M1 has been achieved as the predominant phase using a nominal metal stoichiometry of Mo/V/Te/Nb = 1/0.2/0.17/0.17 and oxalate /molybdenum molar ratios in the range between 0.2 and 0.6.[137] The as-synthesized metal stoichiometry of the M1-rich materials correspond to Mo/V/Te/Nb = 1/0.20–0.27/0.19–0.22/0.19–0.21.

Applying a nominal stoichiometry of Mo/V/Te/Nb = 1/0.3/0.16/0.12, Ueda and coworkers prepared orthorhombic Mo–V–Te–Nb mixed oxides at the standard synthesis conditions in small (70 ml) Teflon-lined autoclaves.[136] The XRD patterns of the annealed product displayed exclusively reflections of the orthorhombic M1 phase. Polycrystalline orthorhombic Mo–V–Te[140] and Mo–V[141–143] oxides have also been prepared operating at the same synthesis temperature, but applying different metal ratios, molybdenum concentrations, and synthesis times, as well as separating undesired phases by manual sorting.[140] By controlling the pH of the precursor solution, a trigonal Mo–V-oxide structure containing the 6-, and 7-membered rings, and the pentagonal bipyramidal units in different ratio has been obtained.[144]

As with the slurry synthesis, preparation parameters, such as temperature, metal-salt concentration, metal stoichiometry, batch size, pressure, and redox potential of the synthesis mixture have a strong impact on resultant phase composition.[84,92,132–135,137] Moreover, different characteristics of a specific autoclave, *e.g.* wall material, volume, and heating/cooling regime have a substantial influence on the crystallization behavior.[92] These factors are well known in synthesis of crystalline alumosilicates,[145,146] but have been less systematically studied for synthesis of transition-metal oxides.[147–150] For a given metal stoichiometry, the impact of reaction temperature and reaction time on M1 formation has been studied.[92] Inhomogeneous distribution of the key elements in the initial gel, which generally contains iso- and heteropolyanions, and giant polyoxometalate clusters in solution together with Nb-rich precipitates, is not necessarily detrimental for formation of phase-pure M1.[92] The structural elements required for crystallization of M1 are established in the course of the hydrothermal process. The reaction could take place by a dissolution–precipitation mechanism involving dissolution or rearrangement of polyoxo metalate building blocks followed by precipitation of an ill-crystallized nanostructured product from solution. Complete reorganization of counterion-containing phases, which is imperative in order to form phase-pure materials, has been shown to be accelerated in presence of superheated water vapor.[92]

The spatial homogeneity of elements in crystalline M1 is reflected in the corresponding precursor. Precursor materials of phase-pure M1 catalysts show long-range order in the [001] direction. Ordering along the other crystallographic directions is less developed, but occurs during subsequent heat treatment of the precursor in an inert gas at 823 K. The presence of residual ammonium containing supramolecular species in the precursor result in formation of phase mixtures in the course of crystallization.[92] Gas-phase transformations, such as the reaction of a monoclinic α-$TeVO_4$ phase with the orthorhombic M1 phase during thermal treatment in inert atmosphere at 678 K have been shown to increase the fraction of M2 in the mixture.[104]

Several attempts have been undertaken to substitute the metals in M1 and M2 under preservation of the M1 or M2 crystal structure, respectively. Mo–V–M oxides with M=Al, Ga, Bi, Sb, Fe, Ti, and Te have been prepared by hydrothermal synthesis by Ueda *et al.*[140,151,152] All synthesized solids showed catalyst particles with rod-shaped morphology and gave a common XRD peak corresponding to 4.0 Å *d*-spacing. However, crystallization of M1 has only been achieved with M=Sb, and Te. The addition of titanium was found to direct the material to form a $(Mo,V,Ti)_5O_{14}$-like phase during heat-pretreatment above 823 K in a N_2 atmosphere.[151] In contrast, restricted substitution of Mo or V in M1 by Ti, Sn, Ge, and W was hypothesized by Millet *et al.*[153] The substitution of Nb by Ta results in a solid isostructural with M1.[90,93,100] M2 was prepared by substituting to various degrees Mo and V with W, Ti, Nb or Fe, and Te with Ce, respectively, applying the slurry method.[128,154]

In summary, despite the availability of sophisticated methods, synthesis of phase-pure M1 is still an ambitious task. Induced by the chemical complexity of the material and due to the multiplicity of synthesis steps that comprise a high number of mutually associated experimental parameters, application of apparent similar synthesis conditions does not necessarily mean that identical products with comparable catalytic properties are obtained. This has to be taken into account when the catalytic behavior of Mo–V–Te–Nb–O_x catalysts is compared. Consequently, accuracy in synthesis and monitoring of the synthesis parameters during preparation is required including also technical factors such as the material of the reaction vessel, or the heat conductivity of wall material, which could result in different phase composition under otherwise similar reaction conditions, *e.g.*, when the reaction is performed in different autoclaves. Deep understanding of the formation mechanism of the key M1 structure in particular during hydrothermal synthesis is still missing. To overcome this disadvantage, the development of *in-situ* techniques is required to monitor the hydrothermal reaction inside the autoclave and to generate quantitative kinetic information. Moreover, crystallization of M1 during heat pretreatment is highly sensitive towards heating and flow rates and the chemical potential of the gas phase. The optimization of redox properties and gas-phase transport phenomena is a rather unexplored area in synthesis of Mo–V–Te–Nb oxides. Improved understanding of the underlying chemistry bears great potential in catalyst optimization.

4.7 Characteristics of M1

4.7.1 Reaction Network and Mechanism

The kinetic experiments performed so far on Mo–V–Te(Sb)–Nb oxides consistently show that propylene is a primary product formed in propane oxidation to acrylic acid.[134,135,155–162] Most of the catalytic studies, however, have been performed on phase mixtures. Therefore, the intrinsic catalytic properties of M1 remain widely unexplored. The reaction starts at temperatures as low as 473 K. Most frequently, reaction temperatures between 623 and 693 K are applied. According to the original patent, the preferred gas hourly space velocity (GHSV) should be chosen between 2000 and 300 h^{-1}, and, consequently, the space velocities applied in the academic literature are generally close to these values. Usually, the feed is highly diluted containing 1–10 vol% propane, a stochiometric propane : oxygen ratio of 1 : 2, 0–60 vol% steam and helium or nitrogen as inert balance gas. Typical products from oxidation of propane over Mo–V–Te–Nb oxides in the presence of steam in the feed are propylene, acrylic acid, acetic acid, CO, and CO_2. Acetone, acrolein, propanal, and propanoic acid are sometimes formed in small amounts. The product distribution strongly depends on reaction conditions and feed composition. In Scheme 4.5, possible reaction pathways are summarized in a formal reaction network (see Chapter 15).

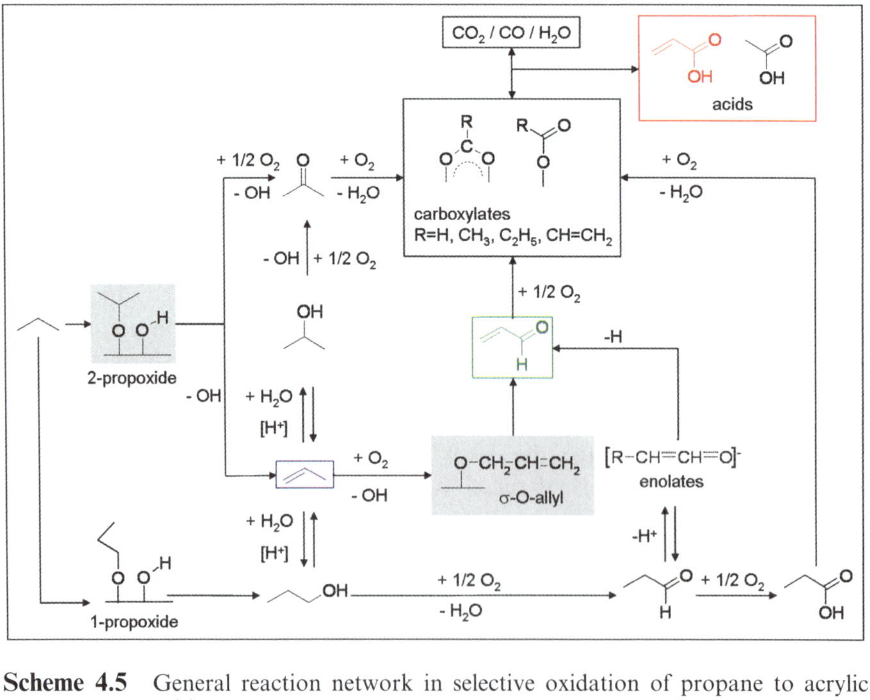

Scheme 4.5 General reaction network in selective oxidation of propane to acrylic acid.

The actual path depends on the redox and acid–base properties of the catalyst that, in turn, is determined by chemical and phase composition of the material and by the reaction conditions. The mechanism differs for phosphates and the various oxide materials. In principle, three different consecutive pathways to acrylic acid might be possible, which may occur in parallel developing a complex network by additional transverse hydration/dehydration, redox, and isomerization reactions. In the first route that results in saturated oxygenates only, the terminal carbon atom of propane C(1) is oxidized to form first 1-propanol that is oxidized further to propanal and propanoic acid. The first reaction step in the second route is the formation of propylene by oxidative dehydrogenation. An isopropoxide species is generally considered to be a surface intermediate, which is decomposed under release of propylene and formation of an OH group by breaking the second C–H bond. Propylene is then stepwise further oxidized to acrolein, *via* an O-allylic intermediate allylic alcohol, and finally acrylic acid. Surface acrylate species are intermediates in formation of acrylic acid, but may also be decomposed by oxidation to acetate and formate species that, in turn, represent precursors for acetic acid, and, finally, CO_2 and CO, respectively. The third pathway involves direct oxidation of propane to 2-propanol again *via* an isopropoxide intermediate, which is further oxidized to acetone. Comparative mechanistic investigations of the total oxidation of alkanes and alkenes on $MgCr_2O_4$ catalysts have shown that the decomposition of the propoxide species to propylene is competitive with its oxidative dehydrogenation to acetone, suggesting that the stability of the propoxide intermediate might be decisive with respect to the selectivity of the catalyst.[163] If cleavage of the first C–H bond is followed by insertion of an oxygen before the second C–H bond at C(2) is broken, the corresponding catalyst is a combustion catalyst rather than a catalyst for oxidative dehydrogenation. On the other hand, propylene formed by oxidative dehydrogenation may readsorb on different sites and electrophilic attack of the propylene double bond by weak Brønsted-acidic OH groups may happen. According to the Markovnikov rule, secondary propoxy species are formed, which likewise undergo further oxidation to acetone. Adsorption of acetone and oxidative breaking of the C–C(=O) bond again result in formation of surface acetate and formate groups, which can be considered as precursor species for both the formation of the corresponding acids or further oxidation or decomposition to CO_2 and CO, respectively.

To confine the reaction network over Mo–V–Te–Nb oxides, transformation of intermediates and products by feeding propylene, 2-propanol, 1-propanol, propanal, acrolein, acetone, acrylic, and acetic acid was investigated by Luo *et al.* at low temperatures (323 and 423 K).[156] The results of the experiments are consistent with a reaction network in which propane is at first transformed into propylene by oxidative dehydrogenation. Mo–V–Te–Nb oxide catalysts are characterized by excellent allylic oxidation capabilities. Accordingly, most of the propylene is then directly oxidized to acrylic acid *via* allylic alcohol and acrolein. Neither ionic or radical allyl species nor allylic alcohol have been detected spectroscopically or isolated as an intermediate over Mo–V–Te–Nb oxide catalysts, respectively, because the oxygen insertion activity seems to be too high and the concentration of such species is practically zero. Only traces of saturated oxygenates were detected, excluding substantial contributions from the pathway *via* 1-propanol. In a parallel path, propylene might be hydrated to isopropanol, which is further oxidized to acetone and acetic acid. The latter pathway seems to be dominant over the more acid–base-driven catalyst VO–H–beta that has been studied for reference in the same work. The hindered internal diffusion in the zeolite seems to contribute additionally to enhanced total oxidation over VO–H–beta. The formation of acetone over Mo–V–Te–Nb oxides is favored at lower temperatures, while acrylic acid becomes the major product by 573 K. Acrylic acid and acetic acid are comparatively stable over Mo–V–Te–Nb oxide and only less than 10% are converted into carbon oxides at 473 K.

Studying the reactivity of products and potential intermediates over Mo–V–Te–Nb oxide at reduced temperatures might involve complications, because the molecules under study exert quite different chemical potential in comparison with the propane/oxygen feed, which could result in changes of oxidation state and chemical composition at the surface of the catalyst. Notwithstanding, similar mechanistic conclusions identifying allylic oxidation of intermediately formed propylene as the predominating pathway and acid-catalyzed hydration of propylene to *iso*-propanol followed by further oxidation to acetone, acetic acid and finally carbon oxides as the main side reaction have been drawn by investigating the trends in product selectivities as a function of propane conversion.[134,158,161] The conversion has been adjusted between 5 and 70% by varying the gas hourly space velocity (GHSV) between 40000 and 500 ml g_{cat}^{-1} h^{-1}. A weak influence of the oxygen content in the feed in the range $1.5 < x_{O2}/x_{propane} < 4$ has been found on the selectivity as a function of propane conversion over eggshell catalysts composed of a typical M1/M2 phase mixture.[158,159] In the latter studies, kinetic coefficients have been calculated based on measurements in a plug-flow reactor operated in integral mode with the option to determine axial concentration profiles along the reactor. The simulations are based on a simplified reaction scheme assuming common formation pathways for the byproducts acetic acid, CO, and CO_2, and first-order rate expressions. According to the results, the oxidative dehydrogenation of propane remains unaffected within the error bar in the investigated range of oxygen concentrations.[158,159] Ueda *et al.* determined a reaction order zero with respect to oxygen and a reaction order of 0.8 with respect to propane by varying the feed composition $C_3H_8/O_2/H_2O/N_2 = 2.5–20/2.5–20/45/37$ at a reaction temperature of 633 K and a GHSV of 9600 h^{-1}.[161] Novakova *et al.* performed kinetic and mechanistic studies in partial oxidation of propane over Mo–V–Sb–Nb oxide catalysts at conversions of propane and oxygen below 10%.[164] Total propane consumption was found to be first order in propane and partial order (0.26) in oxygen in the temperature range between 653 and 712 K at a GHSV of approximately 3300 h^{-1}. They observed that acetic acid shows the behavior of a primary product in propane oxidation and proposed a reaction model that essentially bypasses propylene and assumes primary formation of *iso*-propanol as predominant intermediate in formation of acetic acid over Mo–V–Sb–Nb oxide catalysts.

Water has a substantial effect on the partial oxidation of propane over mixed oxides. Generally, the addition of steam to the feed increases propane conversion and selectivity towards oxygenated products. Depending on the catalyst, the impact of water may differ.[21,23,33,34,78,165,166] Over some catalysts, steam in the feed is indispensable to form acids. The possible impact of water on the catalytic properties in selective oxidation is manifold. Water may change the reaction pathway and modify the catalysts surface simultaneously. It has been speculated that water competes for adsorption sites with propane, propylene, or other intermediates. Furthermore, water has been considered to enhance desorption of acrylic acid from the catalyst surface. Thus, acrylic acid would be preserved from overoxidation.[21,23] Water may promote directly the oxidation of acrolein to acrylic acid by acting as a nucleophilic or oxidizing agent.[1,167] Interference in the reaction mechanism is occasionally reflected in promotion of acetic-acid formation by hydration of propylene to *iso*-propanol and its further oxidation to acetone, acetic acid and CO_x.[23] In wet feed, the concentration of surface hydroxyl groups is increased and the surface acid–base properties are changed. The effect of water on the catalytic properties of V–P–O catalysts has been studied by Landi *et al.*[33,34] The highest yield of acrylic acid was achieved with a steam content in the feed of 15 vol.%. Depending on the steam concentration, reversible changes of the surface acidity were measured by NH_3-TPD, showing a peak in acidity at highest yield to acrylic acid, disclosing similarities with respect to the behavior of V–P–O catalysts in oxidation of butane to maleic anhydride.[26] In addition, the steam improves the crystallinity of V–P–O. Novakova *et al.* investigated the effect of steam in propane oxidation over Mo–V–Sb–Nb oxide catalysts.[165] The rates of formation of acrylic and acetic acid are increased in presence of steam, whereas, in particular, consecutive reactions of propylene towards acetic acid are promoted. The wet reaction atmosphere seems to prevent the Mo–V–Sb–Nb oxide catalyst from irreversible physical changes during propane oxidation. It is speculated that water maintains the metals in a partially reduced oxidation state, and, in this way, keeps the catalyst structure.[165,168] The effect of water is less pronounced with respect to changes in the product composition over highly dispersed vanadia supported on oxides, in which the chemical complexity of the surface is reduced. Generally, water inhibits the rate of alkene formation over VO_x/SiO_2 in oxidative dehydrogenation of ethane,[169] and over VO_x/ZrO_2 in the oxidative dehydrogenation of propane,[170] and increases the relative formation of partial oxidation products over CO_x. Oyama *et al.* explained the influence of steam in the feed by an increased degree of surface hydroxylation in the presence of water vapor that reduces the number of oxygen sites for alkane adsorption and, consequently, the rate.[169] The improved selectivity has been attributed not only to the reduced conversion, but also to the degrading action of water on the surface O/OH ratio. Total oxidation should require larger ensembles of adjacent oxygen sites than partial oxidation and, therefore, is suppressed as the surface O/OH ratio decreases.

4.7.2 Chemical Composition and Active Sites

By varying the chemical composition of Mo–V–Te–(Nb) oxides, Botella *et al.* found a positive correlation between the vanadium content and the space-time yield of acrylic acid in a series of catalysts, which have been prepared by different methods and that were composed of a multitude of phases.[134] Niobium exerts a positive effect on the yield of acrylic acid as well. However, the interpretation of the results was complicated by the diversity of the constituent phases, because the phase composition differs for each catalyst and the chemical composition of the corresponding phases may vary and is unknown. Thus, in addition to M1 and M2, the formation of Nb- and/or V-doped MoO_x, $Mo_{5-x}(V,Nb)_xO_{14}$, or $Te(Mo,V,Nb)_5O_{16}$ has been considered.

Ueda *et al.* studied the role of the elements by comparing the catalytic properties of an orthorhombic ternary Mo–V phase structurally similar to that of M1 with phase-pure M1 composed of Mo, V, and Te, and Mo, V, Te, and Nb, respectively.[110,161,171] The Mo–V oxide shows high activity but poor selectivity in propane oxidation to acrylic acid. Introduction of tellurium resulted in the development of selectivity towards acrylic acid. This effect has been attributed to the formation of tellurium-containing active sites responsible for further oxidation of propylene formed in the initial oxidative dehydrogenation of propane mediated by isolated vanadyl species. Further improvements in the selectivity achieved by addition of niobium have been attributed to spatial isolation of the alkane activating vanadium sites by breaking V–O–V bonds and achieving site isolation as essential feature of oxidation catalysts that are selective also at high conversions.[97] The reduced thermal stability of the Mo–V phase compared to the Te- and Nb-containing phases may, however, constitute a problem with respect to the comparability of the catalytic results.

Significant improvement in the selectivity to acrylic acid was observed when Te, Nb, and Sb oxides were present at submonolayer coverage on the surface of a Mo–V–O catalyst with orthorhombic M1-like structure.[172] Although the original Mo–V–O catalyst was essentially unselective in propane oxidation to acrylic acid, significant increase in the selectivity was found when Te, Nb, and Sb oxides were anchored as surface oxo-species.

The effects of framework and near-surface composition of phase-pure M1 Mo–V–Te–Nb oxide catalysts on their catalytic performance in selective oxidation of propane to acrylic acid has been studied by exploring the phase width of the M1 structure. For this purpose, the metal stoichiometry was varied comprising a Mo content from 51–77 at.%, V content from 12–23 at.%, Te content from 0–15 at.%, and Nb content from 0–23 at.%.[166] By keeping strictly the phase purity, the effect of metal-site occupancy in the M1 framework on surface composition and catalytic behavior was addressed, though there are no clear trends between bulk composition of M1 and the performance of the different M1 catalysts in selective oxidation of propane to acrylic acid. However, the metal stoichiometry at the surface differs from the stoichiometry of the bulk, as evidenced by surface sensitive *in-situ* photoelectron spectroscopy possessing an information depth of approximately 1 nm. Increased percentage of Te and reduced fraction of Mo at the surface of M1 tend to result in higher yields of acrylic acid. Moreover, a clear correlation between the Te:V molar ratio at the surface and the yield of acrylic acid has been found. Higher yields of acrylic acid have been achieved with increasing Te:V ratio, *i.e.* the formation of acrylic acid is favored by progressive dilution of active vanadia sites on the surface with tellurium oxide species. These experiments indicate that the solid-state structure of M1 might be rather a support or a reservoir that creates an active monolayer on its surface (see Chapter 15). These findings reveal close similarities to the dynamic behavior of vanadyl pyrophosphate catalysts in selective oxidation of *n*-butane to maleic anhydride.[173–177]

Only the mixed-oxide catalysts with M=Sb, and Te were active for the oxidation of propane to acrylic acid. The addition of titanium was found to result in a significant increase of the activity for selective oxidation of ethane to ethene and acetic acid, which was ascribed to morphological changes of the catalyst particles and to gain of surface area by addition of titanium.[151] However, during the heat pretreatment above 823 K in a N_2 atmosphere, phase transformation to a $(Mo,V,Ti)_5O_{14}$-like phase with drastic decreases of the surface area and the oxidation activity has been observed. A decrease in activity for oxidation of propane to acrylic acid was also observed by Millet *et al.* when Mo or V in M1 were substituted by Ti or Sn, whereas Ge increased the activity.[153] The substitution of Nb by Ta resulted in a less-efficient catalyst.[100] M2 was prepared by substituting to various degrees Mo and V with W, Ti, Nb or Fe, and Te with Ce.[128,154] W and Ce were found to be able to mimic the role of Mo and Te, respectively, in ammoxidation of propene. W has been considered to be mainly responsible for site-isolation and electronic effects, Ce compensates for the redox effects of Te.

4.7.3 Structure Sensitivity and Termination

The ongoing discussion about the reaction mechanism of direct oxidation of propane to acrylic acid over Mo–V–Te–Nb oxide catalysts afforded models that consider specific metal positions and rigid geometric arrangements of the constituent elements as active sites. Definite occupations of metal-framework M1 positions have been proposed to result in the formation of site-isolated active ensembles on its terminating basal plane that are characterized by close proximity of the required catalytic functions, and, consequently, attributed to high selectivity.[168,178,179] In this respect, the crystallographic positions M3 and M7 (Scheme 4.2) have been considered to be particularly responsible for activation of propane.[179] The argumentation has been inspired by the bronze-like crystal structure and the associated needle- or rod-like morphology of M1, in which the {001} crystal planes (Scheme 4.2) are arranged perpendicular to the length axis of the cylindrical primary catalyst particles (Figure 4.6).

High and stable performance of M1 in selective oxidation has been originally attributed to the distortion of the octahedral units in the M1 crystal structure and on its terminating basal plane arising from the specific arrangement of the pentagonal, hexagonal, and, in particular, heptagonal rings that causes

Figure 4.6 HR-SEM of M1 needles.

strain.[178] Due to the needle shape of the particles, the surface fraction of (001) planes comprises only a minor fraction of the total surface area of the catalyst. Thus, increased activity after grinding of M1 needles has been attributed to the generation of additional (001) surface area.[140,178] However, even though an increased specific surface area was observed after the mechanical treatment, the disadvantage of the grinding process is its well-known impact on the nature and concentration of defects on the overall surface of the catalyst.[180] The "channels" occupied by tellurium, which open up at the basal plane, have been suggested to be of importance in the reduction and reoxidation of the surface during catalysis.[181,182]

Catalytic experiments trying to verify the assignment of the minority {001} faces, which are generally assumed to terminate without restructuring, as the location of active sites gave conflicting results.[183,184] To address the specific catalytic properties of the basal plane, a model catalyst has been prepared that exposes {001} planes preferentially.[183] For this purpose, a batch of crystalline, phase-pure, and well-characterized M1 synthesized hydrothermally in superheated water vapor was divided into two parts. One part was fully coated with silica by silylation using $HSi(OEt)_3$. The other part has been left untreated. Special attention has been paid to a comprehensive microstructural and surface analysis of the model catalysts before and after propane oxidation. The gentle mechanical treatment during pressing and sieving procedures that are commonly required for testing a powdered catalyst in a fixed-bed reactor, caused partial disruption of the silica-coated and the silica-free M1 particles changing the particle-size distribution towards shorter particles for both specimens in a similar manner. Low-energy ion scattering (LEIS) measured in vacuum at room temperature revealed differences in the integral chemical composition of the top-most Mo–V–Te–Nb oxide surface layer of the two catalysts. Due to the structural anisotropy of M1, the {001} planes freshly generated by breaking the silica-covered Mo–V–Te–Nb oxide needles showed enhanced enrichment of tellurium and vanadium compared to the entire M1 surface of the silica-free needles. Irrespective of different terminating lattice planes and surface chemistry, the two catalysts exhibit similar intrinsic reactivity (reactivity normalized to the total Mo–V–Te–Nb oxide surface area) implying that the same metal-oxo clusters are exposed to the reactants under conditions of the formation of acrylic acid on the entire surface of the M1 particles. This would necessarily require mass transport on the surface of the activated catalyst. The lateral surface of the cylindrical particles accounts for the main part of the surface area of M1 (approximately 80%). The stepped morphology of the latter may generate similar metal-oxo arrangements like terminations on the basal plane. Electron microscopy has been applied to verify this assumption.[185] Figure 4.7 represents a STEM image of one M1 crystal viewed in the projection along <001>.

The perimeter of the particle is clearly visible. The termination mode is consistent with a perfectly ordered bulk structure that gives, however, no indication about the surface structure, because the resulting termination is substantially different from an assumed termination along low-indexed planes of crystallographic unit cells (Scheme 4.2). Evidently, the crystal periodicity is broken along the thinnest sections through the mesh of polyhedra, in other

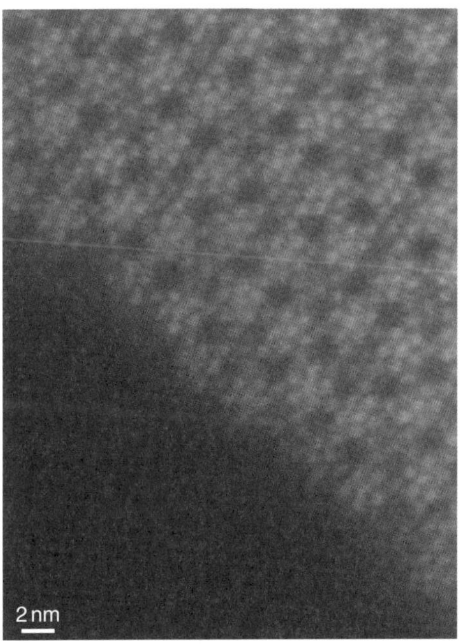

Figure 4.7 STEM image of M1 along the <001> direction.

words, the lateral surface is created along directions with minimum number of bonds to be broken. Building such a termination is shown exemplarily in Scheme 4.6.

Among the many nearest-neighbor lines, a continuous pathway has been chosen arbitrarily. The exact fracture line may vary and may depend on the metal-site occupancy, which, in turn, is controlled by the chemical composition of the synthesis gel and the synthesis kinetics. In any case, the lateral surface of the cylindrical M1 particles is constituted of roughly half-pipes from the formerly closed channels and from wall fragments exposing now the inner surface of the channels, which can be seen clearly in the STEM image (Figure 4.7). These "rain pipes" filled with tellurium oxo-species run probably along the whole prismatic faces, as indicated by the stepped surface morphology seen in the HR-SEM image of Figure 4.6. Consequently, lateral termination, in particular, liberates tellurium oxo-groups that are no longer contained in the channels but can disperse over the whole surface. The termination of M1 evidenced by electron microscopy explains the excess in Te usually detected at the surface of Mo–V–Te–Nb oxide catalysts investigated by XPS in vacuum[83,110,129,132,137,182] and in the presence of the reacting molecules.[166,186] Furthermore, the proposed termination mechanism provides an explanation for the experimental observation that reactive sites exist on all orientations of the anisotropic faces of the M1 crystals.[183]

4.7.4 Surface Dynamics

The outer atoms on the surface of a catalyst largely control the catalytic properties of the solid. The surface of Mo–V–Te–(Nb) oxide catalysts has been studied with low-energy ion scattering (LEIS) in vacuum that gives the atomic composition of the outermost atoms and by adsorption of methanol and allyl alcohol as probe molecules.[184,187] In agreement with XPS studies and with the termination model discussed above, these investigations indicated that the terminating layer of the catalyst is composed differently from the bulk.

The surface of M1 Mo–V–Te–Nb oxide has been studied by *in-situ* photoelectron spectroscopy at reaction temperature of 623 K in the presence of the reacting molecules propane and oxygen, and the additive steam at a total pressure of 20–30 Pa.[166,186] A depth profile of the elemental composition was recorded by varying the kinetic energy between 170 eV and 470 eV separately for each core level. Escape depths comprising 63% of the detected signal of *ca.* 0.8 nm denominated as surface, and *ca.* 1.6 nm denominated as subsurface were probed. *In-situ* valence-band spectroscopy has also been measured at 623 K in the presence of all reactants, revealing a substantial covalent character of the metal–oxygen bonds in M1.[166,186,188] The *in-situ* XPS experiments showed an enhanced mobility of Te with changing feed composition and temperature, and, especially, after addition of water vapor (Figure 4.8).

In the presence of propane at 623 K, the surface concentration of tellurium is reversibly enhanced at the expense of molybdenum, which is even more noticeable in the presence of steam. By analyzing the subsurface concentration, it

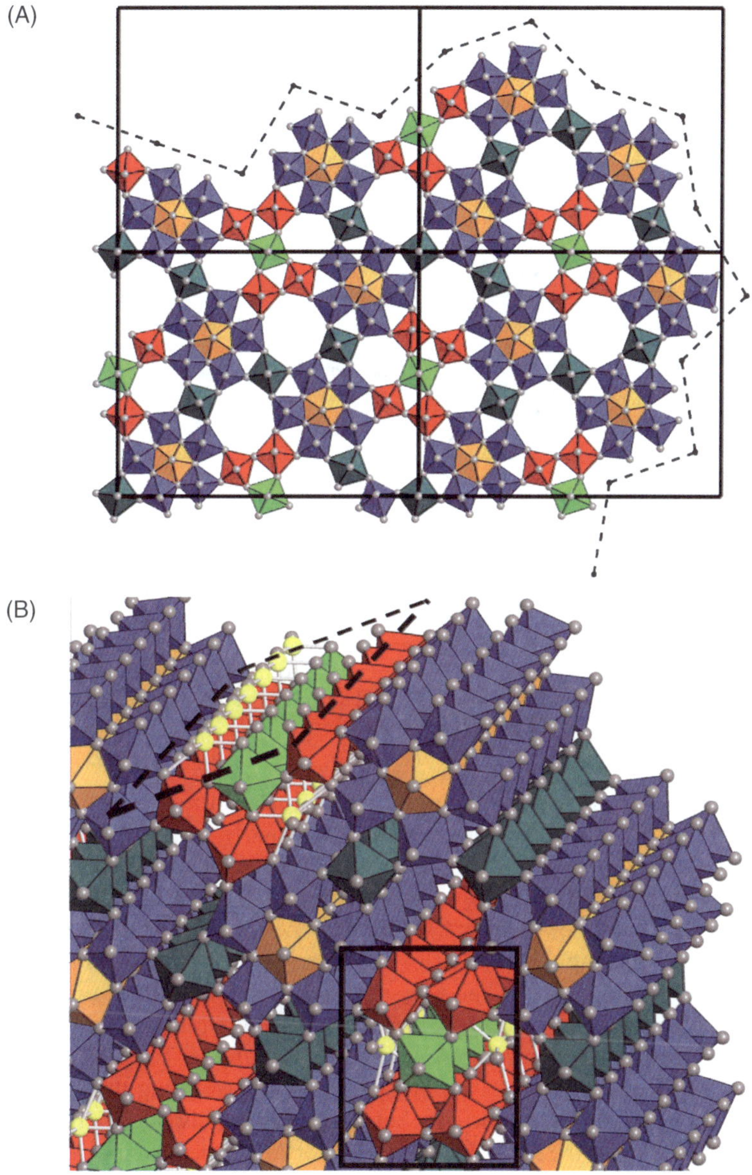

Scheme 4.6 (A) Building a termination model of M1 starting with four unit cells of M1; Representation viewed along *c* onto the *a,b* plane; TeO units in the channels are omitted for clarity; Centroids of hexagonal and heptagonal channels are connected by a network of nearest-neighbor lines between 7 and 9 Å in length and among the many nearest-neighbor lines a continuous pathway is chosen arbitrarily, The part of the model lying on or outside the line is cut away. (B) Termination model of M1: Possible active sites are highlighted with boxes: (1) continuous line: "classical" site on the *a,b* plane; (2) broken line: "nonclassical" site(s) on the lateral termination.

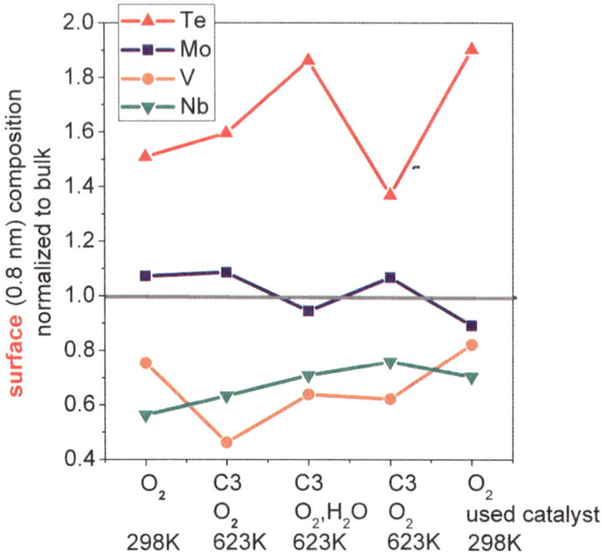

Figure 4.8 Quantification of the elemental composition of a M1 catalyst (ID 1761) by *in-situ* XPS (at.%) at escape depth of *ca.* 0.8 nm; The following reaction conditions were applied: measurement before admission of propane at 298 K and $p(O_2) = 30$ Pa; measurement at 623 K with molar ratio hydrocarbon:$O_2 = 1{:}2$, $p = 20$ Pa; $p(H_2O) = 10$ Pa; an oxygen pressure of 30 Pa was applied in the *ex situ* experiment.

becomes apparent that the bulk supplies tellurium to the surface. These observations reflect the mobility of tellurium, particularly in a wet reaction atmosphere. Enhanced volatility of TeO_2 in water vapor has been attributed to the formation of gaseous $TeO(OH)_2$.[189] The positive effect of steam on the selectivity to acrylic acid may be, therefore, related to the formation of mobile Te-containing clusters through hydrolysis of oligomeric $(TeO)_n$ species in the channels and on the lateral surface of the M1 structure, which migrate over the surface. Thus, restructuring and possible formation of metal tellurates under the hydrothermal reaction conditions may provoke spatial isolation of vanadium containing active sites. Identification the type of oxo-clusters constituting the surface under reaction conditions is highly relevant, but, spectroscopically challenging due to the low surface area and the semiconducting properties of the mixed oxide.

4.8 Functional Model of M1

Nanostructured multimetal-oxide catalysts composed of Mo, V, Te, and Nb show high yields with respect to acrylic acid in partial oxidation of propane, but the productivity is not yet sufficient for an industrial application. Distinguished selectivity towards acrylic acid is unambiguously related to the "M1" crystal structure. The M1 phase unifies sophisticated structural chemistry with considerable chemical complexity that provides with Mo, V, and Te three redox-active elements. Therefore, it is not surprising that approaches to explain the catalytic properties of Mo–V–Te–Nb oxides often refer to specific features of the M1 framework structure. Furthermore, it is assumed that the complicated chemical nature of the mixed-metal oxide is indispensable, *i.e.* all four metals are certainly required. In the following, an attempt is undertaken to combine the present knowledge from the literature and our laboratory to draw a preliminary picture from the material function of the M1 phase in selective oxidation of propane to acrylic acid.

It has certainly been proven that the crystalline M1 phase is essential for sustained catalytic performance of the catalyst. High crystallinity of the metastable phase is obviously required, possibly to prevent segregation of binary phases and formation of unselective sites. Vanadium and tellurium are key elemental constituents contributing to the performance of the catalyst. But, there are a number of observations that clearly indicate substantial structural and compositional differences of M1 bulk and M1 surface:

1. Neither a correlation between the geometric surface area of M1 and the activity in propane oxidation, nor structure sensitivity with respect to the fraction of (001) faces in the catalyst can be observed.
2. Under reaction conditions, the surface composition on average is so strongly different from the bulk composition that the surface termination cannot be a cut through the bulk structure.
3. The yield of acrylic acid increases with increasing Te:V ratio on the surface of M1, and, in contrast, decreases with the Mo surface abundance. This suggests that the active sites do not contain all constituent elements of the M1 phase.

The combined experimental facts imply that the nature of the active sites is not related to a rigid section of the M1 structure. Notwithstanding, the existence of M1 seems to be an essential requirement for the formation and long-lasting operation of the active surface species. Consequently, the role of the M1 phase is not to provide directly the active sites in its structure but to uniquely enable their formation. Due to the partial covalent character of the metal–oxygen bonds in molybdenum-oxide-based mixed systems, M1 can be classified as a stable host structure that liberates some Te_xO_y guest species without structural collapse. It has been shown that tellurium can be removed, possibly reversibly, on demand under the hydrothermal reaction conditions of propane oxidation by changing the concentration of steam in the feed. Furthermore, the specific termination of the lateral surface of the cylindrical M1 particles that generates half-pipes filled with tellurium oxide species contributes to an enrichment of Te on the catalyst surface *a priori* with termination. The M1 system can thus be considered as a support for an active phase that comprises a thickness of about 1 nm and that is composed of isolated V_xO_y moieties embedded in a matrix of Te_xO_y surface species (see Chapter 15). It is assumed that chemical bonds exist between the surface layer and the bulk structure that account for the high stability of the catalyst. Possibly, surface polyoxo species are formed under reaction conditions. The electronic and molecular structure of the active ensembles remain, however, still unknown. Their elucidation would require *in-situ* spectroscopic efforts that overcome the poor spectroscopic accessibility of the material and provide information in *sub*nanometer resolution. How vanadium- and/or tellurium-containing surface oxo species are involved in individual reaction steps of the complex and in detail unknown reaction mechanism also remains to be clarified. In any case, the moderate average productivity of the M1 catalyst implies that active sites are a rare species on the reactive surface. This is conceivable when we assume that at least 2 V atoms need to be in an active site for C–H activation; this number is the minimum configuration to reversibly hold 4 electrons in exchange that are required to activate one oxygen molecule. The observed negative correlation of performance with the main constituting element Mo may be seen as indication for the proposed oligomeric nature of the active site: the more Mo in the structure the more unlikely will be a V–O–V bridge situation.

In summary, adequate surface V_xO_y distribution and abundance of Te_xO_y species represent two parameters that seem to be the key guarantee for high performance of phase-pure and highly crystalline M1 in selective oxidation of propane to acrylic acid. These variables are controlled on the one hand by catalyst synthesis in its whole complexity, which determines not only the bulk composition but also morphology and surface termination, and on the other hand by the chemical potential of the operation. Thus, the difficulties in catalyst optimization become evident. Even though considerable progress has been made for the last 20 years in clarification of the phase composition and phase structures, understanding the catalyst synthesis, improving the reproducibility of the synthesis, and analyzing the physical and chemical properties of the catalyst, no breakthrough has been achieved with respect to performance compared to the original patents filed by Mitsubishi Corp. in the early 1990s. The selectivity seldom exceeds 80% at maximum conversion of 60% so that the magic benchmark of 50% yield has been surpassed only in very rare cases in the patent literature. Critical analysis of the present knowledge accumulated in our laboratory and in the literature leads to the conclusion that further improvement of the catalyst performance demands concerted strategies involving optimization of the operation conditions individually adapted to M1 composition and morphology. Moreover, novel approaches in reaction engineering are required that allow the alignment of the chemical potential over the catalyst bed in the reactor.

Acknowledgements

Our own research on multimetal-oxide catalysts for selective oxidation of propane to acrylic acid has been performed with the participation of many coworkers of the Department of Inorganic Chemistry at the Fritz Haber Institute. Special thanks go to Dr. Almudena Celaya Sanfiz and Dr. Ayyamperumal Sakthivel for synthesis of catalysts, Dr. Frank Girgsdies for profound crystallographic analysis, Dr. Thomas W. Hansen, Gisela Weinberg and Dr. Wei Zhang for electron microscopy, Dr. Sabine Wrabetz for microcalorimetry, and Dr. Detre Teschner and Peter Schnörch for *in-situ* photoelectron spectroscopic studies. We acknowledge the good collaboration with Professor Sharifah Bee Abd Hamid and coworkers, COMBICAT, University Malaya, Kuala Lumpur, Malaysia for screening the catalytic performance. I would like to thank Professor Robert Schlögl for continued discussion and support.

References

1. M. M. Bettahar, G. Costentin, L. Savary and J. C. Lavalley, *Appl. Catal., A: General*, 1996, **145**(1–2), 1–48.
2. M. M. Lin, *Appl. Catal. A: General*, 2001, **207**(1,2), 1–16.
3. E. K. Novakova and J. C. Vedrine, Propane selective oxidation to propene and oxygenates, in *Metal Oxides – Chemistry and Applications*, ed. J. L. G. Fierro, CRC Taylor & Francis, Boca Raton, London, New York, 2006, pp. 413–461.
4. G. Centi, F. Cavani and F. Trifiro, Chapter 5: Control of the Surface Reactivity of Solid Catalysts: New Alkane Oxidation Reactions, in *Selective Oxidation by Heterogeneous Catalysis*, Kluwer Academic, New York, Boston, Dordrecht, London, Moscow, 2001, pp. 203–283.
5. F. Cavani and F. Trifirò, Partial oxidation of C2 to C4 paraffines, in *Basic Principles in Applied Catalysis*, ed. M. Baerns, Springer, Berlin, Heidelberg, New York, 2003, pp. 19–84.
6. M. Beller, B. Cornils, C. D. Frohning and C. W. Kohlpaintner, *J. Mol. Catal. A: Chemical*, 1995, **104**(1), 17–85.
7. R. K. Grasselli and J. D. Burrington, Oxidation of low-molecular-weight hydrocarbons, in *Handbook of Heterogeneous Catalysis (2nd Edition)*, 2008, Wiley-VCH, Weinheim, vol. 7, pp. 3479–3489.
8. T. V. Andrushkevich, *Catal. Rev.: Science & Engineering*, 1993, **35**(2), 213–259.
9. R. K. Grasselli, J. D. Burrington, D. J. Buttrey, P. De Santo, Jr., C. G. Lugmair, A. F. Volpe, Jr. and T. Weingand, *Top. Catal.*, 2003, **23**(1–4), 5–22.
10. G. Mestl, *Top. Catal.*, 2006, **38**(1), 69–82.
11. P. Kampe, L. Giebeler, D. Samuelis, J. Kunert, A. Drochner, F. Haass, A. H. Adams, J. Ott, S. Endres, G. Schimanke, T. Buhrmester, M. Martin, H. Fuess and H. Vogel, *Phys. Chem. Chem. Phys.*, 2007, **9**(27), 3577–3589.
12. D. C. Wadley, M. S. Tam, P. B. Kokitkar, J. E. Jackson and D. J. Miller, *J. Catal.*, 1997, **165**(2), 162–171.
13. J. N. Chheda, G. W. Huber and J. A. Dumesic, *Angew. Chem. Intl. Ed.*, 2007, **46**(38), 7164–7183.
14. F. Cavani, N. Ballarini and A. Cericola, *Catal. Today*, 2007, **127**(1–4), 113–131.
15. D. Vitry, J. L. Dubois and W. Ueda, *J. Mol. Catal. A-Chem.*, 2004, **220**(1), 67–76.
16. A. Godefroy, G. S. Patience, T. Tzakova, D. Garrait and J. L. Dubois, *Chem. Eng. Tech.*, 2009, **32**(3), 373–379.

17. H.-G. Lintz and A. Reitzmann, *Catal. Rev.: Science & Engineering*, 2007, **49**(1), 1–32.
18. L. Chalakov, L. K. Rihko-Struckmann, B. Munder and K. Sundmacher, *Chem. Eng. J.*, 2009, **145**(3), 385–392.
19. W. Ueda and Y. Suzuki, *Chem. Lett.*, 1995(7), 541–542.
20. M. Ai, *Catal. Today*, 2003, **85**(2–4), 193–198.
21. M. Ai, *Catal. Today*, 1998, **42**(3), 297–301.
22. M. Ai, *J. Chem. Soc., Chem. Commun.*, 1986, 786–787.
23. M. Ai, *J. Catal.*, 1986, **101**(2), 389–395.
24. M. Ai, *Catal. Today*, 1992, **13**(4), 679–684.
25. N. Gribot-Perrin, J.-C. Volta, A. Burrows, C. Kiely and M. Gubelmann-Bonneau, *Stud. Surf. Sci. Catal.*, 1996, **101**(Pt. B, 11th International Congress on Catalysis – 40th Anniversary, 1996), 1205–1214.
26. J.-C. Volta, *Catal. Today*, 1996, **32**(1–4), 29–36.
27. Y.-F. Han, H.-M. Wang, H. Cheng and J.-F. Deng, *Chem. Commun.*, 1999, 521–522.
28. S. Marengo, P. Patrono, P. Comotti, G. Galli, P. Galli, M. A. Massucci and M. T. Meloni, *Appl. Catal. A: General*, 2002, **230**(1–2), 219–231.
29. B. Kerler, A. Martin, M. M. Pohl and M. Baerns, *Catal. Lett.*, 2002, **78**(1–4), 259–265.
30. B. Kerler, A. Martin and M. Baerns, *Chem. Eng. Tech.*, 2003, **26**(4), 420–424.
31. A. Martin, B. Kerler, J.-C. Volta and J. Radnik, *J. Mater. Chem.*, 2003, **13**(1), 125–129.
32. G. Landi, L. Lisi and J. C. Volta, *J. Mol. Catal. A: Chemical*, 2004, **222**(1–2), 175–181.
33. G. Landi, L. Lisi and J. C. Volta, *Catal. Today*, 2004, **91–92**(1), 275–279.
34. G. Landi, L. Lisi and J. C. Volta, *Chem. Commun.*, 2003, 492–493.
35. G. Landi, L. Lisi and G. Russo, *J. Mol. Catal. A: Chemical*, 2005, **239**(1–2), 172–179.
36. L. Zeng, H. Jiang and J. Niu, *J. Mol. Catal. A: Chemical*, 2005, **232**(1–2), 119–122.
37. R. Zhao, J. Wang, Q. Dong and J. Liu, *J. Nat. Gas Chem.*, 2005, **14**(2), 88–94.
38. Y. H. Taufiq-Yap, C. S. Saw and R. Irmawati, *Catal. Lett.*, 2005, **105**(1–2), 103–110.
39. Y. H. Taufiq-Yap and C. S. Saw, *Catal. Today*, 2008, **131**(1–4), 285–291.
40. X.-K. Li, W.-J. Ji, J. Zhao, Z. Zhang and C.-T. Au, *Appl. Catal., A: General*, 2006, **306**(1), 8–16.
41. Q. Jiang, J. Zhao, X. K. Li, W. J. Ji, Z. B. Zhang and C. T. Au, *Applied Catalysis, A: General*, 2008, **341**(1–2), 70–76.
42. Y. Takita, S. Kikukawa, Y. Abe, Y. Mizuhara and T. Ishihara, *Nippon Kagaku Kaishi*, 1992(4), 354–360.
43. Y. Takita, H. Yamashita and K. Moritaka, *Chem. Lett.*, 1989, **18**(10), 1733.
44. Z. Wang, W. Wu, G. Liu, G. Mao and D. Kuang, *Acta Petrolium Sinica (Shiyou Xuebao, Shiyou Jiagong)*, 1998, **14**(3), 21–26.
45. Y. Han, H. Wang, H. Cheng, R. Jin and J. Deng, *New J. Chem.*, 1998, **22**(11), 1175–1176.

46. H. Cheng, Y. F. Han and H. M. Wang, *Shiyou Huagong*, 1999, **28**(12), 803–807.
47. T. Komatsu, Y. Uragami and K. Otsuka, *Chem. Lett.*, 1988(11), 1903–1906.
48. N. Dimitratos and J. C. Vedrine, *J. Mol. Catal. A: Chemical*, 2006, **255**(1–2), 184–192.
49. N. Dimitratos and J. C. Vedrine, *Catal. Commun.*, 2006, **7**(10), 811–818.
50. N. Dimitratos and J. C. Vedrine, Selective oxidation of propane on Cs2.5H1. 5PV1WxMo11-xO40 heteropolyoxometallate compounds, in *Catalysis in Application, [Papers presented at the International Symposium on Applied Catalysis]*, Glasgow, United Kingdom, July 16-18, 2003, pp. 145–152.
51. N. Dimitratos, C. D. Pina, E. Falletta, C. L. Bianchi, V. Dal Santo and M. Rossi, *Catal. Today*, 2007, **122**(3–4), 307–316.
52. J. H. Holles, C. J. Dillon, J. A. Labinger and M. E. Davis, *J. Catal.*, 2003, **218**(1), 42–53.
53. H.-S. Jiang, X. Mao, S.-J. Xie and B.-K. Zhong, *J. Mol. Catal. A: Chemical*, 2002, **185**(1–2), 143–149.
54. N. Mizuno, D.-J. Suh, W. Han and T. Kudo, *J. Mol. Catal. A: Chemical*, 1996, **114**(1–3), 309–317.
55. N. Mizuno, M. Tateishi and M. Iwamoto, *Appl. Catal. A: General*, 1995, **128**(2), L165–L170.
56. N. Mizuno, M. Tateishi and M. Iwamoto, *Appl. Catal. A: General*, 1996, **134**(1), 183.
57. W. Ueda, Y. Suzuki, W. Lee and S. Imaoka, *Stud. Surf. Sci. Catal.*, 1996, **101**(Pt. B, 11th International Congress on Catalysis – 40th Anniversary, 1996, Pt. B), 1065–1074.
58. W. Li and W. Ueda, *Stud. Surf. Sci. Catal.*, 1997, **110**(3rd World Congress on Oxidation Catalysis, 1997), 433–442.
59. J.-S. Min and N. Mizuno, *Catal. Today*, 2001, **71**(1–2), 89–96.
60. J.-S. Min and N. Mizuno, *Catal. Today*, 2001, **66**(1), 47–52.
61. M. E. Davis, C. J. Dillon, J. H. Holles and J. Labinger, *Angew. Chem., Intl. Ed.*, 2002, **41**(5), 858–860.
62. C. J. Dillon, J. H. Holles, R. J. Davis, J. A. Labinger and M. E. Davis, *J. Catal.*, 2003, **218**(1), 54–66.
63. C. J. Dillon, J. H. Holles, M. E. Davis and J. A. Labinger, *Catal. Today*, 2003, **81**(2), 189–195.
64. T.-J. Cai, C.-L. Yu, Q. Deng, Z.-S. Peng, Y.-F. Long and X.-M. Lue, *Chin. J. Chem.*, 2004, **22**(8), 831–836.
65. X.-K. Li, J. Zhao, W.-J. Ji, Z.-B. Zhang, Y. Chen, C.-T. Au, S. Han and H. Hibst, *J. Catal.*, 2006, **237**(1), 58–66.
66. W. Li, K. Oshihara and W. Ueda, *Appl. Catal., A: General*, 1999, **182**(2), 357–363.
67. H. Tsuji and Y. Koyasu, *J. Am. Chem. Soc.*, 2002, **124**(20), 5608–5609.
68. H. Hibst, F. Rosowski and G. Cox, *Catal. Today*, 2006, **117**(1–3), 234–241.
69. P. Beato, A. Blume, F. Girgsdies, R. E. Jentoft, R. Schloegl, O. Timpe, A. Trunschke, G. Weinberg, Q. Basher, F. A. Hamid, S. B. A. Hamid, E. Omar and L. Mohd Salim, *Appl. Catal. A: General*, 2006, **307**(1), 137–147.

70. E. M. Thorsteinson, T. P. Wilson, F. G. Young and P. H. Kasai, *J. Catal.*, 1978, **52**(1), 116–132.
71. M. Dieterle, G. Mestl, J. Jaeger, Y. Uchida, H. Hibst and R. Schloegl, *J. Mol. Catal. A: Chemical*, 2001, **174**(1–2), 169–185.
72. O. Ovsitser, Y. Uchida, G. Mestl, G. Weinberg, A. Blume, J. Jaeger, M. Dieterle, H. Hibst and R. Schloegl, *J. Mol. Catal. A: Chemical*, 2002, **185**(1–2), 291–303.
73. J. B. Wagner, D. S. Su, S. A. Schunk, H. Hibst, J. Petzoldt and R. Schoegl, *J. Catal.*, 2004, **224**(1), 28–35.
74. G. A. Zenkovets, G. N. Kryukova, V. Y. Gavrilov, S. V. Tsybulya, V. A. Anufrienko, T. A. Larina, D. F. Khabibulin, O. B. Lapina, E. Roedel, A. Trunschke, T. Ressler and R. Schloegl, *Mater. Chem. Phys.*, 2007, **103**(2–3), 295–304.
75. D. L. Stern and R. K. Grasselli, *J. Catal.*, 1997, **167**(2), 550–559.
76. D. L. Stern and R. K. Grasselli, *J. Catal.*, 1997, **167**(2), 560–569.
77. D. L. Stern and R. K. Grasselli, *Stud. Surf. Sci. Catal.*, 1997, **110**(3rd World Congress on Oxidation Catalysis, 1997), 357–365.
78. A. Kaddouri, C. Mazzocchia and E. Tempesti, *Appl. Catal. A: General*, 1999, **180**(1–2), 271–275.
79. N. Fujikawa, K. Wakui, K. Tomita, N. Ooue and W. Ueda, *Catal. Today*, 2001, **71**(1–2), 83–88.
80. R. K. Grasselli, G. Centi and F. Trifiro, in *4th International Symposium on Uses of Selenium and Tellurium*, Selenium Tellurium Development, 1989, pp. 609–632.
81. T. Ushikubo, H. Nakamura, Y. Koyasu and S. Wajiki, *US Pat.* 5,380,933A, 1995.
82. H. Hinago and S. Komada, *US Pat.* 6,063,728, 2000.
83. M. Baca, A. Pigamo, J. L. Dubois and J. M. M. Millet, *Top. Catal.*, 2003, **23**(1–4), 39–46.
84. P. Botella, E. Garcia-Gonzalez, J. M. Lopez Nieto and J. M. Gonzalez-Calbet, *Solid State Sciences*, 2005, **7**(5), 507–519.
85. T. Ushikubo, K. Oshima, A. Kayou, M. Vaarkamp and M. Hatano, *J. Catal.*, 1997, **169**(1), 394–396.
86. T. Ushikubo, K. Oshima, A. Kayou, M. Hatano and X. Can Li and Qin, Ammoxidation of propane over Mo-V-Nb-Te mixed oxide catalysts, in *Studies in Surface Science and Catalysis*, Elsevier, 1997, Vol. 112, pp. 473–480.
87. M. Aouine, J. M. M. Millet and J. L. Dubois, *Chem. Commun.*, 2001, 1180–1181.
88. P. DeSanto, Jr., D. J. Buttrey, R. K. Grasselli, C. G. Lugmair, A. F. Volpe, Jr., B. H. Toby and T. Vogt, *Zeitschrift fuer Kristallographie*, 2004, **219**(3), 152–165.
89. P. DeSanto, D. J. Buttrey, R. K. Grasselli, C. G. Lugmair, A. F. Volpe, B. H. Toby and T. Vogt, *Top. Catal.*, 2003, **23**(1–4), 23–38.
90. P. DeSanto, D. Buttrey, R. Grasselli, W. Pyrz, C. Lugmair, A. Volpe, T. Vogt and B. Toby, *Top. Catal.*, 2006, **38**(1), 31–40.

91. H. Murayama, D. Vitry, W. Ueda, G. Fuchs, M. Anne and J. L. Dubois, *Appl. Catal. A: General*, 2007, **318**(1), 137–142.
92. A. Celaya Sanfiz, T. W. Hansen, F. Girgsdies, O. Timpe, E. Roedel, T. Ressler, A. Trunschke and R. Schloegl, *Top. Catal.*, 2008, **50**(1–4), 19–32.
93. W. D. Pyrz, D. A. Blom, N. R. Shiju, V. V. Guliants, T. Vogt and D. J. Buttrey, *Catal. Today*, 2009, **142**(3–4), 320–328.
94. W. D. Pyrz, D. A. Blom, T. Vogt and D. J. Buttrey, *Angew. Chem., Intl. Ed.*, 2008, **47**(15), 2788–2791.
95. W. D. Pyrz, D. A. Blom, N. R. Shiju, V. V. Guliants, T. Vogt and D. J. Buttrey, *J. Phys. Chem. C*, 2008, **112**(27), 10043–10049.
96. J. Loub, *Collect. Czech. Chem. Commun.*, 1993, **58**(8), 1717–1738.
97. R. K. Grasselli, *Catal. Today*, 2005, **99**(1–2), 23–31.
98. J. Holmberg, R. K. Grasselli and A. Andersson, *Appl.Catal., A: General*, 2004, **270**(1–2), 121–134.
99. R. K. Grasselli, A. Andersson, D. J. Buttrey, J. D. Burrington, C. G. Lugmair and A. F. Volpe, in *Abstracts of Papers, 228th ACS National Meeting*, Philadelphia, PA, United States, August 22–26, 2004, pp. COLL-245.
100. R. K. Grasselli, D. J. Buttrey, J. D. Burrington, A. Andersson, J. Holmberg, W. Ueda, J. Kubo, C. G. Lugmair and A. F. Volpe, Jr., *Top. Catal.*, 2006, **38**(1–3), 7–16.
101. R. G. Teller, J. F. Brazdil, R. K. Grasselli and W. Yelon, *J. Chem. Soc., Faraday Trans. 1: Physical Chemistry in Condensed Phases*, 1985, **81**(7), 1693–1704.
102. R. K. Grasselli, *Top. Catal.*, 2002, **21**(1–3), 79–88.
103. R. K. Grasselli, *Top. Catal.*, 2001, **15**(2–4), 93–101.
104. D. Vitry, Y. Morikawa, J. L. Dubois and W. Ueda, *Top. Catal.*, 2003, **23**(1–4), 47–53.
105. J. Holmberg, R. K. Grasselli and A. Andersson, *Top. Catal.*, 2003, **23**(1–4), 55–63.
106. J. B. Wagner, O. Timpe, F. A. Hamid, A. Trunschke, U. Wild, D. S. Su, R. K. Widi, S. B. A. Hamid and R. Schloegl, *Top. Catal.*, 2006, **38**(1–3), 51–58.
107. P. Korovchenko, N. R. Shiju, A. K. Dozier, U. M. Graham, M. O. Guerrero-Perez and V. V. Guliants, *Top. Catal.*, 2008, **50**(1–4), 43–51.
108. M. Baca, M. Aouine, J. L. Dubois and J. M. M. Millet, *J. Catal.*, 2005, **233**(1), 234–241.
109. J. M. Oliver, J. M. L. Nieto and P. Botella, *Catal. Today*, 2004, **96**(4), 241–249.
110. W. Ueda, D. Vitry and T. Katou, *Catal. Today*, 2004, **96**(4), 235–240.
111. F. Ivars, B. Solsona, E. Rodríguez-Castellón and J. M. López Nieto, *J. Catal.*, 2009, **262**(1), 35–43.
112. R. K. Grasselli, C. G. Lugmair, A. F. Volpe, Jr., A. Andersson and J. D. Burrington, *Catal. Letters*, 2008, **126**(3–4), 231–240.
113. M. M. Lin, *Appl. Catal. A: General*, 2003, **250**(2), 305–318.
114. T. Ushikubo, I. Sawaki, K. Oshima, K. Inumari, S. Kobayakawa and K. Kiyono, *US Pat.* 5,422,328, 1995.

115. R. I. Maksimovskaya, V. M. Bondareva and G. I. Aleshina, *Eur. J. Inorg. Chem.*, 2008(31), 4906–4914.
116. Y. Sun, J. Liu and E. Wang, *Inorg. Chim. Acta*, 1986, **117**(1), 23–26.
117. S. J. Anderson, *Nature*, 1937, **140**(3550), 850.
118. H. T. Evans, *J. Am. Chem. Soc.*, 1948, **70**(3), 1291–1292.
119. H. T. Evans, *J. Am. Chem. Soc.*, 1968, **90**(12), 3275–3276.
120. J. M. Oliver, J. M. Lopez Nieto, P. Botella and A. Mifsud, *Appl. Catal. A: General*, 2004, **257**(1), 67–76.
121. G. Y. Popova, T. V. Andrushkevich, Y. A. Chesalov, L. M. Plyasova, L. S. Dovlitova, E. V. Ischenko, G. I. Aleshina and M. I. Khramov, *Catal. Today*, 2009, **144**(3–4), 312–317.
122. G. Y. Popova, T. V. Andrushkevich, L. S. Dovlitova, G. A. Aleshina, Y. A. Chesalov, A. V. Ishenko, E. V. Ishenko, L. M. Plyasova, V. V. Malakhov and M. I. Khramov, *Appl. Catal. A: General*, 2009, **353**(2), 249–257.
123. J. C. J. Bart, G. Petrini and N. Giordano, *Zeitschrift für anorganische und allgemeine Chemie*, 1975, **413**(2), 180–192.
124. F. Ivars, B. Solsona, S. Hernandez and J. M. Lopez Nieto, *Catal. Today*, 2010, **149**(3–4), 260–266.
125. X. Tu, N. Furuta, Y. Sumida, M. Takahashi and H. Niiduma, *Catal. Today*, 2006, **117**(1–3), 259–264.
126. G. Y. Popova, T. V. Andrushkevich, G. I. Aleshina, L. M. Plyasova and M. I. Khramov, *Appl. Catal. A: General*, 2007, **328**(2), 195–200.
127. M. M. Lin, *Appl. Catal. A: General*, 2003, **250**(2), 287–303.
128. J. Holmberg, S. Hansen, R. K. Grasselli and A. Andersson, *Top. Catal.*, 2006, **38**(1–3), 17–29.
129. M. Baca and J.-M. M. Millet, *Appl. Catal. A: General*, 2005, **279**(1–2), 67–77.
130. B. Deniau, G. Bergeret, B. Jouguet, J. L. Dubois and J. M. M. Millet, *Top. Catal.*, 2008, **50**(1–4), 33–42.
131. W. Ueda, D. Vitry, T. Kato, N. Watanabe and Y. Endo, *Research on Chemical Intermediates*, 2006, **32**(3–4), 217–233.
132. P. Botella, P. Concepcion, J. M. L. Nieto and Y. Moreno, *Catal. Today*, 2005, **99**(1–2), 51–57.
133. P. Botella, J. M. Lopez Nieto and B. Solsona, *Catal. Lett.*, 2002, **78**(1–4), 383–387.
134. P. Botella, J. M. Lopez Nieto, B. Solsona, A. Mifsud and F. Marquez, *J. Catal.*, 2002, **209**(2), 445–455.
135. P. Botella, B. Solsona, A. Martinez-Arias and J. M. L. Nieto, *Catal. Lett.*, 2001, **74**(3–4), 149–154.
136. D. Vitry, Y. Morikawa, J. L. Dubois and W. Ueda, *Appl. Catal. A: General*, 2003, **251**(2), 411–424.
137. F. Ivars, P. Botella, A. Dejoz, J. M. L. Nieto, P. Concepcion and M. I. Vazquez, *Top. Catal.*, 2006, **38**(1–3), 59–67.
138. N. R. Shiju and V. V. Guliants, *ChemPhysChem*, 2007, **8**(11), 1615–1617.
139. P. Botella, P. Concepcion, J. M. Lopez Nieto and B. Solsona, *Catal. Lett.*, 2003, **89**(3–4), 249–253.

140. W. Ueda and K. Oshihara, *Appl. Catal. A: General*, 2000, **200**(1–2), 135–143.
141. T. Katou, D. Vitry and W. Ueda, *Chem. Lett.*, 2003, **32**(11), 1028–1029.
142. M. Sadakane, K. Kodato, T. Kuranishi, Y. Nodasaka, K. Sugawara, N. Sakaguchi, T. Nagai, Y. Matsui and W. Ueda, *Angew. Chem., Intl. Ed.*, 2008, **47**(13), 2493–2496.
143. W. Ueda, M. Sadakane and H. Ogihara, *Catal. Today*, 2008, **132**(1–4), 2–8.
144. M. Sadakane, N. Watanabe, T. Katou, Y. Nodasaka and W. Ueda, *Angew. Chem., Intl. Ed.*, 2007, **46**(9), 1493–1496.
145. C. S. Cundy and P. A. Cox, *Chem. Rev.*, 2003, **103**(3), 663–702.
146. C. S. Cundy and P. A. Cox, *Microporous Mesoporous Mater.*, 2005, **82**(1–2), 1–78.
147. B. L. Cushing, V. L. Kolesnichenko and C. J. O'Connor, *Chem. Rev.*, 2004, **104**(9), 3893–3946.
148. G. J. A. A. Soler-Illia, C. Sanchez, B. Lebeau and J. Patarin, *Chem. Rev.*, 2002, **102**(11), 4093–4138.
149. A. Michailovski and G. R. Patzke, *Chemistry – A European Journal*, 2006, **12**(36), 9122–9134.
150. R. Kiebach, N. Pienack, W. Bensch, J.-D. Grunwaldt, A. Michailovski, A. Baiker, T. Fox, Y. Zhou and G. R. Patzke, *Chem. Mater.*, 2008, **20**(9), 3022–3033.
151. N. F. Chen, K. Oshihara and W. Ueda, *Catal. Today*, 2001, **64**(1–2), 121–128.
152. K. Oshihara, Y. Nakamura, M. Sakuma and W. Ueda, *Catal. Today*, 2001, **71**(1–2), 153–159.
153. B. Deniau, J. M. M. Millet, S. Loridant, N. Christin and J. L. Dubois, *J. Catal.*, 2008, **260**(1), 30–36.
154. R. Haeggblad, J. B. Wagner, B. Deniau, J.-M. M. Millet, J. Holmberg, R. K. Grasselli, S. Hansen and A. Andersson, *Top. Catal.*, 2008, **50**(1–4), 52–65.
155. M. Lin, T. B. Desai, F. W. Kaiser and P. D. Klugherz, *Catal. Today*, 2000, **61**(1–4), 223–229.
156. L. Luo, J. A. Labinger and M. E. Davis, *J. Catal.*, 2001, **200**(2), 222–231.
157. E. Balcells, F. Borgmeier, I. Grisstede and H. G. Lintz, *Catal. Lett.*, 2003, **87**(3), 195–199.
158. E. Balcells, F. Borgmeier, I. Grisstede, H. G. Lintz and F. Rosowski, *Appl. Catal. A: General*, 2004, **266**(2), 211–221.
159. I. Grisstede, *Thesis "Die partielle Oxidation von Propan zu Acrylsäure"*, University Karlsruhe, 2004.
160. B. Zhu, H. Li, W. Yang and L. Lin, *Catal. Today*, 2004, **93–95**(1), 229–234.
161. W. Ueda, D. Vitry and T. Katou, *Catal. Today*, 2005, **99**(1–2), 43–49.
162. R. Fushimi, S. O. Shekhtman, A. Gaffney, S. Han, G. S. Yablonsky and J. T. Gleaves, *Ind. Eng. Chem. Res.*, 2005, **44**(16), 6310–6319.
163. E. Finocchio, G. Busca, V. Lorenzelli and R. J. Willey, *J. Chem. Soc., Faraday Trans.*, 1994, **90**(21), 3347–3356.
164. E. K. Novakova, J. C. Vedrine and E. G. Derouane, *J. Catal.*, 2002, **211**(1), 226–234.

165. E. K. Novakova, E. G. Derouane and J. C. Vedrine, *Catal. Lett.*, 2002, **83**(3–4), 177–182.
166. A. Celaya Sanfiz, T. W. Hansen, D. Teschner, P. Schnörch, F. Girgsdies, A. Trunschke, R. Schlögl, M. H. Looi and S. B. A. Hamid, *J. Phys. Chem. C*, 2010, **114**(4), 1912–1921.
167. G. Centi, F. Cavani and F. Trifiro, Chapter 8: New aspects of the mechanisms of selective oxidation and structure/activity relationships, in *Selective Oxidation by Heterogeneous Catalysis*, Kluwer Academic, New York, Boston, Dordrecht, London, Moscow, 2001, pp. 363–495.
168. V. V. Guliants, R. Bhandari, J. N. Al-Saeedi, V. K. Vasudevan, R. S. Soman, O. Guerrero-Perez and M. A. Banares, *Appl. Catal. A: General*, 2004, **274**(1–2), 123–132.
169. S. T. Oyama, A. M. Middlebrook and G. A. Somorjai, *J. Phys. Chem.*, 1990, **94**(12), 5029–5033.
170. K. Chen, A. Khodakov, J. Yang, A. T. Bell and E. Iglesia, *J. Catal.*, 1999, **186**(2), 325–333.
171. T. Katou, D. Vitry and W. Ueda, *Catal. Today*, 2004, **91–92**(1), 237–240.
172. V. V. Guliants, R. Bhandari, B. Swaminathan, V. K. Vasudevan, H. H. Brongersma, A. Knoester, A. M. Gaffney and S. Han, *J. Phys. Chem. B*, 2005, **109**(50), 24046–24055.
173. H. Bluhm, M. Hävecker, E. Kleimenov, A. Knop-Gericke, A. Liskowski, R. Schlögl and D. S. Su, *Top. Catal.*, 2003, **23**(1), 99–107.
174. M. Haevecker, R. W. Mayer, A. Knop-Gericke, H. Bluhm, E. Kleimenov, A. Liskowski, D. Su, R. Follath, F. G. Requejo, D. F. Ogletree, M. Salmeron, J. A. Lopez-Sanchez, J. K. Bartley, G. J. Hutchings and R. Schloegl, *J. Phys. Chem. B*, 2003, **107**(19), 4587–4596.
175. M. Haevecker, A. Knop-Gericke, H. Bluhm, E. Kleimenov, R. W. Mayer, M. Fait and R. Schloegl, *Appl. Surf. Sci.*, 2004, **230**(1–4), 272–282.
176. E. Kleimenov, H. Bluhm, M. Haevecker, A. Knop-Gericke, A. Pestryakov, D. Teschner, J. A. Lopez-Sanchez, J. K. Bartley, G. J. Hutchings and R. Schloegl, *Surf. Sci.*, 2005, **575**(1–2), 181–188.
177. G. J. Hutchings, *J. Mater.Chem.*, 2009, **19**(9), 1222–1235.
178. K. Oshihara, T. Hisano and W. Ueda, *Top. Catal.*, 2001, **15**(2–4), 153–160.
179. R. K. Grasselli, D. J. Buttrey, P. DeSanto, J. D. Burrington, C. G. Lugmair, A. F. Volpe, Jr. and T. Weingand, *Catal. Today*, 2004, **91–92**(1), 251–258.
180. B. Kubias, M. J. G. Fait and R. Schlögl, Mechanochemical methods, in *Handbook of Heterogeneous Catalysis*, eds. G. Ertl, H. Knözinger, F. Schüth and J. Weitkamp, 2008.
181. J.-M. M. Millet, *Top. Catal.*, 2006, **38**(1–3), 83–92.
182. J. M. M. Millet, H. Roussel, A. Pigamo, J. L. Dubois and J. C. Jumas, *Appl. Catal. A: General*, 2002, **232**(1–2), 77–92.
183. A. Celaya Sanfiz, T. W. Hansen, A. Sakthivel, A. Trunschke, R. Schloegl, A. Knoester, H. H. Brongersma, M. H. Looi and S. B. A. Hamid, *J. Catal.*, 2008, **258**(1), 35–43.

184. V. V. Guliants, R. Bhandari, H. H. Brongersma, A. Knoester, A. M. Gaffney and S. Han, *J. Phys. Chem. B*, 2005, **109**(20), 10234–10242.
185. W. Zhang, A. Trunschke, R. Schlögl and D. Su, *Angew. Chem. Intl. Ed.*, 2010, **49**(35), 6084–6089.
186. D. Teschner, E. M. Vass and R. Schlögl, Photoelectron spectroscopy of catalytic oxide materials, in *Metal Oxide Catalysis*, eds. S. D. Jackson and J. S. J. Hargreaves, Wiley-VCH: Weinheim, 2009, pp. 243–298.
187. V. V. Guliants, R. Bhandari, A. R. Hughett, S. Bhatt, B. D. Schuler, H. H. Brongersma, A. Knoester, A. M. Gaffney and S. Han, *J. Phys. Chem. B*, 2006, **110**(12), 6129–6140.
188. R. Tokarz-Sobieraj, K. Hermann, M. Witko, A. Blume, G. Mestl and R. Schloegl, *Surf. Sci.*, 2001, **489**(1–3), 107–125.
189. O. Glemser, R. v. Haeseler and A. Müller, *Z. Anorg. Allg. Chem.*, 1964, **329**(1), 51–55.

CHAPTER 5

Ammoxidation of Propylene and Propane to Acrylonitrile

ROBERT K. GRASSELLI

Center for Catalytic Science and Technology, University of Delaware, Newark, DE 19716, USA; Department of Chemistry, Technische Universität München, Lichtenbergstr. 4, D-85747 Garching, Germany

5.1 Introduction and Historic Background

About twenty five per cent of the most important organic chemicals and intermediates used in the manufacture of industrial products and consumer goods are produced by selective heterogeneous oxidation catalysis. Within this group, the selective oxidation and ammoxidation of propylene to strategic intermediates acrolein, acrylic acid and acrylonitrile command a significant role.

The cradle of selective oxidation catalysis of light hydrocarbons resides in Cleveland, Ohio, USA, where at the Research Center of SOHIO (The Standard Oil Corporation of Ohio) the first experiments were conducted in 1952 on the oxidation of propylene and propane. Our underlying concept was to utilize lattice oxygen of selected metal oxides, if judiciously chosen, to convert propylene to acrolein. Acrolein was considered to be a versatile functionalized intermediate and therefore a good target molecule. The basic rationale was that lattice oxygen of properly chosen metal oxides would be a more versatile and selective oxidant than dioxygen in noncatalytic radical homogeneous gas-phase reactions. Based on this premise, a solid oxidant process could be envisioned where a given hydrocarbon could be oxidized by the metal oxide to the desired functionalized product, the metal oxide being reduced in the process, and the

RSC Nanoscience & Nanotechnology No. 19
Nanostructured Catalysts: Selective Oxidations
Edited by Christian Hess and Robert Schlögl

Published by the Royal Society of Chemistry, www.rsc.org

so-reduced solid subsequently reoxidized to its original fully oxidized state by dioxygen in a separate step (regeneration).

With this concept in mind the search began to uncover the desired metal oxide. The first approach was to investigate the 1st-row transition metals. It became soon apparent that these single oxides were too unselective. One exception was copper oxide that under partial reduction became selective towards acrolein and became the basis for the Callahan and Grasselli "Site Isolation Principle".[1] This principle states that in order to achieve selectivity in a catalytic oxidation reaction it is necessary to limit the number of available active lattice oxygens at the active sites of a heterogeneous oxidation catalyst; the sites must be spatially isolated from each other on the catalyst surface. Although acrolein yields of about 10% were achieved with partially reduced copper oxide, the requirements to maintain partial reduction under reaction conditions were too cumbersome for a practical process. For this reason our attention shifted to binary oxides where site isolation could be obtained on the basis of solid-state structure rather than reaction conditions.

A second helpful concept towards the discovery of useful selective metal oxide systems was the postulate that a binary metal oxide must contain metal–oxygens (M–O) of intermediate strength to be active and selective. We hypothesized that too weak M–O bonding would lead to complete combustion, while too tight bonding would lead to inactivity. The main focus of our search centered on binary oxides selected from molybdates, antimonates and vanadates. These three oxide classes can be considered as salts of relatively weak acids. To obtain solids with intermediately strong M–O bonds that release lattice oxygen at modest rates to the reacting hydrocarbon, it became apparent to us that the cationic portion of such a desired solid will have to be chosen from intermediately strong bases or those with an amphoteric character. Both the cationic and anionic portions of the binary oxide must contain elements with variable valences and hence be redox active.

After a broad range of metal oxides were surveyed we identified Cu-, Bi-, Sn- and Sb-molybdates as good candidates for a proposed, propylene to acrolein solid oxidant process. In this process it was envisioned that the hydrocarbon was to be contacted with the solid oxidant whose lattice oxygen was to be selectively removed to generate acrolein. In a separate step, the reduced solid oxide was to be regenerated by air back to its original fully oxidized state. A schematic representation of the SOHIO Fluid-Bed Oxidant Process with external regeneration of the metal oxide is shown in Figure 5.1. Among the four identified metal oxides mentioned above, Bi-molybdate ($Bi_9PMo_{12}O_{52}$-SiO_2) gave the highest acrolein yields ($\sim$55%) and showed the best promise for further study and process development. Although the concept of such an oxidant process proved technically sound, the amount of solids to be circulated was too large to make the process commercially attractive (200 g of solids had to be circulated to produce 1 g of acrolein).

Fortunately, Bi-molybdate turned out to be not only a good solid oxidant but also a good heterogeneous oxidation catalyst. Therefore, propylene and oxygen (air) could be fed simultaneously to the reactor containing this solid,

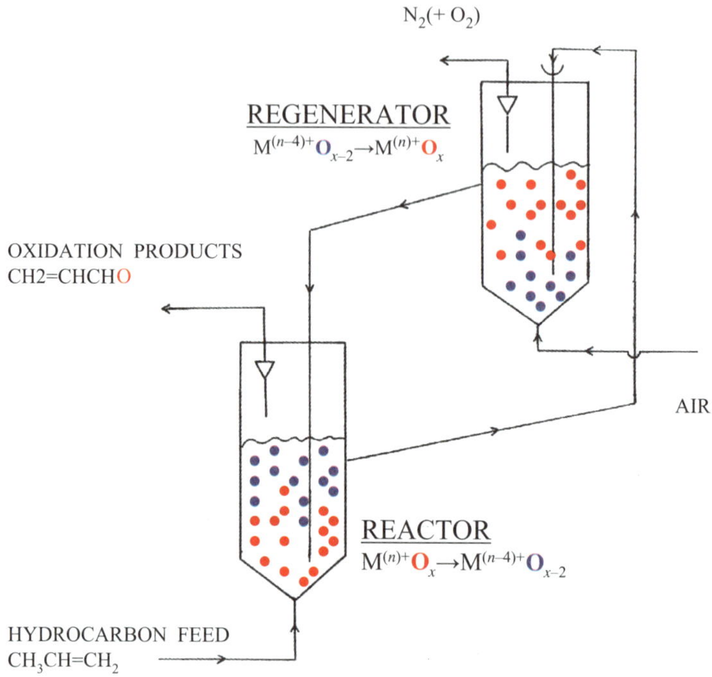

Figure 5.1 SOHIO oxidant process.

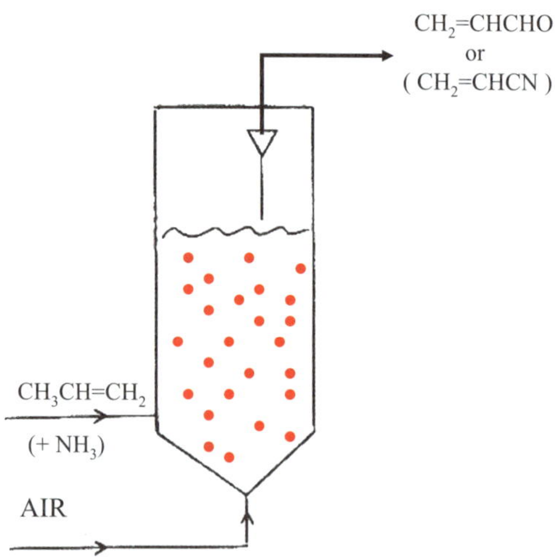

Figure 5.2 SOHIO fluid-bed catalytic process: Oxidation (in the absence of NH_3), ammoxidation (in the presence of NH_3).

producing acrolein continuously without the need of external regeneration (Figure 5.2). Called Catalyst A at SOHIO, with an empirical formula 50%$Bi_9PMo_{12}O_{52}$/50%SiO_2 (supported on silica to gain mechanical strength), it exhibited, for that time, very good productivity ($\sim$55% per pass yield) and also good on-stream stability (about one year). It became the catalyst of choice for the production of acrolein, and was commercialized in 1956 by licensing it to Degussa to produce acrolein on a plant scale. The successful commercialization of this SOHIO acrolein process formed the advent of commercial selective oxidation catalysis of light hydrocarbons.

Since acrolein has a rather limited market, we turned our attention to acrylic acid and acrylonitrile whose markets are much larger. With acrylic acid we soon discovered that a two step fixed-bed process was most advantageous, where in the first reactor 50%$Bi_9PMo_{12}O_{52}$/50%SiO_2 catalyst produces acrolein and in the second stage a 50%$H_4PMo_{12}O_{48}$/50%SiO_2 catalyst converts the acrolein coming from the first reactor directly to acrylic acid. Acrylic acid yields of about 50% were obtained in the mid-1950s. Since then much progress has been made in this area and overall yields of more than 90% acrylic acid are currently achieved by using sophisticated complex mixed-metal catalysts in both stages. Space constraints prevent us to report further on this subject but a recent account of this field, including direct oxidation of the paraffin propane to acrylic acid, can be found in ref. 2. We will limit our discussion here to the ammoxidation of propylene and propane to acrylonitrile.

With our attention, since the mid-1950s, now turned away from acrolein, the ammoxidation of propylene to acrylonitrile became our main research target. And it was our good fortune, that Catalyst A, $Bi_9PMo_{12}O_{52}/SiO_2$, was not only effective for the selective oxidation of propylene to acrolein, but also, in the presence of ammonia, for the ammoxidation of propylene to acrylonitrile (Figure 5.2). Thus, a single catalyst was capable to catalyze the complex transformation of propylene, oxygen and ammonia to acrylonitrile; a major breakthrough in oxidation catalysis. $Bi_9PMo_{12}O_{52}/SiO_2$ became the first industrial ammoxidation catalyst, commercialized by SOHIO in 1960 and thereafter licensed worldwide. This first-generation ammoxidation catalyst gave a 65% acrylonitrile yield in commercial operation. However, this is far from the thermodynamically possible 100%. Therefore, our studies to improve the catalyst, by making it more selective and more productive, continued vigorously throughout the following years and up to now, so that the newest complex mixed-metal oxide ammoxidation catalysts give currently acrylonitrile yields in excess of 80% in commercial units. Simultaneously, the life of the catalyst has been extended from originally one year to more than ten years.

5.2 General

Ammoxidation refers to the interaction of ammonia with an organic material in the presence of oxygen and a suitable catalyst. The organic material can be any hydrocarbon, but is most often an alkene, alkane, or aromatic. Oxygen can be dioxygen, but most commonly air is used. Ammoxidation catalysts are solids, most commonly mixed-metal oxides containing variable-valence elements. They are readily reducible by ammonia and hydrocarbons and readily reoxidizable by dioxygen (air) present in the reaction mixture. In this manner it is the lattice oxygen of the catalyst that interacts with ammonia and the hydrocarbon to produce the functionalized product; the thereby reduced metal oxide is concurrently rapidly reoxidized by dioxygen to its original, fully oxidized state. The catalyst is actually a solid reactant that is continuously reduced and reoxidized, from where the term redox catalyst or redox process is derived.

Ammoxidation of alkenes (eqn (5.1)) is a six-electron oxidation, yielding as the desired product an unsaturated nitrile and water; it is related to the four-electron oxidation of alkenes (eqn (5.2)), yielding unsaturated aldehydes and water; and the two-electron oxydehydrogenation of olefins to diolefins and water (eqn (5.3)):

$$CH_2{=}CRCH_3 + NH_3 + \tfrac{3}{2}O_2 \xrightarrow{\text{catalysts}} CH_2{=}CRCN + 3H_2O \ (R{=}H, \text{alkyl}) \quad (5.1)$$

$$CH_2{=}CRCH_3 + O_2 \xrightarrow{\text{catalysts}} CH_2{=}CRCHO + H_2O \quad (5.2)$$

$$CH_2{=}CHCH_2CH_2R + \tfrac{1}{2}O_2 \xrightarrow{\text{catalysts}} CH_2{=}CHCH{=}CHR + H_2O \quad (5.3)$$

In the ammoxidation process, ammonia is present in the reaction mixture in addition to the hydrocarbon and dioxygen. In the oxidation process the reacting gases are hydrocarbon and dioxygen; no ammonia is present. Oxydehydrogenation refers to the oxidation of alkenes having four or more contiguous carbon atoms in the chain. From first principles, it is expected that ammoxidation is a more difficult and demanding process than oxidation, which is more demanding and difficult than oxydehydrogenation. Although redox catalysts that catalyze these three reaction types are chemically similar, those catalyzing the ammoxidation reaction are generally the most complex and sophisticated ones among them. Hence, a catalyst for the oxydehydrogenation reaction does not necessarily catalyze the ammoxidation reaction. However, ammoxidation catalysts generally catalyze both the oxidation and oxydehydrogenation reactions. Nevertheless, to achieve the highest desired product yields in any given oxidation class, the respective catalyst must be specifically optimized for the reaction in question.

5.3 Propylene Ammoxidation to Acrylonitrile

The commercially most important alkene ammoxidation is that of propylene to acrylonitrile (eqn (5.4)):

$$CH_2{=}CHCH_3 + NH_3 + \tfrac{3}{2}O_2 \xrightarrow{\text{catalysts}} CH_2{=}CHCN + 3H_2O \quad (5.4)$$

SOHIO commercialized in 1960 its revolutionary ammoxidation process based on propylene/ammonia/air (eqn (5.4) and Figure 5.2).[3a] Prior to that invention the BASF (formerly part of I. G. Farben Industries) acetylene/HCN route (eqn (5.5)) was the leading commercial process for the production of

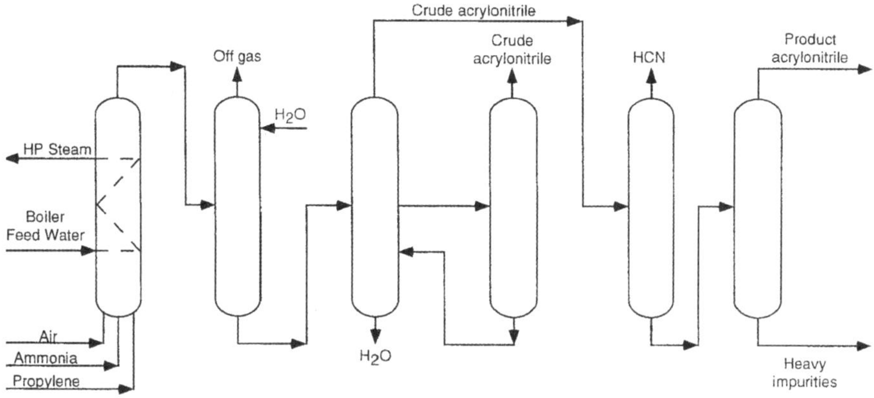

Figure 5.3 The SOHIO acrylonitrile process, flow diagram, (HP = high pressure).[3a,33]

acrylonitrile, and was practised by BASF, Du Pont, American Cyanamid and Monsanto. A similar process based on ethylene oxide/HCN (eqn (5.6)) was practised by BASF, Union Carbide and American Cyanamid.

$$HC{\equiv}CH + HCN \xrightarrow{CuCl+NH_4Cl} CH_2 = CHCN \quad (5.5)$$

$$C_2H_4O + HCN \xrightarrow{\text{base catalysts}} HOCH_2CH_2CN \longrightarrow CH_2{=}CHCN + H_2O \quad (5.6)$$

$$4\,CH_2{=}CHCH_3 + 6NO \xrightarrow{Ag_2O/SiO_2} 4CH_2 = CHCN + 6H_2O + N_2 \quad (5.7)$$

A third obsolete route to acrylonitrile was based on propylene oxidation with NO (eqn (5.7)); reached the pilot plant stage at Du Pont, but was never commercialized. The high costs and environmentally unfriendly properties of the starting materials, as well as difficult process conditions, made all of these processes (eqns (5.5)–(5.7)) noncompetitive with the much more efficient, cost-effective and environmentally friendly SOHIO ammoxidation process (eqn (5.4), Figure 5.3).

By 1970, the last commercial and semicommercial plants using the outdated, highly expensive, process technologies (eqns (5.5)–(5.7)) were shut down. The SOHIO process revolutionized the acrylonitrile industry and accounts currently for all of the acrylonitrile produced worldwide.

5.3.1 The SOHIO Acrylonitrile Process

The revolutionary SOHIO acrylonitrile process (eqn (5.4)) was discovered by SOHIO scientists[3] in the late 1950s, commercialized in 1960 and since 1970 it commands exclusivity in the production of acrylonitrile worldwide. The process has been licensed the world over and about 70 plants in 30 countries are currently using the SOHIO technology. With a current world production of nearly 6 billion kg/year, it is one of the most important selective heterogeneous oxidation processes. This large-scale industrial process produces worldwide $\sim$1 kg acrylonitrile/person/year.

The immediate impact of the SOHIO ammoxidation process resulted in a drastic price reduction of acrylonitrile coupled with an explosive production increase worldwide. This in turn led to the discovery of many new applications of this highly versatile petrochemical intermediate in the areas of fibers, resins, rubbers and specialty products.[4] The versatility of acrylonitrile is reflected in the steady growth of its end-product base, with market share continuing to grow and expansion particularly robust in developing and emerging countries.

With the invention and commercialization of each newer and more efficient catalyst, the acrylonitrile production increased dramatically over the years (Figure 5.4) and continues to grow at an annual rate of $\sim$4%.

The SOHIO ammoxidation process (Figures 5.2 and 5.3) is a single-pass process with about 98% conversion of propylene, producing approximately 1 kg of acrylonitrile per 0.9 to 1.05 kg propylene feed. Useful byproducts are HCN (about 0.1 kg per kg of acrylonitrile produced), which is primarily used as a starting material in the manufacture of methyl methacrylate, and acetonitrile (about 0.03 kg per kg acrylonitrile produced), which is a common industrial solvent used in the rubber industry.

The process uses a fluid-bed reactor wherein propylene, ammonia and air are contacted with a solid, fluid-bed catalyst. World-scale reactors are about 12 m in diameter and contain about 75 000 kg of catalyst, with particle size ranging from 40 to 100 μm. The particle-size distribution is critical for optimum

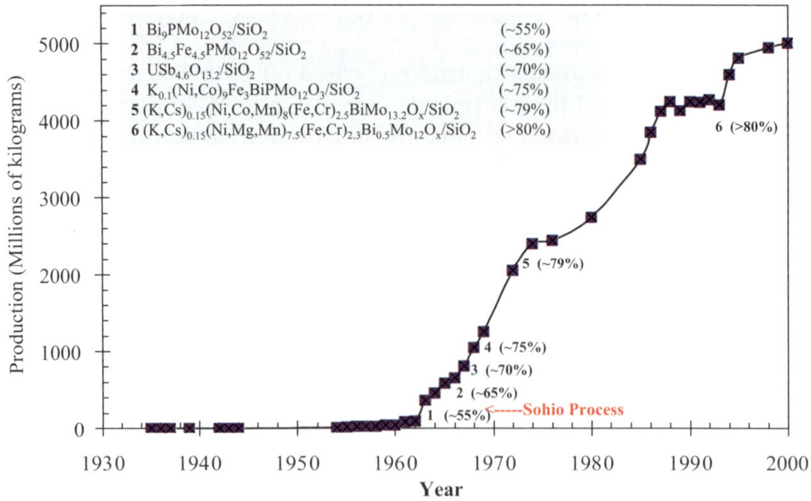

Figure 5.4 World production of acrylonitrile.[5] The numbers 1–6 denote the approximate time (year) when new and more efficient catalysts were commercially introduced by SOHIO (now BP America), the numbers in parentheses the approximate commercially achieved acrylonitrile in-tank yields using the respective catalysts.

fluidization, and a prescribed amount of catalyst fines ($<40\,\mu m$ in diameter) are needed for best fluidization. The reactor bed is maintained at a constant temperature, which ranges between 400 °C and 450 °C depending on the specific catalyst composition used. With today's more sophisticated multicomponent catalysts, the operating temperature is kept at about 420 °C or lower. The operating pressure is essentially atmospheric, ranging between 1.5 and 3 bar. It is desirable to keep the pressure as low as possible, since undesirable byproducts are kinetically higher-order reactions than the acrylonitrile reaction, which is first order in propylene. Therefore, increased pressure favors byproducts and reduces selectivity to acrylonitrile. The superficial linear gas velocity ranges between 0.2 and $0.5\,m\,s^{-1}$. The feed gases enter through two separate inlets, separated in space so that the air enters at the bottom of the reactor through an air grid, propylene and ammonia through a common feed sparger positioned slightly above the air grid (Figure 5.2). The air grid and feed sparger feed directly into the catalyst bed to minimize free space, reducing undesirable homogeneous gas-phase combustion, and preventing a remotely possible explosion. Once the hydrocarbon and ammonia are in intimate contact with the catalyst, uncontrolled combustion cannot occur. In over 50 years of commercial operation, no explosion has ever occurred in any of the ammoxidation reactors producing acrylonitrile from propylene using SOHIO technology.

Commercial fluid-bed reactors contain a large number of steam coils whose purpose it is to conduct the exothermic heat of the reaction out of the reactor, maintain uniform temperature throughout the reactor bed, improve catalyst fluidization, minimize catalyst backmixing and create a pseudoplug-flow condition in the fluid-bed reactor. Thereby, consecutive reactions leading to undesirable byproducts are minimized.

In commercial operation (Figure 5.3) hot reactor effluent is taken overhead, quenched with water in a countercurrent absorber where unreacted ammonia is neutralized with sulfuric acid. The resulting ammonium sulfate can be recovered and used as fertilizer. The absorber off-gas, composed primarily of N_2, CO, CO_2, and unreacted hydrocarbons, is passed through an incinerator to combust the CO and hydrocarbons to CO_2 and water. The acrylonitrile-containing solution from the absorber bottom is passed to a recovery column that yields a crude acrylonitrile stream overhead, also containing HCN produced as a byproduct in the ammoxidation reaction. The column bottoms are passed to a second recovery column to remove water and produce crude acetonitrile. The crude acetonitrile is either incinerated, or refined in the acetonitrile recovery column to produce solvent-quality acetonitrile. Crude acrylonitrile is taken overhead from the acetonitrile recovery column, combined with the main crude acrylonitrile stream, and fed to the lights column to remove HCN and light ends. The bottoms containing the main amount of acrylonitrile are fed to the product column where acrylic fiber quality (99.4% minimum) acrylonitrile product is taken overhead. The high boiling impurity bottoms are deep-welled or preferably incinerated.

The disposal of process impurities has become of ever-increasing environmental importance. Methods of environmentally acceptable disposal include biological treatment, wet air oxidation and incineration. Therefore, catalysts optimized for the highest acrylonitrile yield and simultaneously lowest byproduct formation (*i.e.* acrolein, acrylic acid, oxazole, succinonitrile, *etc.*) are the catalysts of choice under the prevailing industrial environment. Hence, much research is being expended to devise not only the most product-efficient but also the most environment-friendly catalysts.

The most effective commercial catalysts for propylene ammoxidation are complex mixed-metal molybdates or antimonates discovered and developed by SOHIO researchers over the past 50 years (Figure 5.4). From the very onset, intense fundamental studies accompanied exploratory and developmental catalyst research and became (and still are) an indispensable, integral part of the overall effort. By unraveling and defining the principles governing the chemical, physical and structural behavior of the catalytic solids, the knowledge so gained could readily be applied to the design of ever more sophisticated and efficient catalyst systems. This interplay between fundamental and applied research approaches that culminates in commercially viable catalytic systems will become more apparent by reading the account below. Although much has already been accomplished in perfecting the propylene ammoxidation catalysts, the game is not over. There is still ample room for further catalyst improvements since thus far only slightly above 80% of the thermodynamically possible 100% acrylonitrile yield, under commercial operating conditions, has been achieved.

5.3.2 Key Catalytic Functionalities

Effective ammoxidation catalysts are multifunctional and possess several key properties. These are sometimes referred to as the "seven pillars" of selective (amm)oxidation catalysis.[5] The active sites are comprised of at least two vicinal metal oxide moieties having optimum metal–oxygen bond strengths, both of which are readily reducible and reoxidizable. The individual active sites must be spatially isolated from each other to achieve desired product selectivity. They obey the site-isolation principle that was proposed some 50 years ago by Callahan and Grasselli,[1,6] a principle by now well accepted and universally applicable in (amm)oxidation catalysis. The active sites are either themselves capable of dissociating dioxygen and incorporating it into the lattice of the solid catalyst, or they must be located in the vicinity of auxiliary reoxidation sites that contain metals having a facile redox couple. The active sites and the reoxidation sites are generally distinct from each other, but must be electronically and spatially situated in a manner that electrons, lattice oxygen (O_L^{2}) and anion vacancies can readily move between them. Therefore, the catalytic solid must be able to accommodate a certain number of anion vacancies without structural collapse.

5.3.3 Role of Lattice Oxygen – Catalyst as Redox Solid

As already mentioned in the Introduction, ammoxidation catalysts are solid reactants whose lattice oxygen participates in the catalytic process. Lattice oxygen, once removed from the solid catalyst by the process of oxidizing a hydrocarbon to the corresponding nitrile, is continuously replenished by dioxygen from the gas phase. This fact was recognized early in the discovery and development of selective oxidation and ammoxidation catalysts at SOHIO,[7,8] and was based on the working hypothesis that lattice oxygen ($O_L{}^{2-}$) from certain multivalent metal oxides would serve as a more selective and versatile oxidant than gas-phase molecular oxygen. Thus, in the ammoxidation of propylene to acrylonitrile, lattice oxygen first reacts with ammonia to form isoelectronic NH^{2-} surface moieties and water (eqn (5.8)):

$$M^{(n)+}O_x + NH_3 \rightarrow M^{(n-2)+}O_{x-1}NH + H_2O \tag{5.8}$$

This surface species reacts further with propylene and additional lattice oxygen to form acrylonitrile and water (eqn (5.9)):

$$\begin{aligned} &M^{(n-2)+}O_{x-1}NH + CH_2{=}CHCH_3 + 2M^{(n)+}O_x \rightarrow \\ &3M^{(n-2)+}O_{x-1} + CH_2{=}CHCN + 2H_2O \end{aligned} \tag{5.9}$$

The metal oxide that was reduced in this process by six electrons can be reoxidized by dioxygen (eqn (5.10)), thereby reconstituting the active solid oxidant to its original, fully oxidized state:

$$3M^{(n-2)+}O_{x-1} + \tfrac{3}{2}O_2 \rightarrow 3M^{(n)+}O_x \tag{5.10}$$

In this manner, the metal oxide becomes a catalyst when both the reduction and reoxidation processes (*i.e.* redox process) can be carried out simultaneously with propylene, ammonia and dioxygen as the reactants. The overall catalytic process is then simply:

$$CH_2{=}CHCH_3 + NH_3 + \tfrac{3}{2}O_2 \rightarrow CH_2{=}CHCN + 3H_2O \tag{5.11}$$

That lattice oxygen is truly involved in the catalytic processes of selective alkene oxidation and ammoxidation has been amply demonstrated through isotope experiments using $^{18}O_2$ instead of $^{16}O_2$ and observing the isotope distribution of the first-formed products.[8–11] The results show that the first-formed partial oxidation products are rich in the ^{16}O isotope, followed by later-formed products rich in the ^{18}O isotope. This is convincing verification that alkene and ammonia react with the lattice oxygen ($^{16}O^{2-}{}_L$) of the catalyst producing the nitrile (or aldehyde in the absence of ammonia) and water, and that the oxygen of the dioxygen incorporates in a subsequent (*i.e.* reoxidation) step into the solid catalyst. It also independently confirms the validity of the original SOHIO redox catalyst hypothesis mentioned above.

To accomplish the redox process effectively, ammoxidation catalysts must be readily reducible by ammonia and the hydrocarbon to be oxidized; and they must be readily reoxidizable by dioxygen. They must perform a complex sequence of bond-breaking and bond-making processes (*e.g.* in sum 32 bonds in ammoxidation). They must provide a facile pathway to the desired product,

Table 5.1 Thermodynamics of ammoxidation reactions.[12]

Reaction		ΔG° (427 °C) (kJ mol^{-1})
$C_3H_6 + NH_3 + \frac{3}{2}O_2$	$\rightarrow CH_2CHCN + 3H_2O$	–569.4
$C_3H_6 + \frac{3}{2}NH_3 + \frac{3}{2}O_2$	$\rightarrow \frac{3}{2}CH_3CN + 3H_2O$	–595.4
$C_3H_6 + 3NH_3 + 3O_2$	$\rightarrow 3HCN + 6H_2O$	–1144.2
$C_3H_6 + 3O_2$	$\rightarrow 3CO + 3H_2O$	–1275.9
$C_3H_6 + \frac{9}{2}O_2$	$\rightarrow 3CO_2 + 3H_2O$	–1940.8

acrylonitrile, since it is only a kinetically isolatable intermediate while thermodynamically it is a less-stable product than the byproducts acetonitrile, HCN, CO and CO_2. A comparison of the respective free energies of formation, elaborate this point (Table 5.1).[12]

5.3.4 Generalized Mechanism of Alkene Ammoxidation

A generalized mechanistic cycle (Figure 5.5) illustrates the major steps in the sequence an ammoxidation catalyst undergoes from activating ammonia and an alkene, to producing the corresponding nitrile, and then returning to its original, starting state. Ammonia interacts first with the bifunctional active site of the catalyst (having a first call on the active oxygens of the catalytic site, ahead of the alkene), transforming it to an ammoxidation site on which the alkene coordinates to form an adsorbed surface complex. After multiple rearrangements and oxidations, the surface complex is transformed to the nitrile molecule that subsequently desorbs from the catalyst and is released into the gas phase. As a consequence of these transformations a reduced surface site is formed that is subsequently regenerated to its original fully oxidized state by means of lattice oxygen (O_L^{2-}) coming from a separate but adjacent reoxidation site. The latter is capable of dissociating dioxygen to lattice oxygen and sending it to the oxygen-starved reduced surface site that simultaneously transmits lattice vacancies [V] to the reoxidation site. It is apparent that these sites must communicate with each other by residing in a common solid lattice, capable of easy electron, anion vacancies, and lattice oxygen transmission. The solid must be able to accommodate an appreciable amount of anion vacancies (oxygen vacancies) without structural collapse to be an acceptable redox catalyst.

5.3.5 Multifunctional Nature of Active Sites

The bifunctional (or more general multifunctional) nature of active sites of selective oxidation and ammoxidation catalysts has been researched extensively.[13–17] Insights into these catalytically active phases are derived mainly from extensive model compound probe experiments, *in-situ* Raman studies, and excellent structural knowledge (based primarily on X-ray diffraction, neutron diffraction and HREM). By way of example, a schematic representation of the

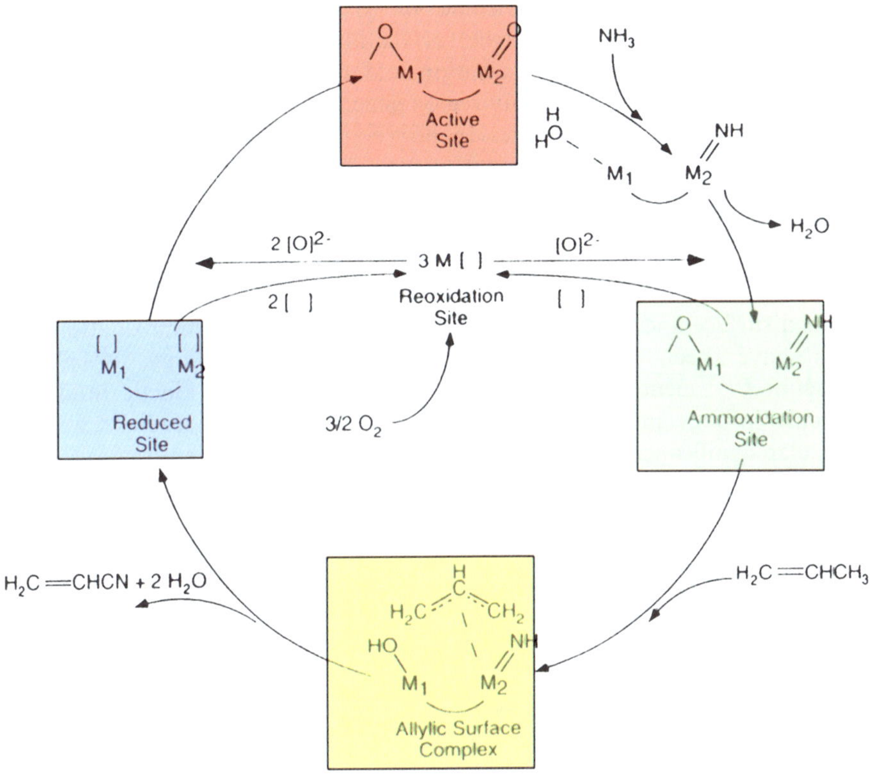

Figure 5.5 Generalized mechanistic cycle for alkene ammoxidation.[3a,33]

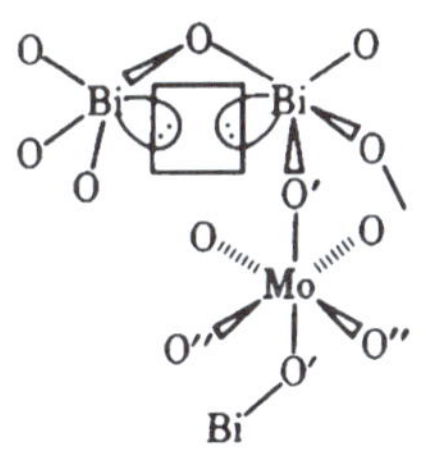

Figure 5.6 Schematic representation of the active site of Bi_2MoO_6.[3a] O′ = oxygen responsible for α-H abstraction. O″ = oxygen associated with Mo; responsible for oxygen insertion into allylic intermediate. □ = proposed center for O_2 reduction and dissociative chemisorption.

catalytically active site of Bi_2MoO_6 (Figure 5.6)[2,13] reveals that it is bifunctional, and that the various catalytic functionalities can be assigned to specific elements and to specific lattice oxygen positions. It is concluded that bridging oxygens Bi–O–Mo that are more strongly associated with Bi than Mo are responsible for the α-hydrogen abstraction of a reacting alkene, while oxygens associated with Mo are responsible for oxygen (Mo=O) and nitrogen (Mo=NH) insertion into the allylic surface intermediate. It is also reasoned that the locus of dioxygen dissociation and its reduction to lattice oxygen occurs at the high electron density region generated by the orbitals of the two lone electron pairs of Bi–O–Bi sites. In more complex multicomponent catalysts the reoxidation sites are associated with such elements as Fe, Ce, U, or Cu, all being more easily reduced than Bi.

5.3.6 Ammoxidation Mechanism (Molybdates, Antimonates)

Detailed surface and solid-state mechanisms of propylene ammoxidation catalysis have been determined using Raman spectroscopy,[18–21] X-ray diffraction,[22,23] neutron diffraction,[24,25] HREM,[26] pulse kinetic studies,[27] molecular probes and isotope studies.[8,13] Molecular probe studies[13] were of particular significance in unraveling the detailed molecular steps that occur on the catalyst surface during the ammoxidation reaction. Mechanisms for the selective ammoxidation of propylene to acrylonitrile over the commercially most significant molybdate-based (Figure 5.7)[14,28] and antimonite-based systems (Figure 5.8)[14] are of interest in order to understand the processes that effective ammoxidation catalysts must perform and to gain some appreciation of how in the future these already sophisticated solids might be further perfected.

The mechanisms for Bi-molybdate catalysts are shown in Figure 5.7 for both ammoxidation and selective oxidation. The reason for showing both mechanisms together is twofold: First, the catalytically active sites in the absence of ammonia are oxidic in nature, and only after the catalyst is contacted with ammonia do isoelectronic nitrogen containing species replace lattice oxygen. In molybdates the replacement occurs through the formation of Mo=NH species, whereas in antimonates Sb–NH–Sb (and in tellurates Te–NH–Te) species are formed. Second, although ammonia is more strongly chemisorbed than propylene, and NH species easily form at reaction temperature on the active sites of the catalysts of choice, any regions of catalyst that might be devoid of ammonia will contain Mo=O, *etc.*, surface species. These will produce acrolein, and because of the elevated temperature compared to selective oxidation process conditions, the acrolein will further react to produce CO_x. It is therefore imperative that at least stoichiometrically sufficient ammonia be fed to the reactors and that gas mixing and catalyst fluidization are good in commercial reactors to achieve optimum acrylonitrile yields.

In bismuth molybdate-based catalysts (Figure 5.7) the first step of the mechanism entails the fast transformation of Mo=O groups to Mo=NH moieties, followed by the chemisorption of propylene through its π electrons to the coordinately unsaturated Mo centers. This is followed by a slow, rate limiting mechanistic step of α-hydrogen abstraction from the propylene by lattice oxygens associated with bismuth to form the σ-allyl Mo-complex, and subsequently the σ-N-allyl Mo-complex. A 1,4-hydrogen shift produces another Mo-surface-complex, that does not desorb but undergoes another

1,4-hydrogen shift after reoxidation of the molybdenum. This reoxidation (*via* lattice oxygen transfer) occurs prior to or simultaneously with the final hydrogen shift, otherwise an overall five-electron reduction of Mo would be required to transform this surface complex to acrylonitrile, which is highly unlikely. After acrylonitrile is formed and desorbed from the catalyst surface, the catalytic cycle is completed by replenishment of lattice oxygen at the reduced site *via* lattice oxygen migration from the reoxidation site and vacancy migration in the opposite direction, culminated by dioxygen dissociation and lattice oxygen reincorporation at the reoxidation site. The six-electron ammoxidation cycle is thereby completed with no net change of the catalyst.

Organometallic complexes, including molybdenum (VI) allyloxo and allylamido, have been synthesized as model compounds[29–31] to mimic the catalytic surface intermediates involved in the oxidation and ammoxidation of propylene to acrolein and acrylonitrile, respectively. The results obtained in these studies are consistent with those we proposed for the actual catalytic surface reactions (eqns. (5.8) and (5.9)), providing independent support of the mechanistic views presented above (Figure 5.7).

Advanced multicomponent bismuth molybdates contain, in addition to Bi and Mo, redox elements (*e.g.* $Fe^{2+/3+}$, $Ce^{3+/4+}$, $Cr^{3+/6+}$) having reduction potentials that lie below those of the α-H-abstracting Bi^{3+} and $(NH)^{2-}$-inserting Mo^{6+}. Therefore, lattice oxygen (O_L^{2-}) can constantly be supplied from these redox moieties to the active Bi–O–Mo centers, keeping the latter in their desired and active, high oxidation state. These redox elements are capable of dioxygen dissociation, its reduction to lattice oxygen, and its incorporation into the catalyst matrix. A schematic mechanism of such a redox cycle for the ammoxidation of propylene over Bi molybdate, containing redox elements M^{n+} (M=Fe, Ce, Cr), is described in the literature.[32,33]

The active sites of antimony-based selective oxidation and ammoxidation catalysts are comprised of bridging oxygens that connect two Sb^{5+} and two Sb^{3+} atoms (Figure 5.8). The Sb^{5+} atoms are the locus of the NH_3 activation and propylene chemisorption and are also associated with the O and NH insertion species. The bridging oxygens of the Sb^{3+} polyhedra perform the rate-limiting α-hydrogen abstraction. The coexistence of $Sb^{3+/5+}$ pairs in active and selective antimonate catalysts has been observed by several groups, and their significance in catalytic reactions realized.[32,33] In contrast to molybdates, the initially formed π-allylic species undergo virtually irreversible insertion of one of the Sb^{5+} bridging oxygens (oxidation) or bridging NH species (ammoxidation) to form the respective σ-species. In ammoxidation the σ-N-allyl-complex rearranges to an allyl amine moiety that remains bonded to the antimony center. After reoxidation and subsequent hydrogen abstraction it produces the desired acrylonitrile (which desorbs) and a reduced site. Reoxidation of the reduced active site occurs, as with molybdates through lattice oxygens coming from a separate, but vicinal reoxidation site, capable of dioxygen dissociation and lattice oxygen reincorporation, whereby anion vacancies migrate in the opposite direction. The catalytic and redox cycles are thereby completed; again

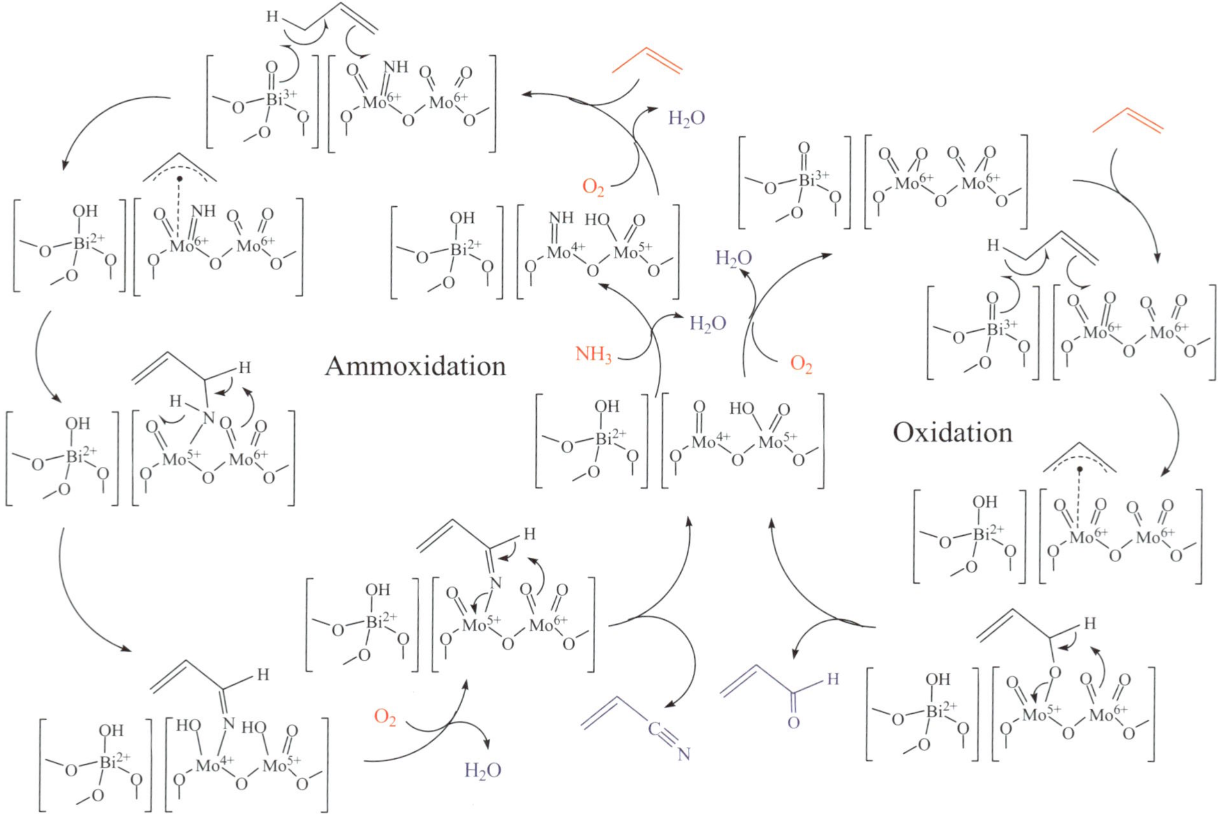

Figure 5.7 Propylene oxidation and ammoxidation mechanisms over bismuth molybdate catalysts.[28]

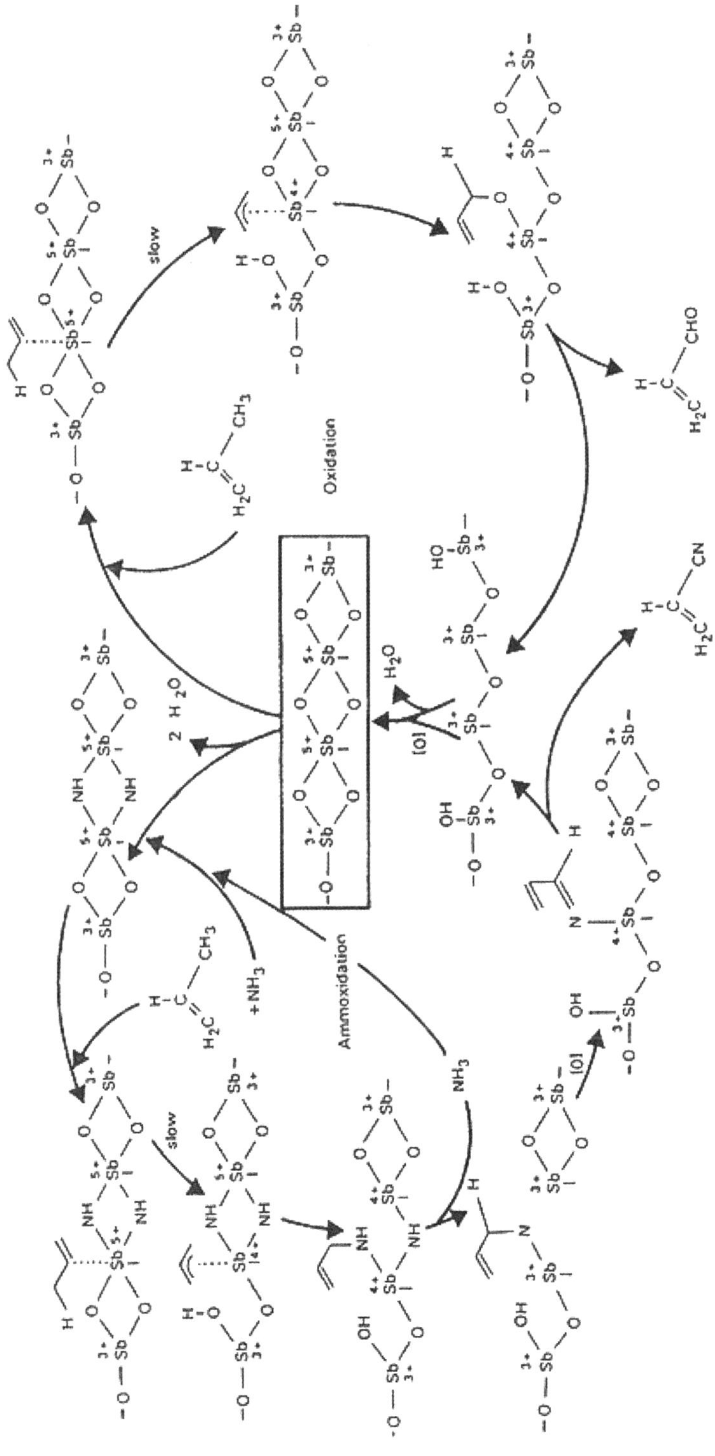

Figure 5.8 Propylene oxidation and ammoxidation mechanisms over antimonate catalysts.[14]

with no net change of the catalyst. Both have been adequately discussed in the literature.[32,33]

In sum, the key catalytic functionalities required of effective ammoxidation catalysts include an α-H-abstraction component (*e.g.* Bi^{3+}, Sb^{3+}, Te^{4+}, Se^{4+}), an ammonia- and alkene-chemisorbing and oxygen- and nitrogen-insertion component (Mo^{6+}, Sb^{5+}), and a redox couple ($Fe^{2+/3+}$, $Ce^{3+/4+}$, $Cr^{3+/6+}$, $U^{5+/6+}$) to enhance lattice-oxygen transfer between the bulk and surface of the catalyst. An interesting empirical correlation of the electronic structure requirements of these elements with the respective functionalities of these elements has been advanced (Table 5.2).[3a] It corroborates molecular probe findings extremely well,[13] and is therefore deemed to be of substantial value in catalyst design.

According to this concept (Table 5.2), the α-H-abstracting element of the active site possesses some electron density on the metal cation of the oxide and in the majority of cases it actually contains a lone pair of electrons. This imparts a partial radical character to the oxygen bonded to the metal in question and thereby facilitates the abstraction of a hydrogen radical from the alkene. Except for uranium, the electronic structure of the cations listed is characterized by a large overlap between their occupied s levels and the oxygen 2p electronic states in the solid state. The unoccupied p levels of the metal lie above the 2p states of oxygen, thereby constituting the conduction band for the solid. The alkene-chemisorbing and oxygen/nitrogen-inserting component is a fully oxidized, coordinately unsaturated cation with the proper metal–oxygen bond strength for partial oxidation of the allylic intermediate. Another common feature of the oxygen-inserting component is its variable valence with at least two stable oxidation states in the solid state. The redox component is a variable-valence metal with a reduction potential greater than that of the oxygen/nitrogen-insertion element. In this manner, the redox component maintains the activity of the catalyst by promoting rapid reoxidation and reconstruction of the surface active sites of the catalyst. The redox element may also serve as the site for catalyst reoxidation by dioxygen. However, it is not sufficient for an effective catalyst to fulfill only these stated electronic requirements; the elements of choice must also reside in an appropriate solid-state matrix of appropriate overall structure.

Table 5.2 Electronic structure of some catalytically active elements.[3a]

α-H abstraction	*Alkene chemisorption/ oxygen insertion*	*Redox couple*	*Example*
Bi^{3+} $5d^{10}6s^26p^0$	Mo^{6+} $4d^05s^0$	Ce^{3+}/Ce^{4+} Fe^{2+}/Fe^{3+}	$Bi_2O_3{\cdot}nMoO_3$ $M^{2+}{}_aM^{3+}{}_b\ Bi_xMo_yO_z$
Te^{4+} $4d^{10}5s^25p^0$	$Mo^{6+}4d^05s^0$	Ce^{3+}/Ce^{4+}	Te_2MoO_7 $(Te_aCe_bMo_y)O_z$
Sb^{3+} $4d^{10}5s^25p^0$	Sb^{5+} $5d^05s^0$	Fe^{2+}/Fe^{3+}	$Fe_xSb_yO_z$
U^{5+} $5f^16d^07s^0$	Sb^{5+} $5s^05p^0$	U^{5+}/U^{6+}	USb_3O_{10}
Se^{4+} $3d^{10}4s^24p^0$	Te^{6+} $5s^05p^0$	Fe^{2+}/Fe^{3+}	$FeSe^{4+}{}_bTe^{6+}{}_cO_x$

5.3.7 Molybdate Catalysts

The most studied and commercially most important ammoxidation catalysts are those based on bismuth molybdates. The catalytically most important simple molybdate phases lie in the compositional range Bi:Mo of 2:3 and 2:1 and are α-$Bi_2Mo_3O_{12}$, β-$Bi_2Mo_2O_9$, and γ-Bi_2MoO_6. The α phase has a Scheelite ($CaWO_4$) derived structure, based on tetrahedrally coordinated Mo, and has a large number (1/3) of cation vacancies.[34] The β structure contains features of both α and γ structures.[35] The molybdenum coordination is tetrahedral, akin to that in the α structure, and it contains cation vacancies (1/9). However, most of the bismuth has coordination similar to that found in the γ structure. The γ phase has the naturally occurring Koechlinite layer structure, and contains octahedrally coordinated Mo.[35] The structures of these phases are shown in Figure 5.9.[26,28,33] The three phases show no mutual solubility and their surface compositions are those of their respective bulk phases. The highest ammoxidation yield is obtained with the β phase, followed by the α phase, with the γ phase being a distant third.[7,36–38]

A plausible explanation for the difference in ammoxidation performance of the three phases resides in the compositional difference of their respective Bi/Mo ratios.[28,33] The α structure has the desired Bi/Mo ratio of 2/3 and provides therefore ample NH_3 and alkene chemisorbing and NH inserting sites (*i.e.* Mo sites). These sites are associated with sufficient vicinal Bi moieties to provide for

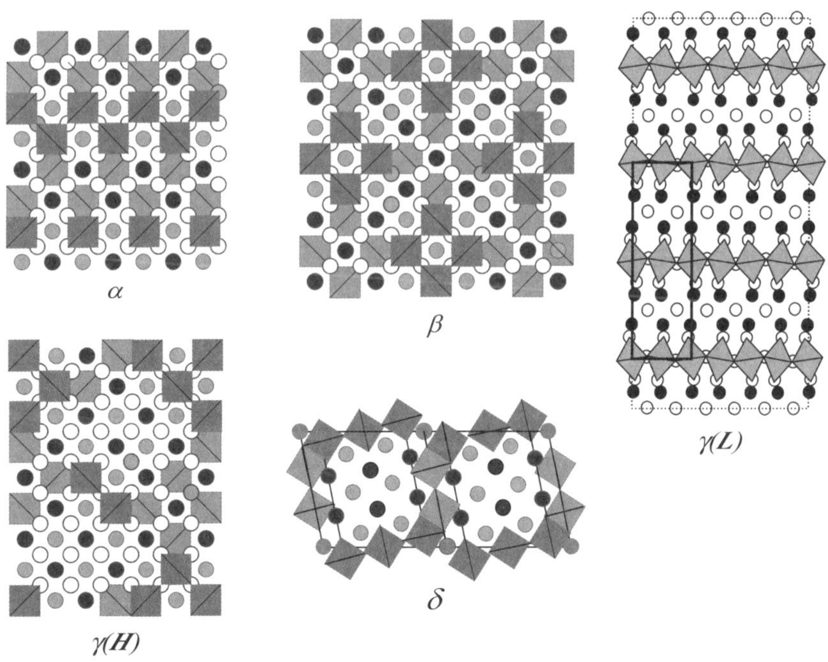

Figure 5.9 Bismuth molybdate phases projected in principal Bi channel direction.[26]

easy NH_3 activation and α-H abstraction. Additionally, the Bi centers are isolated from each other so that undesired overoxidation does not occur. The large number of cation vacancies imparts the structure good pliability and integrity to accommodate anion vacancies created by the ammoxidation process.

The γ phase has, by virtue of its Bi/Mo ratio of 2/1, too much Bi and too little Mo for best ammoxidation performance. It does not have sufficient NH_3 and alkene chemisorbing sites (lack of Mo), and has too many reoxidation (Bi) sites, which are not isolated from each other. This leads to undesired overoxidation. However, owing to its layer structure, with alternating octahedral MoO_2 and Bi_2O_2 layers, it is well suited to be an excellent reoxidation phase. It contains ample dioxygen dissociation sites (Bi sites) and offers facile lattice oxygen transport throughout its structure by means of the Bi_2O_2 channels.

The β phase might be considered a hybrid of the α and γ phases, and thus somewhat more effective in ammoxidation reactions than the α phase. Its Bi:Mo ratio of 1:1 is optimal for having the largest number of active sites. Having tetrahedrally coordinated Mo and cation vacancies similar to the α phase it can easily insert NH into a surface allyl radical. And with Bi coordination similar to the γ phase it provides for easy vacancy migration, dioxygen reincorporation, and hence rapid reoxidation kinetics. Unfortunately, the α phase is too fragile by virtue of its structure to be stable on its own under ammoxidation conditions. Allegedly, it is stable as an overgrowth on both α and γ phases.

In addition to possible overgrowths, interaction between various Bi–Mo phases is possible because of their structural similarity. Epitaxial intergrowth can occur at certain lattice planes providing for interaction and ultimately communication between the phases. It has been observed that physical mixtures of α and γ phases, when thoroughly mixed and in good physical contact with each other, give ammoxidation yields that exceed, over a wide range of compositions, those of the individual phases.[39] It can be reasoned that the active sites of the α phase become more readily reoxidized by the external but proximal cooperation of the γ phase.[3a] The result comprises catalytic synergism, owing to the cooperation of phases having different catalytic functions, thereby becoming simple forerunners of complex multiphase multicomponent catalytic systems of today.[6]

5.3.8 Multicomponent Molybdates

Ammoxidation yields can be substantially improved over those obtained with binary molybdates (*i.e.* Bi molybdates) by incorporating into these molybdates additional elements, particularly redox elements (*e.g.* Fe, Cr, Ce). If substitutions can be made directly into the parent molybdate (*i.e.* depending on the solubility properties of the substituting element), then new and more effective catalysts can be obtained in this manner. Examples of such catalysts are $Bi_{a-x}Fe_xMo_yO_z$ and $Bi_{a-x}Ce_xMo_yO_z$. Single-phase compositions of these systems with improved efficiency over the parent catalysts do exist, however, only over very limited compositional ranges because of solubility limitations.[40,41]

When miscibility problems occur, it is possible to achieve the desired catalytic result by having two immiscible phases, having two different and distinct functions, cooperate with each other by means of a common superstructure, epitaxial overgrowths, or epitaxial lattice matching. An example of the latter is the composition having the empirical formula $M^{2+}{}_aFe^{3+}{}_{10-a}Bi_xMo_yO_z$ (Figure 5.10),[40] where $M^{2+}$is Fe^{2+}, or Fe^{2+} plus Ni^{2+}, Co^{2+}, Mg^{2+}, and/or Mn^{2+}. With this system the acrylonitrile yield is about 18% when there is no Fe^{3+} present, rising sharply to about 82% and staying at that level over a wide range of compositions, then falling off sharply to about 21% when M^{2+} becomes zero.

The catalytic synergism is due to an Fe^{3+}-containing α-Bi-molybdate phase cooperation with an Fe^{2+}-molybdate phase stabilized in an M^{2+} host lattice, where M^{2+} can be Ni, Co, Mg, and/or Mn. The first phase is the catalytically active one, the second the auxiliary reoxidation (or cocatalyst) phase. The latter keeps the first phase highly oxidized, thereby increasing the time-averaged number of working active sites at the surface of the first phase during the catalytic process. The two phases are able to cooperate with each other owing to the existence of mutual, epitaxially well-matched lattice plains (Figure 5.11).[42]

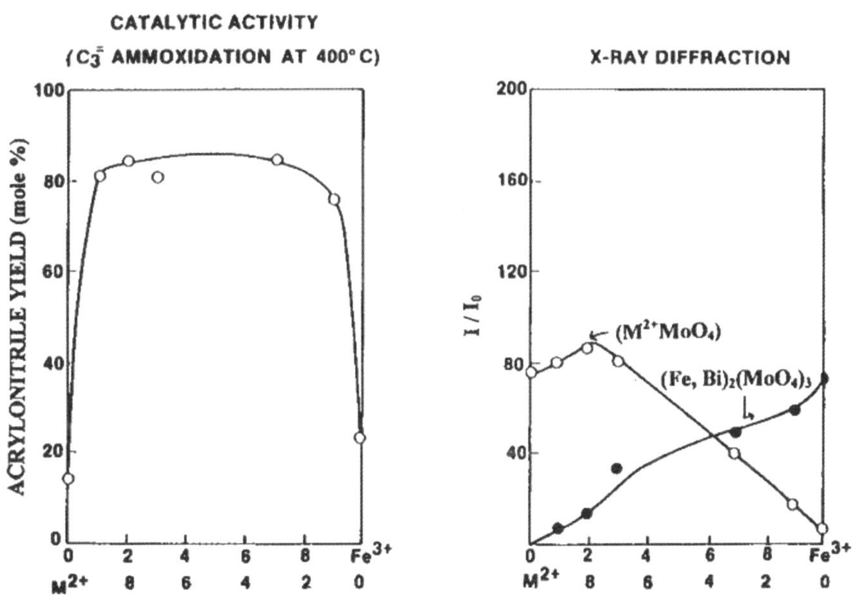

Figure 5.10 Relationship between catalytic activity and phase composition of $M^{2+}{}_aM^{3+}{}_{10-a}Bi_xMo_yO_z$-$SiO_2$.[40]

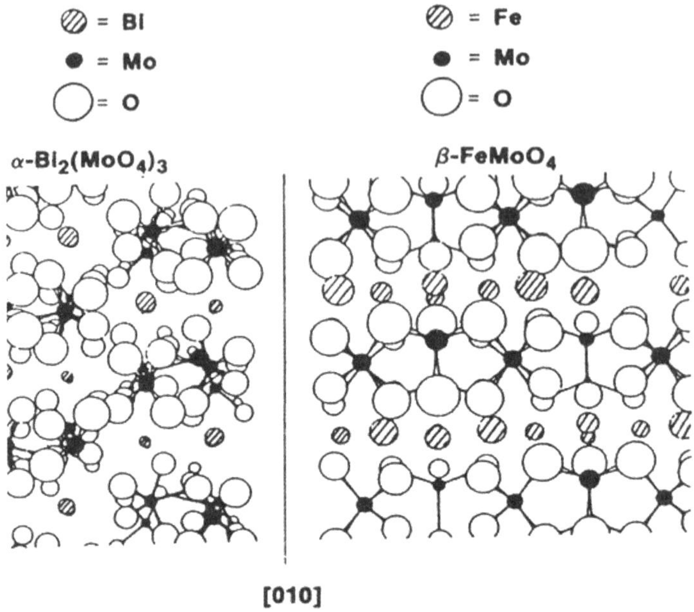

Figure 5.11 Epitaxial match at [010] face of $Bi_2Mo_3O_{12}$ and β-$FeMoO_4$.[42]

Another beautiful example of catalytic phase cooperation is the $Bi_{2-x}Ce_x$-Mo_3O_{12} system, discussed thoroughly in the literature.[33,40,41]

5.3.9 Commercial Molybdate Catalysts

The most important commercial ammoxidation catalysts are metal molybdates. And the most prominent among them are those based on bismuth molybdates and their modifications.

As already mentioned above, the first commercially successful ammoxidation catalyst, introduced industrially by SOHIO in the early 1960s, had an empirical formula $Bi_9PMO_{12}O_{52}$ and was supported on SiO_2 at a 50% level.[7] The composition is written as the salt of a heteropoly acid (*i.e.* phosphomolybdic acid) and as written is only of historic value. In the search for a useful ammoxidation catalyst at SOHIO in the late 1950s, a large number of heteropoly- and isopoly-acid salts were synthesized and tested for the named reaction. The most promising results were obtained with Cu, Sn, Sb, and Bi phosphomolybdates. And among these, the best was the Bi composition shown below (Table 5.3).

Since the ammoxidation reaction was conducted with this catalyst in the range of 450–480 °C, it is obvious that the heteropoly anion cannot retain its integrity under these conditions, and the as-synthesized composition transforms essentially to $4Bi_2Mo_3O_{12}$ (α-phase) + $BiPO_4$. The precise structural location of the phosphorus atoms has never been fully determined, nor has its exact significance. Incorporation of phosphorous prevents whiskering and is

Table 5.3 Propylene ammoxidation catalysts (molybdates).

1. *Early catalysts*	Acrylonitrile *In-Tank Yields*
$Bi_9\ PMo_{12}\ O_{52}$ –SiO_2	~55%
$Fe_{4.5}\ Bi_{4.5}\ PMo_{12}\ O_{52}$ –SiO_2	~65%
2. *Advanced multicomponent catalysts (MCM)*	
$K_a(Ni,Co)_9Fe_3BiPMoO_{12}$ –SiO_2	~75%
$(K,Cs)_a(Ni,Co,Mn)_{9.5}(Fe,Cr)_{2.5}BiMo_{12}O_x$ –SiO_2	~78–80%
$(K,Cs)_a(Ni,Mg,Mn)_{7.5}(Fe,Cr)_{2.3}Bi_{0.5}Mo_{12}O_x$ –SiO_2	>80%

deemed to impart structural stability to the Bi molybdate under ammoxidation conditions. The simple Bi molybdates (α, β, γ), while interesting model compounds (Figure 5.9) do not possess sufficient structural stability by themselves to withstand the rigors of redox catalysis for extended periods of time, which is required of commercial catalysts.

The SiO_2 support was used to impart mechanical strength to the catalyst. With this first-generation 50%$Bi_9PMo_{12}O_{52}$/50%SiO_2 catalyst (SOHIO Catalyst-A) about 55% in-tank acrylonitrile yields were achieved in commercial fluid-bed reactors (Table 5.3), sufficient to displace the then prevalent IG Farben/BASF process based on acetylene and HCN feeds. The SOHIO catalyst and process was such a revolutionary development that it was eagerly licensed by such internationally renowned companies as Mitsubishi, Mitsui Toats, Asahi, Monsanto, American Cyanamid, Du Pont, Montedison, Hoechst, Baeyer, and finally BASF. Ultimately, the total number of licensees exceeded 30 in those early days and has increased substantially since then.

The first improvement over Catalyst-A was accomplished by introducing a redox element into its structure.[43] Thus, 50% $Fe_{4.5}Bi_{4.5}PMo_{12}O_{52}$ – 50% SiO_2[44] became the 2nd-generation commercial catalyst yielding approximately 65% acrylonitrile on a commercial scale (Table 5.3). Higher propylene throughputs were possible because with the introduction of the redox element iron the time-averaged number of the active Bi–O–Mo sites increased. Although this composition was a distinct improvement over the 1st-generation catalyst as to yield and long-term catalyst stability, the system required rigorous control of the feeds, particularly the oxygen partial pressure. At high oxygen partial pressures, a high concentration of Fe^{3+} leads to the formation of free $Fe_2Mo_3O_{12}$ that leads to overoxidation and seriously raises complete combustion (CO_x).

This result led us to rethink the whole redox chemistry of Bi–Fe–Mo–O catalysts and led us to the discovery of the next and very significant improvement in catalyst composition and performance; the first multiphase, multicomponent catalyst.[3a,43] The empirical formula of one of these compositions is $K_{0.1}(Ni,Co)_9Fe_3BiPMo_{12}O_X$, and the active phase is again supported on 50% SiO_2.[45] This catalyst (Catalyst-41) gives typically about 75–79% acrylonitrile yields on the laboratory-bench scale and approximately 75% in commercial-scale operation. The Fe and Bi concentrations have been reduced, and divalent elements such as Ni and Co introduced into the structure. Also, a small amount of an alkali such as potassium was added to increase selectivity and the stability of the catalyst. The result is a very stable and easy to run catalyst, giving higher yields, allowing higher alkene throughputs, and long life. Again on the basis of the earlier discussion, it is apparent that the divalent lattice of $(Ni,Co)MoO_4$ becomes a host for Fe^{2+}. The β-$FeMoO_4$ is iso-structural with the β-phase of $(Ni,Co)MoO_4$, forming a solid solution of $(Fe,Ni,Co)MoO_4$, and thus stabilizing the divalent oxidation state of iron under oxidizing conditions. The so-stabilized divalent iron molybdate is an excellent dioxygen dissociator, incorporator and transmitter of lattice oxygen. Since the divalent molybdate can easily form epitaxial intergrowths with the active Scheelite β-$Bi_{1-X}Fe_X Mo_3O_{12}$ phase, the stage is now set for the active phase to be easily and continuously pumped up to its fully oxidized and catalytically active state during the course of the reaction. The incorporation of the alkali aids primarily the selectivity to acrylonitrile. Partly, it decreases the surface acidity of the catalyst, but more significantly it has a profound effect on the electronic structure and surface topology of the active phase. It aids in creating surface shear domains of desired proportions (site isolation), resulting in enhanced selectivity properties.

The next improvement in the hierarchy of Bi molybdate catalysts came by further fine tuning the foregoing composition. We designated it Catalyst-49, it had an empirical formula of $(K,Cs)_{0.15}(Ni,Co,Mn)_8(Fe,Cr)_{2.5}BiMo_{13.2}O_X$ and was supported on 50% SiO_2.[46] Bench-scale laboratory acrylonitrile yields range between 78% and 82%, with commercial in-tank yields being slightly below or at about 80%. The addition of Mn further stabilizes the β structure of the divalent Ni,Co,Fe-molybdate host, which comprises the regeneration phase. The addition of Cr^{3+} dilutes the Fe^{3+} concentration in the active phase of α-$Bi_{1-X}(Fe,Cr)_XMo_3O_{12}$ and thereby helps prevent undesirable waste formation (site isolation).[1,6] The addition of Cs further tunes the electronic, structural and acid properties of the surface. The overall result of these modifications is a still higher acrylonitrile yield, less-undesirable byproduct formation, and a very rugged catalyst under commercial operating conditions.

The latest catalyst improvement is a still further refinement of the preceding composition. Its empirical formula is $(K,Cs)_{0.15}(Ni,Mg,Mn)_{7.5}$-$(Fe,Cr)_{2.3}Bi_{0.5}Mo_{12}O_x$, and it too is supported on 50% SiO_2.[47] Bench-scale laboratory acrylonitrile yields range between 80% and 85%, with commercial in-tank yields topping the 80% mark. The most significant change from the parent Catalyst-49 composition is the elimination of Co from the divalent host lattice and its replacement with Mg. This substitution further stabilizes the β-molybdate structure of the solid solution $(Fe,Ni,Mg,Mn)MoO_4$ reoxidation phase, making the reoxidation process of the catalyst even more efficient, and at the same time eliminating a potentially environmentally undesirable element (*i.e.* Co).

As already discussed, the original discovery of the first commercial ammoxidation catalyst by SOHIO was based on the catalytic properties of Bi molybdate. The next four decades of intensive research brought about substantial improvements in acrylonitrile yields with the discovery and development of five new generations of commercial ammoxidation catalysts. Remarkably, the most current catalysts are still based on the unique properties of Bi molybdates. Without Bi, all of the other elements mentioned above, taken separately or together, are not capable of catalyzing the ammoxidation reaction.

The only molybdate catalyst of some note that does not contain Bi is Te-molybdate, discovered by Montedison Corporation.[48] Although this catalyst was commercialized on a small scale in Italy, it never achieved significant commercial importance, primarily because of the inherent problem of tellurium volatility. During the redox process of the ammoxidation reaction, tellurium reduces to Te^{4+}/Te^{2+}, whose oxides are rather volatile, leave the reactor, and deplete the catalytically active phase of its α-H-abstracting and NH_3-activating element. Additionally, Te is an environmentally undesirable element to handle in large quantities.

Bi molybdates are neither entirely problem free in commercial reactors, but these are minor compared to those of the Te molybdates. Under redox conditions of the ammoxidation reaction some Mo is lost from the Bi molybdate-based compositions as $MoO_2(OH)_2$ but is captured as MoO_3 on cooling coils in commercial reactors. By "rotating" the cooling coils (*i.e.* shutting some of them off on a periodic basis and letting them warm up), the so-deposited MoO_3 is revaporized from the cooling coils and returned into the fluidized catalyst bed. Most of it is reincorporated into the working catalyst, but some is inadvertently carried out of the reactor and must be processed externally. Although this sublimation loss has been greatly curtailed by the development of the newer catalyst compositions, it still remains a minor problem to be dealt with. A simple expedient that works extremely well in practice is the addition of small, measured amounts of MoO_3 to the reactor bed.[3a,49] The addition must start at the very beginning of a catalyst startup to prevent undesirable catalyst restructuring that could become irreversible and irreparable. However, employing this expedient, catalysts have performed effectively for over 10 years under commercial reaction conditions (Catalyst 49 and Catalyst 49-C).[33]

5.3.10 Antimonate Catalysts

In addition to the molybdates discussed above, the second most important class of ammoxidation catalysts attaining commercial status are those based on antimonates. And among these the most important are those containing at least one of the elements U, Fe, Sn, Mn, or Ce. All of these auxiliary elements possess multiple oxidation states, their redox potential lies above that of the $Sb^{5+/3+}$ redox couple, they provide the reoxidation moieties of the antimonate catalysts. No single structure describes all of the above antimonates; some belong to the rutile, some the trirutile family, and still others possess even more complex structural makeups (*e.g.* U-antimonates).

The most effective antimony-based ammoxidation catalyst is U-antimonate whose active phase we identified to be USb_3O_{10}.[50] It is comprised of a single phase with orthorhombic structure, space group Fddd, $a = 7.346$ Å, $b = 12.72$ Å, $c = 15.40$ Å, $z = 8$,[51,52] and contains all of the essential ammoxidation functionalities within a single phase. The α-H-abstracting and NH_3 activating function is associated with Sb^{3+} moieties, the alkene and NH_3-chemisorbing, O- or NH-inserting function with coordinately unsaturated Sb^{5+} sites, while the reoxidation function is provided by the $U^{5+/6+}$ couple.

Noteworthy is the site isolation of U moieties in the USb_3O_{10} structure as schematically depicted in Figure 5.12.[6] Each U is surrounded by eight Sb neighbors, so that in no instance do U doublets (U–O–U) or larger vicinal U aggregates (U–O–U–O–U, *etc.*) exist. Vicinal U moieties would lead to undesirable overoxidation of the alkenes to waste products (CO_x).

In addition to USb_3O_{10} a second phase with the composition $USbO_5$ can also form depending on the synthesis conditions, calcination temperature and stoichiometry (U/Sb ratio). In contrast to USb_3O_{10}, the $USbO_5$ phase exhibits, because of only limited site isolation of U moieties, a much lower selectivity towards acrylonitrile and produces substantial amounts of CO_x. If all of the antimony is left out in the synthesis, U_3O_8 is produced, which is a ferocious waste former, making solely CO_x. In this phase no site isolation is possible because all of the U sites have only U neighbors. A comparison of USb_3O_{10}, $USbO_5$ and U_3O_8 phases, viewed in the c direction, is shown in Figure 5.12.

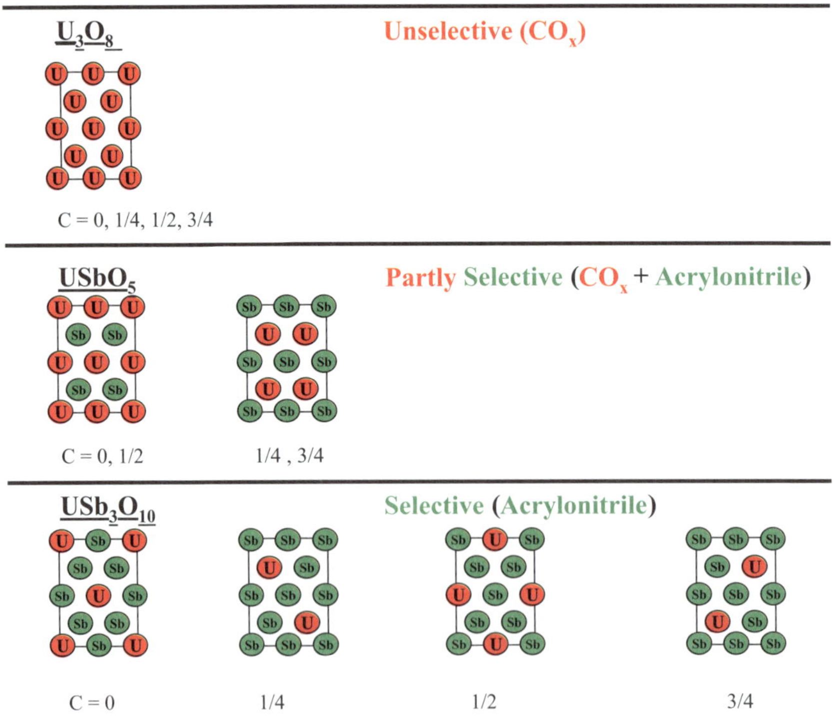

Figure 5.12 Heavy atom positions of USb_3O_{10}, $USbO_5$, and U_3O_8.[52]

Only next-nearest heavy atoms to individual U atoms are shown, illustrating the extent of site isolation attainable by each of the respective phases.

The structure of the catalytically desirable USb_3O_{10} active phase[51,52] is one of the best manifestations of the site-isolation principle of selective (amm)oxidation catalysis.[1,6] It forms the basis for a superb, highly selective catalyst for the production of acrylonitrile under ammoxidation conditions and a superb, highly selective catalyst for the production of acrolein in the absence of ammonia. Therefore, the catalyst has been commercialized.

SOHIO commercialized, in the late 1960s, the first and only U-based ammoxidation catalyst having the empirical formula $USb_{4.6}O_{13.2}$ (*i.e.* USb_3O_{10} – $0.8Sb_2O_4$), supported on 40% SiO_2[50–53] with the designation Catalyst 21. Herewith, the largest, non-nuclear use of uranium was created; of course ^{235}U-depleted uranium was used in the catalyst. The ammoxidation yields of the U-based catalyst[52] exceeded those of the first $Bi_9PMo_{12}O_{52}/SiO_2$ and second-generation $Bi_{4.5}Fe_{4.5}PMo_{12}O_{52}/SiO_2$ catalysts (Figure 5.4). This third-generation system displaced the first two in commercial plants and became a decided success. However, because of the residual minuscule radioactivity and by containing the heavy element U, Catalyst 21 had to be phased out by the late 1970s. It was a great catalyst whose structure and properties we knew well, a system amenable to easy yield improvement through judicial elemental substitutions and doping;[54] however, unfortunately the improvements never went past the pilot-plant stage because of environmental constraints.

It must, however, also be mentioned that antimony-based catalysts are prone to antimony loss during the redox reaction. As was mentioned above, besides the desirable USb_3O_{10} phase a lower antimony containing phase, namely $USbO_5$ also exists that is undesirable since it is much less selective than USb_3O_{10}. If the antimony loss were to be severe, the USb_3O_{10} phase can transform to the undesired $USbO_5$ phase, which would be unacceptable in commercial operation. It is therefore imperative to minimize antimony loss by starting with an excess of antimony in the catalyst (see Catalyst 21 composition above), adding redox enhancers into the composition (*e.g.* Fe-molybdate,[55] physically adding antimony oxide periodically to the fluid-bed reactor, and/or using an engineering expedient of internal reoxidation, *i.e.* raised feed inlet sparger.

Besides U antimonate, Fe antimonates also attained commercial status (Table 5.4). The active phase in these catalysts is ascribed to $Fe_2Sb_2O_7$, having a trirutile structure, a naturally occurring but scarce mineral named tripuhyite.

Table 5.4 Propylene ammoxidation catalysts (antimonates).

1. *Early catalysts*	Acrylonitrile *In-Tank Yields*
$FeSb_{8.6}O_x$ –SiO_2	~65%
$USb_{4.6}O_x$ –SiO_2	~70%
2. *Advanced multicomponent catalysts (MCA)*	
$Na_{0.3}$ (Cu, Mg, Zn, Ni)$_{0-4}$(V,W)$_{0.05-1}$ $Mo_{0.1-2.5}$ $Te_{0.2-5}$ Fe_{10} Sb_{13-20} O_x –SiO_2	~75%

However, to achieve selectivity, it is necessary to add additional antimony to the active phase,[57,58] which fits again the concept of site isolation.[1] Various alternate explanations have also been given, ranging from epitaxy between crystallographically well-matched planes of $Fe_2Sb_2O_7$ (or $FeSbO_4$) and α-Sb_2O_4, leading to intergrowths of these two distinct phases,[59] to epitaxial overgrowths of Sb_2O_4 (*i.e.* suprasurface antimony sites)[60] on $FeSbO_4$. All of these suppositions are consistent with the well-accepted mechanistic concept[14] that NH_3 activation and α-H abstraction of the alkene occur on a Sb^{3+} site, the chemisorption of NH_3 and the alkene, and the NH insertion on an Sb^{5+} site. The underlying Fe antimonate phase provides for dioxygen dissociation, lattice oxygen incorporation and its transfer to the antimonate phase or overlayer, to keep the Sb^{3+}–O–Sb^{5+} catalytic sites in their proper oxidation states during ammoxidation. α-Sb_2O_4 and Sb_2O_5, while capable of ammoxidizing alkenes to their corresponding nitriles, are unable to sustain activity for any practical time. The simple antimony oxide phases reduce rapidly, are unable to dissociate dioxygen and incorporate it as lattice oxygen into their structure; they cannot maintain a sufficiently high oxidation state for effective catalysis. Therefore, incorporation of an effective reoxidation element such as $Fe^{3+/2+}$ into the antimony oxide phase or the placement of an Fe-antimonate phase into close proximity of an antimony oxide phase is necessary. As soon as $Fe_2Sb_2O_7$ appears as a crystallographically discernible phase in the X-ray diffraction pattern of the solid, the solid becomes a catalyst. The ammoxidation properties of Fe antimonates can be further improved by the incorporation of additional selected elements. Nonetheless, Fe-antimonates have the same redox weakness and loss of antimony problem as do U-antimonates. The same preventive measures as discussed above for $USbO_x$ can be applied here.

Other antimonates that behave similarly to Fe antimonates and also have similar structure are those containing as the redox element Mn, V, or Ce. All require excess antimony to achieve selectivity, and all can be further improved by the addition of certain selected elements.

Among the Sb-based catalysts, highly fine tuned, complex multicomponent antimonates (MCA) (Table 5.4)[61] are currently used in commercial units. Most of this work has been performed by Nitto Corporation, while SOHIO/BP (with the exception of $FeSb_{8.6}O_x$–SiO_2 and $USb_{4.6}O_x$–SiO_2) continued to concentrate on the Bi-molybdates, including the currently used complex multicomponent molybdates (MCM), discussed earlier (Table 5.3). Because MCM catalysts are much more resilient under commercial reaction conditions, and because they give higher in-tank acrylonitrile yields than MCA catalysts, the Bi–Mo-based MCM catalysts retain, among these two rivals, by far the more important place in the commercial ammoxidation world (Figure 5.4).

5.4 Propane Ammoxidation to Acrylonitrile

With increasing crude oil prices and the price differential between propylene and propane increasing, the incentive to replace the feed material propylene by propane in the well-established SOHIO acrylonitrile process[3a,4] becomes steadily more attractive. To achieve this goal, an efficient catalyst for the direct conversion of propane to acrylonitrile must be discovered, so that the entire process can take place in a single reactor.

The ammoxidation of propane is an eight-electron oxidation:

$$CH_3CHCH_3 + NH_3 + 2O_2 \xrightarrow{\text{catalysts}} CH_2{=}CHCN + 4H_2O \qquad (5.14)$$

Over the past 15 years considerable research has been done to discover the desired catalyst for this demanding, complex reaction. Table 5.5 summarizes the most important catalysts discovered thus far. To this end, MoVNbTeO[28] discovered by Mitsubishi[62–65] is the most promising candidate, followed closely by MoVNbSbO.[66]

Although acrylonitrile yields in exceed of 60% per pass have been obtained on microreactor scale, they do not suffice to replace the olefin-based SOHIO process. A yield of 70+% is needed. At that level a paraffin-based process becomes feasible; particularly if building a grass roots acrylonitrile plant.

However, these desired, lofty yields have thus far eluded the research community. Towards this goal, it is imperative to understand the needed properties of such a catalyst, with fundamental studies aimed at an understanding of the functionalities required of the chemical makeup of such a solid catalyst being of utmost importance.

In this vain, it is of interest that up to now (Table 5.5) all effective propane ammoxidation catalysts contain at least some vanadium; and that in its highest oxidation state. V^{5+} is an essential component for the activation of the paraffin. Secondly, as with alkenes, essentially all effective paraffin compositions are comprised of either molybdates or antimonates.

Among the molybdates the most prominent is the Mitsubishi catalyst having an empirical formula of $Mo_{0.6}V_{0.16-0.21}Nb_{0.06-0.10}Te_{0.10-0.18}O_x$.[65] An important feature in its synthesis is calcination in N_2 or other reducing atmosphere at about 620 °C. The structure and mechanism of this intriguing catalyst is discussed below. Also-ran molybdate catalysts have the general formula $VMo_xM_yO_z$, with M being Bi, Te, Ga, Nb, Ta, Ce, and Ag.[69,70] Bi-containing compositions have often the Scheelite structure;[71] others are multiphase and are more complex.

Table 5.5 Paraffin ammoxidation catalysts (microreactor yields).

Molybdates	
(a) $VMo_xM_yO_z$	M=Bi, Te, Ga, Nb, Ta, Ce, Ag
(b) $Mo_{0.6}V_{0.16-0.21}Nb_{0.06-0.10}Te_{0.10-0.18}O_x$[65]	55.1% AN yield (86.7% conv., 63.5% sel.)
(c) $Mo_{0.6}V_{0.187}Nb_{0.085}Te_{0.14}O_x$[28]	61.8% AN yield (86% conv., 72% sel.)
(d) $MoV_{0.3}Nb_{0.06}Sb_{0.20}G_{0.30}O_x$[66]	44.7% AN yield (86% conv., 52% sel.)
Antimonates	
(a) $VSb_xM_yO_z$	M=W, Te, Nb, Sn, B, Bi, Al, Ti
(b) $VSb_5W_{0.5}TeSn_{0.5}O_x$[67]	39.1% AN yield (68.8% conv., 56.7% sel.)
Oxy-nitride VAlON[68]	36% AN yield (55% conv., 66% sel.)

Among the antimonates, the most studied are $VSb_xM_yO_z$, where M is most frequently W, Te, Nb, Sr, Bi, Al, and/or Ti.[72] All are of the rutile structure. The highest acrylonitrile yields (Table 5.5) have been attained with a SOHIO-discovered composition having the empirical formula $VSb_5W_{0.5}Te_{0.5}Sn_{0.5}O_x$ supported on SiO_2/Al_2O_3 (39.0% yield, 68.8% conversion, 56.7% selectivity).[72a,73] The highest selectivity to acrylonitrile from propane is reported for a catalyst having the empirical formula $VSb_{3.5}Bi_2Fe_2O_x/Al_2O_3$ (18.9% yield, 21% conversion, 90% selectivity).[74] Interestingly, this Rhone-Poulenc-developed catalyst may be considered a combination of two SOHIO-patented catalysts, namely, VSb_xO_y[72] to convert propane to propylene, and $BiFe_xSb_yO_z$[75] to subsequently convert propylene to acrylonitrile.

The vanadium antimonate catalysts are multiphase, containing $VSbO_4$ with rutile structure and α-Sb_2O_4. Extensive mechanistic studies[43a76,77] show that propane activation is the rate-limiting step with a V–O$^\bullet$ moiety abstracting the first hydrogen from the methylene group of propane, thereby forming propylene as a desorbed primary product and the main precursor of acrylonitrile. The desorbed propylene reacts further, as is the case in propylene ammoxidation, by an Sb^{3+}-O site abstracting an α-hydrogen from the propylene, and an $Sb^{5+}=NH$ site inserting the nitrogen into the chemisorbed allylic surface complex, ultimately yielding desorbed acrylonitrile product. Therefore, V–Sb-oxide catalysts are multifunctional, having V–O, Sb^{3+}–O$^\bullet$, and Sb^{5+}–O moieties in near proximity of each other. The reoxidation of the catalyst occurs by lattice oxygen ($O^{2-}{}_L$) migration from the reoxidation site to the catalyst active site, and anion-vacancy migration in the opposite direction. The reoxidation site contains V=O moieties, which are also responsible for dioxygen dissociation and its incorporation into the lattice, thereby keeping the catalyst lattice oxygen concentration at a steady state in this Mars–van Krevelen redox process.

5.4.1 Structure of MoVNbTeO Catalysts

The MoVNbTeO catalytic system is comprised of three crystalline phases:[28,78] orthorhombic $Mo_{7.8}V_{1.2}NbTe_{0.94}O_{28.9}$ (M1) (*Pba2*: $a = 21.1337$ Å, $b = 26.6440$ Å, $c = 4.01415$ Å, $z = 4$), pseudohexagonal $Mo_{4.67}V_{1.33}Te_{1.82}O_{19.82}$ (*M2*) (*Pmm2*: $a = 12.6294$ Å, $b = 7.29156$ Å, $c = 4.02010$ Å, $z = 4$) and a trace of monoclinic $TeMo_5O_{16}$ ($P2_{1/C}$: $a = 10.0349$ Å, $b = 14.430$ Å, $c = 8.1599$ Å, $\beta = 90.781°$, $z = 1$). A Schematic three-phase section of the MoVNbTeO phase diagram is shown in Figure 5.13.[28] The structures of the M1 and M2 phases respectively are shown in Figures 5.14 and 5.15. The catalytically active and selective centers reside on the surface of the basal plane of the M1 phase and are comprised of an assembly of five metal-oxide octahedra ($2V^{5+}{}_{0.32}/Mo^{6+}{}_{0.68}$, $1V^{4+}{}_{0.62}/Mo^{5+}{}_{0.38}$, $2Mo^{6+}{}_{0.5}/Mo^{5+}{}_{0.5}$) and two tellurium-oxygen sites ($2Te^{4+}{}_{0.94}$). This catalytic assembly is stabilized and structurally isolated from each other (site isolation) by four Nb^{5+} centers, each surrounded by five molybdenum–oxygen octahedra (Figure 5.16). The V^{5+} surface sites, distinguished through their ($V^{5+}=O \leftrightarrow {}^{4+}V^\bullet$–$O^\bullet$) resonance structure, are the

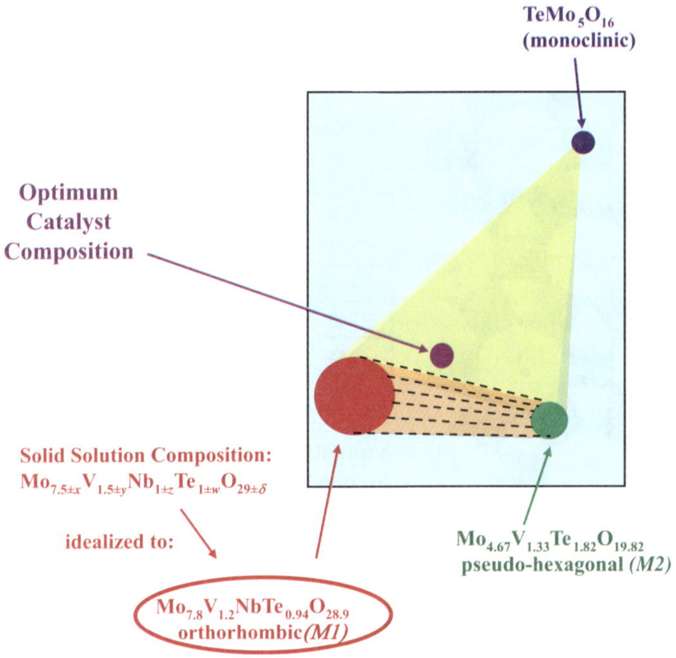

Figure 5.13 Three-phase section from the MoVNbTeO phase diagram.[28]

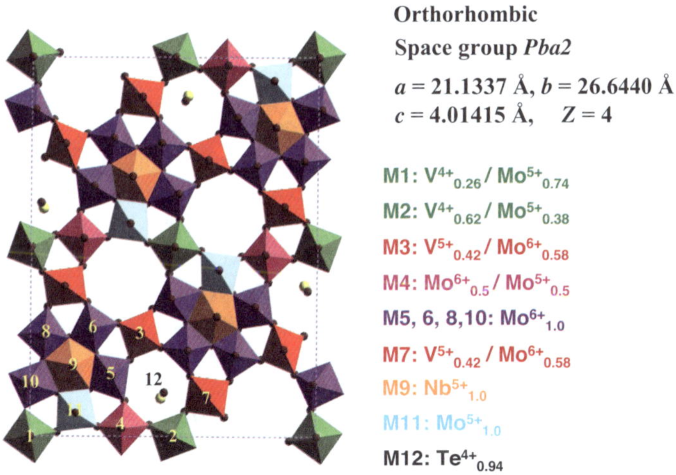

Figure 5.14 Structure of $Mo_{7.8}V_{1.2}NbTe_{0.94}O_{28.9}$ (M1 phase).[28,78]

paraffin activating sites capable of methylene-H abstraction; the Te^{4+} sites (lone pair of electrons) for the α-H abstraction of the chemisorbed propylene molecule, once formed; and the adjacent Mo^{6+} sites for the NH insertion into the π-allylic surface intermediate. Herewith, all key catalytic elements needed to

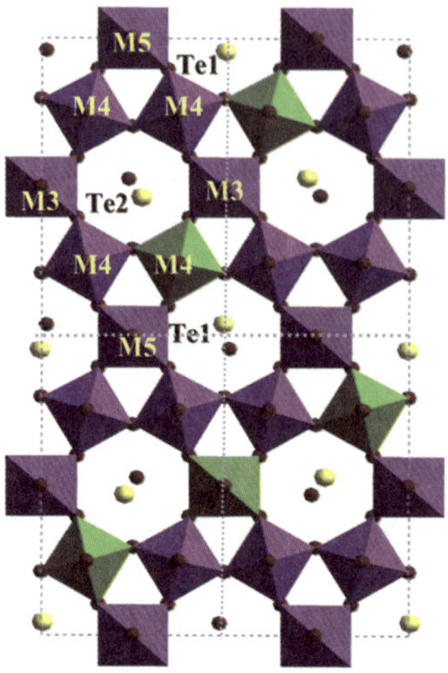

"Pseudo-Hexagonal" (Orthorhombic)
Space group *Pmm2*

a = 12.6294 Å, b = 7.29156 Å, c = 4.02010 Å

Z =4

Site	Occupancy
$Te^{4+}1$	4 x 0.237
$Te^{4+}2$	4 x 0.215
M3	0.54 Mo^{6+} / 0.46 V^{4+}
M4	0.78 Mo^{6+} / 0.22 V^{4+}
M5	1.0 Mo^{6+}

Vanadium is V^{4+}; virtually <u>no</u> V^{5+}
No heptagonal channel (strain)

Figure 5.15 Structure of $Mo_{4.67}V_{1.33}Te_{1.82}O_{19.82}$ (M2 phase).[78,79]

transform propane directly to acrylonitrile are contained, strategically arranged and within bonding distance of each other, at the active center of the M1 phase (Figure 5.16). Schematically, the presumed interacting area of an M1 active center with a reacting propane molecule is illustrated in Figure 5.17. It comprises an M7 (V), M4 (Mo), M12 (Te) and M5 (Mo) site of the active center.

We have calculated[79] the metal site distribution probabilities at the active center of M1, concluding that 44% of the sites should be active and selective for propane ammoxidation, 46% inactive for propane but active for propylene ammoxidation and 10% being waste-forming sites (Figure 5.18). Based on this calculation a theoretical acrylonitrile selectivity of 81% is predicted. This level has not as yet been achieved experimentally; the maximum lies at about 72%. However, if coproduced propylene is added to the acrylonitrile yield, then the combined acrylonitrile and propylene yields give a summed selectivity in the 80% range, at their optimum. The agreement between calculated and experimental values is quite good, giving some credence to the proposed reaction model at the active site of M1.

5.4.2 Symbiosis between M1 and M2 Phases

Under mild operating conditions (low propane conversion and low w/w/h) the M1 phase alone suffices to effectively convert propane directly to acrylonitrile.[28,80–89] Under demanding conditions, akin to commercial requirements (high propane conversion and high w/w/h), symbiosis takes place between the M1 and M2 phases. The M1 phase still continues to be the sole paraffin-activating phase, producing acrylonitrile as it does under mild conditions. However, in addition of acrylonitrile, desorbed propylene is also produced without being converted on the M1 phase to acrylonitrile. The N=H inserting

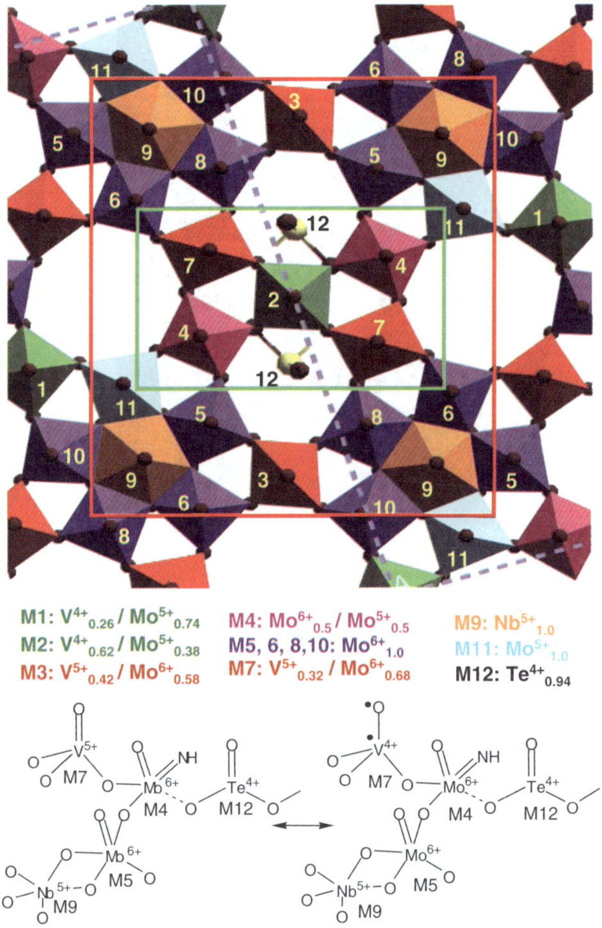

Figure 5.16 Catalytically active center of $Mo_{7.8}V_{1.2}NbTe_{0.94}O_{28.9}$ (M1 phase) in [001] projection, on surface of an M1 basal plane; and chemical drawing illustrating the catalytically active center.[28,78]

function of the M1 phase cannot keep up with the demand created by the high concentration of propylene at the reaction site. The regeneration of N=H inserting sites (or lack thereof) becomes the limiting reaction step at the active site on the surface. Unconverted propylene now desorbs from the M1 surface and enters the gas phase. If M2 phase is in intimate contact with the M1 phase it will chemisorb this free propylene and effectively convert it to acrylonitrile. The M2 phase serves now as a cocatalyst or mop-up phase to the M1 phase. As has been shown convincingly[79,83,84] the M2 phase is incapable of propane activation, lacking V^{5+} sites, but is an excellent propylene ammoxidation catalyst, equal to or better than the M1 phase. Under conditions of symbiosis, a maximum acrylonitrile yield from propane of 61.8% (86% conversion, 72% selectivity at 420 °C) has been achieved with a nominal catalyst composition of $Mo_{0.6}V_{0.187}$ $Nb_{0.085}$ $Te_{0.14}O_x$, comprised of 60% M1, 40% M2 and a trace of

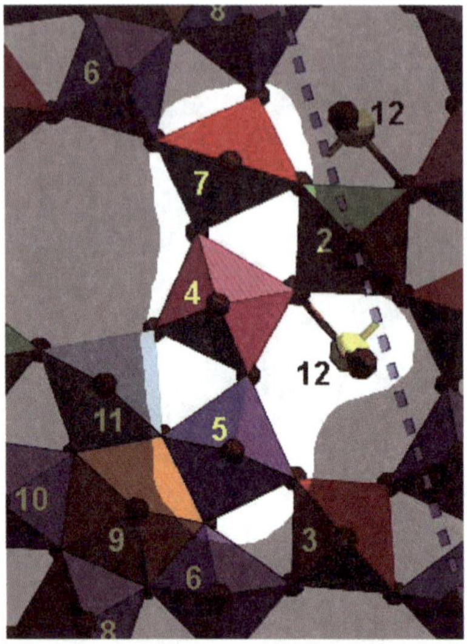

Figure 5.17 Interactive area of a propane molecule with the key catalytic sites of an M1 active center of a MoVNbTeO catalyst. Interacting sites are M7(V), M4(Mo), M12(Te) and M5(Mo).[90]

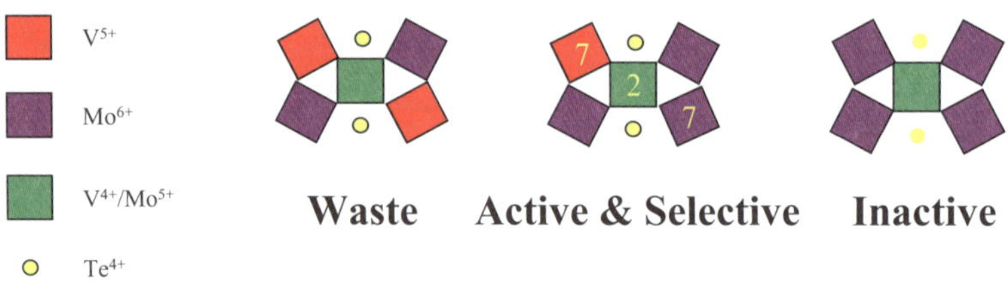

M7	M2	M7	Catalytic Properties
V^{5+} (32)	V^{4+} (62)	V^{5+} (32)	0.06 waste
V^{5+} (32)	V^{4+} (62)	Mo^{6+} (68)	0.14 active & selective
Mo^{6+} (68)	V^{4+} (62)	V^{5+} (32)	0.14 active & selective
Mo^{6+} (68)	V^{4+} (62)	Mo^{6+} (68)	0.28 inactive
V^{5+} (32)	Mo^{5+} (38)	V^{5+} (32)	0.04 waste
Mo^{6+} (68)	Mo^{5+} (38)	Mo^{6+} (68)	0.18 inactive
V^{5+} (32)	Mo^{5+} (38)	Mo^{6+} (68)	0.08 active & selective
Mo^{6+} (68)	Mo^{5+} (38)	V^{5+} (32)	0.08 active & selective

AN Max. Yield: Stat. 81% Exp. 62% Thermo. 100%

Figure 5.18 Distribution probabilities of metal occupancies at the active center of $Mo_{7.8}V_{1.2}NbTe_{0.94}O_{28.9}$ (M1 phase).[79]

$TeMo_5O_{16}$.[28] The catalytic optimum was identified using combinatorial methodology.[89] The ammoxidation performance (acrylonitrile yield) of MoVNbTeO and also MoVTaTeO systems, respectively, as a function of M1 phase concentration is illustrated in Figure 5.19. Both systems exhibit symbiotic yield optima. The acrylonitrile yields of the respective M1 phases are similar. No acrylonitrile is formed with only M2 phase (*i.e.* zero M1 concentration). However, the respective symbiotic optima are vastly different (MoVNbTeO >> MoVTaTeO). Some explanation for this difference is given in the literature.[90]

Based on the observations discussed above, a reaction scheme (Figure 5.20) is proposed for the reaction of propane, ammonia and oxygen (air) on the surface of M1 and M2 catalyst phases. Under mild operating conditions (low w/w/h and low conversion) the M1 phase is entirely self-sufficient in converting propane to acrylonitrile, with the M2 phase being simply a spectator phase. Under demanding reaction conditions (high w/w/h and high conversion) the M1 phase is overworked, cannot regenerate its N=H-insertion functions and permits propylene to desorb unconverted from its surface. Released into the vapor phase, propylene encounters the surface of the M2 phase, if it is in very close proximity, where it readily converts to acrylonitrile because M2 is an efficient olefin-conversion catalyst.

Symbiosis between M1 and M2 phases can take place only if the two phases are in very close contact with each other. This occurs if the two phases are prepared simultaneously in a one-step synthesis[28] or if the two phases are prepared separately, ground separately to fine powders ($\leq 5\,\mu m$) and thereafter

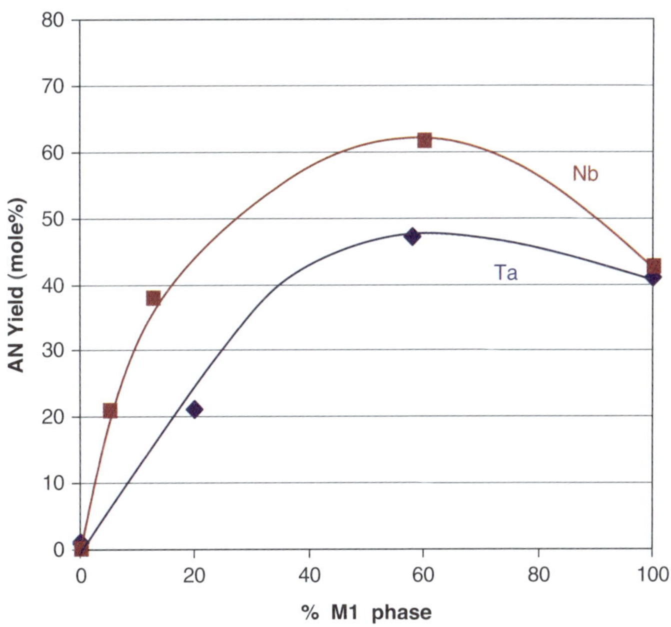

Figure 5.19 Propane ammoxidation performance of MoVNbTeO and MoVTaTeO systems as a function of M1 concentration. Symbiosis at about 60% M1 and 40% M2.[90]

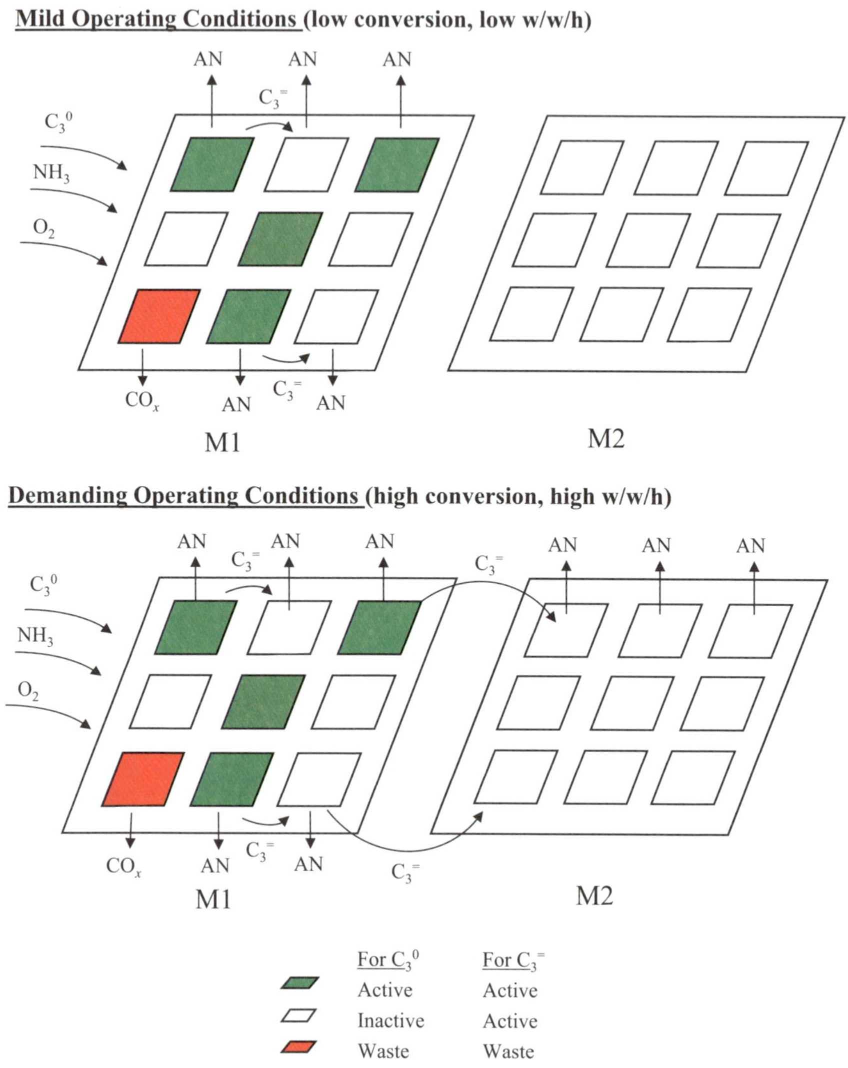

Figure 5.20 Schematic of the propane ammoxidation process on MoVNbTeO catalysts.[79]

thoroughly mixed and compacted.[91,92] As is illustrated in Figure 5.21, no symbiosis is observed when the particles of the individual phases are larger than 5 μm, while symbiosis did occur when the particles were smaller.

5.4.3 Reaction Network and Mechanism

On the basis of extensive experiments and numerous literature studies[28 and references therein] we have proposed a reaction network for the ammoxidation of

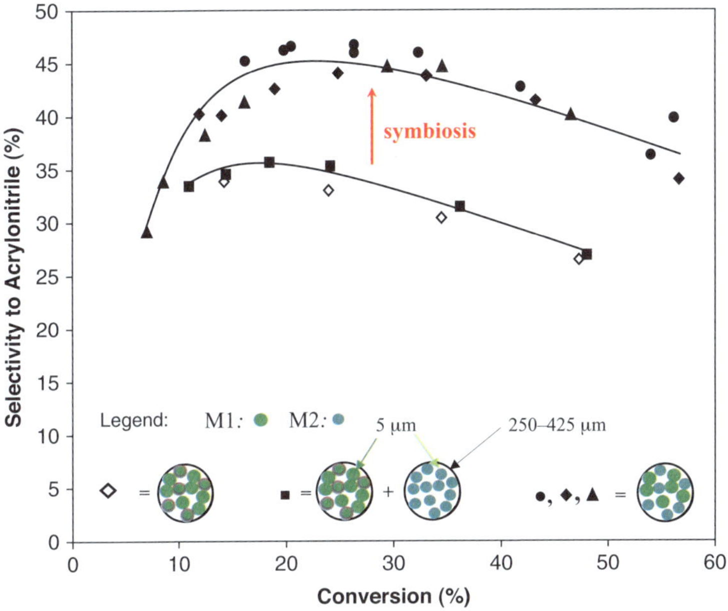

Figure 5.21 Symbiosis (phase cooperation) between M1 and M2 physical mixtures in propane ammoxidation.[91,92] ◇ Neat M1 phase ($Mo_{7.8}V_{2.2}Nb_{1.5}Te_{0.9}O_x$), ground to ~5 μm powder, compacted and sized to 250–425 μm particles; ■ M1 + M2 separately ground to ~5 μm powders, separately compacted and sized to 250–425 μm particles, then mixed; ● M1 + M2 separately ground to ~5 μm powders, mixed, compacted and sized to 250–425 μm particles; ◆ M1 + M2 as in (●) and additionally heated at 550 °C for 1h; ▲ M1 + M2 as in (●) and additionally heated at 600°C for 1 h. Physical mixtures = 50 wt.%M1/50wt.%M2, with surface area ratio of 4:1. Reaction temperature = 380 °C; propane/ammonia/oxygen/argon = 6.1/7.0/18.0/70.2; space velocity = 3.3–26.3 N cm^3/(min g).

propane using M1 and M2 phases (Figure 5.22).[92] In this network, M1 is deemed to be the sole propane-activating phase leading to a surface adsorbed propyl radical $[CH_3\dot{C}HCH_3]_a$ species that further reacts on the M1 active center to an adsorbed propylene $[CH_3CH{=}CH_2]_a$ species. This adsorbed propylene specie reacts further without desorption, with but minimal movement (all key catalytic elements being within a bonding distance of this reaction complex) leading directly to acrylonitrile. Conversely, the M1 adsorbed propylene species can desorb, readsorb on M1 and react further to give acrylonitrile or CO_x; depending on the site it encounters.[79] If M2 is present in close proximity (conditions given above) then the desorbed propylene can react on this phase to acrylonitrile and/or CO_x. Since the M2 phase is allegedly a more effective propylene to acrylonitrile conversion catalyst than the M1 phase ($k_{11} > k_9$), it becomes an important propylene mop-up component of the composite M1 + M2 catalyst system. As already mentioned, the M2 phase is completely

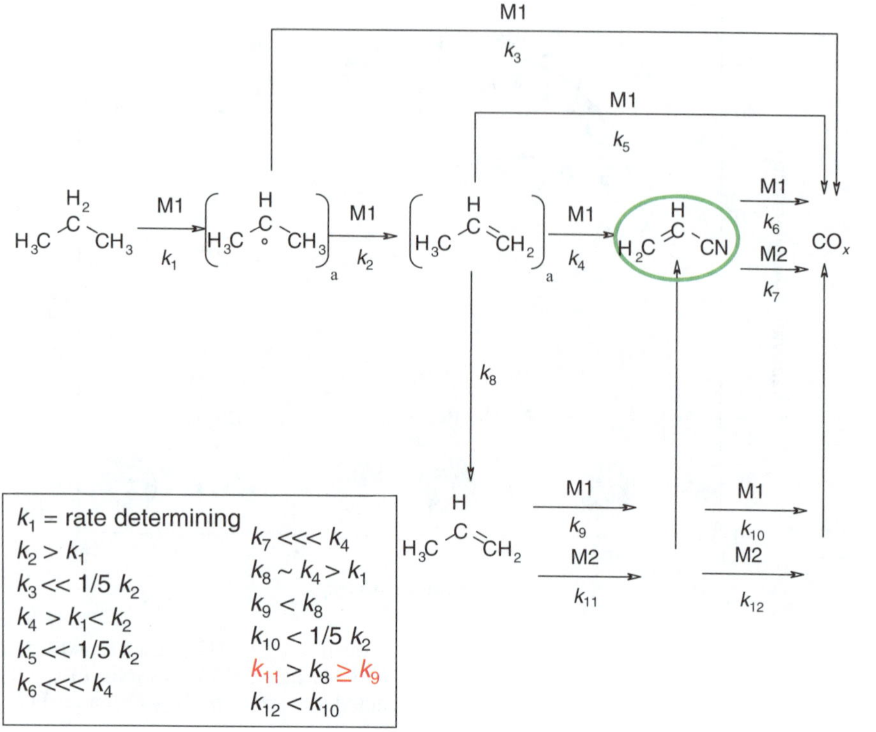

Figure 5.22 Proposed propane ammoxidation network using M1 and M2 as catalysts.[92]

ineffective for propane conversion, having no V^{5+} activating sites. Waste-forming pathways are also indicated in the proposed reaction network.

A detailed reaction mechanism of propane ammoxidation over MoVNbTeO catalysts (M1 phase) is shown in Figure 5.23. All transformations from the activation of propane and ammonia to the ultimate formation of acrylonitrile are possible on this phase. The mechanism is based on structural considerations, catalyst performance,[28] earlier mechanistic literature[93,94] and theory.[95] The activation of propane occurs by methylene-H abstraction on a V^{5+} site (M7) of the M1 active center (Figure 5.16). The radical character of the paraffin-activating surface vanadyl (V=O) oxygen atom, can be explained by the resonance form in which the V assumes a 4+ character, namely $V^{5+}{=}O \leftrightarrow {}^{4+}V^{\bullet}{-}O^{\bullet}$. The activation of paraffins by V–O surface moieties of V–Sb–O_x and V–Mo–O_x catalysts has been suggested by several investigators.[43a,76,77] From classical organic chemistry it is well known that H-atom abstraction by O-radical centers is a facile process. In the present case, the abstraction occurs at the methylene position of the propane molecule to produce the more stable secondary propyl radical. The so-formed propyl radical loses a methyl-H to the oxygen of an adjacent $Te^{4+(6+)}$ at (M12) and produces a propylene molecule.

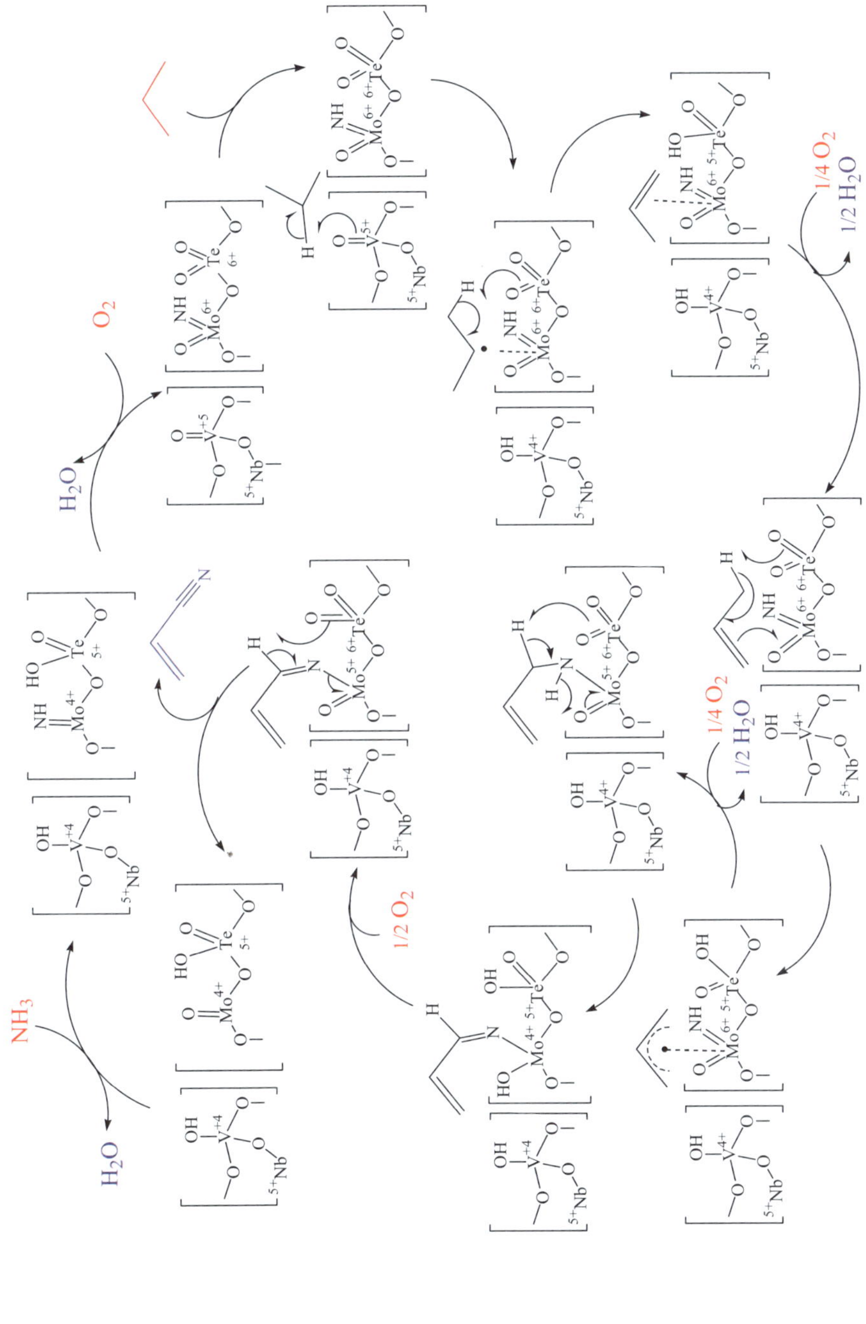

Figure 5.23 Proposed propane ammoxidation mechanism over MoVNbTeO catalysts.[28]

There is some evidence of surface Te^{6+},[91,96,97] which might be ideal for this second H-abstraction. Without desorption, the π-electrons of the now-formed propylene molecule coordinate with an adjacent Mo^{6+} center (M4) while its neighboring Te^{4+} (M12) site abstracts an α-H forming an allyl intermediate. Te^{4+} is superbly suited for this later task, with its lone pair of electrons, akin to Bi^{3+} in Bi-molybdates.[3a,98] A regeneration step restores the reduced Te to Te^{4+} as indicated in Figure 5.17. An N=H of the same (M4) site now inserts into the allyl radical, aided by the spectator oxo group (Mo=O) at (M4), thereby satisfying the conditions of ref. 95. The remaining transformations (1,4 hydrogen shifts, *etc.*) occur as in classical olefin ammoxidation catalysis (Figure 5.7). The subsequent 2-e reduction of the catalyst, in the acrylonitrile-forming step, is now shared by the molybdenum atoms at (M4) and (M5). No desorption of intermediate products (e.g, propylene) is necessary, since it occurs on an active center (Figure 5.16) containing all of the necessary key catalytic elements positioned strategically adjacent to one another. The reaction is essentially concerted because the activated propane molecule is within a bonding distance of all key catalytic functionalities. Since the active centers of the $Mo_{7.8}V_{1.2}NbTe_{0.94}O_{28.9}$ system are multifunctional catalytic assemblies spatially isolated (Figure 5.24) from each other (site isolation) they enjoy high

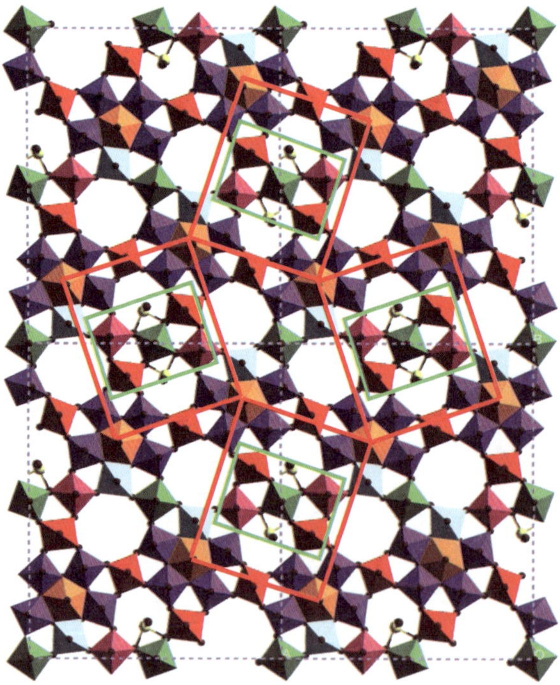

Figure 5.24 2×2 unit cell structure model of $Mo_{7.8}V_{1.2}NbTe_{0.94}O_{28.9}$ in the [001] projection showing four isolated catalytically active centers (site isolation).[28]

selectivity. And because of their multimetallic/multivalent character they also exhibit high activity (*vide supra*).

5.4.4 Future Research

Substantial progress has been made towards achieving commercially acceptable acrylonitrile yields in the ammoxidation of propane, the yields thus far (60 + %) fall, however, short of the 70 + % goal deemed necessary to replace the well-established, currently practiced, propylene-based SOHIO acrylonitrile ammoxidation process.

In the catalyst development area, one of the major improvements comes through enhanced catalyst stability and ruggedness, achieved by moving from MoVNbTeO to MoVNbSbO systems.[99,100] The acrylonitrile yields of the latter catalyst can be improved by selective doping.[66] Yields in excess of 55% have been obtained with these more-stable catalyst systems. Obviously additional work is needed to further improve the yield structure. One possible approach towards this goal is to develop a catalyst capable of activating propane at lower temperatures.

Another hopeful lead is doping with phosphorous. Although the realized substantial yield improvements were demonstrated on MoVNbTeO systems,[101–103] it is likely that this approach might also be applicable to the MoVNbSbO system.

Since symbiosis between certain catalyst phases has definitively been demonstrated (*vide supra*); the opportunity exists to combine two catalysts: one for the activation of the paraffin and the other for the conversion of the intermediately formed olefin to acrylonitrile. Each reaction can be studied separately, which simplifies the research. Of course, the acid test comes when the two systems are combined. In this study it must be kept in mind that the two systems must be chemically compatible, so that they do not poison each other. And they have to be temperature matched at their respective yield optima so that they can operate in a single reactor. With such a hypothetically proposed system it is currently an open question whether the two phases need to be in superclose (nanoscale) contact, possibly preferentially even epitaxially matched within individual solid catalyst particles, or if this postulate is not necessarily a prerequisite for success if the catalysts are to be operated in a fluid-bed reactor (which is the norm for commercial propylene-based ammoxidation processes) where contact between the two phases would be excellent and could be maximized by particle-size and – density manipulations. Some attempts towards this end are being pursued.

Reactor innovations might also contribute to the end goal. For example, it might be possible to achieve synergy between a propane-to-propylene catalyst and a propylene to acrylonitrile catalyst by making the first catalyst denser, so that it resides primarily in the lower part of a fluidized-bed reactor. The second catalyst, if less dense, will therefore preferentially reside in the upper part of the reactor. In this manner two catalysts could cooperate under an essentially plug-flow condition (sieve trays would help). However, since it is difficult to devise an oxidation catalyst that would be selective for propylene from propane conversion at high concentrations of propylene, recycle might be needed, which is not an ideal situation for a commercial process. Conversely, it might be possible to design a propane-to-propylene oxydehydrogenation catalyst that would operate in the presence of ammonia and have no α-H abstracting sites; thereby leaving the propylene unreacted until it reaches the propylene to acrylonitrile conversion catalyst. This might be an interesting and viable approach.

Ultimately, the most rewarding approach, on which my hope rests, will be the discovery and development of an entirely new, sophisticated composition (single or multiphase) that will achieve the propane to acrylonitrile transformation essentially concertedly and be more effective than anything we have seen thus far. One could speculate that it might be comprised of V–Mo–Bi–(Nb)–O, whose structure might be M1, but more likely a variation thereof. Bi and Mo will surely be two of the required elements, Nb (and/or Ta) a likely multipurpose element (site isolator, Mo-catinator and valence inducer). Whether an adequate substitute for V can be found is open for debate; most probably not.

In summary, it is deemed that it will be the synthetic chemist who will provide us with the solution to the puzzle discussed above by devising a new and commercially feasible propane ammoxidation catalyst. It is only a matter of time before this will be accomplished.

References

1. J. L. Callahan and R. K. Grasselli, *AIChE J.*, 1963, **9**, 755.
2. R. K. Grasselli and J. D. Burrington, in *Handbook of Heterogeneous Catalysis*, G. Ertl, H. Knözinger, F. Schüth, J. Weitkamp (ed.), Wiley-VCH Verlag GmbH & Co. KGaA, Weinheim, 2008, Vol. 8, 3479.
3. (a) R. K. Grasselli, *J. Chem. Ed.*, 1986, **63**, 216–221 and references therein; (b) J. D. Idol, US Patent 2 904 580, 1959.
4. (a) R. K. Grasselli, in: *Proceedings of the DGMK Conference*, Hamburg, Germany, 2001, 147; (b) R. K. Grasselli, *La Chimica e l'Industria*, 2001, **83**, 25.
5. R. K. Grasselli, *Top. Catal.*, 2002, **21**, 79.
6. R. K. Grasselli, *Top. Catal.*, 2001, **15**, 93.
7. J. L. Callahan, R. K. Grasselli, E. C. Milberger and H. A. Strecker, *Ind. Eng. Chem. Proc. Res. Dev.*, 1970, **9**, 134–142.
8. R. K. Grasselli and J. D. Burrington, *Adv. Catal.*, 1981, **30**, 133.
9. G. W. Keulks, *J. Catal.*, 1970, **19**, 232.
10. G. W. Keulks, and L. D. Krenzke, *Proceedings 6th International Congress on Catalysis*, The Chemical Society, London, 1976, Vol. 2, p. 806.
11. J. D. Burrington, C. T. Kartisek and R. K. Grasselli, *J. Catal.*, 1980, **63**, 235.
12. D. R. Stull, E. F. Wiestrum, Jr. and G. C. Sinke, *The Chemical Thermodynamics of Organic Compounds*, Wiley, New York, 1969.

13. R. K. Grasselli and J. D. Burrington, *I&EC Res. & Dev.*, 1984, **23**, 393.
14. J. D. Burrington, C. T. Kartisek and R. K. Grasselli, *J. Catal.*, 1984, **87**, 363.
15. T. Otsubo, H. Miura, Y. Morikawa and T. Shirasaki, *J. Catal.*, 1975, **36**, 240.
16. (a) B. Gryzbowska, J. Haber and J. Janas, *J. Catal.*, 1977, **49**, 150; (b) K. Bruckman, J. Haber and T. Wiltowski, *J. Catal.*, 1987, **106**, 188.
17. T. P. Snyder and C. G. Hill, *Catal. Rev.-Sci. Eng.*, 1989, **31**, 43.
18. J. F. Brazdil, L. C. Glaeser and R. K. Grasselli, *J. Catal.*, 1983, **81**, 142.
19. L. C. Glaeser, J. F. Brazdil, M. A. Hazle, M. Mehicic and R. K. Grasselli, *J. Chem. Soc., Faraday Trans. 1*, 1985, **81**, 2903.
20. W. Ueda, Y. Moro-oka and T. Ikawa, *J. Chem. Soc., Faraday Trans. 1*, 1982, **78**, 495.
21. T. Ono and N. Ogata, *J. Chem. Soc., Faraday Trans.*, 1994, **90**, 2113.
22. J. F. Brazdil and R. K. Grasselli, *J. Catal.*, 1983, **79**, 104.
23. A. W. Sleight, *Advanced Materials in Catalysis*, J. J. Barton, R. C. Garten (ed.), Academic Press, New York, 1977, 181.
24. R. G. Teller, J. F. Brazdil, R. K. Grasselli and J. D. Jorgensen, *Acta Crystallogr. C.*, 1984, **40**, 2001.
25. R. G. Teller, J. F. Brazdil, R. K. Grasselli, R. T. L. Corliss and J. Hastings, *J. Solid State Chem.*, 1984, **52**, 313.
26. D. J. Buttrey, D. A. Jefferson and J. M. Thomas, *Philos. Mag. A*, 1986, **53**, 897.
27. J. F. Brazdil, D. D. Suresh and R. K. Grasselli, *J. Catal.*, 1980, **66**, 347.
28. R. K. Grasselli, J. D. Burrington, D. J. Buttrey, P. DeSanto, C. L. Lugmair, A. F. Volpe and T. Weingand, *Top. Catal.*, 2003, **23**, 5.
29. (a) W. A. Nugent, R. J. McKinney, R. V. Kasowski and F. A. van Catledge, *Inorg. Chem. Acta*, 1982, **65**, 91; (b) W. A. Nugent, *Inorg. Chem.*, 1983, **22**, 965; (c) D. M.-T. Chan, W. C. Fultz, W. A. Nugent, C. D. Roe and T. H. Tulip, *J. Am. Chem. Soc.*, 1985, **107**, 251; (d) D. M.-T. Chan and W. A. Nugent, *Inorg. Chem.*, 1985, **24**, 1422.
30. (a) E. A. Maatta and Y. Du, *J. Am. Chem. Soc.*, 1988, **110**, 8249; (b) E. A. Maatta, Y. Du and A. L. Rheingold, *J. Chem. Soc., Chem. Commun.*, 1990, 756.
31. J. Belgacem, J. Kress and J. A. Osborn, *J. Chem. Soc., Chem. Commun.*, 1993, 1125.
32. R. K. Grasselli, J. F. Brazdil, J. D. Burrington, *Proceedings 8th International Congress on Catalysis*, Berlin, Schoen & Wetzel, Frankfurt, 1984, Vol. 5, p. 369.
33. R. K. Grasselli, in: *Handbook of Heterogeneous Catalysis*, Vol. 5, G. Ertl, H. Knoezinger, J. Weitkamp, ed., Wiley-VCH, Weinheim, 1997, p. 2302.
34. W. Jeitschko, *Acta Crystallogr.*, 1973, **B29**, 2074.
35. (a) J. Zemann, *Heidelbeger Beitr. Minerol Petrog.*, 1956, **5**, 139; (b) A. F. V. D. Elzen and G. D. Rieck, *Acta Cryst.*, 1973, **B29**, 2436; (c) B. C. Gates, J. R. Katzer and G. C. A. Schuit, *Chemistry of Catalytic Processes*, McGraw-Hill, New York, 1979, p. 370.

36. I. K. Kolchin, S. S. Bobkov and L. Ya. Margolis, *Netekhimiya 4*, 1964, **2**, 301.
37. I. Matsuura, R. Schut and K. Hirakawa, *J. Catal.*, 1980, **63**, 152.
38. R. K. Grasselli, *Proceedings of the first Joint CIESC-AIChE Meeting*, Beijing, China, 1982, **2**, 731.
39. J. C. Vedrine, G. Coudurier, M. Forissier and J. C. Volta, *Mater. Chem. Phys.*, 1985, **13**, 365.
40. R. K. Grasselli, *Appl. Catal.*, 1985, **15**, 127.
41. (a) J. F. Brazdil and R. K. Grasselli, *J. Catal.*, 1983, **79**, 104; (b) J. F. Brazdil, L. C. Glaeser and R. K. Grasselli, *J. Phys. Chem.*, 1983, **87**, 5485.
42. J. F. Brazdil, M. Mehicic, L. C. Glaeser, M. A. S. Hazle, R. K. Grasselli, *ACS Symp. Ser. 288* (ed. M. L. Deviney, J. L. Gland), 1985, 26.
43. R. K. Grasselli, in *Surface Properties and Catalysis by Non-Metals*, J. Bonelle, B. Delmon, E. Derouane, (eds.), D. Reidel, Dordrecht, 1983, 273 and 289.
44. K. Sennewald, W. Vogt, J. Kandler, R. Sommerfeld, G. Sorbe, US Patent 3 226 422, 1965.
45. (a) R. K. Grasselli, H. F. Hardman, US Patent 3 642 930, 1972; (b) R. K. Grasselli, H. F. Hardman, US Patent 4 503 001, 1985.
46. R. K. Grasselli, D. D. Suresh, H. F. Hardman, US Patent 4139 552, 1979.
47. D. D. Suresh, M. S. Friedrich, M. J. Seely, US Patent 5 212137, 1993.
48. (a) G. Caporali, N. Ferlazzo, N. Giordano, US Patent 3 691224, 1972; (b) J. C. J. Bart and N. Giordano, *J. Catal.*, 1980, **64**, 356; (c) J. C. J. Bart and N. Giordano, *Ind. Eng. Chem. Prod. Res. Dev.*, 1984, **23**, 56.
49. R. K. Grasselli, G. Centi and F. Trifiro, *Appl. Catal.*, 1990, **57**, 149.
50. R. K. Grasselli and J. L. Callahan, *J. Catal.*, 1969, **14**, 93.
51. R. K. Grasselli, D. D. Suresh and K. Knox, *J. Catal.*, 1970, **18**, 356.
52. R. K. Grasselli and D. D. Suresh, *J. Catal.*, 1972, **25**, 273.
53. J. L. Callahan, B. Gertisser, US Patent 3 198 750, 1965.
54. D. D. Suresh, R. K. Grasselli, US Patent 4 315 838, 1982.
55. W. G. Shaw, R. K. Grasselli, H. F. Hardman, US Patent 3 551 470, 1970.
56. J. L. Callahan, E. C. Milberger, R. K. Grasselli, US Patent 3 427 343, 1969.
57. J. L. Callahan, R. K. Grasselli, W. R. Knipple, US Patent 3 338 952, 1967.
58. D. D. Suresh, J. F. Brazdil, R. K. Grasselli, US Patent 4 290 920, 1981.
59. R. G. Teller, J. F. Brazdil and R. K. Grasselli, *J. Chem. Soc., Faraday Trans. 1*, 1985, **81**, 1693.
60. R. Nilsson, T. Linblad and A. Andersson, *J. Catal.*, 1994, **148**, 501.
61. Y. Sasaki, T. Nakamura, Y. Nakamura, K. Moriya, H. Utsumi, S. Saito, US Patent 4 370 279, 1983.
62. M. Hatano, A. Kayo, Europ. Patent 318,295, 1988.
63. T. Ushikubo, K. Oshima, A. Tayo, T. Umezawa, K. Kiyona, I. Sawaki, Europ. Patent 529,853, 1992.
64. H. Hinago, S. Komada, A. K. Kogoyo, US Patent 6,063,728, 2000.
65. T. Ushikubo, *Catal. Today*, 2000, **57**, 331.

66. C. G. Lugmair, J. Zysk, R. K. Grasselli, US Patent appl. 0054869 A1, 2005.
67. A. T. Guttmann, R. K. Grasselli, J. F. Brazdil, US Patent 4 746 641, 1998; A. T. Guttmann, R. K. Grasselli, J. F. Brazdil, US Patent 4 788 317, 1989; A. T. Guttman, R. K. Grasselli, J. F. Brazdil, US Patent 4 797 381, 1989.
68. R. Prada Silvy, P. Grange, *Oil Gas Eur. Mag.*, 2003, **3**, 145; M. Florea, R. Prada Silvy, P. Grange, *Catal. Lett.* 2003, 87, 63.
69. D. D. Suresh, J. F. Brazdil, L. C. Glaeser, US Patent 4760 159, 1988.
70. (a) Y. Moro-oka, N. Miura, N. Fujikawa, Y. C. Kim and W. Ueda, in *Proceedings of the 10th International Congress on Catalysis*, eds. L. Guczi, F. Solymosi, P. Tetenyi, Akademiai Kiado, Budapest, 1992C, 1983; (b) C. Kim, W. Ueda and Y. Moro-oka, *Catal. Today*, 1992, **13**, 673.
71. J. S. Kim and S. I. Woo, *Appl. Catal. A*, 1994, **110**, 207.
72. (a) A. T. Guttmann, R. K. Grasselli, J. F. Brazdil, D. D. Suresh, US Patent 4 746 641, 1988; (b) A. T. Guttmann, R. K. Grasselli, J. F. Brazdil, D. D. Suresh, US Patent 4 788 317, 1988; (c) J. P. Bartek, A. T. Guttmann, US Patent 4 797 381, 1989.
73. R. K. Grasselli, *Catal. Today*, 1999, **49**, 141.
74. G. Blanchard, G. Ferre, US Patent 5 336 804, 1994.
75. J. L. Callahan, R. K. Grasselli, W. R. Knipple, US Patent 3 445 521, 1969.
76. A. Andersson, S. L. T. Andersson, G. Centi, R. K. Grasselli, M. Sanati and F. Trifiro, in: *Proceedings 10th International Congress on Catalysis*, eds. L. Guczi, F. Solymosi, P. Tetenyi, Akademiai Kiado, Budapest, 1992A, 691.
77. H. Bluhm, M. Haevecker, E. Kleimenov, A. Knop-Gericke, A. Liskowski, R. Schloegl and D. Su, *Top. Catal.*, 2003, **23**, 90.
78. P. DeSanto, D. J. Buttrey, R. K. Grasselli, C. G. Lugmair and A. F. Volpe, *Z. Kristallogr.*, 2004, **219**, 152.
79. R. K. Grasselli, D. J. Buttrey, P. DeSanto, J. D. Burrington, C. G. Lugmair, A. F. Volpe and T. Weingand, *Catal. Today*, 2004, **91–92**, 251.
80. T. Ushikubo, K. Kayou and M. Hatano, *Stud. Surf. Sci. Catal.*, 1997, **112**, 473.
81. T. Ushikubo, K. Oshima, T. Numazawa, M. Vaarkamp and I. Sawaki, *Stud. Surf. Sci. Catal.*, 1999, **121**, 339.
82. T. Ushikubo, *Catal. Today*, 2000, **57**, 331.
83. J. Holmberg, R. K. Grasselli and A. Andersson, *Top. Catal.*, 2003, **23**, 55.
84. (a) J. M. M. Millet, H. Roussel, A. Pigamo, J. L. DuBois and J. C. Jumas, *Appl Catal. A Gen.*, 2002, **232**, 77.
85. M. Baca, A. Pagamo, J. L. Dubois and J. M. M. Millet, *Top. Catal.*, 2003, **15**, 39.
86. D. Vitry, Y. Morikawa, J. L. Dubois and W. Ueda, *Top. Catal.*, 2003, **15**, 47.
87. E. Garcia-Gonzalez, J. M. Lopez Nieto, P. Botella and J. M. Gonzalez-Calbert, *Chem. Mater.*, 2002, **14**, 4416.
88. K. Oshihara, T. Hisano and W. Ueda, *Top. Catal.*, 2001, **15**, 153.

89. A. Hagemeyer, B. Jandeleit, Y. Liu, D. M. Poojary, H. W. Turner, A. F. Volpe, Jr. and W. H. Weinberg, *Appl. Catal. A*, 2001, **221**, 23.
90. R. K. Grasselli, D. J. Buttrey, J. D. Burrington, A. Andersson, J. Holmberg, W. Ueda, J. Kubo, C. G. Lugmair and A. F. Volpe, Jr., *Top Catal.*, 2006, **38**, 7.
91. J. Holmberg, R. K. Grasselli and A. Andersson, *Appl. Catal. A: Gen*, 2004, **270**, 121.
92. R. K. Grasselli, *Catal. Today*, 2005, **99**, 23.
93. R. K. Grasselli and J. D. Burrington, *Adv. Catal.*, 1981, **30**, 133.
94. J. D. Burrington, C. T. Kartisek and R. K. Grasselli, *J. Catal.*, 1983, **81**, 489.
95. Y. H. Jang and W. A. Goddard III, *Top. Catal.*, 2001, **15**, 273.
96. M. Aouine, J. L. Dubois and J. M. M. Millet, *Chem. Commun*, 2001, 1180.
97. W. Ueda, D. Vitry and T. Katou, *Catal. Today*, 2005, **99**, 43.
98. J. F. Brazdil and R. K. Grasselli, *J. Catal*, 1980, **66**, 66.
99. H. Hinago, S. Komeda US Patent 6 143 916, 2000.
100. T. Inoue, H. Ishida US Patent 6 514 902, 2003.
101. R. K. Grasselli, in: *Turning Points in Solid-State, Materials and Surface Science*, K. D. M. Harris, P. P. Edwards (ed.), RSC Publishing, Cambridge, 2007, Ch. 34, p. 577.
102. R. K. Grasselli, C. G. Lugmair and A. F. Volpe, Jr., *Top. Catal.*, 2008, **50**, 66.
103. R. K. Grasselli, C. G. Lugmair, A. F. Volpe, Jr., A. Andersson and J. D. Burrington, *Catal. Lett.*, 2008, **126**, 231.

CHAPTER 6

Selective Oxidation of n-*Butane over Vanadium–Phosphorous Oxide*

MOISES A. CARREON[1] AND VADIM V. GULIANTS*[2]

[1] Department of Chemical Engineering, University of Louisville, Louisville, KY 40292, USA; [2] Department of Chemical & Materials Engineering, University of Cincinnati, Cincinnati OH 45221, USA

6.1 Introduction

The low cost and abundance of light alkanes have generated considerable interest in their oxidative catalytic conversion to olefins, oxygenates and nitriles in the petroleum and petrochemical industries.[1–4] Rough estimates place the annual worth of products that have undergone a catalytic oxidation step at \$20–40 billion worldwide.[4] Among these, the 14-electron selective oxidation of *n*-butane to maleic anhydride on vanadium–phosphorus oxide (VPO) catalysts is one of the most fascinating catalytic processes:[4,5]

The selective oxidation of *n*-butane to maleic anhydride is the only industrial process of a selective vapor-phase oxidation of an alkane that uses dioxygen.[5] The demand for maleic anhydride comes principally from the manufacture of pharmaceuticals, lubricating oil additives, unsaturated

RSC Nanoscience & Nanotechnology No. 19
Nanostructured Catalysts: Selective Oxidations
Edited by Christian Hess and Robert Schlögl

Published by the Royal Society of Chemistry, www.rsc.org

polyester resins, agricultural chemicals, and food additives.[6] In 1995, the global production of maleic anhydride was estimated to be ~900 000 tons with a value of $700 million.[4]

Since the discovery of production of maleic anhydride from *n*-butane over VPO catalysts in 1966 by Bergman and Frisch,[7] *n*-butane has been increasingly used instead of benzene as the raw material for maleic anhydride production due to lower price, high availability and low environmental impact.[8] At present, more than 70% of maleic anhydride is produced from *n*-butane.[6] However, productivity from *n*-butane is lower than in the case of benzene due to lower selectivities to maleic anhydride at higher conversions and somewhat lower feed concentrations (<2 mol.%) used to avoid flammability of a process stream. Under typical industrial conditions (2 mol.% *n*-butane in air, 673–723 K, and space velocities of 1100–2600 h^{-1}) the selectivities for fixed-bed production of maleic anhydride from *n*-butane are 67–75 mol.% at 70–85% *n*-butane conversion.[9] A remarkable characteristic of the VPO catalysts is that no support is required in the partial oxidation of *n*-butane. Many studies of *n*-butane oxidation on the VPO catalysts indicated that crystalline vanadyl(IV) pyrophosphate, $(VO)_2P_2O_7$, is present in the most selective catalysts. However, the VPO system is characterized by facile formation and interconversion of a number of crystalline and amorphous V^{III}, V^{IV} and V^{V} phosphates.[9]

The VPO catalysts are prepared by thermal dehydration of its precursor, vanadyl(IV) hydrogen phosphate hemihydrate, $VOHPO_4 \cdot 0.5H_2O$. The catalytic performance of the VPO catalysts depends on (a) $VOHPO_4 \cdot 0.5H_2O$ synthesis method (nature and reagents concentrations, reducing agents and solvents, the reduction temperature and synthesis time), (b) activation and conditioning procedures of the precursor and (c) presence of metal promoters. These factors important for understanding the catalytic behavior of the VPO system in *n*-butane oxidation have been discussed previously in a number of reviews.[9–13]

Herein, we briefly discuss fundamental aspects of VPO catalysis, such as the mechanism of $VOHPO_4 \cdot 0.5H_2O$ formation and its transformation to active and selective VPO catalysts, the active VPO surface, the mechanism of *n*-butane oxidation, the role of promoters and several novel synthesis approaches that have recently shown promise for the molecular engineering of new well-defined VPO phases.

6.2 $VOHPO_4 \cdot 0.5H_2O$ Precursor

6.2.1 Conventional Synthesis Methods

It is well known that the necessary synthesis conditions to obtain an optimal VPO catalyst are the following:[4,9,13–20] (a) synthesis of microcrystalline $VOHPO_4 \cdot 0.5H_2O$ in an alcohol characterized by the preferential exposure of the basal (001) planes, (b) the presence of defects in the stacking of the (001) planes and (c) a slight excess of phosphate with respect to the stoichiometric amount employed in the synthesis (P/V = 1.01–1.10).

Three major conventional synthesis methods have been reported for preparation of the $VOHPO_4 \cdot 0.5H_2O$ precursor:

1. *Aqueous synthesis.* V^V compounds (*e.g.* V_2O_5) are reduced to V^{IV} in aqueous solutions of orthophosphoric acid, followed by evaporation of the solvent to dryness:[21]

$$V_2O_5 + 2NH_2OH \cdot HCl + 2H_3PO_4 \rightarrow 2VOHPO_4 \cdot 0.5H_2O + N_2 + 2HCl + 4H_2O$$

2. *Organic synthesis.* V^V compounds are reduced by an anhydrous alcohol, followed by the addition of anhydrous orthophosphoric acid dissolved in the same alcohol and precipitation of $VOHPO_4 \cdot 0.5H_2O$:[15,18]

$$V_2O_5 + EtCHOHMe + 2H_3PO_4 \rightarrow 2VOHPO_4 \cdot 0.5H_2O + EtCOMe + 2H_2O$$

3. *Model organic synthesis.* V^V orthophosphate dihydrate, $VOPO_4 \cdot 2H_2O$, is first synthesized from V_2O_5 and H_3PO_4 in aqueous medium and then reduced to $VOHPO_4 \cdot 0.5H_2O$ by an alcohol in a separate step:

$$V_2O_5 + 2H_3PO_4 + H_2O \rightarrow 2VOPO_4 \cdot 2H_2O + H_2O$$

$$2VOPO_4 \cdot 2H_2O + EtCHOHMe \rightarrow 2VOHPO_4 \cdot 0.5H_2O + EtCOMe + 3H_2O$$

In general, the *organic route* provides the most active and selective catalysts.[4,9,13,18] XRD patterns indicate that organic precursors are less crystalline and preferentially expose the (001) planes,[13] as manifested in a broader (001) reflection and its lower relative intensity as compared to the intensity of the inplane (130) reflection. All three methods may lead to various hydrated vanadyl(IV) hydrogen-phosphate phases, $VOHPO_4 \cdot nH_2O$ ($n = 0.5$, 1, 2, 3, and 4), which are all precursors of the VPO catalysts. The precursor with $n = 0.5$ ($VOHPO_4 \cdot 0.5H_2O$) produces the best VPO catalysts.[9]

Currently, only organic VPO catalysts are employed industrially in *n*-butane oxidation as the most active and selective. Several steps have been proposed in the formation of the $VOHPO_4 \cdot 0.5H_2O$ in organic medium:[18] (a) the formation of colloidal V_2O_5, (b) the dissolution of V_2O_5 through the formation of V^V-alcoholate species, (c) the reduction of the dissolved V^V-alcoholate species in the liquid phase to solid V_2O_4, and (d) the reaction of V_2O_4 with H_3PO_4 to form $VOHPO_4 \cdot 0.5H_2O$. In both aqueous and organic syntheses, the $VOHPO_4 \cdot 0.5H_2O$ precursor has a P/V ratio higher than the stoichiometric value.[9,15,18] The maximum value corresponds to P/V = 1.1, while the excess phosphate remains in synthesis solution. X-ray photoelectron (XPS) analysis has shown that the excess phosphate is localized at the surface of the vanadyl pyrophosphate catalysts (surface P/V = 1.5–3.0).[22]

6.2.2 Thermal Activation

Several crystalline VPO phases have been observed during the transformation of the $VOHPO_4 \cdot 0.5H_2O$ precursor to the active VPO catalyst[10,11] depending on: the temperature, time, and atmosphere of activation; the morphology of the precursor; the P/V ratio in the precursor; and the presence of defects.

Two different thermal activation procedures have been used to activate the $VOHPO_4 \cdot 0.5H_2O$ precursor:

(a) Activation in an oxygen-free atmosphere at $T > 673\,K$, followed by introduction of the reactant mixture of *n*-butane in air. $VOHPO_4 \cdot 0.5H_2O$ transforms to poorly crystalline $(VO)_2P_2O_7$ during the first step,[15] which can be partially oxidized to V^V orthophosphates after the introduction of the reactant mixture.[12]
(b) Calcination in air at $T < 673\,K$, after which the reactant mixture is introduced.[14,15]

When the $VOHPO_4 \cdot 0.5H_2O$ precursor is heated, the trapped alcohol molecules are released, which creates structural defects, microcracks and increases the surface area. The precursor first transforms into an amorphous phase,[14,22] which can be further dehydrated to crystalline $(VO)_2P_2O_7$ and partially oxidized to V^V orthophosphates once the reactant mixture is introduced. Excess phosphate present in the precursor stabilizes $(VO)_2P_2O_7$ against overoxidation in an oxygen-containing atmosphere.[12] The *in-situ* Raman spectroscopy provided insights into the nature of the transformation of the model organic $VOHPO_4 \cdot 0.5H_2O$ precursor.[23] When the precursor phase was heated in an inert He atmosphere, a continuous loss of local order was observed up to 724 K. At that temperature, fingerprint Raman bands of $VOHPO_4 \cdot 0.5H_2O$ were no longer visible, and instead weak features corresponding to poorly crystalline $(VO)_2P_2O_7$ emerged. The *in-situ* spectra suggested that these two phases did not coexist during the precursor transformation in an inert atmosphere. Similar observations were made during the precursor transformation in a reactive *n*-butane–air environment. The precursor is completely transformed into a mixture of δ-$VOPO_4$ and $(VO)_2P_2O_7$ at 708 K. The redox environment leads to partial oxidation of vanadyl(IV) species into a V(V) orthophosphate. The presence of V(V) orthophosphate is usually responsible for higher activity and lower selectivity of these catalysts as compared to the fresh catalysts activated in an inert atmosphere. It is interesting that the transformation proceeds *via* formation of an amorphous VPO intermediate regardless of the gas environment, and that $VOHPO_4 \cdot 0.5H_2O$ and $(VO)_2P_2O_7$ phases are not observed simultaneously during the transformation. Figure 6.1 shows the different phase transformations in the VPO system.[24] Xue and Schrader further examined the VPO phase transformations using *in-situ* Raman spectroscopy.[25] They found that $(VO)_2P_2O_7$ experienced a structural disorder during prolonged exposure to water vapor at elevated temperatures that resulted in the V–O–P bond breaking, the loss of phosphate and the appearance of bulk vanadium oxide. The thermal transformation of the $VOHPO_4 \cdot 0.5H_2O$ precursor into $(VO)_2P_2O_7$ is considered to be topotactic in the sense that the initial crystal morphology was preserved during the transformation.[26]

6.2.3 Equilibrated VPO Catalyst

After the crystalline $VOHPO_4 \cdot 0.5H_2O$ precursor is transformed into the fresh, poorly crystalline $(VO)_2P_2O_7$ and overoxidized V(V) orthophosphate impurity phases in an oxygen-containing atmosphere, changes in its catalytic performance and physicochemical properties occur during on-stream conditioning in the reactant mixture. A "nonequilibrated" catalyst is termed as "fresh" in contrast to an "equilibrated" catalyst that spent considerable time, *e.g.* 30 days, under catalytic reaction conditions.[4,9,27] A fresh catalyst is more active and less selective in *n*-butane activation, particularly at high conversion, owing to the presence of the V(V) orthophosphate impurities and easier oxidation of V^{IV} to V^{V} at the end of the reactor in a fixed-bed operation as the reactant mixture becomes more oxidizing at lower *n*-butane concentration. The aim of such a conditioning procedure is to reduce V(V) orthophosphate impurity phases in the presence of *n*-butane, crystallize $(VO)_2P_2O_7$ in the VPO catalysts and increase their surface area.[9] Recently, 340×40 nm $VOHPO_4 \cdot 0.5H_2O$ crystallites were transformed and equilibrated into highly crystalline $(VO)_2P_2O_7$ at 663 K in an *n*-butane–air mixture for only 300 h.[28] In this study, no intermediate crystalline phases were formed from these nanosized precursors, whereas α_{II}-$VOPO_4$ and δ-$VOPO_4$ formed (as intermediate phases) from larger $VOHPO_4 \cdot 0.5H_2O$ crystals. Shorter activation times (75 h) of $VOHPO_4 \cdot 0.5H_2O$ in an *n*-butane–air mixture lead to poorly crystalline $(VO)_2P_2O_7$ with limited exposure of the most active and selective (200) plane.[29]

A number of crystalline $VOPO_4$ phases have been detected in the VPO catalysts by different research groups[9,10] that displayed different catalytic behavior during on-stream conditioning (Figure 6.1). Poorly crystalline and amorphous $VOPO_4$ phases present in the catalysts are often undetected by XRD. The presence of such disordered phases may be confirmed by other characterization techniques, such as Raman spectroscopy and ^{31}P spin-echo NMR. The incomplete phase analysis of VPO catalysts is one of the major reasons responsible for several alternative definitions of the equilibrated state and the nature of the active VPO phase.[9–11] In many cases, only short-term VPO conditioning was conducted and the attainment of the steady state was reported after only several hours on-stream.[30–32] Equilibrated organic catalysts have displayed superior performance in *n*-butane oxidation over aqueous catalysts, while the phase composition and crystalline order of $(VO)_2P_2O_7$ in these catalysts after conditioning remains almost identical.[23] No significant changes in the bulk phase composition (XRD) or the surface P/V ratios (XPS) have been observed during VPO conditioning.

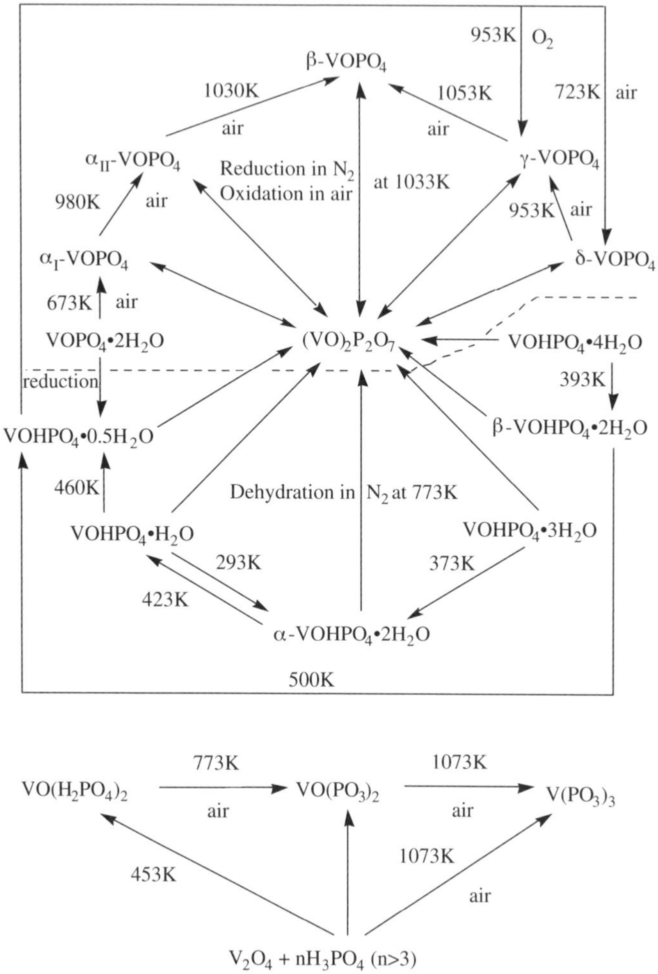

Figure 6.1 VPO phase transformations.[24] The dashed line separates the *reversible redox transformations* to $(VO)_2P_2O_7$ (top) from *irreversible dehydrations* (bottom).

6.2.4 Optimal P/V Ratio

The optimal VPO catalyst composition is characterized by a slight excess of phosphate with respect to the empirical formula of the $VOHPO_4 \cdot 0.5H_2O$ precursor.[15,22] A considerable excess of surface phosphate (P/V = 1.5–3.0) prevents the bulk oxidation of $(V^{IV}O)_2P_2O_7$ to the $V^{V}OPO_4$ phases.[9,33,34] In catalysts with a slight phosphate deficiency (P/V = 0.95), the rates of V^{IV} oxidation and reduction are high.[35] The increased V^{V} content leads to more active but less selective catalysts, while low reducibility of V^{IV} in VPO catalysts with high surface phosphate concentration results in low catalytic activity.

Therefore, catalysts with a slight excess of phosphate (P/V = 1.05) show the right compromise between reducibility and oxidizability needed to obtain both high activity and selectivity in *n*-butane oxidation.

6.3 Active VPO Surface

Bordes suggested that the active sites in *n*-butane oxidation to maleic anhydride are associated with coherent interfaces between slabs of the (100) planes of various $VOPO_4$ phases and the (200) planes of $(VO)_2P_2O_7$ along the (001) and (201) planes, respectively.[24] However, the best $(VO)_2P_2O_7$ catalysts display the lack of other impurity $VOPO_4$ phases. Therefore, this mechanism may be appropriate to explain the catalytic behavior of nonequilibrated or overoxidized VPO catalysts that contain various microcrystalline $VOPO_4$ phases.

On the other hand, Volta *et al.* proposed that the active sites are not associated with interfaces between these crystalline phases.[36] On the basis of comparison between XRD and radial electron distribution data they suggested that the active phase for selective oxidation of *n*-butane consists of a mixture of well-crystallized $(VO)_2P_2O_7$ and an amorphous V^VOPO_4 phase involving corner-sharing VO_6 octahedra, as shown in Figure 6.2(a). This amorphous phase was interpreted by them as a precursor of β-$VOPO_4$, which formed at higher reaction temperatures. Later, Volta *et al.* suggested on the basis of kinetic data as well as XRD and ^{31}P MAS NMR results that domains of γ-$VOPO_4$ supported on a $(VO)_2P_2O_7$ matrix are necessary for selective *n*-butane oxidation.[37] However, XRD, Raman and ^{31}P NMR studies showed that the best catalysts did not contain amorphous or microcrystalline V^V phosphates.[38] Therefore, the mechanism of Volta *et al.* may explain the behavior of only nonequilibrated or overoxidized VPO catalysts that contain $VOPO_4$ phases.[34] Hutchings *et al.*[39] and Volta *et al.*[40] suggested that the actives sites for *n*-butane oxidation to maleic anhydride involve a V^{4+}/V^{5+} couple well dispersed on the surface of a range of VPO phases, *e.g.* $(VO)_2P_2O_7$ for equilibrated VPO catalysts. Hutchings *et al.* also suggested the possibility that the actives sites could be the defects as well as interfaces between microcrystalline $VOPO_4$ phases and well-crystalline $(VO)_2P_2O_7$.

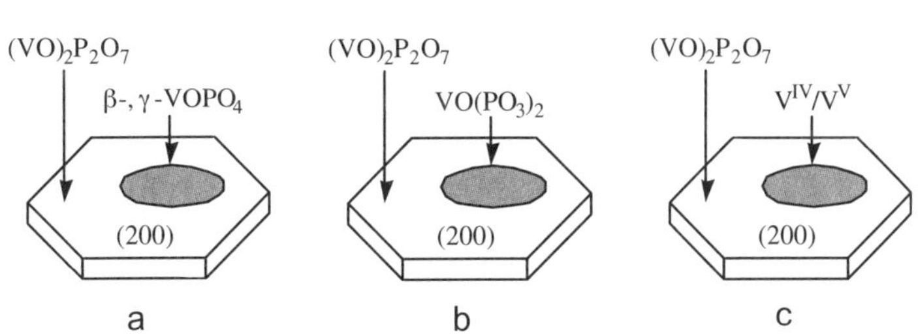

Figure 6.2 Models of active surface of the VPO catalysts: (a) Volta *et al.*,[36,37] (b) Yamazoe *et al.*[30] and (c) Trifirò *et al.*[9]

Yamazoe *et al.* reported $VO(H_2PO_4)_2$ as the precursor of the active and selective phase in *n*-butane oxidation.[30] This precursor transformed to an amorphous $VO(PO_3)_2$ catalyst that was much less active but just as selective as the $(VO)_2P_2O_7$ catalysts as shown in Figure 6.2(b). Volta *et al.*[41] found traces of $VO(PO_3)_2$ in the VPO catalysts, by ^{31}P NMR spin-echo mapping. The authors concluded that the V^{5+} species in strong interaction with the $VO(PO_3)_2$ structure is a controlling factor in the catalytic performance in *n*-butane oxidation to maleic anhydride. However, low activity and high selectivity of $VO(PO_3)_2$-containing catalysts in *n*-butane oxidation was later explained by the presence of either $(VO)_2P_2O_7$ impurity or $VOPO_4$ phases formed during solid-state disproportionation of $VO(PO_3)_2$ into V(III) and V(V) phases.[38]

Trifirò *et al.* attributed the activity of the VPO catalysts in *n*-butane oxidation to vanadyl pyrophosphate, whereas the selectivity to maleic anhydride was associated with the presence of a very limited and controlled amount of V^V sites as shown in Figure 6.2(c).[9] TEM analysis of spent catalysts, as well as the ratio of intensities of the interlayer (200) to inplane (024) reflection in the XRD patterns of $(VO)_2P_2O_7$ catalysts showed a high exposure of the (200) planes in the $(VO)_2P_2O_7$ crystallites. The authors suggested that the active surface is obtained by truncation of the $(VO)_2P_2O_7$ crystals along the (200) plane. Other studies of well-defined model VPO catalysts showed that the best catalysts contained only well-crystalline $(VO)_2P_2O_7$.[4] The active and selective surface sites for *n*-butane oxidation to maleic anhydride were associated with the vanadyl dimers present at the surface (200) planes of $(VO)_2P_2O_7$.[4] On the other hand, the side planes (021) and (001) of $(VO)_2P_2O_7$ have been proposed as *unselective* surfaces in *n*-butane oxidation.[41–43]

Four types of termination of the surface (200) planes in $(VO)_2P_2O_7$: A, B, C, and D have been suggested.[12] Type A termination where only vanadium species are present on the surface is the least probable,[13] since excess phosphate is always present in real catalysts. Such a situation would be characterized by high activity and low selectivity to maleic anhydride,[44] in agreement with the catalytic behavior of VPO catalysts with a $P/V < 1$. Ebner and Thompson proposed the termination of the surface with pendant $H_2P_2O_7^{2-}$ groups, which surround two vanadyl dimers (type B).[44] Such surface termination may be envisioned by replacing every phosphate tetrahedron facing up in the bulk (200) planes of $(VO)_2P_2O_7$ with pyrophosphate groups. In this type of surface termination, the pyrophosphate "barrier" isolates pairs of vanadyl dimers from other pairs, which was proposed to be beneficial for the selectivity to maleic anhydride during *n*-butane oxidation. In the surface (200) planes terminated with HPO_4^- groups (type C), every phosphate tetrahedron facing up in the bulk (200) planes of $(VO)_2P_2O_7$ would correspond to a surface HPO_4^- group. This situation is similar to that of V^VOPO_4, and the vanadium sites in the dimers are more easily accessible from the gas phase than in the type B termination, which is expected to be detrimental for the selectivity to maleic anhydride. In such a case the formation of discrete V^V species or V^VOPO_4 phases can occur.[43] The (200) surface termination with $VO(PO_3)_2$ (type D) occurs when considerable excess of phosphate is present,[41] which significantly lowers the catalytic activity in *n*-butane oxidation. This situation corresponds to the model of active surface proposed by Yamazoe *et al.*[30]

Steady and transient performance experiments of equilibrated VPO catalysts with different P/V ratios have been studied to gain information on the nature of the VPO active surface under reaction conditions.[45] For catalysts having P/V ratios slightly higher than 1, a layer made of vanadium oxide and P_2O_5/H_2O has been suggested as an active site in *n*-butane oxidation to maleic anhydride. Also, both adsorbed and surface lattice oxygen on vanadyl pyrophosphate have been postulated as active sites for the formation of maleic anhydride and CO_x from *n*-butane.[46]

6.4 Proposed Reaction Steps and Active Sites in *n*-Butane Oxidation

Table 6.1 shows the proposed reaction steps for the transformation of *n*-butane to maleic anhydride.[4] For kinetic reasons, none of the intermediates shown in Table 6.1 was detected under typical reaction conditions. The activation of secondary C–H bonds in *n*-butane is the first and rate-determining step. When compared to oxidation of 1-butene on the VPO catalysts, the rate of *n*-butane disappearance is ~20–60 times slower.[9,47] This results in a fast transformation of the intermediates to maleic anhydride before any desorption of the former may occur.

The following observations led to the proposed reaction steps shown in Table 6.1. 1-Butene, *cis/trans*-2-butenes, butadiene and furan have been detected in the oxidation of *n*-butane on the VPO catalysts under very unusual conditions, such as under oxygen deficiency at high *n*-butane concentration and very short contact times,[4] or in high vacuum in a temporal analysis of products (TAP) reactor.[48] Phthalic anhydride has been detected under

Table 6.1 Proposed reaction steps in *n*-butane oxidation to maleic anhydride and polyfunctional nature of VPO catalysts.

Proposed step	*Type of reaction*	*Probe molecule*	*Product*
n-butane → butenes	oxidative dehydrogenation	isobutyric acid cyclohexane	methacrylic acid benzene
butenes → butadiene	allylic oxidation	butene 2,5-dihydrofuran	butadiene furan
butadiene → 2,5-dihydrofuran	1,4-oxygen insertion	benzene naphthalene	maleic anhydride naphthaquinone
2,5-dihydrofuran → furan	allylic oxidation	-------------------	-------------------
furan → γ-butyrolactone	electrophilic insertion	-------------------	-------------------
γ-butyrolactone → maleic anhydride	electrophilic oxygen insertion	methacrolein *o*-xylene	methacrylic acid phthalic anhydride

hydrocarbon-rich conditions (10% mol *n*-butane, 17% O_2).[49] All the intermediates in Table 6.1 (except 2,5-dihydrofuran and γ-butyrolactone) have been observed in the oxidation of butenes and butadiene to maleic anhydride on the VPO catalysts.[9] 2,5-dihydrofuran oxidation to furan occurs over VPO catalysts at much lower temperatures $\sim$533 K with very high selectivity $\sim$90 mol.%, which is probably why 2,5-dihydrofuran has not been observed experimentally during *n*-butane oxidation over the VPO catalysts.[9]

The unique features of the VPO catalysts in carrying out the reaction steps shown in Table 6.1 are (a) the ability to selectively activate *n*-butane during the rate-determining step, (b) fast oxidation of chemisorbed intermediates to maleic anhydride with high selectivity, and (c) the lack of desorption of any intermediate contributing to high selectivity to maleic anhydride. Due to the high rate of oxidation of intermediates on VPO catalysts, their surface concentration is low, which makes it difficult to detect these intermediates by spectroscopic techniques.[9]

The reaction pathway in Table 6.1 shows the single steps and the polyfunctional nature of active sites that are required for the oxidation of *n*-butane to maleic anhydride.[50–53] The polyfunctional nature of the VPO catalysts has been demonstrated by the oxidation of probe molecules in accordance with the steps proposed for *n*-butane oxidation to maleic anhydride.[12,54–56] The high yield of maleic anhydride observed in the oxidation of these molecules provided support for the reaction steps shown in Table 6.1.

Chen and Munson have challenged the proposed reaction pathway shown in Table 6.1.[57,58] They employed a series of fresh VPO catalysts including one $(VO)_2P_2O_7$ and a number of overoxidized amorphous VPO phases possessing bulk P/V ratios of 0.9–1.2, average V oxidation states in the 3.92–4.95 range and ^{13}C-labeled *n*-butane reactant in the absence of dioxygen under non-conventional pseudoflow reaction conditions. They observed that the label in the butadiene, produced from 1,4-^{13}C-*n*-butane, was completely scrambled, but in maleic acid, also produced from 1,4-^{13}C-*n*-butane, it was largely unscrambled. They concluded that butadiene was unlikely to be the primary reaction intermediate for the conversion of *n*-butane to maleic anhydride, because of the observed discrepancy in the amount of label scrambling between maleic acid and butadiene. However, the small amount of label scrambling in maleic acid suggested that both reaction pathways may occur simultaneously to produce maleic acid. The authors observed the formation of much more isotopically labeled ethylene from fully ^{13}C-labeled butane as from [1,4-^{13}C] butane, indicating that ethylene was produced mainly from the two methylene carbons of *n*-butane, while the abstraction and oxidation of the two methyl groups of *n*-butane resulted in the total oxidation of *n*-butane. Further studies employing well-characterized equilibrated VPO catalysts under realistic reaction conditions are needed to support the mechanism proposed by Chen and Munson.

The best VPO catalysts preferentially expose the (200) planes of $(VO)_2P_2O_7$.[4,9–12] Therefore, the models of *n*-butane oxidation on the VPO catalysts proposed to date are based on the hypothetical active sites present on the surface (200) plane. The following hypothetical active sites have been proposed to exist on the surface (200) plane:[12] (a) Brønsted-acid sites, probably -POH groups; (b) Lewis-acid sites, probably V^{IV} and V^{V}; (c) one-electron redox couple, V^{V}/V^{IV}, V^{IV}/V^{III}; (d) two-electron redox couple, V^{V}/V^{III}; (e) bridging oxygen, V–O–V, V–O–P, or VO(P)V; (f) terminal oxygen, $V^{V=O}$, $V^{IV=O}$; and (g) activated molecular oxygen, η^1-peroxo and η^2-superoxo species. The roles of these species according to the proposed mechanisms of *n*-butane oxidation have been reviewed elsewhere.[59]

6.5 Promoted VPO

The VPO catalysts employed industrially are usually promoted with metal compounds to improve their catalytic performance.[4] Hutchings *et al.* have reviewed the nature, the location and the roles of metal promoters in the VPO system.[60,61] However, there is a lack of agreement in the literature concerning the functions of promoters in controlling the morphology, phase and elemental bulk and surface composition, structures and redox properties of VPO catalysts, which define their catalytic performance. Several approaches have been used in the past to incorporate promoters in the VPO lattice.[61–95] These methods include (a) the incorporation of promoters during the $VOHPO_4 \cdot 0.5H_2O$ precursor synthesis, (b) the impregnation of the $VOHPO_4 \cdot 0.5H_2O$ precursor with promoters prior to its thermal activation and (c) promoter impregnation into VPO catalysts. The promoters can be introduced in the VPO host as metals, salts, oxides and phosphates. Metal phosphates incorporated as ortho- and pyrophosphates have a particularly beneficial effect on catalytic performance.[74] The promoters are typically introduced into VPO catalysts in the 0.25–5 wt.% range.[61] Since the typical surface areas of organic VPO catalysts are in the 20–30 m^2/g range, these levels of promoters correspond to the surface coverage of ~0.1 to 2 theoretical monolayers.[60] At loadings below a theoretical monolayer (~2–3 wt.%), a promoting element may form a two-dimensional surface oxide overlayer and partially dissolve in the VPO lattice. The latter is typically observed when soluble promoter salts are introduced directly during the $VOHPO_4 \cdot 0.5H_2O$ synthesis. At higher loadings (>3 wt.%), formation of promoter-containing bulk phases, such as oxides or phosphates, may also occur. We briefly discuss here the findings of recent studies of promoted VPO catalysts with respect to promoter locations and functions in *n*-butane oxidation to maleic anhydride. Table 6.2 summarizes some key studies on the preparation methods and catalytic behavior of promoted VPO catalysts in *n*-butane oxidation to maleic anhydride.

Zazhigalov *et al.* modified the redox properties of the VPO catalysts by incorporating metallic Co in the VPO lattice, which improved desorption of maleic anhydride and enhanced the selectivity to maleic anhydride.[63] Volta *et al.* promoted their VPO catalysts with metallic Co and Fe and studied their catalytic behavior in *n*-butane oxidation.[64–66] They found that both promoters improved the selectivity to maleic anhydride. However, the presence of Fe

Table 6.2 Selected promoted-VPO catalysts in *n*-butane oxidation to maleic anhydride.

Promoter loading	*Approach*	*Promoted catalytic performance*	*Unpromoted catalytic performance*	*Promoter location or function*	*Ref.*
Co 4% wt.	Impregnation of VHP with Co acetate or acac	S(MA) = 60% at C = 80%; 703 K	S(MA) = 30% at C = 80%; 703 K	Co phosphate phase	68,69
Co 1–5% at.	Co acac added to V_2O_5, *i*-ButOH and H_3PO_4 under reflux	S(MA) = 71.4% at C = 25%; 673 K (Co 1%)	S(MA) = 50.3% at C = 25%; 673 K	Stabilizes amorphous VPO phase	71
Co 1.2–4.3% molar	Co acac added to V_2O_5, *i*-ButOH and H_3PO_4 under reflux	S(MA) = 73% at C = 62%; 548 K	S(MA) = 11% at C = 62%; 548 K	Stabilizes phosphorus in the VPO lattice	89
Co 0.5–3% at.	Co acac added to V_2O_5, *i*-ButOH and H_3PO_4 under reflux	S(MA) = 78% at C = 45%; O_2/C_4 = 12 673 K	S(MA) = 72% at C = 40%; O_2/C_4 = 12 673 K	Stabilizes VPO phase	90
Co 1% at.	Co nitrate added to V_2O_5, *i*-ButOH and H_3PO_4 under reflux	S(MA) = 61% at C = 54%; 673 K	S(MA) = 61% at C = 30%; 673 K	Increase of oxygen species originated from active phase V^{4+}	91
Cr 1% at.	Cr nitrate added to V_2O_5, *i*-ButOH and H_3PO_4 under reflux	S(MA) = 60% at C = 43%; 673 K	S(MA) = 61% at C = 30%; 673 K	Increase of oxygen species originated from active phase V^{4+}	91
Mo (1–5%) Cr (0.5–3%) W (1.7–10.5%)	Impregnation and precipitation	C = 80% S(MA) = 84% Mo S(MA) = 70% Cr S(MA) = 40% W 631 K, 700 K, 675 K, respectively	C = 80% S(MA) = 50% 718 K	Increase of Lewis-acid sites and generation of V^{5+} close to V^{4+} center of vanadyl pyrophosphate	95
Fe 1–5% at.	Fe acac added to V_2O_5, *i*-ButOH and H_3PO_4 under reflux	S(MA) = 73.5% at C = 25%; 673 K (1% at. Fe)	S(MA) = 50.3% at C = 25%; 673 K	solid solution in $(VO)_2P_2O_7$	71
Fe 0.1–0.5 molar	Intercalation of Fe acac in vanadyl n-hexylphosphate	S(MA) = 65% at C = 15%; 723 K	Not reported	Fe dispersed in the bulk Enhance redox ability	92

Ce–Fe 5% wt.	Mixed oxide precipitated from nitrates and added to V_2O_5, *i*-ButOH, benzyl alcohol and H_3PO_4 under reflux	Y(MA) = 39–42%; 673 K	Y(MA) = 27%; 673 K	Most likely as surface Ce–Fe oxide	73
Bi–Fe 1% wt Bi/Fe:2	Reflux Bi and Fe nitrates with VHP	S(MA) = 61% at C = 72%; 673 K (Best catalytic performance)	S(MA) = 63% at C = 83%; 673 K	Enables reoxidation and increases oxygen mobility	93
	Addition of BiFe oxide to VHP				
	Mechanochemical				
Bi Bi/V = 0.1	Ball milling of $BiPO_4$ and organic VHP	S(MA) = 74% at C = 88%; 713 K	S(MA) = 67% at C = 73%; 713 K	Bulk $BiPO_4$	
Bi/V = 0.1	Ball milling of Bi_2O_3 and organic VHP	S(MA) = 71% C = 83%; 713 K		Surface $BiPO_4$	
Bi–Mo 1% mol and Mo 1% mol	Mo and Bi alkoxides grafted on preformed vanadium phosphate precursor	S(MA) = 87% at C = 40% (Bi–Mo) S(MA) = 90% at C = 40% (Mo) 663 K	S(MA) = 80% at C = 40% 663 K	Solid solution and Mo surface layers, respectively	94
Mo–Zr–Zn V/P/Zr/Mo/Zn = 1.0/1.2/0.03/0.03/0.03 (at.)	Zr nitrate, Zn acetate and ammonium molybdate added to V_2O_5, HCl and water. Citric acid then added as a complexing agent.	S(MA) = 88% at C = 46%; 673 K	Not reported	Cations occupying vacant surface and bulk sites in VPO phases	82
Nb,V Nb/V = 0.01 (at.)	Organic VHP added to Nb ethoxide in *i*-ButOH under reflux	Y(MA) = 52.5%; 673 K	Y(MA) = 41%; 673 K	Solid solution: $V_{1-x}Nb_xO_2P_2O_{7+x}$	85
Nb, Si, Ti, V Zr 0.25 wt.%	Incipient wetness impregnation of VHP with Nb, Si, Ti, V and Zr alkoxides	S(MA) = 58% at C = 15% for Nb-VPO catalyst; 653 K	S(MA) = 41% at C = 15%; 653 K	Surface enrichment by XPS; may partially form a solid solution	86

acac = acetylacetonate; VHP = $VOHPO_4 \cdot 0.5H_2O$
C = *n*-butane conversion; S(MA) = Maleic anhydride molar selectivity; Y(MA) = Maleic anhydride yield

resulted in increased *n*-butane conversion, while the Co addition led to decreased catalytic activity. They suggested that these promoters affected to a different extent the dispersion of various $VOPO_4$ and $(VO)_2P_2O_7$ phases during precursor activation, which was responsible for the observed behavior. In a different report, Volta *et al.* suggested that Co delayed the transformation of the VPO precursor leading to an enhancement of disorder in the inorganic lattice at lower temperature that stabilized the amorphous VPO phase and led to lower *n*-butane conversion, but higher selectivity to maleic anhydride as compared to the unpromoted VPO catalyst.[67]

Cornaglia *et al.* studied the effect of Co cations on the performance of VPO catalysts prepared by impregnation of $VOHPO_4 \cdot 0.5H_2O$ with cobalt acetate and acetyl acetonate.[68,69] They found that Co added at 4 wt.% significantly improved the yield of maleic anhydride. The superior catalytic performance of their catalysts was associated with an optimum concentration of very strong Lewis-acid sites, very low concentration of isolated V^V centers and the absence of V^V phases. Furthermore, the cobalt phosphate phase detected by XPS aided in stabilizing the excess surface phosphate. The selectivity to maleic anhydride was 60 mol.% for the promoted and 30% for the unpromoted catalyst at 80% *n*-butane conversion at 703 K. Other studies[62,70] suggested that the presence of Co in the VPO phases enhanced the surface enrichment in phosphate and modified the surface acidity, increasing the concentration of strong Lewis-acid sites. Also, it has been suggested that Co stabilizes phosphorous in the VPO lattice[89,90] and increases the amount of oxygen species originated from the active phase V^{4+},[91] resulting in highly selective catalysts to maleic anhydride. Recently, Cr has been employed as a VPO promoter, improving *n*-butane conversion.[91] The authors suggested that a higher amount of oxygen species associated with V^{4+} were responsible for the higher *n*-butane conversions. It has been suggested that Mo and Cr can generate V^{5+} sites close to the V^{4+} center of vanadyl pyrophosphate and increase the concentration of Lewis-acid sites, leading to highly selective catalyst to maleic anhydride.[95]

Hutchings *et al.* reported the effect of Co and Fe ions added during the $VOHPO_4 \cdot 0.5H_2O$ synthesis on *n*-butane oxidation to maleic anhydride.[71] At low levels (1–5 at.%), both Co and Fe significantly enhanced the selectivity and intrinsic activity in maleic anhydride formation expressed as moles of maleic anhydride/m^2 h. For instance, the selectivity to maleic anhydride at 25% *n*-butane conversion was $>$63 mol.% for the promoted phases and only 50 mol.% for the unpromoted VPO catalyst at 673 K. The authors suggested that Co was insoluble in the $(VO)_2P_2O_7$ phase. Although Co was found to form a solid solution in $VOHPO_4 \cdot 0.5H_2O$, it migrated to a disordered V^{4+}–V^{5+} phosphate phase during thermal activation of $VOHPO_4 \cdot 0.5H_2O$. Therefore, the authors proposed that Co present at the surface functions as a promoter for the disordered VPO phases indicating that these disordered phases may play an important role in selective oxidation of *n*-butane. On the other hand, Fe ions may be soluble in the $(VO)_2P_2O_7$ lattice and, therefore, function as an electronic promoter for this phase. It has been suggested that Fe incorporated in VPO can enhance redox activity[92] and increase oxygen mobility.[93] Satsuma *et al.*[72]

reported a novel approach for the introduction of promoters into VPO phases. Lamellar vanadyl benzylphosphate (LVBP) was used as a host material and an Fe complex, $Fe(acac)_3$, as a guest promoter compound. It was found that $Fe(acac)_3$ was successfully inserted into LVBP by heating it in a toluene solution of $Fe(acac)_3$. It was confirmed that $Fe(acac)_3$ was uniformly dispersed in the interlayer space of LVBP without significant disruption of the lamellar structure. However, *n*-butane oxidation over these lamellar VPO phases has not been investigated.

VPO catalysts promoted with a mixed Ce–Fe oxide (5 wt.%) were investigated for *n*-butane oxidation.[73] It was found that the promoted catalyst exhibited higher conversion and selectivity to maleic anhydride than the unpromoted catalyst during *n*-butane oxidation in the absence of gas-phase oxygen. The maleic anhydride yield for the promoted catalysts was much higher (39–42 mol.%) than for the unpromoted VPO catalyst (27 mol.%) at 673 K. The authors suggested that the introduction of Ce–Fe oxides improved the redox performance of VPO catalysts by increasing the lattice oxygen activity. The authors suggested that Ce leads to the formation of highly selective oxygen species, while Fe increases the reactivity of oxygen species.

Zazhigalov *et al.* studied the incorporation of alkali and alkaline-earth metal ions in the VPO lattice.[74] They found that Li, Na, K, Cs, Be, Mg, Ca and Ba cations present at different concentrations easily donated electrons to the VPO lattice with P/V ratios of 1.07 and 1.20, leading to increased negative charge on lattice oxygen atoms and enhancing the rate of *n*-butane oxidation. The presence of these promoters caused an increase of the surface P/V ratio and corresponding changes of the surface acidity. However, the preparation of a catalyst characterized by high activity in *n*-butane oxidation and high selectivity to maleic anhydride still requires fine tuning of the basicity of surface oxygen atoms to facilitate the activation of *n*-butane and the surface acidity to control the residence times of the reaction intermediates.

Bi is one of the most efficient promoters of VPO catalysts, which was used to improve the selectivity to maleic anhydride in *n*-butane oxidation.[61,75,76] Zazhigalov *et al.* found that the incorporation of Bi at $Bi/V = 0.1$ increased the surface phosphate concentration and enhanced the acidity and the strength of acid sites of the VPO catalysts that improving their catalytic performance in *n*-butane oxidation.[77,78] In particular, the selectivity to maleic anhydride improved with an increase in the Brønsted acidity. Haber *et al.* further studied the effect of Bi on the catalytic performance of organic VPO catalysts.[79] Most notably, Bi-promoted VPO catalysts with significantly enhanced selectivity to maleic anhydride were obtained by mechanochemical treatment of bulk $BiPO_4$ and Bi_2O_3 with $VOHPO_4 \cdot 0.5H_2O$. They observed that Bi_2O_3 formed a surface Bi phosphate phase in the VPO catalysts due to phosphate migration to the Bi_2O_3 phase and that $BiPO_4$ ball milled together with $VOHPO_4 \cdot 0.5H_2O$ was finely dispersed as a bulk phase in the $BiPO_4$-$(VO)_2P_2O_7$ catalyst. In addition, the mechanochemical treatment produced thinner $VOHPO_4 \cdot 0.5H_2O$ platelets and increased the surface area of promoted catalysts. The presence of the surface Bi phosphate phase in the Bi_2O_3-$(VO)_2P_2O_7$ catalyst was claimed to be responsible for the slightly lower selectivity to maleic anhydride displayed by this catalyst (Table 6.2). Lopez-Nieto *et al.* found that the incorporation of Bi in the VPO catalysts led to a modification in the surface properties that resulted in improved catalytic behavior.[80] They claimed that the incorporation of Bi stabilized the desirable $(VO)_2P_2O_7$ structure, led to an increase in the specific surface area and a decrease in the O 1s-electron binding energy, which enhanced the rate of *n*-butane oxidation. On the other hand, Taufiq-Yap *et al.* found that Bi incorporation into the $(VO)_2P_2O_7$ lattice resulted in the surface area increase as well as the reduction of the V oxidation state (from +4.24 to +4.08).[81] However, no catalytic performance for the Bi–VPO catalyst was reported. Bi–Fe[93] and B–Mo[94] promoted VPO catalysts have been suggested to act as electronic promoters in *n*-butane oxidation (Table 6.2).

Ji *et al.* promoted VPO catalysts by a mixture of Mo, Zr and Zn ions, finding this particular system to be highly selective for the partial oxidation of *n*-butane to maleic anhydride.[82] The combination of these three species created a synergistic effect that enhanced the performance of the promoted VPO catalysts. The selectivity to maleic anhydride of 88 mol.% at 46% *n*-butane conversion at 673 K was reported for the Mo–Zr–Zn–VPO catalysts. However, no comparison was made with unpromoted VPO catalysts in their study. The authors suggested that these cations were incorporated in the vacant surface sites as well as the bulk VPO lattice. Pt,[83] La–Bi[80] and Zr[80] were among other recently investigated VPO promoters. Pt was incorporated as H_2PtCl_6 into the layered $VOHPO_4 \cdot 0.5H_2O$ precursor during its synthesis in aqueous and organic media.[83] However, the catalytic performance of the Pt–VPO catalyst was demonstrated only for the hydrogenation of nitrobenzene and the oxidation of tetrahydrofuran. La–Bi and Zr promoters were introduced as nitrates during organic $VOHPO_4 \cdot 0.5H_2O$ precursor synthesis in butanol.[80] Although the catalytic behavior of the La–Bi and Zr-promoted catalysts in *n*-butane oxidation was not reported, these catalysts were active for the oxidation of ethane to ethylene.

Nb is another highly attractive promoter for the VPO catalysts. Volta *et al.* have demonstrated that Nb is an effective promoter of the activity of the VPO catalysts in *n*-butane oxidation to maleic anhydride.[84,85] Nb increases the number of defects at the surface of $(VO)_2P_2O_7$ platelets, *e.g.* O-vacancies, and facilitates the initial C–H activation of *n*-butane. These sites mainly correspond to coordinatively unsaturated cations acting as Lewis-acid sites of low acidity. Furthermore, Volta *et al.* proposed that the observed improvement in catalytic properties resulted from a redox effect of Nb on the V^{5+}/V^{4+} balance. They proposed that the incorporation of Nb into the VPO framework occurs *via* substitution of V^{4+} by Nb^{5+} in VO^{2+}, *i.e.* $(VO)_2P_2O_7 + Nb^{5+} = V_{1-x}Nb_xO_2P_2O_{7+x}$. The observed yield of maleic anhydride was 52.5 mol.% for the Nb-promoted VPO catalyst and only 41 mol.% for the unpromoted catalyst at 673 K. Guliants *et al.* introduced the Nb, Si, Ti, V and Zr promoters at 0.14–0.26 theoretical monolayer surface coverage into model organic $(VO)_2P_2O_7$ catalysts possessing thin platelet crystal morphology that preferentially exposed the surface (200) planes.[86] They found that the promoted $(VO)_2P_2O_7$ catalysts containing surface niobia species were highly active and selective catalysts in *n*-butane oxidation to maleic anhydride. The Nb-promoted $(VO)_2P_2O_7$ catalyst containing only 0.26 monolayer NbO_x surface coverage displayed 58 mol.% selectivity to maleic anhydride at 15% *n*-butane conversion, while the unpromoted catalyst at the same conversion was characterized by 41 mol.% selectivity at 653 K. The authors proposed that the surface acidity (*i.e.* Nb species and -POH groups at superstoichiometric P/V ratios) were mainly responsible for the improved catalytic performance of the Nb-promoted $(VO)_2P_2O_7$ catalyst. They further suggested that the promoter elements may partially form a solid solution.

Hutchings *et al.* reported *n*-butane oxidation over $(VO)_2P_2O_7$ catalysts promoted with group 13 elements, B, Al, Ga, In.[87] These promoters were found to change the morphology of the $VOHPO_4 \cdot 0.5H_2O$ precursor from the platelet-like to rosette-like. Although all of these promoters functioned as modifiers of the VPO crystal morphology, only In- and Ga-promoted catalysts displayed improved catalytic performance in *n*-butane oxidation to maleic anhydride. However, the catalytic role of these elements is currently poorly understood. Aluminum-phosphate-promoted VPO catalysts displayed improved catalytic performance during *n*-butane oxidation and shortened equilibration time as compared to conventional VPO catalysts.[88] Aluminum phosphate was incorporated during the synthesis of the $VOHPO_4 \cdot 0.5H_2O$ precursor. The morphology of the $VOHPO_4 \cdot 0.5H_2O$ precursor was influenced by the presence of the aluminum phosphate. The authors found that both $(VO)_2P_2O_7$ and the original amorphous aluminum phosphate crystallized during the equilibration of the promoted $VOHPO_4 \cdot 0.5H_2O$ precursor. However, no mixed V–Al phosphate phase was observed. Moreover, the reasons for the improvement in the catalytic properties of the promoted system were not discussed.

The results summarized in Table 6.2 and briefly discussed in this section demonstrate the beneficial effects of promoting the VPO system for the partial oxidation of *n*-butane to maleic anhydride. However, the specific roles of promoters in modifying the morphology, phase and elemental bulk and surface compositions, structures and redox properties of the VPO catalysts at present are poorly understood. Improved fundamental understanding of the VPO promoter effects will enable rational design of VPO catalysts with enhanced catalytic performance in *n*-butane oxidation to maleic anhydride. Therefore, detailed studies of several classes of well-defined promoted VPO catalysts containing promoters (a) in solid solution with the VPO lattice, (b) as surface species and (c) nanosized oxides or phosphates, *etc.*, are expected to provide critical fundamental insights into the specific roles of key promoter species in selective oxidation of *n*-butane.

6.6 Alternative Synthesis Strategies to VPO Catalysts

Conventional synthesis methods offer limited control over desirable phase and elemental (bulk and surface) compositions, preferential exposure of active and selective surface planes, surface areas and pore structures of VPO catalysts, which define their catalytic performance in selective oxidation of *n*-butane.

For instance, conventional methods produce VPO catalysts with relatively low surface areas $<20\,m^2/g$ and a limited choice of crystal morphologies. Reported meso- and macroscale self-assembly approaches conducted, respectively, in the presence of surfactant mesophases[96–98] and colloidal sphere arrays[99] are highly promising for the molecular engineering of novel catalytic mixed-metal oxides. These novel methods offer the possibility to control surface and bulk chemistry (V oxidation state and P/V ratios), wall nature (amorphous or nanocrystalline), morphology, pore structures and surface areas of mixed-metal oxides. Furthermore, these novel catalysts represent well-defined model systems that are expected to lead to new insights into the nature of the active and selective surface sites and the mechanism of *n*-butane oxidation. In this section, we describe several promising synthesis approaches to VPO catalysts, such as the self-assembly of mesostructured VPO phases, the synthesis of macroporous VPO phases, intercalation and pillaring of layered VPO phases, other methods, and supported VPO.

6.6.1 Mesostructured VPO Phases

Figure 6.3 illustrates the mesoscale self-assembly pathway for mesostructured metal oxides possessing a 2D-hexagonal pore structure. This general route was initially proposed for mesostructured silicate phases. The formation of highly ordered mesostructures containing cylindrical mesopores takes place *via* charge-matching interactions between the inorganic species at the interface with a pre-existing hexagonal lyotropic liquid-crystal phase. Then, the template removal leads to highly ordered mesoporous structures.[96–98]

Mesostructuring of VPO phases offers the possibility to design novel VPO catalysts with high surface areas, controlled porosities, tunable surface compositions and improved catalytic performance. Few reports exist on the synthesis of mesostructured VPO phases. Iwamoto *et al.* reported the synthesis of mesostructured hexagonal vanadium–phosphorus oxide phases using alkyltrimethyl ammonium surfactants.[100] However, the mesostructure in these phases was lost upon calcination. Doi and Miyake reported the synthesis of a hexagonal mesostructured VPO compound from $VOHPO_4 \cdot 0.5H_2O$ by surfactant

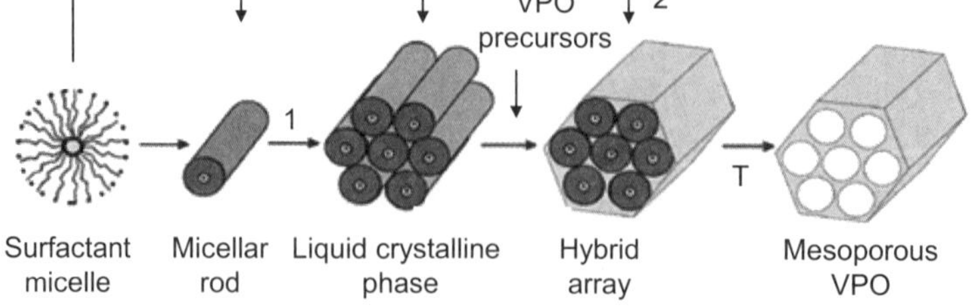

Figure 6.3 Formation of 2D-hexagonal *mesoporous VPO* by liquid-crystal templating mechanism.

intercalation and subsequent hydrothermal treatment.[101] However, these materials suffered from poor thermal stability and low phosphorus content that are detrimental for their potential applications in *n*-butane oxidation. Amoros *et al.* described the synthesis of hexagonal mesostructured oxovanadium phosphates.[102] However, the surfactant removal from their vanadium phosphates resulted in the collapse of the mesostructure. Mizuno *et al.* described the synthesis of hexagonal, cubic and lamellar mesostructured vanadium–phosphorus oxides that lost the structural order upon thermal treatment.[103] Datta *et al.* synthesized mesostructured lamellar VPO phase employing hexadecylamine as a template. However, no attempts were made to remove the surfactant.[104]

Carreon and Guliants reported hexagonal, cubic and lamellar VPO phases, which displayed improved thermal stability, desirable chemistries (*i.e.* the P/V ratios and vanadium oxidation states), and pore structures for the partial oxidation of *n*-butane.[105,106] These VPO phases displayed the selectivities to maleic anhydride up to ~40 mol.% at 673K at ~10% *n*-butane conversion.[107] A conventional organic VPO catalyst containing well-crystallized vanadyl(IV) pyrophosphate, the proposed active and selective phase for *n*-butane oxidation to maleic anhydride, displayed the selectivities to maleic anhydride 50 mol.% under the same reaction conditions. The low yields observed for mesoporous VPO catalysts confirmed the critical role of the vanadyl pyrophosphate phase $(VO)_2P_2O_7$ in catalyzing the oxidation of *n*-butane to maleic anhydride. Therefore, the amorphous nature of the mesoporous VPO samples may be responsible for the observed poor maleic anhydride yields. On the other hand, the mesoporous VPO phases were thermally unstable under *n*-butane oxidation conditions. The BET surface area measurements of the fresh and used catalysts indicated the collapse of the mesoporous structures with concomitant decrease of the surface area from ~190 to 15 m^2/g. Low *n*-butane conversions observed at even short reaction times (<1 h) suggested that the mesoporous structure collapsed very quickly under reaction conditions. Furthermore, the XRD diffraction pattern of the used catalyst showed no reflections at low 2ϑ angles confirming the mesostructure collapse. Only dense VPO phases such as γ-$VOPO_4$ were observed. The incomplete crosslinking of the mesoporous VPO framework, its low Tamman temperature and redox properties are mainly responsible for its limited thermal stability.

6.6.2 Macroporous VPO Phases

Figure 6.4 shows the macroscale templating of bulk mixed-metal oxide phases in the presence of colloidal sphere arrays consisting of three steps. First, the interstitial voids of the monodisperse sphere arrays are filled with metal oxide precursors. In the second step, the precursors condense and form a solid framework around the spheres. Finally, the spheres are removed by either calcination or solvent extraction leading to the formation of 3D ordered macroporous structures.[99]

Macroporous VPO phases with tunable phase compositions, pore structures and morphologies have been synthesized by employing monodisperse

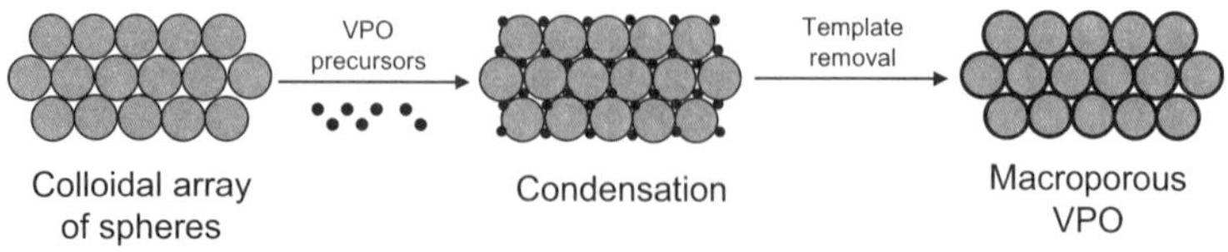

Figure 6.4 Synthesis of ordered *macroporous VPO* phases.

polystyrene sphere arrays as templates.[108,109] Colloidal polystyrene spheres were ordered into close-packed arrays by sedimentation and centrifugation. Depending on the choice of VPO sources and template removal method, various crystalline VPO phases were obtained, *e.g.* $(VO)_2P_2O_7$, γ-$VOPO_4$,-$VOHPO_4 \cdot 0.5H_2O$. The macroscale-templated synthesis produced VPO phases with unprecedented high surface areas $\sim$75 m^2/g, desirable macroporous structures, optimal bulk compositions (P/V$\sim$1.1), desirable vanadium oxidation states (4.1–4.4) and preferential exposure of the surface (200) planes of $VO_2P_2O_7$, the proposed active and selective phase for the partial oxidation of *n*-butane. The catalytic performance of macroporous VPO catalysts was evaluated in the partial oxidation of *n*-butane to maleic anhydride.[110] The observed yield of maleic anhydride was >50 mol.% for the macroporous VPO catalyst that displayed ordered arrays of 380 nm pores and walls composed of 20-nm crystallites of $(VO)_2P_2O_7$. Under similar reaction conditions, the yield of maleic anhydride was $\sim$40 mol.% for a conventional organic VPO catalyst. The ordered open pore structures composed of nanocrystalline $(VO)_2P_2O_7$ walls and high surface areas (>40 m^2/g) in macroporous VPO catalysts enhanced the access of reactant molecules to the active and selective surface that led to improved catalytic performance. These ordered macroporous VPO phases are highly promising for potential applications in selective catalytic oxidation, representing model catalytic systems that will enable improved fundamental understanding of the relationships between the molecular structure and catalytic properties of a broad range of industrially relevant catalytic systems.

6.6.3 Intercalation and Pillaring of Layered VPO Phases

Guliants *et al.* reported new precursors to vanadyl(IV) pyrophosphate catalysts for *n*-butane oxidation based on the structural modification of the layered $VOHPO_4 \cdot 0.5H_2O$ precursor.[111–114] The $VOHPO_4 \cdot 0.5H_2O$ precursor was intercalated with *n*-alkylamines in neat amine that resulted in a significant expansion of the interlayer spacing as well as disordered stacking of the vanadyl(IV) hydrogen phosphate layers. However, this structural disorder in the intercalated precursor phases led to an improved catalytic performance of the VPO catalysts in *n*-butane oxidation after their thermal transformation.[111–114] Vanadyl(IV) phosphite and alkylphosphonates containing vanadyl(IV) dimers in their layered structures transformed into vanadyl(IV) pyrophosphate catalysts with higher surface areas (35 m^2/g) and higher selectivity to maleic anhydride than conventional organic VPO catalysts.[112,113] Furthermore, these new VPO precursor phases transformed into vanadyl(IV) pyrophosphate catalysts at lower temperatures than the conventional organic precursor (*i.e.* 550–650 K). Thermally stable pillared VPO catalyst with 17.0 Å interlayer spacing was also reported, which was obtained by the intercalation of mixed vanadyl(IV) phosphite–phosphate phases with aminopropyltriethoxysilane (APS).[114] These intercalated phases were synthesized by contacting the hemihydrate VPO precursor with a two-fold excess of 0.5 M APS in anhydrous ethanol at room temperature for 48 h. The pillared phase was obtained by calcining the APS-intercalated phase in nitrogen at 673 K. The catalytic performance of the pillared VPO catalyst was reported for *n*-butane oxidation to maleic anhydride at 673 K.[114] The steady-state selectivity to maleic anhydride of 25 mol.% at 45% *n*-butane conversion was achieved after 6 days in stream. The low selectivity to maleic anhydride observed for the pillared VPO catalyst confirmed that the presence of $(VO)_2P_2O_7$ is critical for selective *n*-butane oxidation to maleic anhydride.

6.6.4 Other Synthesis Methods of Dense VPO

Other novel synthesis approaches to VPO phases have been reported. Hutchings *et al.* have reported a new synthesis route for the VPO catalysts using supercritical CO_2 as an antisolvent.[115,116] CO_2 can be liquid at <42 bar at 293 K or a supercritical fluid ($T_c = 304$ K, $P_c = 72$ bar). This route produced amorphous microspheroidal VPO phases, which were more active (intrinsic activity expressed as moles of maleic anhydride produced per 1 m^2 of catalyst per h) than crystalline $(VO)_2P_2O_7$ catalyst in *n*-butane oxidation to maleic anhydride. Furthermore, these novel phases did not require an extensive activation period to achieve full catalytic activity. These observations, however, contradict a number of published studies that reported that the presence of crystalline vanadyl(IV) pyrophosphate was critical for superior activity and selectivity in *n*-butane oxidation.

VPO catalysts consisting of dominant $(VO)_2P_2O_7$ and $VOPO_4$ impurity phases were prepared using ultrasound- and microwave-irradiation methods.[117] The ultrasound method produced a catalyst in a relatively short time that had similar composition and surface area to those prepared by conventional organic method. In this method, V_2O_5 and H_3PO_4 were mixed with 2-butanol and benzyl alcohol immersed in a water bath and the mixture was subjected to ultrasound irradiation. On the other hand, the VPO catalyst obtained in a microwave synthesis (same precursors and solvents as in the ultrasound method) showed different surface morphology as compared to the products of conventional synthesis. However, the catalytic performance of these novel VPO phases in *n*-butane oxidation was not reported.

Relatively high surface area VPO catalysts have been prepared in the presence of polyethylene glycols (PEG).[118] The resultant catalysts possessed the specific surface areas of 41–54 m^2/g, with an average particle size of 100–200 nm and crystalline $(VO)_2P_2O_7$ phase, displaying improved catalytic performance for the partial oxidation of *n*-butane to maleic anhydride. For instance, the yields of maleic anhydride >65 mol.% were observed for the VPO catalysts synthesized in the presence of PEG, while the VPO catalysts obtained in the absence of PEG displayed maleic anhydride yields <55 mol.% under similar reaction conditions. Most likely the PEG acts as a pore former increasing the specific surface area of the resultant VPO phases. In the same study[118] high surface area VPO catalysts were prepared by a ball-milling mechanochemical route. The ball-milling process not only increased the surface area of the VPO catalysts from 15 to 38 m^2/g, but also changed the local environment and the redox behavior of the active surface sites, which resulted in enhanced performance. The ball-milled catalysts displayed the selectivity to maleic anhydride of 59 mol.% at 93% *n*-butane conversion at 698 K. Similar selectivity was observed for conventional organic VPO catalysts at somewhat lower *n*-butane conversion (83%). The authors suggested that the ball-milling treatment led to a higher reactivity of the lattice oxygen associated with the V^{5+} and V^{4+} species. The enhancement of catalytic performance was attributed in this study to the increased exposure of the (200) planes of $(VO)_2P_2O_7$ as a direct result of ball milling.

6.6.5 Supported VPO

In principle, supported VPO catalysts offer several advantages over unsupported catalyst, such as higher mechanical strength, larger surface area to volume ratio of the active species, and better heat transfer properties. Several high surface area oxide supports have been used in the past for VPO catalysts in *n*-butane oxidation, including Al_2O_3,[119] SiO_2,[120] ZrO_2,[121] TiO_2,[122] and periodic ordered mesoporous silicas (SBA-15,[123,124] MCM-41,[123,125]). Selected supported VPO catalysts for *n*-butane oxidation are briefly discussed here. Taufiq-Yap *et al.* reported VPO supported catalysts on γ-Al_2O_3 displaying maleic anhydride selectivity as high as 39% at *n*-butane conversion of 34% (673 K).[119] The catalytic performance was highly dependent on the formed VPO crystalline phase. Samples calcined for 6 and 10 h produced the most active and selective supported catalysts consisting of pure $(VO)_2P_2O_7$. Larger calcination times (>30 h) led to β-$VOPO_4$, which was detrimental for catalytic performance. Au *et al.* studied fumed SiO_2 supported VPO for the partial oxidation of *n*-butane.[120] The authors found that at 653 K these catalysts showed *n*-butane conversions of 33–40% and maleic anhydride selectivity of 65–87%. The highly uniform dispersion of VPO in the support and the presence of $(VO)_2P_2O_7$ as the predominant phase was suggested as the key parameters for the good catalytic performance of the VPO-fumed SiO_2 catalysts. TiO_2-supported VPO catalysts have been employed in *n*-butane oxidation displaying low yields in the 10–18% range at 493–648 K.[122] The amorphous nature of the precipitated VPO phase and the strong VPO–support interaction were responsible for the poor catalytic performance observed in these catalysts.

Recently, Au *et al.* studied VPO supported phosphated ZrO_2 catalysts for *n*-butane oxidation.[121] These catalysts displayed maleic anhydride selectivity as high as 68.8% at *n*-butane conversion of 89.1% at 673 K (~61% maleic anhydride yield), the best catalytic performance achieved so far among supported VPOs. The authors attributed the superior catalytic performance of these phosphated ZrO_2-supported VPO catalysts to several factors including the presence of crystalline $(VO)_2P_2O_7$ and small fraction of vanadyl orthophosphate species on amorphous ZrO_2, a high coverage of the VPO phase, and a unique VPO–support synergetic interaction. Periodic ordered mesoporous MCM-41 and SBA-15 have been employed as supports for VPO phases.[123–125] Maleic anhydride yields as high as ~47% and ~51% at 673 K have been reported for MCM-41[123] and SBA-15[124]-supported VPO, respectively. The high surface area, high thermal stability, high capacity for loading active species, and low limitation for internal mass transfer make MCM-41 and SBA-15 attractive supports for VPO phases.

6.7 Conclusions

More than four decades have passed since Bergman and Frisch disclosed that VPO catalysts can selectively oxidized *n*-butane to maleic anhydride. The selective oxidation of *n*-butane to maleic anhydride is still the only industrial process of a selective vapor-phase oxidation of a light alkane. The VPO system has been intensively studied over the last two decades and significant fundamental insights into the mechanism of *n*-butane oxidation to maleic anhydride have been obtained. Crystalline $(VO)_2P_2O_7$ displaying preferential exposure of the surface (200) planes and a slight excess of phosphate ($P/V = 1.01–1.10$) is considered as the most active and selective phase for *n*-butane oxidation to maleic anhydride. The catalytic performance of VPO catalysts in this reaction can be improved by the addition of promoters. However, there is still a lack of agreement in the literature concerning specific roles of these promoters in controlling the morphology, phase and elemental bulk and surface compositions, structures and redox properties of VPO catalysts, which define their catalytic performance. Several novel synthesis approaches have recently shown promise for the molecular engineering of new well-defined VPO catalysts that will enable improved fundamental understanding of the relationships between the molecular structure and catalytic properties of a broad range of industrially relevant catalytic systems. Finely tuned pore sizes, metal-oxide wall structures and compositions, high surface areas and enhanced accessibility of the active and selective surface sites in these novel phases are highly attractive for improving the catalytic performance in oxidation of light alkanes. Future studies are expected to focus on the use of these model catalysts to unravel remaining mysteries about the nature of the active and selective VPO phase and surface sites, the mechanism of *n*-butane oxidation and the role of promoters. These insights are critical for the development of molecular surface structure–activity and –selectivity relationships and for the establishment of rational rules of improved VPO catalysts for the selective oxidation of *n*-butane to maleic anhydride.

References

1. D. J. Hucknall, *Selective Oxidation of Hydrocarbons*, Academic Press, London, 1974.
2. S. T. Oyama, A. N. Desikan and J. W. Hightower, *Catalytic Selective Oxidation*, ACS Symposium Series 523, American Chemical Society, Washington, DC, 1993.
3. J. Haber, *Stud. Surf. Sci. Catal.*, 1997, **110**, 1.
4. G. Centi, F. Cavani and F. Trifiro, *Selective Oxidation by Heterogeneous Catalysis. Fundamental and Applied Catalysis*, Kluwer Academic/Plenum Publishers, New York, 2001.
5. F. Cavani and F. Trifirò, *Appl. Catal.*, 1992, **88**, 115.
6. *Encyclopedia of Chemical Technology*, J. I. Kroschwitz, M. Howe-Grant, ed., Wiley, New York, 1991.
7. R. I. Bergman and N. W. Frisch, US Patent 1966, 3, 293, 268.
8. S. Irvin-Monshaw and A. Klein, *Chem. Eng.*, 1989, **96**, 35.
9. G. Centi, F. Trifirò, J. R. Ebner and V. M. Franchetti, *Chem. Rev.*, 1988, **88**, 55, and refs. therein.
10. P. Ruiz and B. Delmon, *Catal. Today*, 1987, **1**, 2, and refs. therein.
11. G. Centi, *Catal. Today*, 1993, **16**, 1, and refs. therein.
12. F. Cavani and F. Trifirò, *Catalysis*, 1994, **11**, 246, , and refs. therein.
13. J. R. Ebner and M. R. Thompson, In *Structure-Activity and Selectivity Relationships in Heterogeneous Catalysis*, R. K. Grasselli, A. W. Sleight, ed., Elsevier, Amsterdam, 1991, 31.
14. R. M. Contractor, J. R. Ebner and M. J. Mummy, In *New Developments in Selective Oxidation*, G. Centi, F. Trifirò, ed., Elsevier, Amsterdam, 1990, 553.
15. H. S. Horowitz, C. M. Blackstone, A. W. Sleight and G. Teufer, *Appl. Catal.*, 1988, **38**, 193.
16. I. Matsuura, *Catal. Today*, 1993, **16**, 61.
17. T. Okuhara and M. Misono, *Catal. Today*, 1993, **16**, 61.
18. L. M. Cornaglia, C. A. Sanchez and E. A. Lombardo, *Appl. Catal.*, 1993, **95**, 117.
19. E. Bordes, *Catal. Today*, 1993, **16**, 27.
20. D. Ye, A. Satsuma, A. Hattori, T. Hattori and Y. Murakami, *Catal. Today*, 1993, **16**, 113.
21. T. Shimoda, T. Okuhara and M. Misono, *Bull. Chem. Soc. Jpn.*, 1985, **58**, 2163.
22. L. M. Cornaglia, C. Caspani and E. A. Lombardo, *Appl. Catal.*, 1991, **74**, 15.
23. V. V. Guliants, S. A. Holmes, J. B. Benziger, P. Heaney, D. Yates and I. E. Wachs, *J. Mol. Catal. A*, 2001, **172**, 265.
24. E. Bordes, *Catal. Today*, 1987, **1**, 499, and refs. therein.
25. Z.-Y. Xue and G. L. Schrader, *J. Phys. Chem. B*, 1999, **103**, 9459.
26. C. C. Torardi, Z. G. Li, H. S. Horowitz, W. Liang and M.-H. Whangbo, *J. Solid State Chem.*, 1995, **119**(2), 349.

27. S. Albonetti, F. Cavani, F. Trifiro, P. Venturoli, G. Calestani, M. Granados and J. L. G. Lopez; Fierro, *J. Catal.*, 1996, **160**(1), 52.
28. H. Imai, Y. Kamiya and T. Okuhara, *J. Catal.*, 2008, **255**, 213.
29. Y. H. Taufiq-Yap and C. S. Saw, *Catal. Today*, 2008, **131**, 285.
30. H. Morishige, J. Tamaki, N. Miura and N. Yamazoe, *Chem. Lett.*, 1990, 1513.
31. N. Yamazoe, H. Morishige, J. Tamaki and N. Miura, In *New Frontiers in Catalysis*, L. Guczi *et al.*, ed., Elsevier, Amsterdam, 1993, 1979.
32. H. Igarashi, K. Tsuji, T. Okuhara and M. Misono, *J. Phys. Chem.*, 1993, **97**, 7065.
33. M. O'Connor, F. Dason and B. K. Hodnett, *Appl. Catal.*, 1990, **64**, 161.
34. M. Guilhoume, M. Roullet, G. Pajonk, B. Grzybowska and J. C. Volta, in *New Developments in Selective Oxidation by Heterogeneous Catalysis*, P. Ruiz, B. Delmon, ed., Elsevier, Amsterdam, 1992, 255.
35. F. Cavani, G. Centi and F. Trifirò, *La Chimica & L'Industria, Milan*, 1992, **74**, 182.
36. G. Bergeret, M. David, J. P. Broyer, J. C. Volta and G. Hecquet, *Catal. Today*, 1987, **1**, 37, and refs. therein.
37. N. Harrouch-Batis, H. Batis, A. Ghorbel, J. C. Vedrine and J. C. Volta, *J. Catal.*, 1991, **128**, 248.
38. V. V. Guliants, J. B. Benziger, S. Sundaresan, I. E. Wachs, J. M. Jehng and J. E. Roberts, *Catal. Today*, 1996, **28**, 275.
39. G. J. Hutchings, C. J. Kiely, M. T. Sananes-Schulz, A. Burrows and J. C. Volta, *Catal. Today*, 1998, **40**(2–3), 273.
40. M. T. Sananes-Schulz, A. Tuel, G. J. Hutchings and J. C. Volta, *J. Catal.*, 1997, **166**(2), 388.
41. J. C. Volta, K. Bere, Y. J. Zhang and R. Olier, In *Catalytic Selective Oxidation*, J. Hightower, T. Oyama, ed., ACS, Washington, DC, 1993, 217.
42. I. Matsuura, M. Yamazaki, In *New Developments in Selective Oxidation*, G. Centi, F. Trifirò, ed., Elsevier, Amsterdam, 1990, 563.
43. T. Okuhara, K. Inumaru and M. Misono, In *Catalytic Selective Oxidation*, J. Hightower, T. Oyama, ed., ACS, Washingon, DC, 1993, 156.
44. J. R. Ebner and M. R. Thompson, *Catal. Today*, 1993, **16**, 51.
45. N. Ballarini, F. Cavani, C. Cortelli, M. Ricotta, F. Rodeghiero, F. Trifiro, C. Fumagalli and G. Mazzoni, *Catal. Today*, 2006, **117**, 174.
46. J. Gascon, R. Valenciano, C. Tellez, J. Herguido and M. Menendez, *Chem. Eng. Sci.*, 2006, **61**, 6385.
47. M. Misono, K. Miyamoto, K. Tsuji, T. Goto, N. Mizuno and T. Okuhara, In *New Developments in Selective Oxidation by Heterogeneous Catalysis*, G. Centi, F. Trifirò, ed. Elsevier, Amsterdam, 1990, 605.
48. J. T. Gleaves, J. R. Ebner and T. C. Kuechler, *Catal. Rev. – Sci. Eng.*, 1988, **30**, 49.
49. N. Ballarini, F. Cavani, C. Cortelli, F. Gasparini, A. Mignani, F. Pierelli, F. Trifiro and C. Fumagalli, *Mazzoni, Catal. Today*, 2005, **99**, 115.
50. J. March, *Advanced Organic Chemistry*, Wiley, New York, 1992.

51. F. Cavani and F. Trifiro, *Appl. Catal. A*, 1997, **157**(1–2), 195.
52. S. O. Shekhtman, G. S. Yablonsky, S. Chen and J. T. Gleaves, *Chem. Eng. Sci.*, 1999, **54**(20), 4371.
53. Z.-Y. Xue and G. L. Schrader, *J. Catal.*, 1999, **184**(1), 87.
54. P. A. Agaskar, L. DeCaul and R. K. Grasselli, *Catal. Lett.*, 1994, **23**, 339.
55. S. J. Puttock and C. H. Rochester, *J. Chem. Soc., Faraday Trans. 1*, 1986, **82**, 2773.
56. S. J. Puttock and C. H. Rochester, *J. Chem. Soc., Faraday Trans. 1*, 1986, **82**, 3013.
57. B. Chen and E. J. Munson, *J. Am. Chem. Soc.*, 1999, **121**, 11024.
58. B. Chen and E. J. Munson, *J. Am. Chem. Soc.*, 2002, **124**(8), 1638.
59. V. V. Guliants, M. A. Carreon, *Vanadium-Phosphorus-Oxides: from Fundamentals of n-Butane to Synthesis of New Phases on Catalysis*, Vol. 18, RSC Publishing, Cambridge, 2005, 1.
60. G. J. Hutchings, *Appl. Catal.*, 1991, **72**, 1.
61. G. J. Hutchings and R. Higgins, *J. Catal.*, 1996, **162**(2), 153.
62. G. J. Hutchings and R. Higgins, *J. Catal.*, 1996, **162**(2), 153.
63. V. A. Zazhigalov, J. Haber, J. Stoch, A. I. Pyatnitzkaya, G. A. Komashko and V. M. Belousov, *Appl. Catal. A*, 1993, **96**, 135.
64. F. Ben Abdelouahab, R. Olier, M. Ziyad and J. C. Volta, *J. Catal.*, 1995, **157**, 687.
65. M. T. Sananes-Schulz, F. Ben Abdelouahab, G. J. Hutchings and J. C. Volta, *J. Catal.*, 1996, **163**, 346.
66. V. Martin, J. M. M. Millet and J. C. Volta, *J. Therm. Anal. Calorim.*, 1998, **53**(1), 111.
67. S. Sajip, J. K. Bartley, A. Burrows, C. Rhodes, J. C. Volta, C. J. Kiely and G. J. Hutchings, *Phys. Chem. Phys.*, 2001, **3**, 2143.
68. L. Cornaglia, S. Irusta, E. A. Lombardo, M. C. Durupty and J. C. Volta, *Catal. Today*, 2003, **78**(1–4), 291.
69. L. M. Cornaglia, C. R. Carrara, J. O. Petunchi and E. A. Lombardo, *Catal. Today*, 2000, **57**(3–4), 313.
70. G. J. Hutchings, C. Kelly, M. T. Sananes-Schulz, A. Burrows and J. C. Volta, *Catal. Today*, 1998, **40**, 273.
71. S. Sajip, J. K. Bartley, A. Burrows, M. T. Sananes-Schulz, A. Tuel, J. C. Volta, C. J. Kiely and G. J. Hutchings, *New J. Chem.*, 2001, **25**(1), 125.
72. A. Satsuma, Y. Kijima, S. Komai, Y. Kamiya, E. Nishikawa and T. Hatttori, *Catal. Today*, 2001, **71**(1–2), 161.
73. S. Shen, J. Zhou, F. Zhang, L. Zhou and R. Li, *Catal. Today*, 2002, **74**, 37.
74. V. A. Zazhigalov, J. Haber, J. Stoch, I. V. Bacherikova, G. A. Komashko and A. I. Pyatnitskaya, *Appl. Catal. A*, 1996, **134**(2), 225.
75. N. Yamazoe H. Morishige, J. Tamaki and N. Miura, *Stud. Surf. Sci. Catal.*, 1993, **75**, 1989.
76. P. Ruiz, Ph. Bastians, L. Caussin, R. Reuse, L. Dasa, D. Acosta and B. Delmon, *Catal. Today*, 1993, **16**, 99.
77. V. A. Zazhigalov, J. Haber, J. Stoch and E. V. Cheburakova, *Catal. Commun.*, 2001, **2**, 375.

78. V. A. Zazhigalov, *Kinetics and Catal.*, 2002, **43**(4), 514.
79. J. Haber, V. A. Zazhigalov, J. Stoch, L. V. Bogutskay and I. V. Batcherikova, *Catal. Today*, 1997, **33**, 39.
80. B. Solsona, V. A. Zazhigalov, J. M. Lopez Nieto, I. V. Bacherikova and E. A. Diyuk, *Appl. Catal. A*, 2003, **249**, 81.
81. Y. H. Taufiq-Yap, K. P. Tan, K. C. Waugh, M. Z. Hussein, I. Ramli and M. B. Abdul Rahman, *Catal. Lett.*, 2003, **89**, 87.
82. L. Xu, X. Chen, W. Ji, Q. Yan and Y. Chen, *React. Kinetics Catal. Lett.*, 2002, **76,2**, 335.
83. A. Datta, M. Agarwal, S. Dasgupta and R. Y. Kelkar, *J. Mol. Catal. A*, 2003, **198**, 205.
84. P. G. Pries de Oliveira, J. G. Eon, M. Chavant, A. S. Riche, V. Martin, S. Caldarelli and J. C. Volta, *Catal. Today*, 2000, **57**(3–4), 177.
85. A. M. Duarte-Farias, W. A. Gonzalez, P. G. Oliveira, J.-G. Eon, J.-M. Herrmann, M. Aouine, S. Loridant and J. C. Volta, *J. Catal.*, 2002, **208**, 238.
86. V. V. Guliants, J. B. Benziger, S. Sundaresan, I. E. Wachs and A. M. Hirt, *Catal. Lett.*, 1999, **62**(2–4), 87.
87. J. C. Sanchez, J. A. Lopez-Sanchez, R. P. K. Wells, C. Rhodes and G. J. Hutchings, *J. New J. Chem.*, 2001, **25**(12), 1528.
88. S. Holmes, L. Sartoni, A. Burrows, V. Martin, G. J. Hutchings, C. Kiely and J. C. Volta, *Stud. Surf. Sci. Catal.*, 2000, **130B**, 1709.
89. N. Govender, H. B. Friedrich and M. J. van Vuuren, *Cat. Today*, 2004, **97**, 315.
90. A. Cruz-Lopez, N. Guilhaume, S. Miachon and J. A. Dalmon, *Cat. Today*, 2005, **107–108**, 949.
91. Y. H. Taufiq-Yap, *J. Natur. Gas Chem.*, 2006, **15**, 144.
92. Y. Kamiya, Y. Kijima, T. Ohkura, A. Satsuma and T. Hattori, *Appl. Catal. A*, 2003, **253**, 1.
93. C. K. Goh, Y. H. Taufiq-Yap, G. J. Hutchings, N. Dummer and J. Bartley, *Catal. Today*, 2008, **131**, 408.
94. K. Kourtakis and P. L. Gai, *J. Molec. Catal. A*, 2004, **220**, 93.
95. B. T. Pierini and E. A. Lombardo, *Catal. Today*, 2005, **107–108**, 323.
96. C. T. Kresge, M. E. Leonowicz, W. J. Roth, J. C. Vartuli and J. S. Beck, *Nature*, 1992, **359**, 710.
97. J. S. Beck, J. C. Vartuli, W. J. Roth, M. E. Leonowicz, C. T. Kresge, K. D. Schmitt, C. T.-W. Chu, D. H. Olson, E. W. Sheppard, S. B. McCullen, J. B. Higgins and J. L. Schlenker, *J. Am. Chem. Soc.*, 1992, **114**, 10834.
98. C. T. Kresge, M. E. Leonowicz, W. J. Roth, J. C. Vartulli (Mobil Oil Corp.) US Patent 1992, 5098684.
99. A. Stein and R. C. Schroden, *Curr. Opin. Solid State Mater. Sci.*, 2001, **5**, 553.
100. T. Abe, A. Taguchi and M. Iwamoto, *Chem. Mater.*, 1995, **7**, 1429.
101. T. Doi and T. Miyake, *Chem. Commun.*, 1996, 1635.
102. J. E. Haskouri, M. Roca, S. Cabrera, J. Alamo, A. Beltran-Porter, D. Beltran-Porter, M. D. Marco and P. Amoros, *Chem. Mater.*, 1999, **11**, 1446.

103. N. Mizuno, H. Hatayama, S. Uchida and A. Taguchi, *Chem. Mater.*, 2001, **13**, 179.
104. S. Dasgupta, M. Agarwal and A. Datta, *J. Mater. Chem.*, 2002, **12**, 162.
105. M. A. Carreon and V. V. Guliants, *Microp. Mesopor. Mater.*, 2002, **55**(3), 297.
106. M. A. Carreon and V. V. Guliants, *Catal. Today*, 2003, **78**, 303.
107. M. A. Carreon, V. V. Guliants, F. Pierelli and F. Cavani, *Catal. Lett.*, 2004, **92**(1–2), 11.
108. M. A. Carreon and V. V. Guliants, *Chem. Commun.*, 2001, 1438.
109. M. A. Carreon and V. V. Guliants, *Chem. Mater.*, 2002, **14**(6), 2670.
110. M. A. Carreon, PhD. Thesis. University of Cincinnati, 2003.
111. V. V. Guliants, J. B. Benziger and S. Sundaresan, *J. Catal.*, 1995, **156**(2), 298.
112. V. V. Guliants, J. B. Benziger, S. Sundaresan, I. E. Wachs and J.-M. Jehng, *Chem. Mater.*, 1995, **7**(8), 1493.
113. V. V. Guliants, J. B. Benziger and S. Sundaresan, *Chem. Mater.*, 1995, **7**(8), 1485.
114. J. B. Benziger, V. V. Guliants and S. Sundaresan, *Catal. Today*, 1997, **33**(1–3), 49.
115. G. J. Hutchings, J. K. Bartley, J. M. Webster, J. A. Lopez-Sanchez, D. J. Gilbert, C. J. Kiely, A. F. Carley, S. M. Howdle, S. Sajip, S. Caldarelli, C. Rhodes, J. C. Volta and M. Poliakoff, *J. Catal.*, 2001, **197**, 232.
116. G. J. Hutchings, J. A. Lopez-Sanchez, J. K. Bartley, J. M. Webster, A. Burrows, C. J. Kiely, A. F. Carley, C. Rhodes, M. Havecker, A. Knop-Gericke, R. W. Mayer, R. Schlogl, J. C. Volta and M. Poliakoff, *J. Catal.*, 2002, **208**, 197.
117. U. R. Pillai, E. Sahle-Demessie and R. S. Varma, *Appl. Catal. A*, 2003, **252**(1), 1.
118. X. Wang, L. Xu, X. Chen, W. Ji, Q. Yan and Y. Chen, *J. Mol. Catal. A*, 2003, **206**(1–2), 261.
119. Y. H. Taufiq-Yap, L. K. Leong and R. Irmawati, *J. Nat. Gas Chem.*, 2007, **16**, 266.
120. C. Y. Xiao, X. Chen, Z. Y. Wang, W. J. Ji, Y. Chen and C. T. Au, *Catal. Today*, 2004, **93–95**, 223.
121. R. M. Feng, X. J. Yang, W. J. Ji, Y. Chen and C. T. Au, *J. Catal.*, 2007, **246**, 166.
122. R. A. Overbeek, P. A. Warringa, M. J. D. Crombag, L. M. Visser, A. J. van Dillen and J. W. Geus, *Appl. Catal. A*, 1996, **135**, 209.
123. X. K. Li, W. J. Ji, J. Zhao, Z. Zhang and C. T. Au, *Appl. Catal. A*, 2006, **306**, 8.
124. X. K. Li, W. J. Ji, J. Zhao, Z. Zhang and C. T. Au, *J. Catal. A*, 2006, **238**, 232.
125. W. Y. Nie, Z. Y. Wang, W. J. Ji, Y. Chen and C. T. Au, *Appl. Catal. A*, 2003, **244**, 265.

CHAPTER 7

Routes to Methacrylic Acid via *Partial Oxidation*

STEPHAN A. SCHUNK AND NADINE BREM

hte Aktiengesellschaft, Kurpfalzring 104, 69123 Heidelberg, Germany

7.1 Routes to Methacrylic Acid *via* Partial Oxidation

Methacrylic acid and its ester methylmethacrylate are two important monomers for high-performance polymer applications. In this chapter we will try and highlight pathways that use partial oxidation reactions for the synthesis of both monomers and shed some light on the current understanding of the structure–property relationships for the catalytic materials employed for the respective processes. In contrast to typical oxygenates of the C_2-, C_3-, C_4- and C_8-value chains there are a number of technical routes to methacrylic acid (MAA) that are currently employed on an industrial basis or that have been brought to a maturity that is close to commercialization. Due to the large number of approaches we will mainly focus on selected routes that are specific for partial oxidation reactions to methacrylic acid and omit catalyst systems and routes that are currently successfully employed in the C_3-, C_4- and C_8-value chains and are considered potential candidate systems for routes to methacrylic acid as well.

7.2 The Industrial Importance of Methacrylic Acid

In order to put the various routes to methacrylic acid into perspective it is worthwhile taking a look at the commercial applications and forecast on growth. The world consumption of methacrylic acid was as high as 2.5 million

RSC Nanoscience & Nanotechnology No. 19
Nanostructured Catalysts: Selective Oxidations
Edited by Christian Hess and Robert Schlögl

Published by the Royal Society of Chemistry, www.rsc.org

metric tons in 2005. Approximately 85% of this amount goes into applications like construction, automotive applications or equipment manufacture. These markets are very dependent on general economic conditions, so the demand for methacrylic acid follows the patterns of the leading world economies. The growth rates of methacrylic acid are among the rather high ones for unsaturated monomers: a rate of 5% was achieved from 2002 to 2005 with a constant forecast on growth until 2010. In particular, the demand for polymethyl-methacrylate based polymers in high-tech applications like flat-screen television and liquid-crystal displays leads to this constant growth rate compensating stagnation in commodity applications for the polymers based on methacrylic acid.[1]

Interestingly, the commercial paths of access to the monomer for the production of polymethyl-methacrylate are very diverse and largely depend on the companies that have developed the routes. Worldwide, a capacity of over 30 methyl-methacrylate plants were operational in 2004 the corresponding plant sizes range from 10 000 to 360 000 metric tons per year.[2] Currently, five different routes are used with commercial relevance[2]:

- The acetone cyanohydrin route starts from acetone and hydrogen cyanide (or alternatively from acetone cyanohydrin) and follows a reaction sequence of dehydration, hydrolysis and esterification to methyl-methacrylate.
- In Asia a route is followed where iso-butene is converted to methacrolein by gas-phase oxidation and in a second gas-phase oxidation step to methacrylic acid, followed by esterification.
- The BASF route also takes advantage of the second oxidative transformation of methacrolein to methacrylic acid and esterification, only the source of methacrolein is different. It is obtained by hydroformylation of ethylene to propionaldehyde and consecutive condensation with formaldehyde to methacrolein.
- The Asahi route takes advantage of the oxidation of isobutene to methacrolein in the gas phase. Methacrolein is recovered as liquid and subjected to liquid phase oxidation and simultaneous oxidation over Pd/Pb containing catalysts. As an alternative to isobutene tert-butyl alcohol can be utilized as educt to produce methacrolein in the first step.
- The Mitsubishi route: In principle, the Mitsubishi route operates similar to the acetone cyanohydrin route but has a very refined recycle strategy: acetone and HCN are reacted and hydrolyzed to form α-hydroxy-isobutyramid. This product is brought to reaction with CO and methanol under pressure and elevated temperatures to form formamide and methyl-α-hydroxyisobutyrat. Formamide can be dehydrated to HCN for recycle and methyl-α-hydroxyisobutyrat is also dehydrated to methyl-methacrylate.

In this chapter we will mostly focus on the technologies and catalysts that are used for oxidative gas-phase transformations namely the oxidation of methacrolein to methacrylic acid and also address new developments that involve oxidative functionalization of isobutane and isobutene to methacrolein and methacrylic acid and the oxidative dehydrogenation of iso-butyric acid to

Table 7.1 Heteropoly compounds reported as useful catalysts for the respective partial oxidation reactions.[3–18]

Heteropolyacid	*Catalyzed reaction*
$H_3PMo_{12}O_{40}$ *n* H_2O	Isobutane → methacrolein/MAA[3]
$Cs_xH_{3-x}PMo_{12}O_{40}$	Isobutane → methacrolein/MAA[4]
$Cs_2Fe_{0.2}H_{0.4}PVMo_{12}O_{40}$ $Fe_{0.85}H_{0.45}PMo_{12}O_{40}$	Isobutane → methacrolein/MAA[5]
$Cs_{2.5}Ni_{0.08}H_{0.34}PMo_{12}O_{40}$	Isobutane → methacrolein/MAA[6]
$Cu_{0.3}Cs_{1.4}P_{1.5}Mo_{12}V_{0.5}As_{0.4}O_x$	Isobutane → methacrolein/MAA[7,8]
$(NH_4)_3PMo_{12}O_{40}$	Isobutane → MAA[9,10]
$Cs_{1.6}H_{2.4}P_{1.7}Mo_{11}V_{1.1}O_{40}$	Isobutane → MAA[11]
$H_{3+x}PMo_{12-x}V_xO_{40}$	Methacrolein → MAA[12]
$H_5PMo_{10}V_2O_{40}$	Methacrolein → MAA[13]
$Cs_xH_{3-x+y}PMo_{12-y}V_yO_{40}$	Methacrolein → MAA[14]
$Cs_x(NH_4)_yH_zPMo_{12}O_{40}$	Methacrolein → MAA[15]
$[PMo_{12}O_{40}]^{3-}$ different counterions	Isobutyraldehyde → methacrolein[16]
$H_4PVMo_{11}O_{40}$	Isobutyraldehyde → MAA[17,18]

methacrylic acid, the latter of which have not yet been commercialized. From the materials point of view we will be discussing in depth the use of heteropolyacids for the partial oxidation reactions of interest. Heteropolyacids have proven to be very useful with regard to oxidation routes related to the production of methacrylic acid.

Table 7.1 shows a range of partial oxidation reactions for the value chain reported above.

7.3 Heteropolyacids: Multifunctional Oxidic Materials with Superior Qualities as Oxidation Catalysts

Heteropolyacids are a unique class of materials with regard to their structural properties and their capabilities as catalysts for a large variety of organic reactions – especially as oxidation catalysts – a feature that we will focus on in the context of this chapter.[19–36] Heteropolyacids can be described as nanoscale clusters of the early transition metals, built up of corner-, edge- and surface-sharing metal–oxygen octahedrons. Acidification of at least one kind of mononuclear oxo-anion results in condensed isopolyanions ($[M_mO_y]^{n-}$) or, respectively, heteropolyanions ($[X_xM_mO_y]^{n-}$).[37–39] M represents here a transition metal in its highest oxidation state, which forms M–O–π bonds in heteropolyacids with the empty d-orbitals. The main elements known to form heteropolyacids in a variety of structures are tungsten and molybdenum, tantalum, vanadium, niobium or mixtures of these elements. Tungsten was also the first element that led to the structural elucidation of the nature of heteropolyacids by Keggin.[40] The Keggin structure is one of the most prevalent forms of occurrence of a wide variety of related structures well described in a review.[41] In the context of this work we want to focus on heteropolyacids that are used as oxidation catalysts and appear in the form of the Keggin structure. This structure family of Keggin-type molybdenum-containing heteropolyacid anions possess a good mix of thermal stability, redox behavior, high jet-tunable acidity and suitable oxidizing properties towards organic substrates.

In the classical Keggin structure the heteroatom is located in the center of the structure and is tetrahedrally coordinated to four oxygen atoms that are surrounded by the so-called addenda atoms. Groups of three metal–oxygen octahedra (so-called M_3 triplets), coupled together by shared edges, are attached to each of the tetrahedrally coordinated oxygen atoms. The ensembles of the four M_3 triplets are connected to each other by shared corners through oxygen atoms. This structural arrangement leads to a situation where each metal atom is connected to a tetrahedral oxygen atom, a terminally bound oxygen atom extending away from the center and two types of bridging oxygen: oxygen atoms within an M_3 triplet and the corner-shared oxygens between triplets. The interatom distance between tungsten and oxygen of approximately 0.17 nm is rather short and implies that terminally bound oxygen forms a tungsten–oxygen (or, respectively, a molybdenum–oxygen) double bond.[42]

A large number of the elements of the periodic table can function as central heteroatoms in Keggin anions.[37] Typical candidates that can also be synthetically easily employed are phosphorous, silicon, germanium or arsenic. In most cases, the exchange of the central heteroatom will change properties of the heteropolyacid like net charge due to the valency of the heteroatom, acidity, thermal stability or redox reactivity.[34,37] In the following, our interest is mainly focused on heteropolyacids with phosphorous as the central heteroatom and molybdenum as addenda atoms, as the combination of phosphorous and molybdenum provides the most suitable blend of properties that render heteropolyacids useful materials for partial oxidation reactions.

The counterions that are required to balance the charge of the macroanions can be protons, ammonia, organic molecules like amines, alkali metals or other metal cations like transition metals. From the point of view of their catalytic properties these counterions are also of vital importance for the catalytic behavior and other physicochemical properties of the heteropolyacid. In the following, we will discuss the influence of the counterions on the catalytic properties for the respective catalytic reaction. From a preparative point of view it is generally accepted that during the precipitation step of heteropolyacid salts the counterions and even mixtures of counterions like caesium and protons typically are not evenly distributed. What one finds by analytical methods after the precipitation step are generally gradients in the metal distribution. Nevertheless, the counterions will be statistically distributed over the volume of the whole solid by thermal activation even under mild conditions.[37,67] The reason for the easy diffusion of even large cations through the solid heteropolyacid is their pseudoliquid behavior. As a bonus, the diffusion of the cations is greatly propelled when water is present.

It is vital to add that the addenda atoms like the six-valent Mo or W can be partly replaced against each other and against metals of different valency.[43] If the replacement is done by atoms of different charge like pentavalent V, Nb or

Ta the net charge of the heteropolyacid will change and thus the number of counterions required to keep the charge balance in the structure as well. Typically, replacement of addenda atoms in heteropolyacids takes place as random substitution; there is an agreement in literature that for the systems Mo/W, Mo/V and V/W substitution takes place as so-called "positional isomerism", meaning that substitution results in addenda atoms at different positions of the Keggin moiety. This positional isomerism has been proven by NMR studies.[37,44] In some cases, pure structurally well-defined isomers can be synthesized with a selected number of metals, disubstituted isomers of tungsten-based Keggin structures can be obtained with Sn^{4+}, Ti^{4+}, V^{5+} and Mo^{6+}, trisubstituted compounds of tungsten-based Keggins are only known for V^{5+} and Mo^{6+}.[45–49] In many other cases mixtures of the isomers are obtained, it is therefore important to remember that the stochiometric formula is not in every case equivalent or even translates to the microstructure of the Keggin anion. Very little is known about the thermal stability of these substituted compounds in comparison to unsubstituted Keggin structures. We will come back to this point, when we discuss the structure of the active catalyst and deactivation mechanisms.

An additional factor adding to the complexity of the heteropoly compounds that has to be taken into account is the development of hierarchical structures that evolve beyond the primary structure of the Keggin anion with its surrounding cations that we have discussed so far. Generally speaking, the domain size that is usually considered as "secondary structure" is the unit cell and its structural arrangement of polyanion, counterions and other molecules like coordinated water or organic structures. When speaking about the "tertiary structure" generally nanocrystallites of the size of 10 nm are addressed (see Figure 7.1).[34] These nanocrystallites aggregate to porous aggregates; their particle size, the resulting surface area and porosity also govern a lot of the catalytic properties of the material.[50]

The macroionic nature of the heteropolyacids and their susceptibility to absorb molecules like water and organic entities even in conditions like exposure in gas-phase catalysis render these materials unique and put them from the point of view of their properties on the borderline between inorganic solid oxides and reactive ionic liquids. In a number of cases heteropolyacids have been observed to exhibit catalytic behavior beyond pure surface chemistry: cases where pseudoliquid behavior in catalytic reactions including absorption of educts and their reaction in the bulk of the heteropolyacid are predominant have been reported.[50]

Heteropolyacids can display rather small specific surface areas from $10\,m^2/g$ and little or no inherent porosity up to large surface areas of over $120\,m^2/g$ with substantial micro- and mesoporosity.[34,51,52] The amount of surface area that the materials display is largely dependent on the composition of the heteropolyacids, the route *via* which the materials have been prepared and especially thermal treatment steps at elevated temperatures that the materials have been exposed to, including catalytic test procedures. Unanimity exists that the occurrence of porosity, especially micro- and mesoporosity, is connected to the

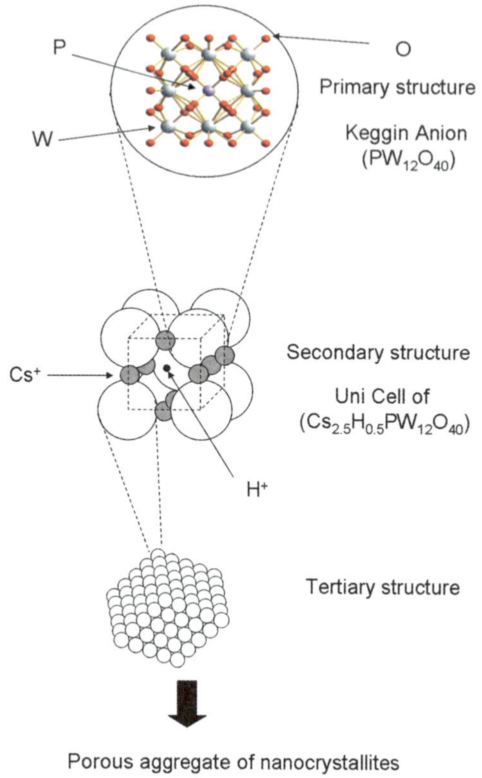

Figure 7.1 Primary, secondary and tertiary structure of HPAs.[34]

introduction of large cations like NH_4^+, K^+ or Cs^+ into the heteropolyacid structure. The amount of porosity and the pore-size distribution is largely dependent on the amount of cations introduced in the structure. Berndt reports that a maximum surface area for molybdophosphoric acid can be obtained for the inclusion of 3 equivalents of caesium in the structure.[67] For Loadings of caesium below 2 or over 3.5 equivalents a dramatic decrease in surface area is observed, especially after thermal treatment. This decrease in surface area, especially for loading of caesium below 2 equivalents, is associated with an increase in the amount of structural defects leading to a decomposition of the Keggin structure and a filling of the pores with decomposed heteropolyacid fragments (see Figure 7.2).

This finding is supported by the fact that the addition of 3 equivalents of caesium to molybdophosphoric acid renders it very thermostable and structural data indicate that the number of defects in the structure is smallest with the presence of 3 equivalents of alkali. With more or less caesium in the structure especially thermal treatment will lead to decomposition of the Keggin anions to heteropolyacids with an equivalent of 3 caesium atoms products of decomposition like lacunary heteropolyacids or fragments thereof. Decomposition

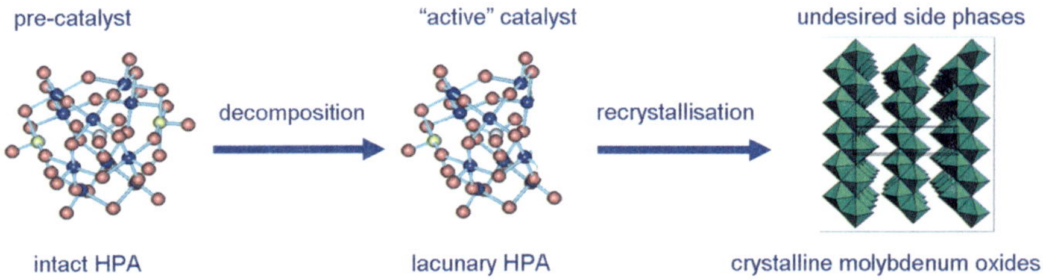

Figure 7.2 Decomposition of structurally intact HPA catalysts to fragmented HPAs and MoO_3.

in the presence of organic species or under reducing conditions favors the decompositional process, the general observation here is that the decomposition temperature is massively reduced.

In the view of the authors the question arises as to whether the initial pore structure is determining for the catalytic behavior of the heteropolyacid species or whether or not it is an accompanying effect of the amount of the alkali content. We feel that the work of Berndt *et al.*[67] has the largest relevance but further discussion is beyond the scope of this review, the reader is recommended to study the work in the references cited above for an overview of controversial information.

7.4 Cs-Containing Molybdophosphoric Keggin Structures: The Industrial Workhorse for the Oxidation of Methacrolein to Methacrylic Acid

Keggin-type heteropolyacids are the most frequently used compounds for industrially relevant oxidation processes with oxygen as terminal oxidant.[53] Apart from the target reaction of this chapter, the oxidation of methacrolein to methacrylic aid, a number of alternative oxidative modifications of alkanes, like the oxidation of propane or of substituted aromatics, can be performed with this catalyst system.[54–57] We will focus on the oxidation of methacrolein to methacrylic acid in this chapter and expand the scope by looking into related heteropolyacids that can be used for the oxidation of iso-butane to methacrolein and methacrylic acid.

The typical heteropolyacids that come into play for the reaction of interest are molybdenum based with phosphorus as central heteroatom. Molybdovanadophosphate systems are most highly valued for the oxidation of methacrolein, the classical counterion these catalysts is caesium. A typical formula for these materials is $H_{4-x}Cs_xPVMo_{11}O_{40} \times yH_2O$ ($x = 1$–4). Generally, one can say that the redox potential increases from W^{VI} to Mo^{VI} to V^V, while tungsten-based materials are in many cases useful for dehydrogenation reactions, the oxo-functionalization is best achieved by molybdenum-based Keggin atoms and can be enhanced with adequate levels of vanadium in combination with transition elements. A number of transition metals are discussed with regard to their beneficial properties at low concentrations, copper, iron, nickel, manganese and cobalt are the most widespread examples; typically they are introduced at a level of $H_{0.08-1.5n}Cs_{2.5}M^{n+}{}_{0.08}PVMo_{11}O_{40} \times yH_2O$. The most relevant "dopant" elements for industrial applications are copper and iron, while for copper it seems pretty evident that it acts as counterion for the molybdovanadophosphate, iron leaves more room for speculation of its location. It is not so obvious what the role of these transition elements is and although a number of in-depth analyses exist, there is still more need for in depth research on that topic.[4,58–60] Copper and iron are identified as additives that help to promote the redox behavior of the heteropolyacid, but their dopant properties go far beyond that, a range of properties is affected by transition metals that are important for the catalytic reaction like acid–base properties or textural properties such as secondary or tertiary structure. For both copper and iron another beneficial effect on the redox behavior of the heteropolyacid is attributed to the fact that both elements help reduce the catalyst and lower the overall oxidation state of molybdenum from Mo^{VI} towards Mo^{V}, rendering the whole catalytic system more active and selective with regard to the target reaction. Several reports also indicate that the addition of amines like pyridinum or quinolinum have similar effects on the oxidation state of the catalyst through partial reduction and render it also useful in alkane oxidation.[61,62]

There are a number of synthetic routes by which HPAs can be synthesized, the most prominent routes, which are also industrially executed proceed *via* the precipitation of soluble salts of all precursors. An exemplary procedure of the industrial synthesis of a caesium-containing catalyst is described in the patent EP 1 142 638A1.[63] For the synthesis ammonium paramolybdate and ammonium metavanadate are dissolved (high concentration of the precursors) with stirring in water of 333 K, followed by addition of pyridine and phosphoric acid and a solution of nitric acid, caesium nitrate and copper nitrate. By heating and stirring the aqueous mixture is concentrated and the obtained clay-like precipitate is molded into columns with 5 mm diameter and 6 mm height. These columns are dried at 393 K for 15 h and calcined in a gaseous nitrogen stream at 703 K for 4 h, followed by an air stream at 673 K for 2 h.

The major issue with heteropolyacid catalyst of the type described above and still one of the greatest challenges in industrial practice is the massive aging behavior that renders this catalyst useful for as commercial catalyst in an industrial plant only for periods up to 24 months. As the catalyst price is largely governed by the pricing of caesium, the catalyst is a very costly material to replace and the frequent turnover of the plant in annual or biannual mode adds also to the price of the monomer produced by gas-phase oxidation. Marosi *et al.*[15,64] set a milestone in the understanding of the deactivation mechanism through *in-situ* X-ray diffractometric techniques, the findings have been confirmed by many other groups using a range of *in-situ* and *ex-situ* techniques.[65–69]

Marosi *et al.* studied the thermal and structural stability of the Keggin-type catalysts $CsH_2PMo_{12}O_{40}$ in the reaction of methacrolein to methacrylic acid.[70,71] The essential difference to previous studies conducted is that the catalyst was characterized by using *in-situ* XRD equipment in which the catalyst could be exposed to realistic test conditions from an industrial viewpoint. Under reaction conditions, in the presence of methacrolein, water and oxygen, a migration of the Mo-oxo-species into the intracrystalline pore system of the crystal lattice could be monitored, at the same time changes in the activity and selectivity were observed. The activity and selectivity of the catalyst change drastically within the first 120 h on time of stream: with increasing formation of Mo-oxo-species the selectivity raises from 63.6 to 68.4% and the conversion level of 56% to 58.5%. After 320 h on-stream the selectivity is found to decrease further to 54% and the conversion level rises further to 63%. Table 7.2 summarizes the performance results of caesium-containing and caesium-free catalysts. It can clearly be seen that the presence of caesium can be beneficial for both the selectivity and lifetime of the catalyst.

What is the microstructural reason behind the catalyst deactivation and can the stabilizing effect of caesium be understood? The *in-situ* XRD-analysis of Marosi and the work of others by analysis with various *in-* and *ex-situ* techniques proved that vanadyl and molybdenyl species are expelled from the Keggin anions under reaction conditions and lead to the formation of lacunary heteropolyacids in the company of monomeric species of molybdenum- and vanadium-oxide entities of different valency (see Figure 7.2).

As a result, an increasing amount of side phases become visible due to oligomerization reactions; hepta- or octamolybdates are formed and finally MoO_3 as a detrimental side phase is detectable *via* various analytical techniques. Once the side phases have increased to a certain amount, degradation in performance of the catalyst is clearly visible. The addition of caesium to a certain extent may stabilize the heteropolyacid under reaction conditions and make a partially caesium-substituted heteropolaycid clearly superior over an unsubstituted one. Still, the dilemma with the catalyst system is that stabilization with caesium is only possible to a certain extent, as beyond a certain equivalent of caesium, the catalyst system loses its activity and selectivity in the target reaction. The optimum amount of caesium for purely molybdenum-based heteropolyacids from the point of view of their performance is around two equivalents of Cs. With a caesium content below two equivalents per Keggin unit the stability of the resulting material is rather low, caesium contents around and above three

Table 7.2 Summary of the undoped and Cs-doped $HPMo_{12}O_{40}$ catalyst. Conversion, selectivity and lifetime are compared after 200 h on-stream. Lifetime is defined as the starting point of decreasing conversion and selectivity.[15]

Catalyst	*BET surface [m²/g]*	*Conversion [%]*	*Selectivity MAA [%]*	*Lifetime [h]*
$H_3PMo_{12}O_{40}$	8	36.4	50.2	~15
$Cs(NH_4)_2PMo_{12}O_{40}$	5	63.7	56	~200
$Cs_2(NH_4)PMo_{12}O_{40}$	3	50.7	54	>380

will render the catalyst inactive for the partial oxidation reaction. From the study it was concluded that caesium also leads to a partial stabilization of the structural disintegration products of the heteropolyacid; the depolymerization is also accompanied by the formation of the thermodynamically more stable, but catalytically inactive caesium salts of the HPA with high caesium content. The presence of caesium also delays the condensation of MoO_3-type oxides and contributes to prolonged catalyst life in this way. In any case though, the final disintegration and structural reorganization product of the heteropolyacid decomposition is MoO_3, an oxide phase not well suited for the target reaction.

From a practical point of view it is important to notice that the disintegration and also the partial recondensation process is strongly affected by the absolute temperature, the heating rate and the presence and amount of water in gas feed. It is important to notice that the presence of water does not lead to a stabilization of the intact Keggin structure. In contrast, the hydrolysis of the Keggin anions seems to be enhanced compared to a situation where no water is contained in the gas atmosphere. This observation is of high importance and practical relevance: water is added to the feed in industrial partial oxidation of methacrolein to methacrylic acid in considerable amounts. Hence, under industrial conditions an active heteropolyacid-derived catalyst is inherently unstable and cannot contain intact Keggin anions as active species. Catalytic partial oxidation conditions lead to a pronounced structural reorganization as described above. Therefore, heteropolyacids in the form that they are employed as industrial catalysts today have to be understood as defined molecular precursor compounds, leading to an active and selective catalyst that is partially depolymerized. The formation of the active phase of the heteropolyacids and the aging of the catalyst are connected. As the relative stability of the depolymerized catalyst under reaction conditions is low, and recondensation leads to the formation of MoO_3 over hepta-, octamolybdates and heteropolyacids with high caesium content, the resulting aged catalyst materials degrades in performance and becomes finally useless for the production of methacrylic acid from an economic point of view.

There are two approaches in the reported literature that intend to overcome the degradation of the heteropolyacid by the use of bronze structures, which are suitable catalysts for the oxidation of acrolein to acrylic acid.[72–74] Although interesting results could be achieved with this approach, caesium-containing heteropolyacids are still unrivalled in their performance.

7.5 SbO_x/Pt Catalysts as an Interesting Materials Class for the Selective Oxidation of Isobutane and Isobutene to Methacrolein

Platinum metals are generally not regarded as selective oxidation catalysts in oxidation reactions. It is even more surprising that a platinum-containing material could be identified as active species for the selective oxidation of isobutane and isobutene to methacrolein (see Figure 7.3).[75–78]

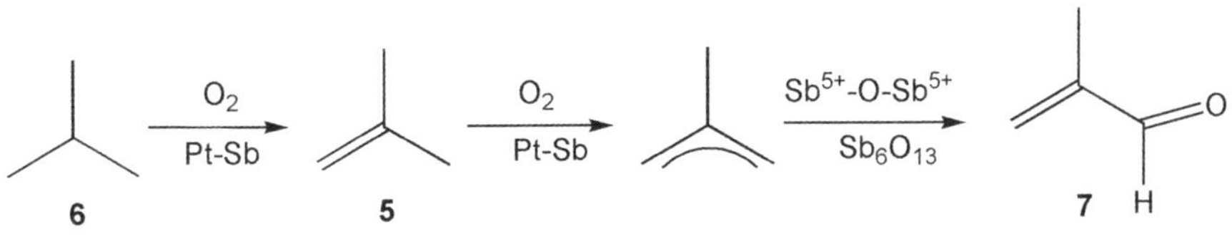

Figure 7.3 Reaction sequence of isobutane oxidation using Pt/SbO_x. (i-C_4H_{10} → i-C_4H_8 → i-C_4H_7 → methacrolein).[75]

The catalytic system is prepared by hydrolysis of an aqueous solution of $SbCl_5$ with an aqueous ammonia solution. After drying and calcination the obtained antimony oxide is impregnated with an acetone solution of $Pt(acac)_2$ and again dried and calcined. If and how much chloride remains after calcination in the catalytic system is not reported.

Under calcination conditions the SbO_x is mainly transformed to Sb_6O_{13}, which consists of a defect pyrochlore-type structure based on corner-shared SbO_6 octahedra with Sb^{5+}–O–Sb^{5+} chains along the [110] axis. The remaining oxygens and Sb^{3+} are occupying vacancies built by the octahedral network and are loosely connected in the crystal structure.[76,77]

The materials show high selectivities to products of value only at low conversions. Nevertheless, in the oxidation reaction of isobutane selectivities of 57% to methacrolein and selectivities of 80–90% to the sum of methacrolein and isobutene could be achieved. The catalytic system is more selective for the oxidation of isobutene: when the olefin is used as educt a selectivity of 90% to methacrolein can be reached. This can be taken as an indication that the initial step in the oxidative dehydrogenation of isobutene determines the overall yield for products of value. Without the presence of platinum the dehydrogenation reaction of isobutane to isobutene does not proceed at all, but simultaneously at higher Pt loadings the material becomes unselective especially at higher conversions and catalyzes the oxidation of hydrocarbons to CO_x.

As a consequence, the sum selectivity of isobutene and methacrolein is highly dependent on the Pt loading of SbO_x (see Table 7.3). Comparing the results of the oxidation of isobutane, an optimum for Pt loadings is achieved between 0.2 and 0.35 wt%, where methacrolein selectivities between 43–52% can be obtained and sum selectivities between 87 and 89% for olefin and unsaturated aldehyde could be obtained. Further oxidation of methacrolein to MAA was not observed, with this class of catalysts.

XRD analysis and Raman spectroscopy revealed Sb_6O_{13} as the main oxide compared to α-Sb_2O_3 in the catalyst before and after the catalytic experiments.[75–78] The authors come to the conclusion that the Sb_6O_{13} structure is responsible for the selective oxidation of isobutene to methacrolein, while the reduced α-Sb_2O_3 catalyzes total oxidation to CO_x. The Pt particles are surrounded or covered by Sb suboxides and small highly dispersed O=SbO_y, which are thought to have the role of a suitable modifier for the inhibition of

Table 7.3 Dependency of yield and selectivity on different Pt loadings for the catalytic oxidation of isobutane using the catalytic system Pt/SbO$_x$ (773 K, 0.3 g catalyst, 2400 h^{-1}, 20% isobutane, 4% oxygen, balance gas helium).[75]

	Yield [%]			*Selectivity [%]*	
Pt [wt%]	*MAL*	*Isobutene*	*CO_2*	*MAL*	*Isobutene*
0	0.0	0.0	0.0	–	–
0.2	0.15	0.15	0.0	43.8	43.3
0.35	0.38	0.27	0.02	52.1	37.5
0.5	0.98	0.43	0.24	56.	24.9
0.72	0.59	0.51	0.53	35.1	30.4

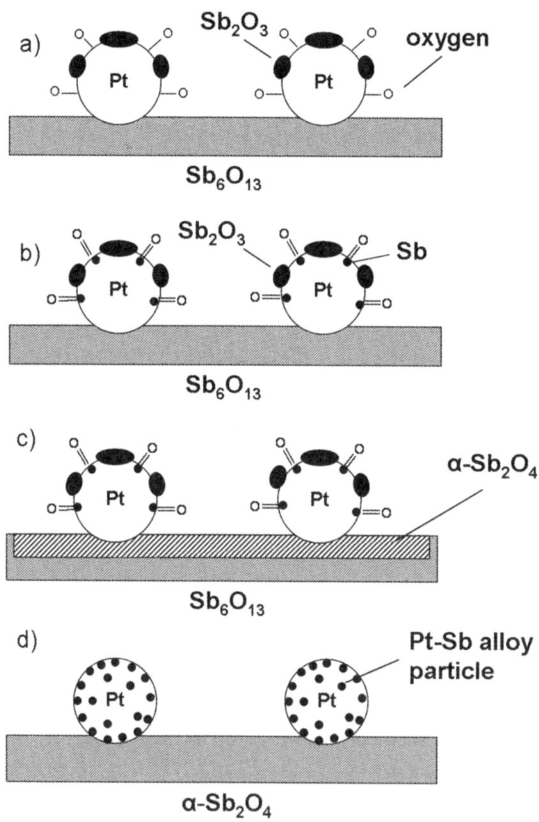

Figure 7.4 Structural changes of Pt/SbO$_x$ during the reaction of isobutene to methacrolein. (a) Phase after calcination before the reaction starts (b) selective phase (c) nonselective phase (d) reduced phase.[75–78]

the total oxidation of organic entities on the Pt nanoparticles (see Figure 7.4). Pt and its respective surface oxides promote the redox cycle of Sb_6O_{13} and lower the temperature for the selective oxidation of isobutene.

Table 7.4 Promoter effect on M/SbO_x for isobutane oxidation (773 K, 0.3 g catalyst, 2400 ml h^{-1}, 20% isobutane, 4% oxygen, balance gas helium).[75]

Metal (wt%)	*Pt (0.5)*	*Pd (0.3)*	*Rh (0.3)*	*Ir (0.5)*	*Cu (0.2)*	*Ag (0.2)*
O_2 conversion [%]	37.5	1.1	2.0	23.3	0.0	0.0
MAL selectivity [%]	56.6	0.0	0.0	0.0	–	–

Without the presence of Pt particles, SbO_x is inactive for the selective oxidative dehydrogenation of isobutane to isobutene but converts the dehydrogenated intermediate, isobutene, to methacrolein with high selectivities of up to 80%. Under isobutane-rich reaction conditions (20% isobutene, 4% oxygen) the Sb_6O_{13} surface is heavily reduced to the nonselective α-Sb_2O_4 phase which causes a decrease in the methacrolein selectivity to less than 20% and simultaneous increase in the total oxidation to CO_x.

The use of alternative metals like rhodium, palladium, iridium, copper or silver instead of platinum on SbO_x did not lead to oxygenates like methacrolein in the oxidation of isobutane (773 K, 0.3 g catalyst, 2400 ml h^{-1}, 20% isobutane, 4% oxygen, balance gas helium). Iridium, palladium and rhodium convert isobutane only to CO_x. Copper- and silver-doped SbO_x were not reactive at all under the conditions tested (see Table 7.4).[75]

7.6 Modified Heteropolyacid Catalysts for the Selective Oxidation of Isobutane to Methacrolein and Methacrylic Acid

As we discussed before, the use of pyridine for the modification of heteropoly compounds ($H_3PMo_{12}O_{40}$) leads to a reduced state of the heteropolyacid during thermal treatment or exposure to organic feed that is beneficial for selective oxidation reactions.[7,8,61] The resulting catalyst systems generally show significantly higher selectivities to target products like MAA compared to the unmodified heteropoly compounds (Keggin-type). As the catalytic activity of these organic/inorganic composite systems is highly dependent on the stability of the reduced state of the resulting heteropolyacid that is achieved by incorporation of the pyridinium ion in the heteropoylacid structure, it is important to find compositionally suitable HPAs for pyridinium modification.[61] From the report of the authors it can be concluded that the pyridinium decomposes during the catalytic oxidation reaction, rendering the heteropolyacid system highly reduced, but stable towards full reduction and decomposition.

Based on the high stability, selectivities of 50% to MAA, and 22% to acetic acid, can be achieved at a conversion of 22% under the following reaction conditions: atmospheric pressure, 3 g catalyst, 2.2 mol% isobutane, 13.7 mol.% oxygen, 33.5 mol.% water, flow rate: 30 ml/min, temperature range 553–613 K.

The acidic form of the heteropoly compound, $H_3PMo_{12}O_{40}$, without pyridinium is much less active (7% conversion of isobutane at the same reaction conditions) and less selective to the target product MAA (11% selectivity) with almost constant selectivity of the fragmentation product acetic acid (21% selectivity).

The catalytic system of interest is prepared by precipitation of the heteropolyacid $H_3PMo_{12}O_{40}$ in aqueous phase and pyridine at 308–313 K under stirring. After evaporation at 353 K and drying at 393 K under nitrogen for 8 h the sample is activated by heat treatment up to 673 K under nitrogen, the completion of the activation procedure is visible by a change of the color from pale yellow to dark green and accompanied by an increase in the specific surface area. In contrast, nonactivated pyridinium salts exposed to catalytic test conditions are inactive for the target reaction. Activation temperatures higher than 743 K lead to degraded catalyst materials that are less-selective catalysts, regarding MAA and acetic acid. It can be observed that for the best-performing samples only one third of the pyridine of the preparation remains in the solid after the pretreatment. Together with the remaining pyridinium molecules the heteropolyacid builds up an orthorhombic phase, which is constructed of pyridinium ions and oxygen-deficient heteropoly anions with a large fraction of reduced molybdenum (V).

Capua *et al.* studied the selective oxidation of isobutane into methacrolein and methacrylic acid over a copper-containing Mo–V–P HPA catalyst (Keggin-type), which was also modified with pyridinium.[61,62] The synthesis of this catalytic system ($Cs_3HCu_{0.12}pyridine_{0.6}PMo_{11}VO_x$) is similar to the synthesis methods mentioned above. The reactants are slowly added one by one in stirred, deionized water, heated up to 353 K for 90 min and are evaporated, dried and calcined at 723 K under nitrogen for 3 h. The performance of the catalytic system was explored at low reaction temperatures and accordingly at low isobutane conversions, several intermediates and final oxidation products were determined. Table 7.5 shows the product selectivities after 4 h under reaction conditions (12.3% isobutane, 6.4% oxygen, 75.5% helium (containing 1% krypton), 5.8% water; contact time 2 s). Using GC/MS and ^{1}H and ^{13}C NMR spectroscopy, it was possible to identify a large quantity of byproducts seldom observed before in the literature, like cyclic products as results of Diels–Alder-type reactions. Further compounds, like maleic acid, isobutanol and formaldehyde derivates could be detected by NMR of the condensed product stream (selectivities between 1.7×10^{-4} and 6.6×10^{-1} %). With the assistance of the spectroscopic analysis methods employed it was possible to highlight the phenomena of skeletal isomerization that occurred to a certain fraction during the oxidation of isobutane. The analysis shows that C–H activation on these solids may also involve reaction pathways that go beyond the side products of "classical" fragmentation during oxidative degradation.

Interestingly, the use of an arsenium-containing Keggin-type heteropoly acid catalyst, $Mo_{12}V_{0.5}P_{1.5}As_{0.4}Cu_{0.3}Cs_{1.4}O_x$, enhances the selective oxidation of isobutane to MAA over isobutene and methacrolein in an acceptable region for an industrial application (productivity of 6.4 mmol/h/g_{cat} methacrolein

Table 7.5 Product selectivities of the oxidation of isobutane at 593 K and 1 bar. The conversion of isobutane was 1.7% and that of oxygen 10.2% ($Cs_3HCu_{0.12}pyridine_{0.6}PMo_{11}VO_x$, 12.3% isobutane, 6.4% oxygen, 75.5% helium (containing 1% krypton), 5.8% water; contact time 2 s).[62]

Product	*Selectivity [%]*
Propene	1.6
Isobutene	5.3
CO	17.5
CO_2	25.7
Acetaldehyde	0.9
Propionaldehyde	0.01
Acetone	2.1
Acrolein	0.6
Methacrolein	25.4
Acetic acid	5.5
Propionic acid	0.3
Isobutyric acid	0.07
Acrylic acid	1.2
Methacrylic acid	14.1

and MAA).[7,8] The productivity of the catalytic system can be increased by increasing temperature or partial pressure of isobutane or by variation of the contact time. At temperatures of 640 K a maximum of the productivity was observed with a sum selectivity of methacrolein and MAA of 45% at isobutane conversion levels of 22% (25 vol.% isobutane, 25 vol.% oxygen, 15 vol.% water, balance nitrogen, contact time 5.4 s). Kinetic experiments prove that the formation of acetic acid and CO_x is caused by overoxidation of MAA as well as by total oxidation of isobutane. The formation of carbon monoxide, however, shows a clear dependency on the formation of acetic acid and acrylic acid. By determination of the reaction rates it was found that isobutene and methacrolein react 500 and, respectively, 120 times faster than isobutane, while the overoxidation of MAA is about two times faster than isobutane oxidation. In the oxidation reaction the reactivity of methacrolein itself is not influenced by the presence of isobutane, but the presence of isobutane has a significant effect on the reactivity of MAA and lowers the overoxidation of MAA.[8]

7.7 Formation of Methacrolein and Methacrylic Acid by Oxidative Dehydrogenation of Isobutyric acid and Isobutyraldehyde using Iron Phosphate Catalysts

As highlighted above in the first paragraphs, several routes to methacrylic acid may be commercially interesting. The oxidative dehydrogenation of isobutyric acid to methacrylic acid has received a lot of attention and, although pilot-plant trials were successful, has until now not found its way to a commercial application.

One of the most exciting studies about the oxidative dehydrogenation reaction of isobutyric acid (IBA) was the one published by Ai and coworkers who were using iron phosphates with different ratios of P/Fe, up to 1.5.[79,80] Depending on the preparation method and the P:Fe ratio different structures could be identified by XRD analysis ($NH_4FeP_2O_7$, $Fe(PO_3)_3$, $FeH_2P_2O_7$, $Fe_2P_2O_7$, $FePO_4$), accompanied by unidentified compounds. These unidentified compounds are formed by transformation of amorphous compounds to $FePO_4$ after calcination and by reduction of $FePO_4$ to $Fe_2P_2O_7$ during the reaction.[79]

Within a certain P:Fe ratio, the preparation method of the catalysts employed in this publication is not really decisive for the catalyst performance: with a P:Fe ratio 1.0–1.5 a MAA selectivity between 75–80 mol% with the isobutyric conversion of up to 99% could be obtained (see Table 7.6; 21.5 mmol/h isobutyric acid, 16.5 mmol/h oxygen, 324 mmol/h steam, 350 mmol/h nitrogen, 653 K, contact time 2.0 s).[79] Raising or decreasing the P:Fe ratio resulted in lower product selectivities, although catalysts with ratios below 1.0 possessed larger specific surface areas. A second dependence of the performance could be identified: the calcination temperature of the material. An increase in the calcination temperatures led to decreasing surface areas and decreasing activity. The main products observed during oxidative dehydrogenation of isobutyric acid on these catalyst systems are MAA as the main and target product as well as acetone, propene and CO_x as side products.

An important general observation together with the performance of the catalyst is associated with the state of oxidation of iron: for materials in the catalytic test (P:Fe ratio 1.2) the valence state of iron changed from Fe^{3+} of the fresh catalyst to reduced Fe^{2+} [$Fe_2P_2O_7$] *via* a bluish black intermediate with a ratio of mixed valency $Fe^{2+}/(Fe^{2+}+Fe^{3+})$ of about 0.5.[81]

The most common method, described in the literature and patents, for the synthesis of iron phosphate catalysts is the precipitation method, where often further promotors employed as nitrates (*e.g.* alkali, Sb, Sn) are added. After mixing an aqueous iron-containing precursor compound (in most cases nitrate salts) with an aqueous solution of phosphoric acid or ammonium phosphate

Table 7.6 Conversion and selectivities of the iron phosphate catalyst (P:Fe ratio 1.2) at 673 K, 21.5 mmol/h IBA, 16.5 mmol/h oxygen, 324 mmol/h water and 350 mmol/h nitrogen).[79]

			Selectivity (mol%)			
Catalyst uses [g]	*Contact time [s]*	*IBA conversion [%]*	*MAA*	*Acetone*	*Propene*	*CO_x*
0.5	0.09	63.0	80.5	10.2	1.0	8.0
0.7	0.13	725	80.8	9.6	0.9	8.4
1.0	0.19	82.7	80.3	9.0	0.8	9.9
1.5	0.28	89.6	78.7	9.6	0.7	11.1
2.0	0.38	96.5	77.5	9.5	0.7	12.4
2.5	0.47	95.6	77.2	9.9	0.7	11.9

and optionally further dopants, the mixture is evaporated, dried and calcined under defined atmospheres like air at 673–773 K. Typically the P:Fe ratios range from unity to 1.5.[82–92]

Ai *et al.* employed various ions (*e.g.* Cs^+, K^+, Li^+, Sr^{2+}, Mg^{2+}, Ce^{3+}, La^{3+}, Ag^+, Mn^{2+}, Nb^{5+}, V^{5+}, Te^{6+}, Bi^{3+}, Pb^{2+}) in search of suitable promotors for the iron phosphate catalysts during the synthetic steps, to study effects on the performance of the oxidative dehydrogenation of isobutyric acid to MAA.[93] The best performance with MAA yields up to 65–75 mol% and MAA selectivities up to 80 mol% were obtained by Pb^{2+} promoted catalysts; especially with a P:Pb:Fe atomic ratios in the range of 1.2–1.3/0.15/1 to 1.3–1.4/0.3/1. Ai *et al.* concluded that Pb^{2+} as promotor, which is found in the catalyst in the form of the phase $Pb_2P_2O_7$, plays a role in suppressing the consecutive oxidation of the produced MAA to CO_x.[93] However, the significant influence of the lead as beneficial dopant described by Ai *et al.* is not confirmed in the patent literature, where lead was used as promoter in comparable examples.[80]

The iron phosphates have the disadvantage that during a long period on-stream under reaction conditions the catalysts get less active and show a decrease in selectivity. A two-step method of regeneration is patented by the Ashland Oil Inc.[94] In the first step the flow of organic feed (isobutyric acid) is stopped, while continuing the flow of steam-air and steam-oxygen to build up an oxidizing atmosphere. In the second step the atmosphere is changed to a mild reducing one by using an organic feed (isobutyric acid) without air or oxygen and kept for a period of time. After repetition of these two steps several times the efficiency of the catalytic system is improved significantly.

Another typical solution to overcome the degradation of the catalyst is a single-step regeneration: the organic feed is stopped, while the flow of steam-air is continued; even better performance restoration is obtained with mixtures of steam and oxygen, at operating conditions over a couple of hours. However, the two-step method is superior and achieves a higher efficiency.

An interesting comparison was published by Ai *et al.* for a range of partial oxidation reactions and oxidative dehydrogenation reactions with the target product methacrylic acid between three different catalyst classes: (1) a heteropolyacid $H_3PMo_{12}O_{40}$ (and its caesium derivates), (2) vanadylpyrophosphate (V–P) and (3) iron phosphate.[95] Concerning the oxidation reaction of isobutene, the V–P catalyst was performing comparably to the heteropolyacid ($Cs_2HPMo_{12}V_2O_{40}$) with regard to activity and selectivity to MAA. In the oxidation of methacrolein the heteropolyacid, $H_3PMo_{12}O_{40}$, and the V–P catalyst have similar selectivities to MAA (40 mol%) and similar conversions could be reached (80 mol%) while the iron catalyst is clearly less selective. For the dehydrogenation reaction of isobutyric acid, however, the $FePO_4$ catalyst shows a much higher selectivity to MAA compared to the heteropolyacid and the V–P compound at similar conversions between 80–83% but requires much higher temperatures (see Table 7.7). The results show that the iron phosphate catalyst is the only effective catalyst for oxidative dehydrogenation reactions of compounds with a tertiary carbon atom in the alpha position, like isobutyraldehyde or isobutyric acid.

Table 7.7 Results obtained for the oxidation of isobutyric acid using different heteropolyacids and phosphates of transition metals as catalysts.[95]

Catalyst	T *[K]*	*Ratio O_2/ $C_4H_8O_2$*	*Contact time [s]*	*Conversion [%]*	*Selectivity [mol%]* MAA	*Acetone*	*Propene*	CO_x
$FePO_4$	673	0.8	0.1	63	80.5	10	1.0	8.5
	673	0.8	0.2	83	80.3	9	1.0	8.3
V–P	538	14.5	4.6	60	55.0	19	24.0	2.0
	548	14.5	4.6	80	50.0	16	25.0	9.0
$H_3PMo_{12}O_{40}$	518	14.5	2.5	60	60.0	29	3.0	8.0
	533	14.5	2.5	80	62.0	25	2.0	11.0
$Cs_2HPMo_{12}O_{40}$	503	14.5	2.5	50	60.0	30	3.0	7.0
$Cs_2HPMo_{12}V_2O_{40}$	493	14.5	2.5	50	69.0	25	2.0	9.0

Similar to the dehydrogenation reaction of isobutyric acid by the P:Fe catalyst (ratio 1.2), Muneyama *et al.* studied the oxidative dehydrogenation of isobutyraldehyde to methacrolein using the same catalytic system.[77,81] As described above, also in this reaction the iron phosphate was reduced from Fe^{3+} in the fresh catalyst to iron (II) pyrophosphate [$Fe_2P_2O_7$] *via* an unidentified intermediate phase. The reactions were carried out with a P:Fe catalyst (ratio 1.2) in a similar regime as the oxidative dehydrogenation of isobutyric acid with a continuous-flow system at an atmospheric pressure of 22.5 mmol/h isobutyraldehyde, 20.8 mmol/h oxygen, 1066 mmol/h water, 350 mmol/h nitrogen at 673 K and different contact times (1–3 s) (1.5 mol% isobutyraldehyde, 1.4 mol% oxygen, 73 mol% water, 24 mol% nitrogen).[96] The contact time was varied by changing the amount of catalyst. During the reaction the main products were methacrolein, CO and CO_2, but no methacrylic acid was formed. In comparison to the dehydrogenation reaction of isobutyric acid, acetone and propene, the main byproducts of the latter reaction, were not observed during the oxidation of isobutyraldehyde.

References

1. S. Bizzari, *SRI Report on methyl-methycrylate*, 2006.
2. *PERP Program – Methyl Methacrylate New Report Alert*, July 2006.
3. N. Ballarini, F. Candiracci, F. Cavani, H. Degrand, J.-L. Dubois, G. Lucarelli, M. Margotti, A. Patinet, A. Pigamo and F. Trifirò, *Appl. Catal. A-Gen*, 2007, **325**, 263.
4. N. Mizuno, M. Tateishi and M. Iwamoto, *J. Catal.*, 1996, **163**, 87.
5. M. Langpape and J. -M. M- Millet, *Appl. Catal. A-Gen.*, 2000, **200**, 89.
6. N. Mizuno, M. Tateishi and M. Iwamoto, *Appl. Catal. A*, 1994, **118**, L1–L4.
7. G. P. Schindler, C. Knapp, T. Ui and K. Nagai, *Top. Catal.*, 2003, **22**, 118.
8. G. P. Schindler, T. ui and K. Nagai, *Appl. Catal. A*, 2001, **206**, 183.
9. F. Cavani, R. Mezzogori, A. Pigamo, F. Trifiró and C. R. Acad. Sci. Pairs, *Série IIc, Chimie/Chemistry*, 2000, **3**, 523.

10. F. Cavani, R. Mezzogori, A. Pigamo, F. Trifiró and E. Etienne, *Catal. Today*, 2001, **71**, 97.
11. L. Jalowiecki-Duhamel, A. Monnier, Y. Barbaux and G. Hecquet, *Catal. Today*, 1996, **32**, 237.
12. US4565801, 1986, Sumitomo Chemical Company.
13. H. Kim, J. C. Jung, D. R. Park, S.-H. Baeck and I. K. Song, *Appl. Catal. A*, 2007, **320**, 159.
14. L. M. Deusser, J. C. Petzold and J. W. Gaube, *Ind. Eng. Chem. Res.*, 1998, **37**, 3230.
15. L. Marosi and C. O. Areán, *J. Catal.*, 2003, **213**, 235.
16. J. Hu and R. C. Burns, *J. Catal.*, 2000, **195**, 360.
17. G. Mestl, T. Ilkenhans, D. Spielbauer, M. Dieterle, O. Timpe, J. Kröhnert, F. Jentoft, H. Knözinger and R. Schlögl, *Appl. Catal. A*, 2001, **210**, 13.
18. T. Ilkenhans, B. Herzog, T. Braun and R. Schlögl, *J. Catal.*, 1995, **153**, 275.
19. E. Cadot, C. Marchal, M. Fournier, A. Tezé, G. Hervé, Role of Vanadium in Oxidation Catalysis by Heteropoly Anions, in: *Polyoxometalates* M. T. Pope, A. M. Müller (ed.), Kluwer Academic, Dordrecht, 1994.
20. M. T. Pope and A. Müller, *Angew. Chem. Int. Ed.*, 1991, **30**, 34.
21. M. Misono, *Catal. Today*, 2005, **100**, 95.
22. M. A. Barteau, J. E. Lyons and I. K. Song, *J. Catal.*, 2003, **216**, 236.
23. T. Okuhara, N. Mizuno and M. Misono, *Appl. Catal. A-Gen.*, 2001, **222**, 63.
24. C. Marchal-Roch and J.-M. Millet, *Cr. Acad. Sci. II C*, 2001, **4**, 321.
25. F. Cavani, *Catal. Today*, 1998, **41**, 73.
26. A. Corma, *Curr. Opin. Solid. St. M.*, 1997, **2**, 63.
27. N. Mizuno and M. Misono, *Curr. Opin. Solid. State Mater.*, 1997, **2**, 84.
28. T. Okuhara, N. Mizuno and M. Misono, *Adv. Catal.*, 1996, **41**, 113.
29. N. Mizuno and M. Misono, *J. Mol. Catal.*, 1994, **86**, 319.
30. M. Misono, *Stud. Surf. Sci. Catal.*, 1993, **75**, 69.
31. Y. Ono, M. Taguchi, S. Gerile, S. Suzuki and T. Baba, *Stud. Surf. Sci. Catal.*, 1985, **20**, 167.
32. G. M. Maksimov, *Russ. Chem. Rev.*, 1995, **64**, 445.
33. T. Okuhara, N. Mizuno and M. Misono, *Adv. Catal.*, 1996, **41**, 113.
34. N. Mizuno and M. Misono, *Chem. Rev.*, 1998, **98**, 199.
35. P. Souchay, *Polyanions and Polycations*, Gauthiers Villars, Paris, 1963.
36. H. T. Evans, *Perspectives Struct. Chem.*, 1971, **4**, 1.
37. M. T. Pope, *Heteropoly and Polyoxometalates*, Springer, Berlin, 1983.
38. K.-Y. Lee and M. Misono, *Handbook of Heterogeneous Catalysts*, Vol. 5, Wiley-VCH, Weinheim, 1997.
39. J. B. Moffat, *Metal-oxygen Clusters: The Surface and Catalytic Properties of Heteropoly Oxometalates*, Kluwer, New York, 2001.
40. J. F. Keggin, *Proc. Roy. Soc. London*, 1934, **A144**, 75.
41. M. T. Pope and A. Müller, *Angew. Chem. Int. Ed.*, 1991, **30**, 34.
42. J. W. Linett, *J. Chem. Soc.*, 1961, 3796.
43. D. P. Smith and M. T. Pope, *Inorg. Chem.*, 1973, **12**, 331.

44. A. Pöppl, P. Manikandan, K. Köhler, P. Maas, P. Strauch, R. Böttcher and D. Goldfarb, *J. Am. Chem. Soc.*, 2001, **123**, 4577.
45. M. T. Pope, S. E. O'Donell and A. R. Prados, *J. C. S. Chem. Comm.*, 1975, **22**.
46. M. T. Pope, S. E. O'Donell, R. A. Pardos, Inorganic Compounds with unusual Properties, American Chemical Society, 1976, p. 85, ACS Adv. Chem. Ser. 150.
47. R. Massart, R. Contant, J. M. Fruchart, J. P. Ciabrini and M. Fournier, *Inorg. Chem.*, 1977, **16**, 2916.
48. P. J. Domaille, *J. Am. Chem. Soc.*, 1984, **106**, 7677.
49. P. J. Domaille and W. H. Knoth, *Inorg. Chem.*, 1983, **22**, 818.
50. M. Misono, *Chem. Comm.*, 2001, **1141**.
51. S. Soled, S. Miseo, G. B. McVicker, J. E. Baumgartner, W. E. Gates, A. Gutierrez and J. Paes, *Advanced Catalysts and Nanostructured Materials*, Academic Press, San Diego, CA, 1996, 435–451.
52. M. A. Schwegler, P. Vinke, M. van der Eijk and H. van Bekkum, *Appl. Catal. A-Gen.*, 1992, **80**, 41.
53. T. Okuhara, N. Mizuno and M. Misono, *Adv. Catal.*, 2001, **41**, 443.
54. M. Misono and N. Nojiri, *Appl. Catal.*, 1990, **64**, 1.
55. I. V. Kozhevnikov, *Catal. Rev. Sci. Eng.*, 1995, **37**, 311.
56. L. Weismantel, J. Stockel and G. Emig, *Appl. Catal.*, 1996, **137**, 129.
57. T. Okuhara, N. Mizuno and M. Misono, *Adv. Catal.*, 1996, **41**, 113.
58. M. Langpape and J.-M. Millet, *Appl. Catal. A-Gen*, 2000, **200**, 89.
59. B. B. Bardin and R. J. Davis, *Appl. Catal. A-Gen*, 1999, **185**, 283.
60. M. Ai, *Appl. Catal. A-Gen*, 1982, **4**, 245.
61. W. Li and W. Ueda, *Catal. Lett.*, 1997, **46**, 261.
62. A. Di Capua, J.-L. Dubois and M. Fournier, *J. Mol. Catal. A: Chem.*, 2007, **263**, 62.
63. EP 1 142 638, 2001, Nippon Shokubai Co., Ltd.
64. L. Marosi, G. Cox, A. Tenten and H. Hibst, *J. Catal.*, 2000, **194**, 140.
65. F. C. Jentoft, S. Klokishner, J. Kröhnert, J. Melsheimer, T. Ressler, O. Timpe, J. Wienold and R. Schlögl, *Appl. Catal. A-Gen*, 2003, **256**, 291.
66. G. Mestl, T. Ilkenhans, D. Spielbauer, M. Dieterle, O. Timpe, J. Kröhnert, F. Jentoft, H. Knözinger and R. Schlögl, *Appl. Catal. A-Gen*, 2001, **210**, 13.
67. S. Berndt, *Dissertation*, Technical University of Berlin, Berlin, 1999.
68. N. Ballarini, F. Candiracci, F. Cavani, H. Degrand, J.-L. Dubois, G. Lucarelli, M. Margotti, A. Patinet, A. Pigamo and F. Trifirò, *Appl. Catal. A-Gen.*, 2007, **325**, 263.
69. C. Marchal-Roch, N. Laronze, N. Guilou, A. Teze and G. Herve, *Appl. Catal. A-Gen.*, 2000, **199**, 33.
70. L. Marosi and C. Otero Aréan, *J. Catal.*, 2003, **213**, 235.
71. L. Marosi, G. Cox, A. Tenten and H. Hibst, *J. Catal.*, 2000, **194**, 140.
72. H. Böhnke, J. C. Petzoldt, B. Stein and J. W. Gaube, *Stud. Surf. Sci. Catal.*, 2000, **130**, 1745.
73. J. B. Wagner, D. Sheng Su, S. A. Schunk, H. Hibst, J. Petzoldt and R. Schlögl, *J. Catal.*, 2004, **224**, 28.

74. S. A. Schunk, A. Sundermann and H. Hibst, *Catal. Today.*, 2008, **137**, 36.
75. T. Inoue, K. Asakura and Y. Iwasawa, *J. Catal.*, 1997, **171**, 184.
76. T. Inoue, K. Asakura and Y. Iwasawa, *J. Catal.*, 1997, **171**, 457.
77. T. Inoue, S. T. Oyama, H. Imoto, K. Asakura and Y. Iwasawa, *Appl. Catal. A*, 1997, **165**, 183.
78. T. Inoue, S. T. Oyama, H. Imoto, K. Asakura and Y. Iwasawa, *Appl. Catal. A*, 2000, **191**, 131.
79. M. Ai, E. Muneyama, A. Kunishige and K. Ohdan, *J. Catal.*, 1993, **144**, 632.
80. J.-M. M. Millet, *Catal. Rev. – Sci. Eng.*, 1998, **40**(1&2), 1.
81. M. Ai, E. Muneyama, A. Kunishige and K. Ohdan, *Catal. Lett.*, 1994, **24**, 355.
82. EP00057320B1, 1982, The Standard Oil Company.
83. GB 02090589A, 1982, Ashland Oil Inc.
84. US03975301, 1976, Eastman Kodak Company.
85. US04298755, 1981, Ashland Oil Inc.
86. US04364856, 1982, The Standard Oil Company.
87. US04366088, 1982, Ashland Oil Inc.
88. US04390725, 1983, Ashland Oil Inc.
89. US04410727, 1983, Ashland Oil Inc.
90. US04410729, 1983, Ashland Oil Inc.
91. US04473707, 1984 The Standard Oil Company.
92. US04499302, 1985, Montedison S. p. A.
93. M. Ai, E. Muneyama, A. Kunishige and K. Ohdan, *Appl. Catal. A*, 1994, **109**, 135.
94. US4359401, 1981, Ashland Oil Inc.
95. M. Ai, *J. Mol. Catal. A: Chem.*, 1996, **114**, 3.
96. E. Muneyama, A. Kunishige, K. Ohdan and M. Ai, *J. Mol. Catal.*, 1994, **89**, 371.

Section II
Other Selective Oxidations

CHAPTER 8

Gold as a Selective Oxidation Catalyst

GRAHAM J. HUTCHINGS

Cardiff Catalysis Institute, School of Chemistry, Cardiff University, Cardiff, CF10 3AT, UK

8.1 Introduction

The recent discovery that supported nanoparticles of gold are effective catalysts for reactions at low temperatures has to be regarded as one of the most fascinating recent observations in chemistry. Until recently, most researchers would not have considered investigating the possibility that gold could be an effective catalyst for any reaction, let alone the oxidation of carbon monoxide at temperatures well below room temperature. Even today many consider gold to be a nonreactive metal, and indeed in its bulk form it is. However, when divided to the nanoscale, to just a few atoms, gold demonstrates remarkable activity for oxidation and reduction reactions. Consequently, a large number of experimental and theoretical studies are being undertaken to try to elucidate the nature of this interesting catalytic activity. This recent research has been reviewed by Haruta,[1–5] Bond and Thompson,[6,7] Bond,[8] Thompson,[9] Freund and coworkers,[10] Cortie,[11] Hashmi,[12] Hutchings,[13–15] Hashmi and Hutchings,[16] Goodman,[17] Gates,[18] Corma,[19] Rossi,[20] Willock[21] and indeed a whole issue of Chemical Reviews has been devoted to the topic of the new chemistry of gold. It should be noted that copper and silver (both in the same triad of the periodic table as gold) are used in many large-scale catalytic processes, and it has been known for many years that the

RSC Nanoscience & Nanotechnology No. 19
Nanostructured Catalysts: Selective Oxidations
Edited by Christian Hess and Robert Schlögl

Published by the Royal Society of Chemistry, www.rsc.org

preparation of active catalysts with copper and silver requires the metal to be well dispersed on a support. Hence, there should not be so much surprise when the same is observed for gold, but because gold was considered to be inert few experimental studies explored this area until the seminal studies of Haruta[22] who discovered the high activity of gold for CO oxidation at subambient temperature. This new discovery brings with it the opportunity that gold, in an appropriate form, is perhaps the most interesting metal in the Periodic Table with respect to its potential to act as a catalyst.

Bond and coworkers[23] in early studies demonstrated that very small gold particles supported on silica could give interesting catalytic performance for hydrogenation, but until very recently the use of gold as selective hydrogenation catalyst has received little attention.[24] In the 1980s there were two significant observations that completely changed this perception and highlighted the special attributes of gold as a heterogeneous catalyst:

- the discovery that supported Au catalysts are very active for low-temperature CO oxidation;[22]
- the prediction that Au would be the best catalyst for ethyne hydrochlorination[25] and the experimental validation that this prediction was correct.[26]

Both of these early experimental demonstrations showed that gold was not just an exceptional catalyst; for these two applications it was shown to be, by far, the best. The observation that Au could oxidize CO at very low temperatures has caught the imagination of many researchers worldwide, and today many groups are active in this field. Since the 1980s publications involving gold as a catalyst have increased exponentially and even when normalized as a proportion of all scientific publications the number of papers concerning catalysis by gold has seen an increase by 2–3 orders of magnitude during this period. In this chapter the use of gold as an oxidation catalyst will be described in an introductory manner.

8.2 Gold Catalysis: the Early Years

While the current fascination with gold catalysis for both homogeneous and heterogeneously catalyzed systems continues, we should consider how this change in fortune for gold has come about. For many years researchers had tried to use gold as a catalyst but it was found to be relatively inactive. Indeed, in 1972 it was stated[27] "Although the catalytic properties of gold are surpassed by those of the Group VIII metals, especially palladium and platinum, possible applications of gold in catalytic processes have been widely studied", and even as late as 1985 Schwank concluded[28] "in spite of its low intrinsic activity, gold can influence the activity and selectivity of group VIII metals". This was for many scientists, at that time, the general conclusion concerning the catalytic prospects for gold, at best it might improve the activity of other more-active metals. This is of course a very comfortable conclusion, after all gold is a very stable element and the chemistry of gold as represented in the inorganic texts of the time was severely limited. However, Bond and coworkers[23] subsequently showed that nanocrystalline gold could be effective for the hydrogenation of dienes. This is a key publication as the key finding was that very small Au particles displayed activity. Unfortunately, gold was not shown to be more effective than well-dispersed palladium or platinum and as Pt and Pt catalysts were well established, gold was not viewed as a viable alternative, and so the idea of gold catalysis was not taken further at that time.

In the early to mid-1980s there were two significant observations that completely changed this perception and highlighted the special attributes of gold as a heterogeneous catalyst. These discoveries were made independently by Haruta and Hutchings,[29] namely the discovery that supported Au catalysts are very active for low-temperature CO oxidation,[22] and the prediction that Au would be the best catalyst for ethyne hydrochlorination[25,26] In both these cases gold was observed to be the most active catalyst, by far, and these discoveries lay at the heart of the new impetus on gold catalysis. However, real interest in gold catalysis did not materialize until the late 1990s, *i.e.* not until over a decade after the initial discoveries; but after the late 1990s the publications and patents concerning gold catalysis have been growing exponentially even when normalized against the general growth in all scientific publications. The reasons for the delay in the development of this new field may be complex. For example, ethyne hydrochlorination is a difficult reaction to study in the laboratory, and with inexpensive oil prices persisting until recently, interest in this reaction has been waning industrially. It is only now that coal is once again competitive with oil that interest in the production of VCM by ethyne hydrochlorination again becomes commercially viable. Also, the preparation of supported gold catalysts can present difficulties and the introduction of standard reference catalysts has eased this situation.[30] However, low-temperature CO oxidation is a facile reaction to replicate and once the key aspects of catalyst preparation had been mastered, the investigation of gold catalysts for this reaction has been the driving force behind the recent research.

The observation that gold can catalyze low-temperature CO oxidation has continued to fascinate most active researchers in the field of gold catalysis and much of this effort has been focused on trying to answer two key questions: (a) what is the mechanism by which supported gold nanocrystals oxidize CO and (b) what is the nature of the active site and is it the same for all supported gold catalysts?

The efficacy of gold for CO oxidation may well find commercial application, and a recent review by Corti *et al.*[31] highlights that this may occur both in the removal of CO[32,33] in respiratory protection as well as in fuel cells.[34] However, there exist possibilities for gold to be used as catalysts in chemical synthesis rather than in total or preferential oxidation, and this is the topic that will be explored in the next sections of this chapter. While this chapter will concentrate on the use of gold in selective oxidation reactions, it should be noted that supported gold catalysts are also highly effective for selective hydrogenations. Following on from the early work by Bond and coworkers,[23] where gold was found to be effective for the hydrogenation of dienes, gold has been found to selectively oxidize an α β-unsaturated aldehyde to the unsaturated alcohol, in preference to the saturated aldehyde.[24,35,36] Indeed, very high specificities can be achieved of *ca.* 85%.[35] Recently, there have been some remarkable developments in which α β-unsaturated nitrocompounds are hydrogenated with molecular hydrogen to give oximes.[37–39] This is an application where previously only noncatalytic stoichiometric reactions using bulky hydrogen donors had been found useful. In particular, the synthesis of cyclohexanone oxime is of central importance in the manufacture of nylon,[37] and this new application of gold catalysis is of significant interest for this application. In addition, gold–palladium alloys have been shown to have very high activities and selectivities for the direct hydrogenation of molecular oxygen to form hydrogen peroxide.[40,41] Hence, it can be considered that gold catalysis should have a very bright commercial future.

8.3 Very Small Gold Clusters are Active in Oxidation Reactions

The nature of the active site for CO oxidation by gold has provoked considerable debate. Bond and Thompson[6] initiated this debate by proposing a model where Au atoms at the interface between the Au particle and the oxide are the active oxidation centers. They based this proposal on the basis of experimental data gleaned from the literature they reviewed. The periphery of a gold nanoparticle exposes active sites that interact with the support, and the smaller the particle diameter the higher the ratio of these peripheral atoms to bulk atoms, and hence their potential relative importance increases. A key question that has provoked considerable debate has been aimed at addressing the size of the active gold nanocrystals in oxidation catalysts.

One of the early benchmark studies was by Goodman and coworkers.[42] Using a combination of STM and spectroscopy they considered that the unusual reactivity could be due to quantum size effects of the very small gold particles. Lopez *et al.*[43] assessed the role of the particle size in the activation of CO in detail and they collated a large number of published results. They compiled a plot of relevant data (Figure 8.1) and this showed that the activity of gold nanoparticles increased dramatically as the particles get smaller. Hence, they concluded that the most important effect controlling reactivity is related to the availability of many low-coordinated gold atoms on the small particles, and this in turn is related to the particle diameter.

A subsequent model study by Goodman and coworkers[44] addressed the role played by these peripheral atoms by designing well-ordered monolayers and bilayers of gold atoms that completely cover the TiO_2 support that clearly do not comprise any of these peripheral sites. They found that the bilayers were more active than the monolayers (Figure 8.2(a)), indicating that the catalysis can proceed on the extended gold surface so created. Goodman[42,45] has previously shown that the particles present in an active Au/TiO_2 catalyst closely resemble the model systems used in this later study[42,44,45] (Figure 8.2(b)).

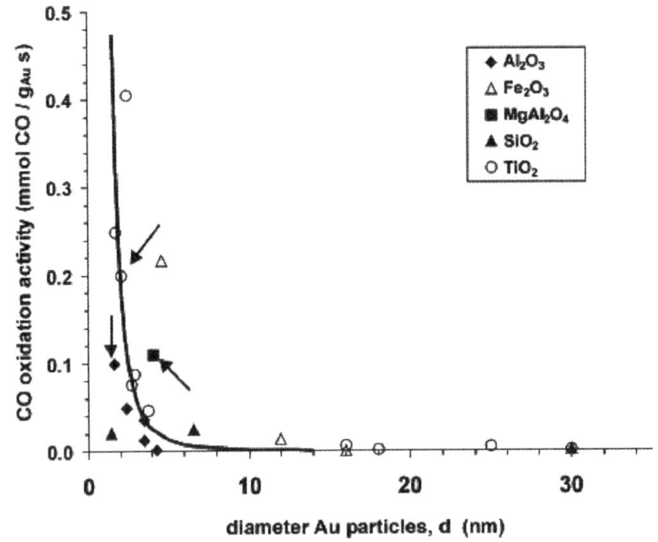

Figure 8.1 Measured activities (in mmol CO/(g Au s)) for CO oxidation at 273 K over different Au-based catalysts as a function of the average particle size (*d*, in nm). Supports are indicated by the symbol shape. Open symbols are used for reducible supports; solid symbols for irreducible supports. The curve shows a $1/d^3$ guide to the eye, showing that the activity of gold catalysts is approximately proportional with the number of low-coordinated atoms at the corners of the gold particles. The three points arrowed are original data presented by Lopez *et al.*,[43] all other data points taken from the literature by the authors.[43]

In active supported gold oxidation catalysts it is apparent that very small particles are active, but the central problem is that real catalyst samples comprise a broad range of particle diameters (see Figure 8.3).[46] Although the mean particle size in the active catalysts can be quite narrow. For example, in Figure 8.3(b) the mean particle size was found to be 1.1 ± 0.5 nm, there is still a broad range of structures present. However, clearly, the activity is not associated solely with nanoparticles in the size range 2–4 nm, which was until recently considered to be the optimal nanoparticle size range. It is more likely that real catalysts prepared with mean particle sizes in this size range happen to produce a number of very active smaller particles. It is only now that the use of aberration-corrected STEM is helping us to determine that very small particles are present in active catalysts.

Recently, aberration-corrected STEM has been used to unravel which structures in a supported gold catalyst are associated with high activity for Au supported on FeOOH, which is known to be a very active catalyst for CO oxidation. This catalyst has been previously studied in great detail[47] and some intriguing results were reported concerning the link between catalyst performance and catalyst drying conditions were reported. A pair of 2.9 at% Au/FeO$_x$ samples (denoted A, B) were prepared from the same coprecipitated precursor.

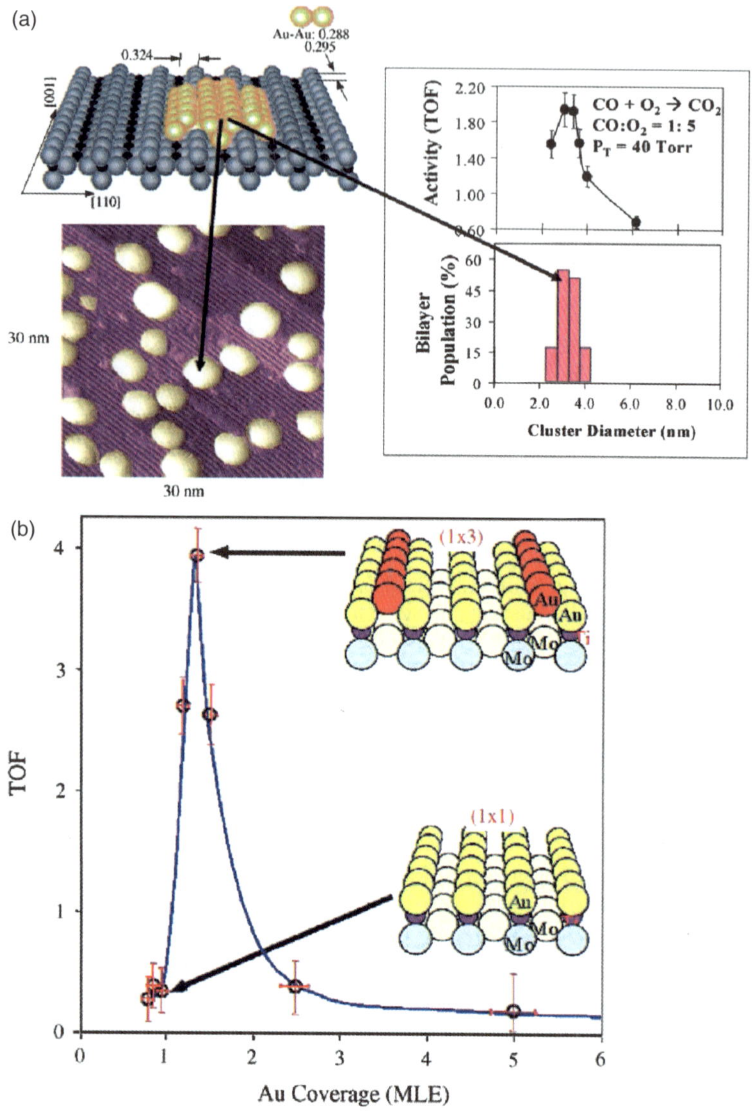

Figure 8.2 (a) An STM image (bottom left) and schematic (upper left) of Au bilayer clusters on TiO_2(1 1 0) whose population are dominant (lower right) for a Au/TiO_2 catalyst most active for CO oxidation (upper right). The activity (at 350 K) is expressed as turnover frequency (TOF) in units of CO_2 molecules produced per Au site per second.[44] (b) Room-temperature CO oxidation activity as a function of Au coverage on Mo(1 1 2)–(8×2)-TiO_x. $CO:O_2 = 2:1$; total pressure = 5 Torr. Schematic models for the (1×1)- and (1×3)-Au/TiO_x surfaces are shown in the inserts; the Au structure was determined by LEED 30. Au coverage is defined as: 1 monolayer equivalent (MLE) = 1.4×10^{15} atoms cm^{-2}.[44]

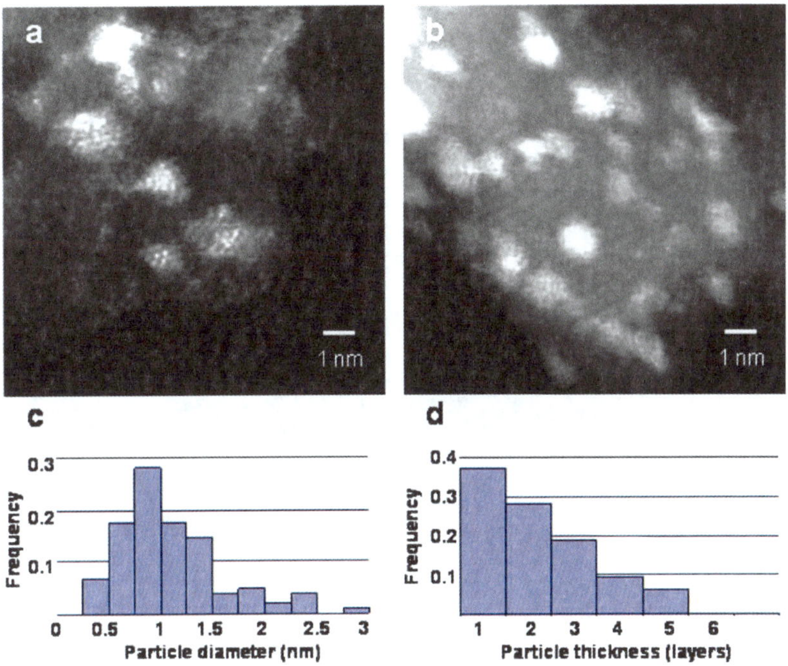

Figure 8.3 High-magnification *Z*-contrast micrographs showing 10 wt% loaded Au on anatase, obtained by Pennycook and coworkers,[46] after two stages of preparation. (a) In the oxidized precursor state following deposition–precipitation of Au, many individual Au atoms are sharply resolved. (b) In the most-active form, after mild reduction in 12% H_2 at 423 K, individual neutral Au atoms are not resolved. (c) Histogram of the distribution of 103 Au particle diameters in the reduced sample. Single atoms and particles over 10 nm were not included. Mean particle size was 1.1 $\pm$0.5 nm. (d) Histogram of the approximate distribution of 32 selected Au-particle thicknesses in the reduced sample determined from relative intensities of *Z*-contrast images.

Sample A, was dried in a tube furnace in static air, while B was dried in flowing air, both at 120 °C for 16 h. The Au loading in each was identical and the underlying support was disordered FeOOH with similar surface areas (*ca.* 190 $m^2 g^{-1}$). X-ray energy-dispersive spectroscopy (XEDS) analysis and high-angle annular dark-field (HAADF) imaging experiments indicated that both samples contained 2–15 nm Au particles, with mean particle sizes of 5.4 nm for A and 7.0 nm for B (Figure 8.4). If particle size was the key factor governing the activity of supported gold catalysts then these two samples should have similar activity. However, catalytic testing of the samples using standard conditions (total flow rate 66 000 vol gas/vol catalyst h; 0.5 vol% CO), revealed that B achieved 100% CO conversion at 25 °C, whereas A gave only trace CO conversion (*i.e.* $<1\%$). In our initial study[47] we were unable to determine the origin of this striking effect due to resolution and sensitivity limitations of

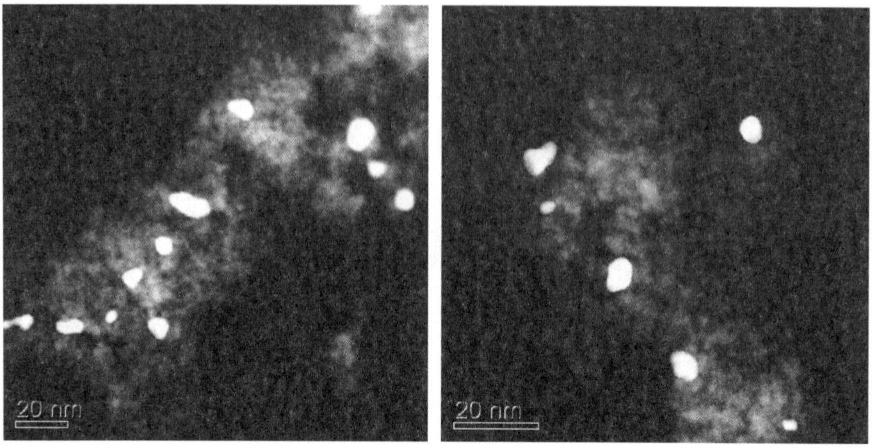

Figure 8.4 Low-magnification aberration-corrected HAADF-STEM images from (a) the inactive, tube furnace catalyst (Sample A) and (b) the highly active, GC oven catalyst (Sample B). The Au particle-size distribution in both samples appears to be very similar at this magnification.[47]

the characterization techniques available. Recently, we re-examined these samples using a state-of-the-art aberration-corrected STEM.[48] Using this higher magnification, (Figure 8.5) it is possible to determine that the Au particle-size distribution and morphology in these samples is quite different. Both samples do indeed contain similar numbers of larger (2–15 nm) Au particles (Figure 8.4). However, it is now possible to image individual gold atoms. Both samples contain a significant number of individual Au atoms (Figure 8.5, indicated by white circles) dispersed on the iron oxide surface. However, the active sample B also contained a large number of subnanometer Au clusters (Figure 8.5, circled in black), which were absent in the inactive sample A. These Au clusters, the majority of which were 0.2–0.5 nm in diameter, contain at most only a few Au atoms, but this study shows clearly that it is these very small species that are active in gold catalysis for CO oxidation. In a subsequent part of this study[48] we demonstrated that it is the bilayer nanoclusters containing around 10 atoms, rather than the smaller monolayer structures, that are the active species. The key observation is that very small clusters can be retained in real catalysts. This study provides a valuable link between model and real catalyst studies. The real challenge now facing materials scientists is how to synthesize and stabilize these structures in high concentrations, since there population based on the total number of gold atoms present in real catalysts is very small.

8.4 Oxidation of Alcohols

The investigation of a range of oxidation reactions using gas- or liquid-phase reagents to produce valuable intermediates or products for the chemicals industry has become a major research target. The reasons for this are many,

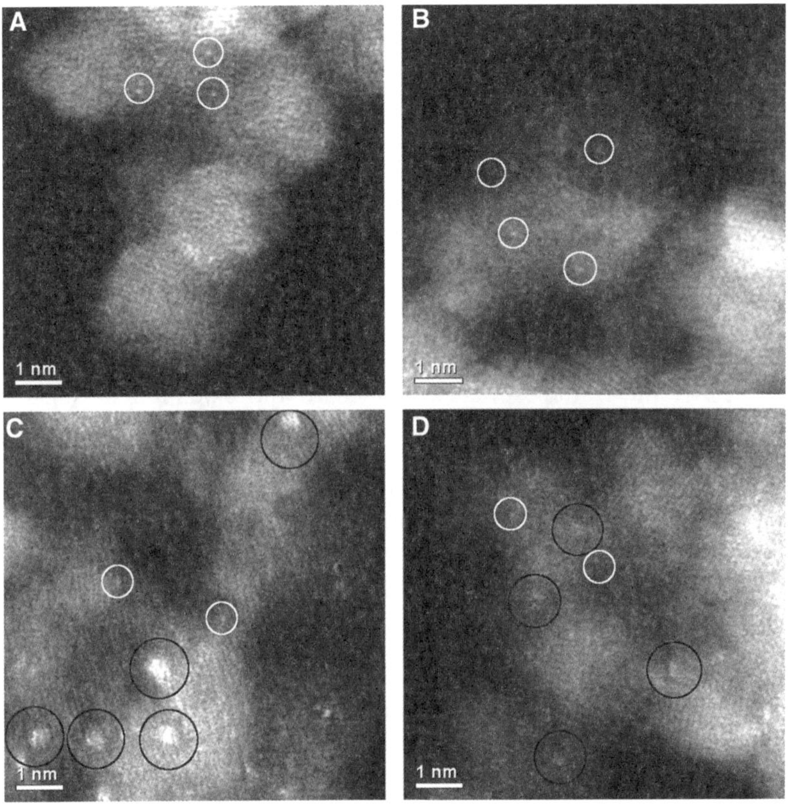

Figure 8.5 High-magnification aberration-corrected STEM-HAADF images of the inactive (A) (a and b) and active (B) (c and d) Au/FeO_x catalysts acquired with the aberration-corrected JEOL 2200FS. The white circles indicate the presence of individual Au atoms, while the black circles indicate sub-nm Au clusters consisting of only a few atoms. Note the presence and image intensity difference of two distinct cluster-types: in (c) there are 0.5-nm, higher contrast clusters whereas in (d) 0.2–0.3-nm low-contrast clusters dominate. This indicates the presence of bilayer and monolayer sub-nm Au clusters in the active catalyst.[48]

but, in particular, it is the drive for greener processes based on the realization that reaction selectivity is the most important factor in chemical synthesis, as making nondesired byproducts can now be very costly. There has therefore been a drive to identify atom-efficient oxidation reactions that can be carried out with molecular oxygen, rather than an activated form of oxygen, using mild solvent-free reaction conditions. The oxidation of alcohols and polyols to chemical intermediates represents a demanding target.

Rossi, Prati and coworkers[49–51] were the first to clearly demonstrate in the seminal studies that supported gold nanoparticles can be very effective catalysts for the oxidation of alcohols, including diols to the corresponding acid, for example 1,2 propane diol formed lactic acid by oxidation of the primary alcohol group rather than the secondary alcohol group, which might have been expected to be more reactive. The presence of base (typically NaOH) was found to be essential for the observation of activity, and consequently sodium salts of the acids were formed as products. The base was considered to be essential for the first hydrogen abstraction, and this is a significant difference between the supported gold catalysts and Pd and Pt catalysts that are effective in acidic as well as basic conditions. The catalysts were Au/carbon and they were observed to be effective for a range of substrates, including alcohols, diols and polyols such as sugars. In addition, the catalysts were shown to be effective with gas-phase reactants and in this case no base addition is required.[52] However, in this case the products formed were aldehydes rather than the sodium salt of the acids that was the dominant product when NaOH was added as the base. Subsequently, these studies were extended to the oxidation of sugars and similar high catalytic efficiency for the oxidation of glucose and sorbitol has been observed.[53,54] More recently, they have reported the synergistic effect of the addition of Pd or Pt to the Au/carbon catalysts for the selective oxidation of D-sorbitol to gluconic acid.[55] Interestingly, Rossi and coworkers[56] have shown that "naked" gold colloidal particles can be very effective catalysts for the oxidation of glucose to gluconic acid. They demonstrated that the initial rates for these nonsupported particles were identical to the rates observed with Au/carbon catalysts operated under the same conditions; hence, confirming that the support is of limited importance in the origin of the catalyst activity in these oxidation reactions. The support–catalyst interaction is, however, essential for the observation of a stable catalyst system. Subsequently, Mertens *et al.*[57,58] have also shown that colloidal gold can catalyze the oxidation of 1,2-diols. Recently, Tsunoyama *et al.*[59] have shown that benzyl alcohol can be oxidized using oxygen in aqueous media with gold nanoclusters stabilized on polymers.

Christensen and coworkers[60,61] have made a number of significant advances in the direct oxidation of primary alcohols using supported gold nanocrystals and they have concentrated on decreasing the amount of base present in these oxidations. They have shown that gold can catalyze the oxidation of aqueous solutions of ethanol to give acetic acid in high yields.[60] This provides a potential new route to a key commodity chemical that is based on a biorenewable feedstock using a substantially green-technology approach. Recently, this group have shown that methyl esters of a broad range of primary alcohols can be produced using a similar approach.[61]

The use of Au/carbon catalysts were extended by Carrettin *et al.*[62–64] for the oxidation of glycerol. Glycerol is a highly functionalized molecule that is readily available from biosustainable sources, for example it can be obtained as a byproduct of the utilization of rape seed and sunflower crops. This makes glycerol a particularly attractive starting point for the synthesis of intermediates, and a large number of products can potentially be obtained from glycerol oxidation and this complexity represents both an opportunity and a problem. In particular, it was found that Au supported on graphite can oxidize glycerol to glycerate with 100% selectivity using dioxygen as the oxidant under relatively mild conditions with yields approaching 60%.[62–64] Under comparable conditions, supported Pd/C and Pt/C always gave other C_3 and C_2 products in addition to glyceric acid and, in particular, also gave some C_1 byproducts. For the Au/carbon catalyst, it was observed that the selectivity to glyceric acid and the glycerol conversion were very dependent upon the glycerol/NaOH ratio. In general, with high concentrations of NaOH, exceptionally high selectivities to glyceric acid can be observed. However, decreasing the concentration of glycerol, and increasing the mass of the catalyst and the concentration of oxygen, leads to the formation of tartronic acid *via* consecutive oxidation of glyceric acid. It is apparent that, with careful control of the reaction conditions, 100% selectivity to glyceric acid can be obtained with 1 wt% Au/C.

In the initial studies with supported gold catalysts it was noted that alcohol oxidation could be achieved under mild conditions when base was present.[51] However, it has been known for some time that highly dispersed metal catalysts, *e.g.* Pd and Ru could effectively catalyze the oxidation of alcohols to aldehydes and ketones. One of the most significant advances in the field of alcohol oxidation has been the observations of Corma and coworkers[65,66] showing that an Au/CeO_2 catalyst is active for the selective oxidation of alcohols to aldehydes and ketones and the oxidation of aldehydes to acids. In these studies the catalysts are active at relatively mild conditions, without the addition of a solvent, using O_2 as oxidant without the requirement for the addition of NaOH to achieve high activity. This last factor is most significant and the omission of NaOH permits the use of green reaction conditions for these oxidation reactions. The results were shown to be comparable to, or higher than, the highest activities that had been previously observed with supported Pd catalysts.[67] The catalytic activity was ascribed to the Au/CeO_2 catalyst stabilizing a reactive peroxy intermediate from O_2. We have previously shown that supported Au–Pd alloys are efficient catalysts for the direct synthesis of H_2O_2 from H_2 oxidation by O_2 at low temperatures.[40,41] In particular, Au–Pd/TiO_2 catalysts were very selective for hydrogen-peroxide synthesis. Hydroperoxy species are considered to be involved in this H_2O_2 formation process, and because hydroperoxy species are key reagents/intermediates in the oxidation of alcohols,[68] we reasoned that these catalysts should also be effective for the oxidation of alcohols. This we found to be the case[69] (Table 8.1), and the catalysts previously found to be effective for hydrogen-peroxide synthesis were all effective for the oxidation of benzyl alcohol, which we used as a test reactant. TiO_2-supported catalysts were found to be the most selective and these were examined in greater detail for the oxidation of benzyl alcohol at 100 °C with O_2 as oxidant in the absence of solvent (Figure 8.6). The Au–Pd/TiO_2 catalysts were very active for this reaction, and the selectivity to benzaldehyde was >96%, with the only byproduct being benzyl benzoate. In contrast, Pd/TiO_2 also produced toluene and benzene as byproducts, and the Au/TiO_2 catalyst produced a significant amount of an acetal product. The selectivity of the Au/TiO_2 catalyst for benzaldehyde decreased with the time on line, but further oxidation of the acetal byproduct led to an increase of the final selectivity in benzaldehyde. The effect of adding Au to a Pd/TiO_2 catalyst is apparent in these initial studies. Although the Pd/TiO_2 catalyst has a high

Table 8.1 Comparative data for benzyl alcohol oxidation and hydrogen-peroxide synthesis.[a,69]

	Benzyl alcohol oxidation					
	Conversion/%		*Benzaldehyde selectivity/%*		*Benzaldehyde productivity*[c]/	*H_2O_2 productivity/*
Catalyst	*0.5 h*	*8 h*	*0.5 h*	*8 h*	*mol/(h kg$_{cat}$)*	*mol/(h kg$_{cat}$)*
2.5%Au–2.5% Pd/Al_2O_3	2.6	83.3	90.5	86.6	174	23
2.5%Au–2.5% Pd/TiO_2	3.7	74.5	95.2	91.6	165	64
2.5%Au–2.5% Pd/SiO_2	3.6	35.7	97.3	88.0	76	80
2.5%Au–2.5% Pd/Fe_2O_3	3.6	63.4	74.9	66.4	102	16
2.5%Au–2.5% Pd/C	2.9	69.2	53.9	46.4	78	30
2.5%Au/TiO_2	0.6	15.3	96.7	63.9	24	(<2) nd[b]
2.5%Pd/TiO_2	13.4	60.1	51.3	54.4	79	24

[a]Comparative results obtained for the oxidation of benzyl alcohol after 8 h reaction and for H_2O_2 synthesis for 0.5 h. The oxidation of benzyl alcohol was carried out at 373 K temperature, 0.2 MPa O_2 pressure and 1500 rpm stirrer speed. The H_2O_2 synthesis was carried out under the conditions described in refs. 26 and 27.
[b]Very low rate <2 mol-H_2O_2/h/kg$_{cat}$
[c]Calculated for 8 h reaction

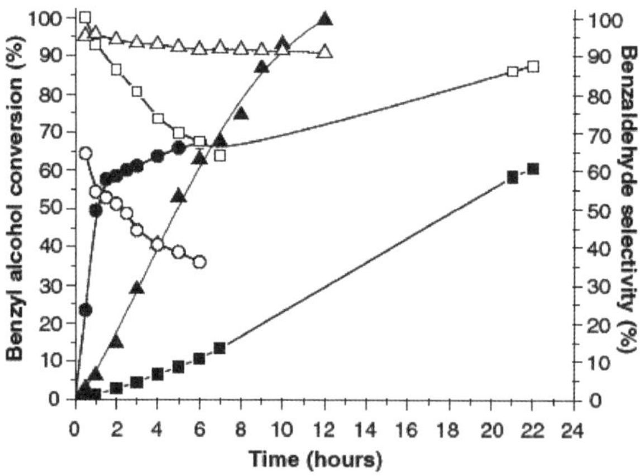

Figure 8.6 Benzyl-alcohol conversion and selectivity in benzaldehyde with the reaction time at 373 K, 0.1 MPa O_2 pressure: (■) Au/TiO_2, (●)Pd/TiO_2, (▲) Au–Pd/TiO_2; solid symbols – conversion, open symbols – selectivity (benzyl alcohol (40 ml), catalyst (200 mg), 100 °C, 0.2 MPa O_2).[69]

initial activity, and the addition of Au decreases the activity, the Au–Pd/TiO_2 catalyst retained high selectivity to benzaldehyde at high conversion rates, a feature not observed with the supported pure-Au and pure-Pd catalysts.

Carbon mass balances were 100%, and no carbon oxides were formed for Au–Pd/TiO_2 or Au/TiO_2 catalysts, but carbon oxides were formed with the supported Pd catalysts. Most importantly, no leaching of gold or palladium was observed with these catalysts, and it should be noted that these reactions were carried out using the catalysts prepared by calcination at 400 °C, which had been shown to be totally stable during the direct formation of hydrogen peroxide and to possess a core–shell structure with a gold rich core and a palladium-rich shell.

The Au–Pd/TiO_2 catalysts were investigated with a range of substrates and conditions (Table 8.2). It is clear that these catalysts are highly reactive for a very broad range of alcohols. As noted previously, both Kaneda and coworkers[67] and Corma and coworkers[65] have shown that supported Pd and Au monometallic catalysts are highly effective for the oxidation of 1-phenylethanol under solvent free conditions at 160 °C with a pO_2 of 0.1 MPa. Under these conditions, the Pd/HAP and Au/CeO_2 catalysts gave TOFs of 9800 and 12 500 h^{-1} for 1-phenylethanol. Under these conditions, the Au–Pd/TiO_2 catalyst, gave a TOF of 269 000 h^{-1}. This represents a significant enhancement on the TOFs of the single metals. The Au–Pd/TiO_2 catalyst is also effective for a range of straight-chain, benzylic, and unsaturated alcohols (Table 8.2) and, in particular, for the oxidation of primary alcohols, such as butan-1-ol and octan-1-ol, and high TOFs are observed.

8.5 Alkene Epoxidation

Supported gold catalysts comprising nanocrystalline metallic gold have been found to be particularly effective for the epoxidation of alkenes as long as a sacrificial reductant, typically H_2 is present to aid the activation of molecular oxygen. Most research has focused on the oxidation of propene to propene oxide that is a major research target, as this is a commodity chemical used in the manufacture of polyurethane and polyols. Recent studies by Lambert and coworkers[70] are particularly promising and they have shown that supported catalysts can give selectivities of *ca.* 50% at 0.25% conversion, in the absence of a sacrificial reductant, but the selectivity declines rapidly with increasing conversion. At present, the most promising commercial route for the synthesis of propene oxide is based on the reaction between propene and hydrogen peroxide using the titanium silicalite TS-1 as catalyst. However, the relative costs of hydrogen peroxide and molecular oxygen make a process based on molecular oxygen far more preferable, but this requires the solution to the major problem of reaction selectivity.

Haruta and coworkers were the first to demonstrate the potential of supported gold catalysts for the epoxidation of propene with dioxygen in the presence of H_2 as a sacrificial reductant. H_2 permits the activation of O_2 at relatively low temperatures permitting the selective oxidation of propene to occur.[71,72] Haruta found that Au/TiO_2, prepared using the deposition precipitation, was selective for propene epoxidation and the catalysis was associated with an intimate contact between hemispherical gold nanocrystals (2–5 nm

Table 8.2 Comparison of the catalytic activity for alcohol oxidation to the corresponding aldehyde. Catalyst is 2.5% Au–2.5% Pd/TiO_2 unless noted otherwise; substrates oxidized without solvent unless specified; catalyst mass varied to give metal concentrations indicated; TOF measured after first 0.5 h reaction.[69]

		Reaction conditions		*[metal] 10^{-5} mol/l*		
Entry	*Alcohol*	T/*K*	P/*Pa*×*10^5*	*Au*	*Pd*	*TOF/h^{-1}*
1	Benzyl alcohol	373	2	63.5	118	607
2	Benzyl alcohol*	373	2	63.5	0	213
3	Benzyl alcohol†	373	2	0	118	2200
4	Benzyl alcohol	373	1	2.1	3.9	6190
5	Benzyl alcohol	373	2	2.1	3.9	6440
6	Benzyl alcohol	373	5	2.1	3.9	6190
7	Benzyl alcohol	373	10	2.1	3.9	5950
8	Benzyl alcohol	383	1	2.1	3.9	14 270
9	Benzyl alcohol	393	1	2.1	3.9	26 400
10	Benzyl alcohol	433	1	2.1	3.9	86 500
11	1-Phenylethanol	433	1	1.8	3.2	269 000
12	3-Phenyl-1-propanol	433	1	2.1	3.9	2356
13	Vanillyl alcohol≠	363	1	21.6	40.6	10
14	Cinnamyl alcohol‡	363	1	21.6	40.6	97
15	Octan-1-ol	433	1	2.5	4.7	2000
16	Octan-2-ol	433	1	2.5	4.7	0
17	Octan-2-ol/Octan-1-ol	433	1	2.1	3.9	0
18	Octan-3-ol	433	1	2.1	3.9	10 630
19	1-Octen-3-ol	433	1	2.1	3.9	12 600
20	Crotyl alcohol	433	5	2.1	3.9	12 600
21	Butan-1-ol	433	5	2.1	3.9	59 30
22	1,2-Butanediol	433	1	2.1	3.9	1520
23	1,4-Butanediol	433	1	2.1	3.9	104 200
24	Benzyl alcohol#	433	1	2.1	3.9	12 500
25	Benzyl alcohol††	433	1	2.1	0	12 400
26	Benzyl alcohol‡‡	433	1	0	3.9	24 800
27	Benzyl alcohol¶	433	1	2.4	4.5	36 500
28	Benzyl alcohol§	433	1	0	3.6	37 600
29	1-Phenylethanol§	433	1	0	3.1	11 600

*2.5%Au/TiO_2
†2.5%Pd/TiO_2
≠0.2 mol/l in toluene as solvent
‡0.2 mol/l in water as solvent
#2.5%Au–2.5%Pd/HAP prepared by impregnation of HAP with $HAuCl_4 \cdot 3H_2O$ and $PdCl_2$
††2.5%Au/HAP prepared by impregnation of HAP with $HAuCl_4 \cdot 3H_2O$
‡‡2.5%Pd/HAP prepared by impregnation of HAP with $PdCl_2$
¶2.5%Au-2.5%Pd/TiO_2 prepared with the method of Kaneda *(6)* using TiO_2 as support
§0.2%Pd/HAP prepared using the method of Kaneda *(6)* using HAP as support

in diameter) and the TiO_2 support. Initial selectivities were low but promising and improvement were made by using different titanium-containing supports including TS-1, Ti-zeolite β, Ti-MCM-41 and Ti-MCM-48.[73–86]

It is clear that all these studies confirm that with H_2 as a sacrificial reactant Au/TiO_x catalysts are very effective for this epoxidation. Many of the early studies concentrated on the use of TS-1 as a support as it is known that this material is selective for the epoxidation of propene with H_2O_2 as the oxidizing species.[73] Haruta and coworkers[74] found the Au/TS-1 catalysts produced more propanal than propene oxide. In contrast, Moulijn[75] and coworkers clearly showed that Au/TS-1 catalysts could be very selective to the formation of propene oxide, and that the catalysts were stable under reaction conditions. Recent mechanistic studies by Nijhuis and coworkers[76,77] show that a bidentate propoxy species is formed as an intermediate, and that the role of the gold nanoparticles in the catalytic reaction is not only to provide the peroxy species but is also involved in establishing the bidentate propoxy intermediate.[77] Nijhuis and coworkers[78] have also recently shown that water plays a role in decreasing the deactivation of these gold catalysts. Studies by Delgass and coworkers[79,80] have confirmed the findings concerning the role of the bidentate propoxy species. In particular, it has been observed that the gold loading requires careful tuning, as had previously been noted by Haruta and coworkers,[81] to obtain high selectivities and activities[87] and when normalized for gold content propene oxide yields of 350 g/h g_{Au} are achievable at 200 °C for a 0.01 wt% Au/TS-1 catalysts with Si/Ti ratio of 500. In these catalysts the active species are considered to be gold species with diameters much smaller than 2 nm.

Haruta and coworkers have met the challenge of poor H_2 utilization and have shown that using mesoporous titanosilicates as a support[87] reasonably efficient H_2 consumption was attained together with high propene oxide yields (93 g/h kg_{cat} at 160 °C) and observing propene oxide selectivities of >90%, at propene conversions of *ca.* 7% together with a hydrogen efficiency of 40%.[83]

In a different approach, using liquid-phase reactions at low temperature, Choudhary and coworkers,[88–92] and subsequently Yin *et al.*[93] have shown that gold nanoparticles supported on a range of oxides are active for the epoxidation of styrene using tertiary butylhydroperoxide (TBHP) in greater than stoichiometric amounts. Selectivities to styrene oxide of *ca.* 50% are readily achieved using this approach.

Subsequently, it was demonstrated that the hydroperoxy species could be used in catalytic amounts with oxygen being the oxidant. Using Au/graphite as a catalyst for alkene oxidation[94] it was found that selective epoxidation could be attained. The initial experiments investigated the oxidation of cyclohexene using 1% Au/C with oxygen using polar solvents (*e.g.* water) in a stirred autoclave reactor or a stirred nonpressurized glass reactor. Even under relatively mild conditions (60–80 °C), only CO_2, formic acid and oxalic acid were formed with up to 100% cyclohexene conversion (Table 8.3), but, crucially, no C_6 products were observed. Experiments involving the potential reaction of expected C_6 products with water as solvent under identical conditions indicated that had these been formed they would have been observed. It is therefore clear that in polar solvents the oxidation of the alkene does not proceed *via* the epoxide. It was at this stage that we carried out a key experiment in the absence of solvent but in the presence of molecular oxygen, in a stirred autoclave reactor. In this case, 2-cyclohexene-1-one and 2-cyclohexen-1-ol were observed.

Table 8.3 Effect of solvent on selective oxidation of cyclohexene using 1% Au/carbon[94]

		Product selectivity					
Solvent	*Conv (%)*	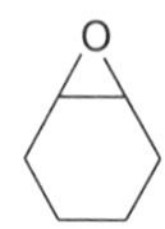	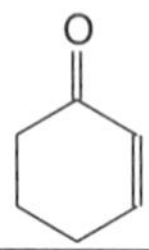	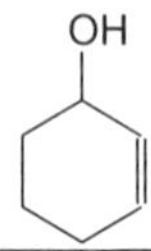	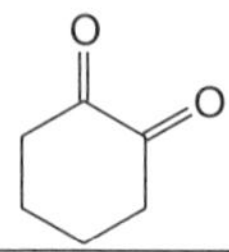	*ΣC_6 Sel*[a]	*Yield*[b]
Water	100[c]	0	0	0	0	0	0
Methanol	27.1[c]	0	0	0	0	0	0
THF	5.8[c]	0	0	0	0	0	0
Hexane	26.1	tr	tr	tr	0	0	0
Toluene	29.1	tr	35.1	25.1	0	60.2	17.5
1,4-Dimethylbenzene	53.5	0	12	0	43.5	55.5	29.7
1,3,5-Trimethylbenzene	8	tr	78.1	Tr	0	78.1	6.2
1,2,3,5-TMB[d]	29.7	50.2	26.3	0	0	76.3	22.7
1,2,4,5-TMB[d]/1,4-Dimethylbenzene	23.1	26	42	9.1	0	77.1	17.8
Quinoline	33.2	0	10.5	0	0	10.5	3.5
1,4-Difluorobenzene	29.1	0	47.1	26.8	0	73.9	21.5
Hexafluorobenzene	15.8	8.9	36.1	2.5	0	47.5	7.5

Reaction conditions: 1% Au/C (0.22 g), C_6H_{10} (0.012 mol), 80 °C, 24 h, solvent (20 ml). For apolar solvents tertiary butyl hydroperoxide (5 mol% based on C_6H_{10}) was added.
[a]Total selectivity to C_6 partial oxidation products
[b]Total yield of C_6 partial oxidation products
[c]Products exclusively CO_2, HCOOH and oxalic acid
[d]Tetramethylbenzene

This demonstrates that the Au/C catalyst was able to selectively oxidize an alkene, although the products observed are those of allylic oxidation, and are not associated with electrophilic epoxidation. We subsequently investigated apolar solvents and found that selective oxidation could be observed, provided a small catalytic amount of a peroxy initiator was added at the start of the reaction. In the absence of the gold catalyst, but in the presence of the catalytic amount of the radical initiator, again no conversion was observed, confirming that the nanocrystalline gold catalyst is necessary. Small amounts of Bi were found to promote the selectivity for the partial oxidation products.[94] This opens up an interesting possibility in the design of supported nanocrystalline gold catalysts; as yet this is a subject area that has received very little attention, but is a topic that could produce new exciting discoveries.

One of the remarkable features of the preliminary results on the gold-catalyzed oxidation of alkenes was the large solvent effect on the reaction rate and the observation that only in aromatic solvents was oxidation controllable so that the alkene (cyclohexene) was not converted into carbon dioxide, formic acid and oxalic acid (Table 8.3). Indeed, the solvent can be used to tune or control the selectivity to a range of products (Table 8.3). The oxidation of *cis*-cyclooctene under solvent-free conditions (Table 8.4) was also investigated and it was found that with *cis*-cyclooctene as substrate, high selectivities can be achieved even when the amount of the peroxy initiator (TBHP) was decreased to as low as

Table 8.4 *cis*-Cyclooctene oxidation with molecular oxygen in the absence of a solvent.[a,94]

Catalyst	TBHP (g)	Con (%)	Selectivity (%) 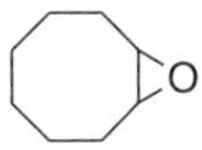	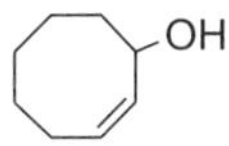	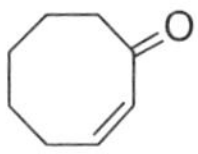	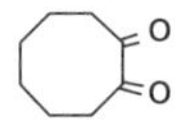	Σ_{sel} C_8
1%Au/graphite	0.12	7.9	81.2	9.3	4.1	0.5	95.1
1%Au/graphite	0.02	7.1	79.2	6.8	3.0	0.5	89.5
1%Au/graphite	0.002	1.3	82.6	7.4	2.1	0.6	92.7
No catalyst	0.008	2.0	Trace	0.0	0.0	0.0	–
Graphite	0.008	2.3	Trace	0.0	0.0	0.0	–

[a]Reaction conditions: 0.12 g catalyst, *cis*-cyclooctene (10 ml, 0.066 mol), 80 °C, 24 h, in a stirred autoclave with O_2 3 bar. TBHP = tertiary butylhydroperoxide.

0.002 g per 10 ml substrate, representing a cyclooctene:TBHP molar ratio of 300 and a molar ratio of epoxide to peroxide of 32. With our nonoptimized system, yields of 35 mol product per mol Au per h (1.9 mol per kg catalyst per h) can be achieved, which is comparable to the previous reports for propene epoxidation by Haruta and coworkers.[84] The high selectivity for the epoxide, particularly for cyclohexene and *cis*-cyclooctene, demonstrates that direct oxidation of the carbon–carbon double bond is occurring with this catalyst system. At present, the results show that the gold catalysts have significant potential for selective epoxide formation rather than the competing allylic oxidation. This characteristic is promising and worthy of further evaluation.

Subsequently, Caps and coworkers[95–97] have studied this experimental approach in detail for the oxidation of stilbene, for which they have proposed a radical mechanism. This mechanism is consistent with the initial observation that catalytic amounts of a peroxy species were required to observe selective epoxidation. Deng and Friend[98] showed using model studies with a Au (111) surface that styrene could be directly epoxidized with oxygen giving 53% styrene oxide. Most recently, Lambert and coworkers[99] have shown that very small Au_{55} nanocrystals supported on carbon are active catalysts for styrene oxidation with oxygen but only minor selectivity to the epoxides was observed and the major product was benzaldehyde. This recent study shows that by tailoring the size of the gold nanoparticles improved reactivity can be observed.

8.6 Concluding Remarks

There is general acceptance in the wider chemical community that supported nanocrystalline gold can be extremely reactive for both oxidation and reduction reactions. In many cases, gold is found to be one of the best catalysts. In this chapter the recent advances concerning the use of gold nanoparticles for the selective oxidation of alcohols and alkenes have been outlined as a means of providing an introduction to the fascinating field of gold catalysis. Indeed, very high reactivities and selectivities can be observed for these demanding reaction using gold catalysts.

References

1. M. Haruta, *Catal. Today*, 1997, **36**, 153.
2. M. Haruta and M. Date, *Appl. Catal. A*, 2001, **222**, 427.
3. M. Haruta, *CATTECH*, 2002, **6**, 102.
4. M. Haruta, *Chemical Record*, 2003, **3**, 75.
5. M. Haruta, *Gold Bull.*, 2004, **37**, 27.
6. G. C. Bond and D. T. Thompson, *Catal. Rev.-Sci. Eng.*, 1999, **41**, 319.
7. G. C. Bond and D. T. Thompson, *Gold Bull.*, 2000, **33**, 41.
8. G. C. Bond, *Gold Bull.*, 2004, **34**, 117.
9. D. T. Thompson, *Appl. Catal. A*, 2003, **243**, 201.
10. R. Meyer, C. Lemaire, Sh. K. Shaikutdinov and H.-J. Freund, *Gold Bull.*, 2004, **37**, 72.
11. M. B. Cortie, *Gold Bull.*, 2004, **37,12**.
12. A. S. K. Hashmi, *Gold Bull.*, 2004, **37**, 51.
13. G. J. Hutchings, *Gold Bull.*, 1996, **29**, 123.
14. G. J. Hutchings, *Gold Bull.*, 2004, **37**, 37.
15. G. J. Hutchings, *Catal. Today*, 2005, **100**, 55.
16. A. S. K. Hashmi and G. J. Hutchings, *Angew. Chem. Int. Ed.*, 2006, **45**, 7896.
17. M. Chen and D. W. Goodman, *Chem. Soc. Rev.*, 2008, **37**, 1860.
18. J. C. Fierro-Gonzalez and B. C. Gates, *Chem. Soc. Rev.*, 2008, **37**, 2127.
19. A. Corma and H. Garcia, *Chem. Soc. Rev.*, 2008, **37**, 2096.
20. C. Della Pina, E. Falletta, L. Prati and M. Rossi, *Chem. Soc. Rev.*, 2008, **37**, 2077.
21. R. Coquet, K. L. Howard and D. J. Willock, *Chem. Soc. Rev.*, 2008, **37**, 2046.
22. M. Haruta, T. Kobayashi, H. Sano and N. Yamada, *Chem. Lett.*, 1987, **16**, 405.
23. P. A. Sermon, G. C. Bond and P. B. Wells, *J. Chem. Soc., Faraday Trans. 1*, 1979, **75**, 385.
24. J. E. Bailie and G. J. Hutchings, *Chem. Commun.*, 1999, 2151.
25. G. J. Hutchings, *J. Catal.*, 1985, **96**, 292.
26. B. Nkosi, N. J. Coville and G. J. Hutchings, *J. Chem. Soc., Chem. Commun.*, 1988, **71**.
27. G. C. Bond, *Gold Bull.*, 1972, **5**, 11.
28. J. Schwank, *Gold Bull.*, 1985, **18**, 1.
29. G. J. Hutchings and M. Haruta, *Appl. Catal., A*, 2005, **291**, 2.
30. http://www.utilisegold.com/uses_applications/catalysis/reference_catalysts/.
31. C. W. Corti, R. J. Holliday and D. T. Thompson, *Top. Catal.*, 2007, **44**, 331.
32. L. Brey, T. E. Wood, G. M. Buccellato, M. E. Jones, C. S. Chamberlain and A. R. Siedle, WO Patent 2005/030382 (2005), 3M Innovative Properties Co.
33. J. T. Brady, L. A. Brey, G. M. Buccellato, C. S. Chamberlain, J. S. Huberty, A. R. Siedle and T. E. Wood, WO Patent 2006/074126 A2 (2006), 3M Innovative Properties Co.

34. G. C. Bond, C. Louis and D. T. Thompson, *Catalysis by Gold*, Imperial College Press, London, 2006.
35. J. E. Bailie, H. A. Abdullah, J. A. Anderson, C. H. Rochester, N. V. Richardson, N. Hodge, J.-G. Zhang, A. Burrows, C. J. Kiely and G. J. Hutchings, *Phys. Chem. Chem. Phys.*, 2001, **3**, 4113.
36. P. Claus, *Appl. Catal. A*, 2005, **291**, 222.
37. A. Corma and P. Serna, *Science*, 2006, **313**, 332.
38. A. Corma, C. González-Arellano, M. Iglesias and F. Sánchez, *Angew. Chem. Int. Ed.*, 2007, **46**, 7820.
39. A. Corma, P. Serna, P. Concepcion, Patricia and J. J. Calvino, Jose, *J. Am. Chem. Soc.*, 2008, **130**, 8748.
40. P. Landon, P. J. Collier, A. J. Papworth, C. J. Kiely and G. J. Hutchings, *Chem. Commun.*, 2002, 2058.
41. P. Landon, P. J. Collier, A. F. Carley, D. Chadwick, A. J. Papworth, A. Burrows, C. J. Kiely and G. J. Hutchings, *Phys. Chem. Chem. Phys.*, 2003, **5**, 1917.
42. M. S. Chen and D. W. Goodman, *Science*, 2004, **306**, 252.
43. N. Lopez, T. V. W. Janssens, B. S. Clausen, Y. Xu, M. Mavrikakis, T. Bligaard and J. K. Norskov, *J. Catal.*, 2004, **223**, 232.
44. M. S. Chen and D. W. Goodman, *Science*, 2004, **306**, 252.
45. T. V. Choudhary and D. W. Goodman, *Appl. Catal. A*, 2005, **291**, 32.
46. S. N. Rashkeev, A. R. Lupini, S. H. Overbury, S. J. Pennycook and S. T. Pantelides, *Phys. Rev. B*, 2007, **76**, 35438.
47. G. J. Hutchings, M. S. Hall, A. F. Carley, P. Landon, B. E. Solsona, C. J. Kiely, A. Herzing, M. Makkee, J. A. Moulijn, A. Overweg, J. Carlos Fierro-Gonzalez, J. Guzman and B. C. Gates, *J. Catal.*, 2006, **242**, 71.
48. A. A. Herzing, C. J. Kiely, A. F. Carley, P. Landon and G. J. Hutchings, *Science*, 2008, **321**, 1331.
49. L. Prati and M. Rossi, *J. Catal.*, 1998, **176**, 552.
50. F. Porta, L. Prati, M. Rossi, S. Colluccia and G. Martra, *Catal. Today*, 2000, **61**, 165.
51. L. Prati, *Gold Bull.*, 1999, **32**, 96.
52. S. Biella and M. Rossi, *Chem. Commun.*, 2003, 378.
53. M. Comotti, C. Della Pina, R. Matarrese, M. Rossi and A. Siani, *Appl. Catal. A*, 2005, **291**, 204.
54. P. Beltrame, M. Comotti, C. Della Pina and M. Rossi, *Appl. Catal., A*, 2006, **297**, 1.
55. N. Dimitratos, F. Porta, L. Prati and A. Villa, *Catal. Lett.*, 2005, **99**, 181.
56. M. Comotti, C. Della Pina, R. Matarrese and M. Rossi, *Angew. Chem. Int. Ed.*, 2004, **43**, 5812.
57. P. G. N. Mertens, M. Bulut, L. E. M. Gevers, I. F. J. Vankelecom, P. A. Jacobs and D. E. De Vos, *Catal. Lett.*, 2005, **102**, 57.
58. P. G. N. Mertens, I. F. J. Vankelecom, P. A. Jacobs and D. E. De Vos, *Gold Bull.*, 2005, **38**, 157.

59. H. Tsunoyama, H. Sakurai, Y. Negishi and T. Tsukuda, *J. Am. Chem. Soc.*, 2005, **127**, 9374.
60. C. H. Christensen, B. Jorgensen, J. Raas-Hansen, K. Egeblad, R. Madsen, S. K. Klitgaard, S. M. Hansen, M. R. Hansen, H. C. Andersen and A. Riisager, *Angew. Chem. Int. Ed.*, 2006, **45**, 4648.
61. I. S. Nielsen, E. Taarning, K. Egeblad, R. Madsen and C. H. Christensen, *Catal. Lett.*, 2007, **116**, 35.
62. S. Carrettin, P. McMorn, P. Johnston, K. Griffin and G. J. Hutchings, *Chem. Commun.*, 2002, 696.
63. S. Carrettin, P. McMorn, P. Johnston, K. Griffin, C. J. Kiely, G. A. Attard and G. J. Hutchings, *Top. Catal.*, 2004, **27**, 131.
64. S. Carrettin, P. McMorn, P. Johnston, K. Griffin, C. J. Kiely and G. J. Hutchings, *Phys. Chem. Chem Phys.*, 2003, **5**, 1329.
65. A. Abad, P. Concepcion, A. Corma and H. Garcia, *Angew. Chem. Int. Ed.*, 2005, **44**, 4066.
66. A. Corma and M. E. Domine, *Chem. Commun.*, 2005, 4042.
67. K. Mori, T. Hara, T. Mizugaki, K. Ebitani and K. Kaneda, *J. Am. Chem. Soc.*, 2004, **126**, 10657.
68. R. A. Sheldon and J. K. Kochi, *Metal-Catalyzed Oxidations of Organic Compounds*, Academic Press, New York, 1981.
69. D. I. Enache, J. K. Edwards, P. Landon, B. Solsona-Espriu, A. F. Carley, A. A. Herzing, M. Watanabe, C. J. Kiely, D. W. Knight and G. J. Hutchings, *Science*, 2006, **311**, 362.
70. O. P. H. Vaughan, G. Kyriakou, N. Macleod, M. Tikhov and R. M. Lambert, *J. Catal.*, 2005, **236**, 401.
71. T. Hayashi, K. Tanaka and M. Haruta, *J. Catal.*, 1998, **178**, 566.
72. M. Haruta and M. Date, *Appl. Catal. A*, 2001, **222**, 427.
73. T. A. Nijhuis, T. Visser and B. M. Weckhuysen, *J. Phys. Chem. B*, 2005, **109**, 19309.
74. B. S. Uphade, S. Tsubota, T. Hayashi and M. Haruta, *Chem. Lett.*, 1998, 1277.
75. T. A. Nijhuis, B. J. Huizinga, M. Makkee and J. A. Moulijn, *Ind. Eng. Chem. Res.*, 1999, **38**, 884.
76. T. A. Nijhuis, T. Q. Gardner and B. M. Weckhuysen, *J. Catal.*, 2005, **236**, 153.
77. T. A. Nijhuis, T. Visser and B. M. Weckhuysen, *Angew. Chem. Int. Ed.*, 2005, **44**, 1115.
78. T. A. Nijhuis and B. M. Weckhuysen, *Chem. Commun.*, 2005, 6002.
79. N. Yap, R. P. Andres and W. N. Delgass, *J. Catal.*, 2004, **226**, 156.
80. A. Zwijnenburg, M. Makkee and J. A. Moulijn, *Appl. Catal. A*, 2004, **270**, 49.
81. C. Qi, T. Akita, M. Okumura, K. Kuraoka and M. Haruta, *Appl. Catal. A*, 2003, **253**, 75.
82. B. Taylor, J. Lauterbach and W. N. Delgass, *Appl. Catal. A*, 2005, **291**, 188.
83. A. K. Sinha, S. Seelan, M. Okumura, T. Akita, S. Tsubota and M. Haruta, *J. Phys. Chem. B*, 2005, **109**, 3956.

84. B. Chowdhury, J. J. Bravo-Suarez, M. Date, S. Tsubota and M. Haruta, *Angew. Chem. Int. Ed.*, 2006, **45**, 412.
85. E. Sacaliuc, A. M. Beale, B. M. Weckhuysen and T. A. Nijhuis, *J. Catal.*, 2007, **248**, 235.
86. J. Lu, X. Zhang, J. J. Bravo-Suárez, K. K. Bando, T. Fujitani and S. T. Oyama, *J. Catal.*, 2007, **250**, 350.
87. A. K. Sinha, S. Seelan, S. Tsubota and M. Haruta, *Angew. Chem. Int. Ed.*, 2004, **43**, 1546.
88. N. S. Patil, B. S. Uphade, P. Jana, S. K. Bhargava and V. R. Choudhary, *Chem. Lett.*, 2004, **33**, 400.
89. N. S. Patil, B. S. Uphade, D. G. McCulloh, S. K. Bhargava and V. R. Choudhary, *Catal. Commun.*, 2004, **5**, 681.
90. N. S. Patil, B. S. Uphade, P. Jana, S. K. Bhargava and V. R. Choudhary, *J. Catal.*, 2004, **223**, 236.
91. N. S. Patil, R. Jha, B. S. Uphade, S. K. Bhargava and V. R. Choudhary, *Appl. Catal. A*, 2004, **275**, 87.
92. N. S. Patil, B. S. Uphade, P. Jana, R. S. Sonawane, S. K. Bhargava and V. R. Choudhary, *Catal. Lett.*, 2004, **94**, 89.
93. D. Yin, L. Qin, J. Liu and C. L. Y. Jin, *J. Molec. Catal. A*, 2005, **240**, 40.
94. M. D. Hughes, Y.-J. Xu, P. Jenkins, P. McMorn, P. Landon, D. I. Enache, A. F. Carley, G. A. Attard, G. J. Hutchings, F. King, E. H. Stitt, P. Johnston, K. Griffin and C. J. Kiely, *Nature*, 2005, **437**, 1132.
95. P. Lignier, F. Morfin, S. Mangematin, L. Massin, J.-L. Rousset and V. Caps, *Chem. Commun.*, 2007, **186**.
96. P. Lignier, S. Mangematin, F. Morfin, J.-L. Rousset and V. Caps, *Catal. Today.*, 2008, **138**, 50.
97. P. Lignier, F. Morfin, L. Piccolo, J.-L. Rousset and V. Caps, *Catal. Today.*, 2007, **122**, 284.
98. X. Deng and C. M. Friend, *J. Am. Chem. Soc.*, 2005, **127**, 17178.
99. M. Turner, V. B. Golovko, O. P. H. Vaughan, P. Abdulkin, A. Berenguer-Murcia, M. S. Tikhov, B. F. G. Johnson and R. M. Lambert, *Nature*, 2008, **454**, 981.

CHAPTER 9

Ethylene Epoxidation over Silver Catalysts

VALERII I. BUKHTIYAROV[1] AND AXEL KNOP-GERICKE*[2]

[1] Boreskov Institute of Catalysis SB RAS, Lavrentieva ave. 5, Novosibirsk, 630090, Russia; [2] Fritz-Haber-Institut der Max-Plank-Gesellschaft, Department of Inorganic Chemistry, Faradayweg 4-6, 14195 Berlin, Germany

9.1 Introduction

One of the most important and successful industrial processes of selective oxidation is the silver-catalyzed epoxidation of ethylene with molecular oxygen. The reaction discovered by Lefort is one of the major petrochemical processes converting several billion US dollars annually due to the importance of ethylene oxide (EO) as a versatile chemical intermediate (see Chapter 16).[1–3] EO is used in the synthesis of ethylene glycol, which is an important component of antifreeze agents, polyester fabrics, surfactants and detergents. The process uses a silver catalyst supported on α-Al_2O_3. The principal challenge associated with the epoxidation process is the prevention of the total oxidation of ethylene, since the competition between this undesired reaction and the desired epoxidation reaction

$$C_2H_4 + \tfrac{1}{2}O_2 \rightarrow C_2H_4O + 25\,\text{kcal/mol} \qquad (9.1)$$

$$C_2H_4 + 3O_2 \rightarrow 2CO_2 + 2H_2O + 340\,\text{kcal/mol} \qquad (9.2)$$

RSC Nanoscience & Nanotechnology No. 19
Nanostructured Catalysts: Selective Oxidations
Edited by Christian Hess and Robert Schlögl

Published by the Royal Society of Chemistry, www.rsc.org

determines the selectivity of EO production. The formation of carbon dioxide is highly exothermic, and the process is running at low conversions (<15%) to achieve high selectivity and to control the heat released by combustion. Another reaction, which declines the selectivity of the process, is the secondary oxidation of ethylene oxide:

$$C_2H_4O + \tfrac{5}{2}O_2 \rightarrow 2CO_2 + 2H_2O \tag{9.3}$$

Due to these side reactions, the selectivity towards ethylene oxide on unpromoted silver is about 40–50%, however, it can be improved up to 80–85% by a promotion of commercial catalyst with chlorine and alkali or alkaline-earth elements.[1–3] Since the improvement of 1% selectivity is estimated to produce a beneficial effect of several tens of millions of dollars annually, efforts to improve the selectivity further by optimizing the catalyst and the process conditions are still continuing.[4]

The practical importance of the process has attracted and continues to attract a great attention of researcher to study both ethylene epoxidation and oxygen adsorption over various silver samples. The number of original papers devoted to these topics is so numerous (simple internet search results in numbers approaching 1000 or even exceeding this number) that they could not be cited in one chapter. The papers published till 1987 were excellently reviewed by van Santen and Kuipers who considered both surface-science and catalytic studies related to the mechanism of ethylene epoxidation.[5] Simple citation of the chapter titles: "The characterization of oxygen-containing adsorbates on silver by surface science", "reactivity of the silver surface", "The mechanism of ethylene epoxidation" indicates that the reviews have highlighted the aspects of the silver-catalyzed epoxidation chemistry. The authors concluded that selectivity towards C_2H_4O is determined by the chemical state of adsorbed oxygen and the inability of a catalyst to activate the carbon–hydrogen bond in ethylene. As consequence, the ability of silver to produce weakly bound electrophilic oxygen, which does not activate C–H bond, explains its uniqueness as a catalyst for this reaction.[5]

The results of surface-science studies published in the subsequent ten years (after van Santen's review) were later analyzed in a critical review of Serafin *et al.*[6] The authors discussed the experimental results related to progress in elucidating and understanding the basic reaction steps and the mechanistic roles of promoters and moderators. For example, they have declared that such long-debated questions as the importance of molecular oxygen (*vs.* atomic species) has been closed in this decade. At the same time, the authors emphasized that in spite of good progress in the field, the "pressure-gap" problem complicates the use of the experimental data to explain all aspects of the silver-catalyzed epoxidation process. Indeed, ethylene epoxidation in industry is performed in a cooled shell-and-tube recycle reactor at temperatures and pressures ranged in 473–573 K and 1–3 MPa, respectively, whereas most atomistic studies of this reaction by various physical methods have been performed in UHV conditions or at pressures below 10^{-4} millibar.[1,2] *In-situ* techniques developed in the

1990s to solve the "pressure-gap" problem offer the possibility to obtain detailed information on the state of the catalyst under operating conditions. Microkinetic modeling of the kinetic data is another effective way to make further progress in elucidation of the mechanism of ethylene epoxidation.[7–9] The model developed by Stolze and coauthors proposed ethylene oxide to be formed as a result of the reaction between electrophilic or covalent oxygen and adsorbed ethylene that proceeds *via* an oxametallacycle as reaction intermediates.[7,8] Formation of both electrophilic oxygen and adsorbed ethylene are promoted by oxide oxygen, which creates $Ag^{\delta+}$ sites and is often termed as nucleophilic or ionic oxygen. The oxametallacycle is a common intermediate in epoxidation, ethylene combustion, and ethylene oxide combustion. Both the uptake of atomic surface oxygen and the surface reaction between adsorbed ethylene and atomic oxygen as rate-limiting steps are also suggested by Linic and Barteau.[9] However, the models disagree in the definition of the active phase of the catalyst and the oxygen species active in epoxidation.

Thus, the ethylene epoxidation over silver continues to be a very real and attractive topic both for fundamental study and for optimization of the catalyst and the process. In this chapter we will review the well-known and recently published literature data about the system. The main attention has been paid to aspects of mechanisms of the reactions, to factors determining the selectivity and to measures that can help to improve the catalysts performance.

9.2 Mechanisms of Ethylene Epoxidation

9.2.1 The Nature of Ethylene Epoxidizing Oxygen Species

Numerous papers devoted to study the oxygen adsorption on silver have been motivated by the suggestion that the nature of oxygen species is the parameter governing various routes of ethylene oxidation. There were two principally different concepts.

The first concept, which has been originally developed by Sachtler *et al.* and later supported by Campbell and Boronin with coauthors associated the epoxidation route with molecularly adsorbed oxygen, and the route of total oxidation – with atomic species.[10–15] The basis of this idea is the well-known experimental fact of activity of organic peracids or hydroperoxides in homogeneous liquid-phase oxidation of ethylene to ethylene oxide – so-called Prilezhaev reaction.[16]

An additional argument in favor of the "molecular oxygen" concept originated from Campbell's papers was the nonobservation of ethylene oxide among the products of titration of adlayer of atomic oxygen with ethylene. Indeed, only CO_2 was registered mass-spectrometrically when ethylene contacts oxygen precovered silver surfaces even for the highest oxygen coverage ($\Theta = 0.67$ ML) produced by the application of a high-pressure cell and subsequent fast transfer of the sample to UHV conditions of analyzer chamber. The formation of atomically adsorbed oxygen was identified by X-ray photoelectron spectroscopy (XPS) due to the appearance of an O1s feature at 528.2 ± 0.2 eV.

On the other hand, molecular oxygen, which is observed only at low temperatures, desorbs from the silver surfaces below room temperature, whereas catalysis proceeds at $T > 200\,^{\circ}C$.[17–19] In spite of the lack of spectroscopic data related to the presence of molecular oxygen under reaction conditions, this concept has dominated since 1990. Its followers have explained this discrepancy by the assumption that the equilibrium concentration of the molecularly adsorbed oxygen is enhanced at pressures of real catalysis by seven orders of magnitude compared to pressures used in spectroscopic investigations.

However, the results of a number of papers where reactivity of oxygen adlayers towards ethylene were investigated using temperature-programmed reaction (TPR) spectroscopy with analysis of TPR products by gas chromatography or mass spectrometry[20–26] disagreed with the "molecular-oxygen" concept. Indeed, silver surfaces with preadsorbed oxygen and ethylene retained their activity in ethylene epoxidation even after evacuation to UHV at room temperature. This behavior has been explained by the assumption that only atomic oxygen, which is stable on the silver surfaces at ambient temperatures, is responsible for epoxidation of ethylene to ethylene oxide.

A practical implication of this assumption is the possibility to exceed the selectivity level of 86% (when molecular oxygen participates in the epoxidation route, the reaction scheme (1–3) gives the maximum selectivity value of 6/7).[27] And indeed, selectivity of 90% has been recently reported for the best industrial catalysts. As consequence, the concepts proclaiming that atomic oxygen epoxidizes ethylene to ethylene oxide became dominant in recent years.

One of the initiators of this concept, Khasin, declared that the same atomic oxygen species is responsible both for epoxidation and for total oxidation of ethylene, since the specific route depends on the bond strength of oxygen atoms to the silver surface, which has been shown to decrease with increasing the oxygen coverage.[20,21] Indirect support of this concept was provided in the papers suggesting that both the selective and nonselective pathways proceed *via* the common intermediate–surface oxametallacycle.[28] The reaction intermediates will be discussed in more detail in the next section. Nevertheless, the concept about constant nature of atomic oxygen participating both in epoxidation and combustion contradicts with the results of Campbell's papers where reactivity of oxygen atomically adsorbed at Ag(110) surface towards ethylene was tested.[12,17] Using a high-pressure reactor for O_2 adsorption, he could increase oxygen coverage up to $\Theta = 0.67$ and decrease the bond strength of oxygen with the surface by 7%. However, CO_2 remained as the only product in this case.

The modified "atomic-oxygen" concept has been developed by van Santen and colleagues.[22,23] It connects the reduction of the bond strength induced by $\Theta(O_{ads})$ rising with changing of the coordination number of adsorbed oxygen atom. Multicoordinated strongly bound oxygen atoms produced at low coverages are active in total oxidation of ethylene, whereas monocoordinated atoms, which are appeared only at high Θ in the presence of subsurface oxygen, are weakly bound and active in epoxidation of ethylene to ethylene oxide. In elaboration of this idea the authors have later suggested that the less-coordinated oxygen atoms have lower charge. Based on earlier published isotope experiments, they also proposed that oxygen atoms with higher charge are active in cleavage of C–H bond, which is the first step in total oxidation, but less-charged oxygen participates in a electrophilic attack of the π-bond in ethylene with the formation of the epoxide cycle.[5,20] This allowed Lambert to designate the former species as nucleophilic oxygen and the latter as electrophilic oxygen. In addition, Lambert's group has further developed the idea about the effect of oxygen charge state on the pathway of ethylene oxidation.[25] They concluded that total oxidation is initiated by charge transfer to adsorbed oxygen, but epoxidation – from oxygen to silver.

It should be noted, however, that for a long time, a basis for this concept was absent, since different oxygen species atomically adsorbed at the silver surfaces were not detected. Reliable data about formation of two atomic oxygen species has been obtained using XPS that has identified two O1s features at 528.2 ±0.2 eV and 530.2 ±0.2 eV after O_2 adsorption at low temperatures on Ag(111) and Ag(100).[29–31] The assignment of the former feature to oxygen atomically adsorbed on silver is beyond question, since it has been repeatedly reported in the literature.[5,17,18] The nature of the second feature is more debatable. Indeed, this feature has been earlier attributed to carbonate species, which are stable on the silver surfaces up to $T = 420$ K and indeed are characterized by an O1s feature with binding energy of 530.5 eV.[5,17–19] Boronin attributed the feature to molecularly adsorbed oxygen identified on the basis of ultraviolet photoelectron spectroscopy (UPS) data.[15] However, other authors have rejected both possibilities either due to the absence of any CO_2 peak in TPD spectra, which are much more sensitive than XPS to any carbonate species decomposed during the sample heating ($T_{des} = 420$ K), or due to the absence of the vibration at 78 meV in high-resolution electron energy-loss spectroscopy (HREES) characteristic of molecular O–O vibration as opposed to the vibration of the O–Ag bond at 39 meV.[25,29–31] They assign the O1s feature at 530.2 ±0.2 eV to atomically adsorbed oxygen species.[29–31] Moreover, the interpretation of angular-dependent XPS data to the depth concentration profiles leads the authors to the conclusion that the oxygen species with BE(O1s) = 530.0 eV adsorbs at the Ag(111) surface, while the species with BE(O1s) = 528.2 eV inserts at least in the top metal layer. The formation of surface silver oxide has been concluded in the latter case.[29]

Bao *et al.* have shown from a Raman study for the first time that three different oxygen species can be distinguished on single-crystalline Ag(111), Ag(110) and microfacetted Ag(110) samples. The samples were exposed to reaction mixtures of oxygen and ethylene. The pressure and temperature ranges are close to those used for industrial oxidation of ethylene using Ag catalysts. Depending on the orientation of the Ag surface and on the history of its preparation and temperature cycling (300–900 K), three Raman lines for three Ag–O species were found, assigned to adsorbed oxygen, subsurface oxygen and oxygen dissolved in the bulk of silver.[32] The formation of two atomic oxygen species under more realistic conditions was observed for the first time after treatment of silver foil with the mixture of ethylene and oxygen at $P \sim 1$ mbar and $T \geq 150$ °C.[26] Such high-pressure treatment also caused the formation of a considerable amount of carbon and oxygen in near-surface layers of the silver foil. It has also been suggested that namely oxygen species with BE(O1s) = 530.0 eV participates in the ethylene epoxidation even in the absence of subsurface oxygen.[33,34] This suggestion was based on the results of TPR experiments that had shown that ethylene oxide is produced as the result of ethylene interaction with oxygen-precovered silver foil only in the case when both oxygen species are registered with XPS.[26,33] In full agreement with literature data it was shown that for a surface containing only one oxidic-type species characterized by BE(O1s) = 528.2 eV, only CO_2 was formed in TP reaction with ethylene.[12,17] Such behavior of the oxygen species allowed the authors to use the terms "electrophilic" and "nucleophilic" oxygen for O_{530} and O_{528} features, respectively. It has also been shown that if the silver surface was covered by the electrophilic oxygen only, it did not adsorb ethylene – neither CO_2 nor C_2H_4O was observed in the subsequent TPR spectra.[26,33] This fact has been explained by the assumption that the nucleophilic oxygen creates the centers for ethylene adsorption (Ag^+ ions) and then adsorbed ethylene reacts with the electrophilic oxygen giving the target product – ethylene oxide.[26] This activation mechanism of adsorbed ethylene has been proposed by van Santen based on theoretical considerations.[5] This picture was supported by isotope experiments. Ethylene interaction with isotope-labeled oxygen layers (nucleophilic oxygen was labeled by ^{18}O, and electrophilic oxygen was labeled by ^{16}O) resulted in an ethylene oxide TPR signal with $m/z = 29$ ($CH^{16}O$ fragment), but a signal with $m/z = 31$ ($CH^{18}O$ fragment) was absent.[33]

It should be noted, however, that most previous studies of silver catalysts for ethylene epoxidation with surface-science techniques have been performed under *ex-situ* conditions after the reaction mixture was pumped to obtain UHV conditions. It is evident that the "pressure-gap" problem made it doubtful to apply the numerous surface-science data to elucidation of the nature of active species. Roberts *et al.* did the first measurements in order to bridge the gap in X-ray photoelectron emission experiments. Ag foil was investigated in 0.5 mbar of oxygen. At 473 K two oxygen species could be distinguished.[35] Later, combined *in-situ* XPS and PTRMS (proton-transfer reaction mass spectrometry) studies of ethylene oxidation over silver foil were performed.[36,37] The highest pressure available for XPS (1–2 mbar) is sufficient for observation of ethylene oxide among the reaction products.[12,13,25] The main advantage of PTRMS, which was used to measure the yield of ethylene oxide, is its high sensitivity to organic molecules.[38] Furthermore, it does fragment the molecules to be analyzed to a lower extent compared to a quadrupole mass spectrometer, which simplifies the analysis of the mass spectra. Thus, in these experiments characterization of the composition of the catalyst surface was combined with catalytic measurements.

PTRMS data shown in Figure 9.1 indicates that for this system we indeed deal with the pressure-gap problem. No PTRMS signal of ethylene oxide is observed when silver interacts with reaction mixture at low pressures, while increase in the pressure causes the appearance of the corresponding signal in PTRMS spectra at $T = 420$ K and with enhanced intensity at $T = 470$ K.

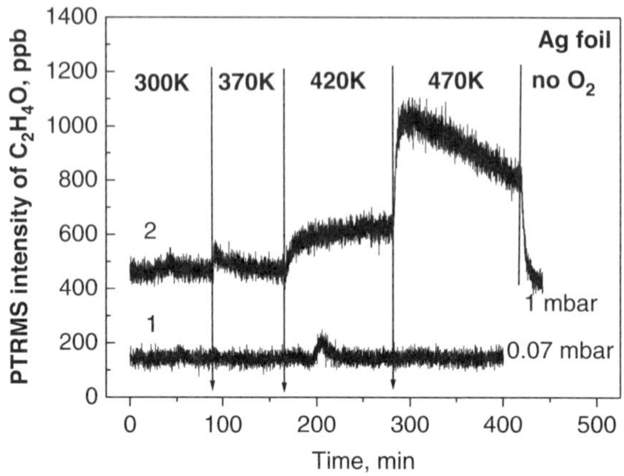

Figure 9.1 Variation of the PTRMS signals of ethylene oxide with temperature, measured in a flowing reaction mixtures at (1) $P(C_2H_4) = 0.0065$ mbar and $P(O_2) = 0.065$ mbar and at (2) $P(C_2H_4) = 0.1$ mbar, $P(O_2) = 0.95$ mbar in the presence of a polycrystalline silver foil. The last part of the high-pressure curve was measured after stopping the oxygen flow.[37]

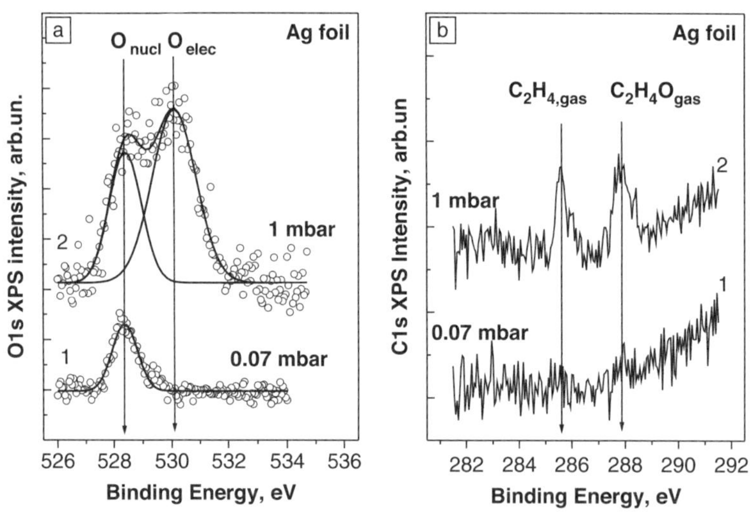

Figure 9.2 O1s (a) and C1s (b) spectra of a silver foil measured at 420 K in a flowing reaction mixture at (1) $P(C_2H_4) = 0.0065$ mbar and $P(O_2) = 0.065$ mbar and at (2) $P(C_2H_4) = 0.1$ mbar, $P(O_2) = 0.95$ mbar.[37]

This result indicates that the used conditions allow the activation of the silver foil for production of the target product (C_2H_4O) of ethylene oxidation. Pressure-induced changes are also observed in XP spectra measured under the same conditions (Figure 9.2).

One can see that the main difference in O1s spectra measured at the higher pressure at 420 K (Figure 9.2(a)) is the presence of the feature originating from electrophilic oxygen. Indeed, only nucleophilic oxygen is observed at low pressures. These results indicate that the presence of nucleophilic oxygen, even in rather high concentrations, is not sufficient to produce a surface that is active for ethylene epoxidation.

Increase in the pressure to 1 mbar also results in the appearance in C1s spectra of two features that have been assigned to gas-phase ethylene (285.8 eV) and ethylene oxide (287.9 eV). XPS observation of gas-phase ethylene oxide is of great significance because it allows unambiguous identification of the PTRMS signal at $m/z = 45$ (Figure 9.1) as ethylene oxide, but not as acetaldehyde, which is also discussed as byproduct of ethylene oxidation. Technically, these isomers exhibit very similar mass spectra, but principally different XP spectra with one C1s peak for identical carbon atoms in ethylene oxide and two C1s peaks from a carbon atom in the methyl group and a carbon atom in the C=O bond of acetaldehyde, CH_3COH.

A set of similar measurements have been performed for different conditions. Figure 9.3, for example, shows transformation of O1s spectra (a) depending on the ratio of O_2/C_2H_4 partial pressures at constant partial pressure of O_2 and (b) depending on the catalyst temperature at constant pressure and ratio of reagents.

When oxygen is admitted to the high-pressure chamber, the nucleophilic oxygen (BE(O1s) = 528.3 eV) is formed on the silver surface (Figure 9.3(a), curves 1 and 2). Introduction of ethylene even at the smallest partial pressure ($P(C_2H_4)/P(O_2) = 1{:}10$) reduces the O1s signal from nucleophilic oxygen and leads to the appearance of a new peak at 530.0 eV (Figure 9.3(a), curve 3) which is due to electrophilic oxygen, as discussed above. An increase in the partial pressure of ethylene up to $P(C_2H_4)/P(O_2) = 1{:}3$ and further to 1:1 leads to a reduction of the surface concentration of the nucleophilic oxygen and an increase in the coverage by the electrophilic oxygen. In parallel, the yield of ethylene oxide was measured with PTRMS, and the dependence of the C_2H_4O yield on the abundance of nucleophilic and electrophilic oxygen is plotted in Figure 9.4. One can see a clear correlation between the abundance of the electrophilic oxygen and the yield of C_2H_4O: the higher the O1s/Ag3d$_{5/2}$ atomic ratio for this oxygen, the higher the partial pressure of ethylene oxide. The revealed linear correlation can be interpreted as participation of the electrophilic oxygen in the production of ethylene oxide.

Although the dependence of ethylene oxide yield *versus* the oxygen abundance for the nucleophilic oxygen is more complicated (Figure 9.4), the general trend is a reduction of the yield of ethylene oxide if the surface concentration of the nucleophilic oxygen is raised. This result can be explained by earlier reported activity of the nucleophilic oxygen in total oxidation of ethylene only.

The O1s spectra change dramatically if they are measured at lower or at higher temperatures than 420 K (Figure 9.3(b), curve 2). At 370 K, the silver surface is characterized by a single O1s peak at 530.5 eV (Figure 9.3(b), curve 1),

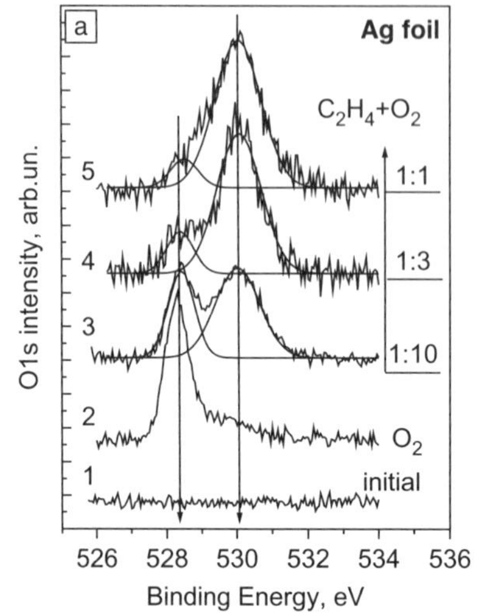

Figure 9.3 O1s spectra measured in flowing reaction mixture: (a) at 420 K, $P(O_2) = 0.33$ mbar and various $P(C_2H_4)$: 2 – 0 mbar (pure O_2); 3 – 0.03 mbar (1:10); 4 – 0.1 mbar (1:3); 5 – 0.24 mbar (1:1). The spectra from initial, "clean" silver surface measured in UHV (1) are also shown for comparison. (b) at $P(C_2H_4) = 0.1$ mbar, $P(O_2) = 0.25$ mbar and at various temperatures: 1 – 370 K, 2 – 420 K, 3 – 470 K. Spectrum (4) was measured at 470 K after the oxygen flow was stopped. The difference spectrum between the measurements in the reaction mixture and in pure ethylene (3-4) shows peaks that exist on the surface only under reaction conditions.[36]

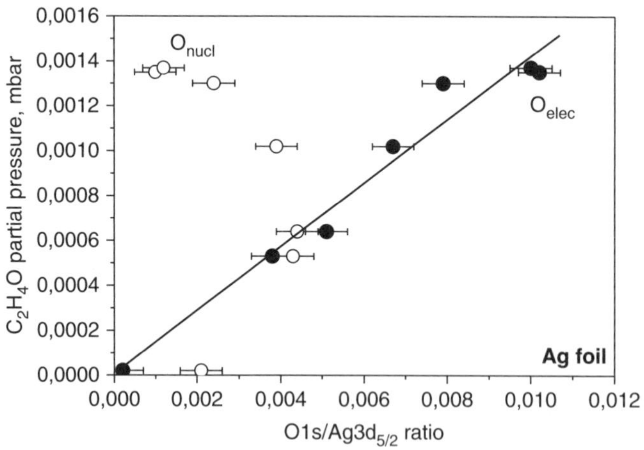

Figure 9.4 The yield of ethylene oxide as a function of the abundance of the nucleophilic (open symbols) and the electrophilic (full symbols) oxygen. All measurements were performed at 420 K and at various partial pressures of oxygen and ethylene (see Figure 9.3(a)).[36]

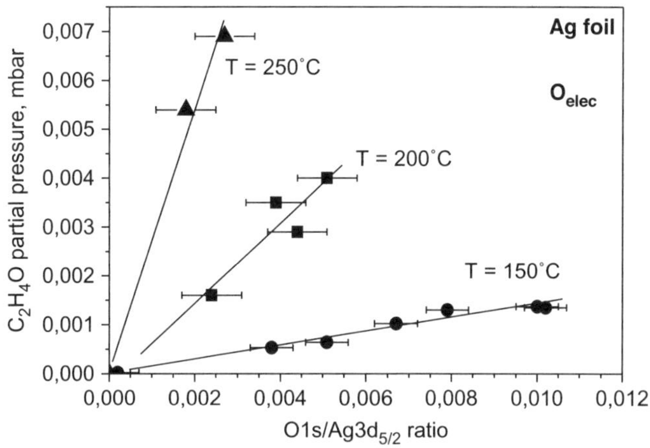

Figure 9.5 The yield of ethylene oxide as a function of the abundance of the electrophilic oxygen measured at different temperatures: 150 °C (circles), 200 °C (squares) and 250 °C (triangles).[36]

while heating the sample to 470 K makes the shape of the O1s spectrum so complex that the existence of more than two O1s features can be concluded. And indeed, switching the reaction mixture flow to pure ethylene retains two peaks at 529.1 and 530.6 eV (Figure 9.3(b), curve 4). The difference spectrum of the O1s spectra measured at 470 K in the reaction mixture and in pure ethylene (Figure 9.3(b), curves 3 and 4) shows two O1s peaks that are absent in the spectrum taken in pure ethylene. These peaks have the same binding energies as nucleophilic (528.5 eV) and electrophilic (530.0 eV) oxygen observed in the spectrum at 420 K (Figure 9.3(b), curve 2). Analysis of C1s and Ag3d spectra measured in the same experiment allowed assignment of the additional peaks to the carbonate species (530.5 eV at 370 K), to the so-called O_γ oxygen (529.1 eV at 470 K) and to subsurface oxygen embedded in octahedral holes (530.5 eV at 470 K).[36] The location of the last two oxygen species in the subsurface region is in line with their inactivity towards ethylene. An additional criterion, which reinforced the assignment, was the temperature range in which they occurred. Schubert *et al.* have shown that dissolved atomic oxygen from the bulk replenishes surface species *via* subsurface oxygen. The interconversion of all three species at the high reaction temperatures required to overcome the barriers for the formation and motion of the various atomic oxygen species limits the overall selectivity of the formaldehyde production.[48] TDS measurements confirmed the existence of three atomic oxygen species located on the surface (T_{des} (desorption temperature) = 575 K) in the interface (T_{des} = 900 K) and as bulk-dissolved species (T_{des} = 700 K).[39]

Identification of various surface and subsurface species makes it possible to understand the variation of activity with temperature. The low activity of silver at $T < 420$ K is caused by the presence of carbonates at the silver surface that reduce the available silver surface area for the catalytic reaction or even completely block it. Accumulation of the embedded oxygen species (O_γ and O_{ss}) explains the decrease of the rate of ethylene-oxide formation with time observed for $T \geq 470$ K (Figure 9.1, curve 2).

In order to calculate the contributions of the electrophilic and nucleophilic oxygen, O1s spectra measured at $T > 420$ K were analyzed with respect to their different components. In order to obtain the dependencies of the ethylene oxide yield on the electrophilic oxygen abundance for $T > 420$ K (Figure. 9.5), the corresponding O1s spectra were deconvoluted by separated components and the contributions of the electrophilic oxygen were calculated. In full agreement with our expectation a linear correlation between catalytic properties of silver in epoxidation and the abundance of the electrophilic oxygen was observed for all temperatures, with the slope of the correlation being dependent on temperature.

Suggesting that the Arrhenius factor is the main reason for the variation of the correlation slope (A_T) with temperature, the activation energies of the ethylene epoxidation have been estimated:[36]

$$E_a = \ln\left(\frac{A_{T_2}}{A_{T_1}}\right) \times \frac{R}{\left(\frac{1}{T_2} - \frac{1}{T_1}\right)} \tag{9.4}$$

It has been established that in the temperature range from 150 °C to 200 °C the E_a value is 58 kJ per mole, while further raising the temperature corresponds to the value of 47 kJ/mol. These values and the tendency of E_a to decrease with increasing temperature are in perfect agreement with the behavior of activation energy for the epoxidation step reported in previous experimental and theoretical studies.[7,13,25] These data evidence again that namely the electrophilic oxygen epoxidizes ethylene to ethylene oxide.

The differences in nature of the nucleophilic and electrophilic oxygen species have been investigated in detail using a combination of physical methods: angular-dependent XPS, UPS, Auger electron (AES) and X-ray absorption near-edge structure (XANES) spectroscopy.[29,32,33] Here, only the data concerning the charge property of the different oxygen species will be discussed. The analysis of this problem should start with the question why the O1s feature of the electrophilic oxygen has such a high binding energy compared to the corresponding value of the nucleophilic oxygen (Figure 9.2(a)). It is pointed out that the cause should be based on an atomic model, since the molecular nature of the electrophilic oxygen as one of the most probable reasons for its high binding energy value can be fully ruled out by XANES experiments.[40] The first step in this direction should be the estimation of contributions of relaxation effect R to the variation in binding energy.[42] Hohlneicher *et al.* proposed to use the following relation to take into account these effects:[43]

$$\Delta\beta = \Delta[2E_b(i) - E_b(k) + E_A(kii)] = 2\Delta R \tag{9.5}$$

where $\Delta\beta$ is the variation of the modified Auger parameter, $E_b(i)$ and $E_b(k)$ are the binding energy values for outer shell (i) and core level (k) that are involved in

Table 9.1 Experimental energy shifts (eV) ΔE_b(O1s), ΔE_b(O2p) and ΔE_A($OKL_{23}L_{23}$), and contributions of the initial $\Delta\varepsilon$(O1s) and final ΔR state in the variation of O1s binding energy for different oxygen species on silver.[41]

Type of O_{ads}	*O1s*	*O2p*	*OKLL*	*Δ(O1s)*	*Δ (O2p)*	*Δ (OKLL)*	ΔR	*Δε(O1s)*
Ag_2O	528.9	2	513.9	0	0	0	0	0
Nucleophilic	528.3	1.9	514.5	–0.6	–0.1	0.6	0.5	0.1
Electrophilic	530.4	3.3	511.9	1.5	1.3	–2.0	–0.45	–1.05

the Auger transition $E_A(kii)$. The corresponding values were measured for the electrophilic and nucleophilic oxygen and were compared with the values from bulk Ag_2O oxide, which was used as a reference (Table 9.1).[44] From the measured energy shifts we calculated ΔR and, finally, the "true" variation of orbital energy ($\Delta\varepsilon$) according to:

$$\Delta\varepsilon = -[\Delta E_b + \Delta R] \tag{9.6}$$

One can see that the shift of the O1s binding energy for the nucleophilic oxygen compared to the value for bulk Ag_2O (–0.6 eV) is determined essentially by the relaxation contribution. The subtraction of the relaxation shift from the total variation of O1s binding energy leads to a very small difference (0.1 eV) in orbital energy of the nucleophilic oxygen and oxygen in the structure of bulk Ag_2O. This is a strong additional confirmation of the chemical similarity of the oxygen–silver interaction in three-dimensional silver oxide and in the nucleophilic oxygen species, which can be considered as surface silver oxide.

The contribution of the final state relaxation effect to the variation of the O1s binding energy is also rather considerable for the electrophilic oxygen (Table 9.1). Approximately half of the shift of E_b(O1s) originates from the relaxation and only half of the shift is determined by the initial state effect, *i.e.* by the changes in the electron configuration of oxygen atoms. The increase in the orbital energy for a negatively charged ion means the decrease in the charge of electrophilic oxygen.[42]

Additional confirmation of a considerable decrease in the charge on the electrophilic oxygen originates from the analysis of the Auger spectra. Ascarelli and Moretti revealed a good correlation between the separation of the $OKL_{23}L_{23}/OKL_{23}L_1$ oxygen lines and their relative intensities (Figure 9.6(a)).[45] This figure also shows the data for the nucleophilic and electrophilic oxygen (Figure 9.6(a), triangles) taken from ref. 33.

The experimental data are in good agreement with the published correlation, indicating a significant variation of the ionicity in the silver–oxygen interaction for the oxygen species studied. To make a quantitative estimation of this variation, we juxtaposed the $E_{kin}(OKL_{23}L_{23}) - E_{kin}(OKL_{23}L_1)$ values for the nucleophilic and electrophilic oxygen (Figure 9.6(b)) with the dependence taken from ref. 45. In this dependence, the Auger line separation is connected with the atomic charge on oxygen calculated using the Pauling's electronegativity scale.[46] One can see that in good agreement with the O1s orbital energy

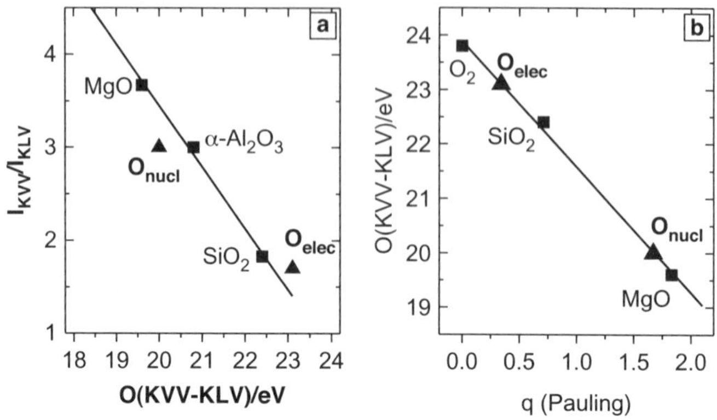

Figure 9.6 Relationships of the spacing of the $OKL_{23}L_1$ and $OKL_{23}L_{23}$ Auger lines with their relative intensities (a) and atomic charge on the oxygen calculated using Pauling's electronegativity scale (b) for the oxygen species on the silver surface (Figure 9.3(b)) and for a number of binary oxides.[41]

variation (Table 9.1), the atomic charge on the electrophilic oxygen is lower by ~1 unit than that on the nucleophilic O_{nucl}. Earlier, the stabilities of two atomic oxygen species with different charges (oxidic, O^{2-}, and oxyradical anion, O^-) were demonstrated in theoretical calculations by Carter and Goddard.[47]

Summarizing the data presented in this part of the chapter, we can conclude that two oxygen species atomically adsorbed on the silver surface exist under reaction conditions. The electrophilic oxygen (530 eV) is involved directly in the epoxidation, and can be described as an O^- anion adsorbed at the surface. It has been also established that the nucleophilic oxygen (528 eV), which has a structure like surface silver oxide (Ag^+–O^{2-}–Ag^+) is active in the total oxidation of ethylene only. In general, this picture is quite close to the concept proposed in ref. 5.

The relevant species of the Ag–O system and their most important properties are summarized in Table 9.2.

The oxygen species assigned as O(4×4) is known as the oxygen-induced reconstruction of the Ag(111) facet. It is formed at rather low temperatures. Basically it is formed as soon as Ag carbonates decompose. This species is not stable under reaction conditions. Since it is an oxygen-induced reconstruction of the Ag(111) surface it is localized at the surface. The Ag–O bonding is characterized by a charge transfer, indicated by the formation of Ag^+ states.

Table 9.2 Binding energies, temperature behavior and location of oxygen species on Ag surfaces.

	BE (eV)	*Temp behavior*	*"Location"*	*O/Ag^{d+}*
$O_{p(4\times4)}$	528.2	Stable $200 < T < 300°C$	Surface	1.7
O_{nucl}	528.5	Formed at 200° C	Surface	0.5
O_{elec}	530.0	Formed at $T > 150°C$	Surface	<<0.5
O_γ	529.7	Predominant at $T > 400°C$	Subsurface bulk	<0.5
O_β	530.6	Formed at $T > 250°C$	bulk	<<0.5

At the same temperature as the O(4×4) the nucleophilic oxygen species is formed. It is characterized by a binding energy of 529.1 eV and surface located. The charge transfer is less pronounced compared to the O(4×4) species. The O_γ species forms at a temperature of about 400 °C, which is significantly higher than the reaction temperature. Therefore, this species is less relevant for the ethylene epoxidation reaction. The species is characterized by a binding energy of 529.7 eV. In the case of small particles the species might form at lower temperature in the subsurface regime. In this case it has a deactivating impact on the reaction. The charge transfer is even lower that for the O_{nucl} species. At binding energies around 530.6 eV two different species appear in the XP spectra. They do form at temperatures around 250 °C and they do have different locations. The species assigned O_β or O_{ns} presents dissolved oxygen in the Ag bulk, whereas the electrophilic oxygen O_{elec} is atomically adsorbed on the Ag surface. The charge transfer of Ag to oxygen is the lowest for these two species compared to all species listed.

The registered binding energies give just a rough value of the binding energy, since the binding energy might depend on the abundance of other oxygen species and on the morphology of the samples. Therefore, the listed binding energies should be considered just as approximate values.

9.2.2 The Transformation of Oxygen Species on Ag Catalysts

The formation and transformation of oxygen species on bulk and supported silver samples will be discussed in this chapter. They depend on the morphology of the sample. On silver foil electrophilic and nucleophilic oxygen species are readily indicated by the XPS maxima at 530.5 eV and 528.4 eV, respectively, as shown in Figure 9.7. Ag particles supported on C (application of a carbon

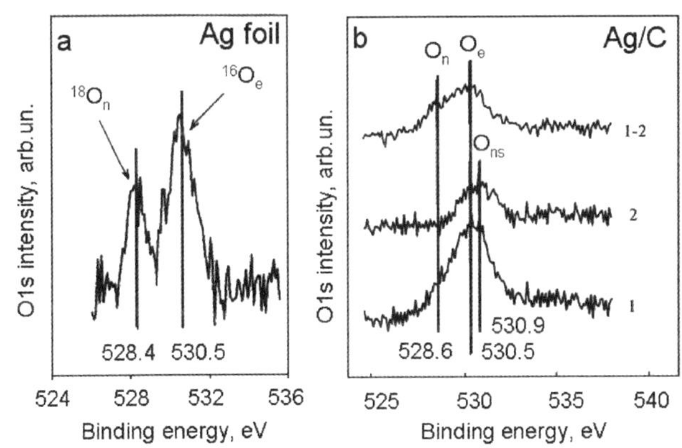

Figure 9.7 O1s XP spectra of adsorbed oxygen on (a) silver foil and (b) Ag/C. O_n, O_{el} and O_{ns} assign nucleophilic, electrophilic and near-surface oxygen, respectively. (1) after O_2 adsorption, (2) after titration by H_2 and (1–2) difference spectrum.[89]

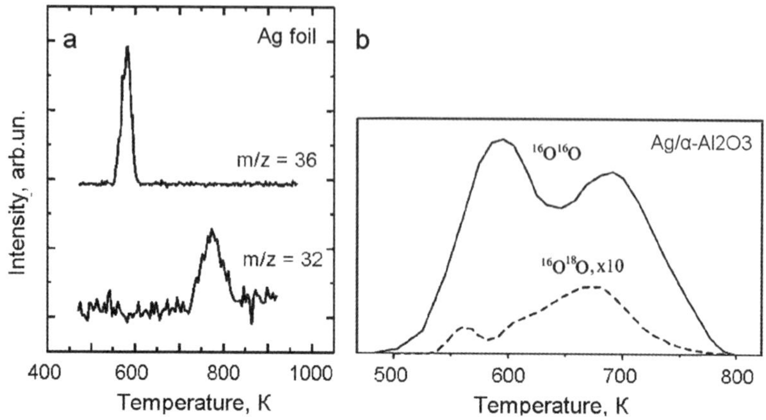

Figure 9.8 TPD spectra of oxygen adsorbed on (a) silver foil and (b) Ag/α-Al_2O_3 ($d = 1000$ Å): (a) O_2 adsorption from the $C_2H_4 + {}^{16}O_2$ mixture at 470 K and 1 mbar with subsequent adsorption of $^{18}O_2$ at the same temperature. (b) adsorption of $^{18}O_2$ at 473 K and 5×10^{-2} Pa with the subsequent adsorption of $^{16}O_2$ at 473 K and 50 mbar.[50]

support instead of alumina allows the measurement of O1s features from oxygen species adsorbed on silver particles) show a further component at higher binding energies. The related oxygen species was assigned as O_β or O_{ns} (near-surface oxygen).[48,49]

When in the case of an Ag foil the electrophilic oxygen is labeled with ^{16}O and the nucleophilic oxygen is labeled with ^{18}O the thermal desorption spectrum does not show isotopic mixing (Figure 9.8). Nucleophilic oxygen desorbs at 600 K as $^{36}O_2$, while the electrophilic oxygen desorbs at 780 K as $^{32}O_2$. The fact that desorption of $^{32}O_2$ was observed at temperatures higher than the reaction temperature might be an indication that the sample did contain a reasonable amount of dissolved oxygen O_{ns}. Analogous experiments performed with supported particles Ag/α-Al_2O_3 show isotopic mixing at around 680 K indicated by the observation of mixed oxygen $^{34}O_2$. These XPS and TPD data indicate the differences in formation of oxygen species on bulk and supported silver samples.

A deep insight into the formation of the electrophilic oxygen species was obtained by steady-state isotopic transient kinetic analysis (SSITKA).[51,52] Figure 9.9 shows the change of $^{18}O_2$ containing reaction products induced by an abrupt change of the gas phase.

The formation of ethylene oxide ($C_2H_4{}^{18}O$) occurs with an induction period relatively to the formation of carbon dioxide as can be seen clearly in Figure 9.9(c). This finding shows that ethylene epoxide and carbon dioxide, are formed from different states of oxygen. This is in agreement with the above-mentioned different chemical functionalities of electrophilic and nucleophilic oxygen in ethylene epoxidation. Since the transfer of the labeled oxygen to electrophilic oxygen takes a longer time, one might assume that electrophilic oxygen is formed from nucleophilic rather than *via* the dissociate adsorption of molecular

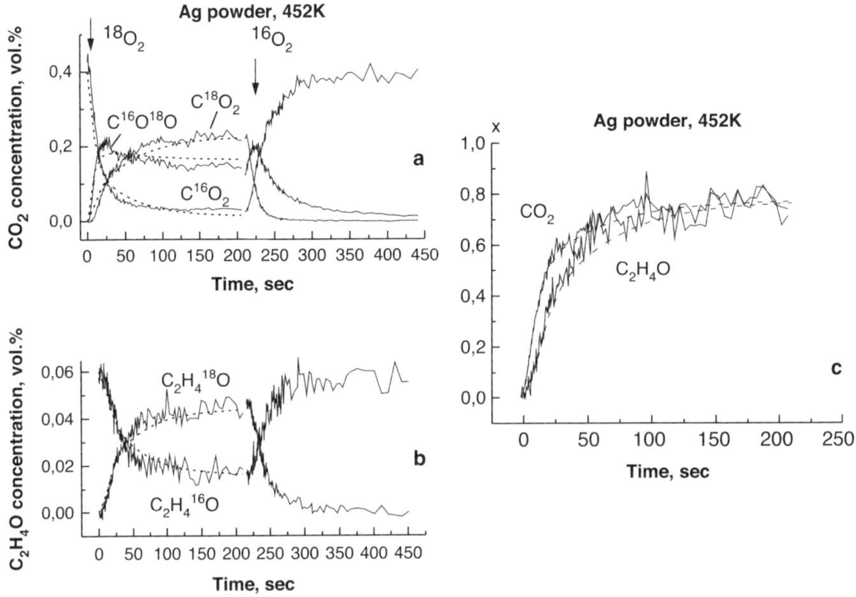

Figure 9.9 Concentrations of (a) CO_2 and (b) C_2H_4O and (c) the isotopic fraction x of ^{18}O in CO_2 and C_2H_4O *vs.* time obtained on silver powder (0.5 g) at 452 K. The gas feed consists of $2.2C_2H_4 + 1.2O_2 + 96.6He$. The flow rate was $0.4\,cm^3/s$. The arrows show the moment when the reaction mixture with ^{18}O was replaced by the mixture of the same composition with ^{16}O. Solid lines show the experiment and dashed lines show the calculations.[89]

oxygen. Since the amount of exchangeable oxygen in silver powder, Ag/α-Al_2O_3 ($d = 1000$ Å) and Ag/α-Al_2O_3 ($d = 220$ Å) are 1.6, 2.9 and 6 monolayers, respectively, a substantial part of the exchangeable oxygen is located in the near-surface layer of silver. As a consequence, one should assume that O_β, or even O_γ, can contribute to the formation of electrophilic oxygen. The participation of oxygen species adsorbed on the support can be excluded, because the portion of the α-Al_2O_3 surface of the samples is about the same. It is very well known that silver adsorbs oxygen above monolayers amounts.[53] The induced reconstruction causes a slow relaxation of the reaction rate.[55] An increase of the exchangeable oxygen is correlated to a substantial increase of the duration of the transient process, as shown in Figure 9.10.

The ultradispersed silver ($d = 200$ Å) has not attained the steady-state value for the isotopic fraction of X_{CO2} even after 30 min. A decrease in the silver particle size increases the amount of the consumed oxygen and the steady state is attained after longer times. An increase in the dispersity of silver increases the number of defects and the portion of even layers decreases. Therefore, one might assume, that O_β is localized in the defect-rich phase (small particles) or close to long defects (grain boundaries). A quantitative analysis of the isotopic curves allowed the authors to suggest the mechanism shown in Scheme 9.1.[52,54]

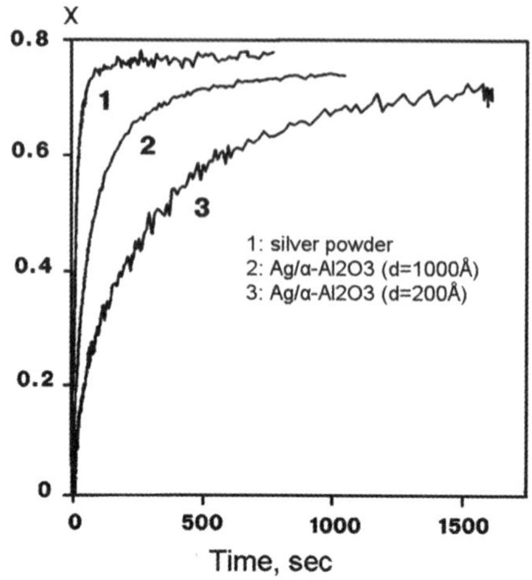

Figure 9.10 Isotopic fraction of ^{18}O in carbon dioxide as a function of time on-stream at 459 K after step-like admission of the reaction mixture containing $^{18}O_2$ onto reduced silver catalysts with various dispersities: (1) silver powder, (2) $Ag/\alpha\text{-}Al_2O_3$ ($d = 1000$ Å) and (3) $Ag/\alpha\text{-}Al_2O_3$ ($d = 200$ Å).[89]

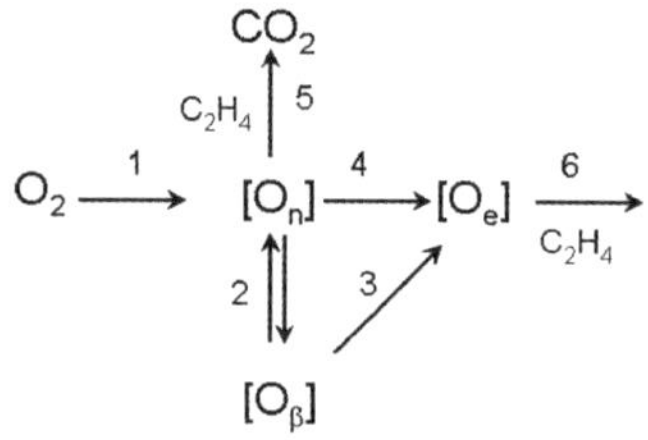

Scheme 9.1 Oxygen transformations during the ethylene oxidation on silver catalysts.[54,89]

Nucleophilic oxygen O_n, which is active in the total oxidation only, is formed by the dissociative chemisorption of molecular oxygen indicated by step 1 in Scheme 9.1. Electrophilic oxygen, which is active in the selective oxidation of ethylene to ethylene oxide, can be formed directly from nucleophilic oxygen or it is generated *via* near-surface oxygen O_β (steps 2, 3 in Scheme 9.1). The exchange rate of O_β and O_n is high and comparable to the adsorption rate of molecular oxygen. The isotopic experiments described above cannot be modeled adequately if step 4 is not considered. In this case, the calculated isotopic curves $X_{CO2}(t)$ and $X_{C2H4O}(t)$ would almost merge. If only step 4 is included in the calculation (formation of O_{el} *via* O_n) the shift of the $X_{C2H4O}(t)$ line relative to $X_{CO2}(t)$ is much larger than in the measurement. Therefore the authors conclude that electrophilic oxygen can be formed directly from nucleophilc oxygen or it is formed *via* near-surface oxygen O_β. The calculations reveal that the rate of formation of electrophilic oxygen *via* nucleophilic oxygen is twice as high as its generation *via* near-surface oxygen.[89] In supported silver catalysts the amount of near-surface oxygen increases with dispersity, because the number of defect states is high. Therefore, the formation of epoxidizing oxygen occurs *via* near-surface oxygen. On bulk catalysts electrophilic oxygen is formed basically from nucleophilic oxygen.

In this context, it is possible to construe the XPS and TDS data shown in Figure 9.7 and Figure 9.8. The amounts of near-surface oxygen, which is formed on imperfect surfaces of dispersed silver, form buffers where intensive isotopic mixing takes place during thermodesorption. For silver powder with micrometer-scale particles, the rate of oxygen transformation into O_β is 20 times higher than that of the transformation into O_{el}.[89] In the case of bulk silver, the transfer of the isotopic label occurs *via* step 4 only. A $^{34}O_2$ MS signal was not observed in TDS measurements of foils or single crystals after isotopic adsorption. This is due to the spatial separation of nucleophilic and electrophilic oxygen. The formation of electrophilic oxygen requires defect states. Therefore, the generation of O_{el} on a single crystal of foil requires much energy. The formation of electrophilic oxygen was observed at high oxygen partial pressure ($\geq$1 torr) and temperature (>550 K) or under milder conditions if abetted by the energy of the chemical reaction. Electrophilic oxygen is localized most probably at defects on the surface. Unlike supported silver the amounts of the two oxygen species covering the surface of a Ag foil or a single crystal are comparable. Thus, the number of neighboring O_n and O_{el} species is much lower than the amount of overall adsorbed oxygen. During the thermal desorption, either O_n fails to transform into O_{el}, since the sites are occupied, or it desorbs before diffusing to defect regions on the surface.[89]

9.2.3 The Nature of the Key Surface Intermediate

DRIFTS measurements of the interaction of ethylene with silver did not reveal any stable intermediate on the pathway from ethylene to ethylene oxide beside the π-complex of ethylene with Ag^+.[55] Since bonding and antibonding orbitals of the π-complex of ethylene are populated, it cannot react with electrophilic oxygen. Therefore, the π-complex has to be activated before the addition of an oxygen atom to the ethylene molecule. The formation of the π-complex is probably necessary to fix ethylene on the silver surface for a certain period of time, which is sufficient to add electrophilic oxygen to the ethylene molecule. This mechanism is supported by data of mixing of *cis* and *trans* isomers of dideuteroethylene.[56–58] The *cis* isomer of ethylene was selectively oxidized and after reaction a *cis*:*trans* ratio close to the equilibrium was detected. During the time an oxygen atom is added to the ethylene, this rotates around the single C–C bond, which is impossible for molecules with a double bond or a cyclic structure. The formation of the π-complex causes a decrease of the electron density of the C–C bond in the ethylene molecule. Since the activating oxygen atom is less electrophilic than an elementary oxygen atom, epoxidation is favored. Therefore, in some papers epoxidizing oxygen is assigned as $O^{\delta-}$.[59–61]

Recently the formation of oxametallacycle as an intermediate in the ethylene epoxidation over silver was suggested in the literature.[9] In this study, ethylene epoxide was adsorbed onto Ag(111) at various temperatures. The authors expected to activate ring opening of the ethylene epoxide and the formation of a stable and detectable intermediate. The Ag(111) surface was investigated by temperature-programmed desorption, high-resolution electron energy-loss spectroscopy (HREELS) after ethylene epoxide adsorption. In addition, density-functional-theory-based *ab initio* calculations were executed.[9] The HREELS spectra of ethylene epoxide adsorbed on Ag(111) show significant differences after adsorbing at 250 K compared to adsorption at 110 and 140 K. TPD spectra gave indications that at 250 K a reaction takes place on Ag(111), with structures at 860, 1090, 1475 and in the range between 2850–3000 cm^{-1}. These structures are in very good agreement with the DFT calculations by the same authors and with IR measurements performed during the oxidation of ethylene in a batch recirculation reactor over a silica-supported silver catalyst by Force and Bell.[62] The authors conclude that the ethylene epoxide undergoes a ring opening to produce a stable surface intermediate. The results mentioned here deliver evidence that this intermediate is a surface oxametallacycle, which reacts at 300 K to reform ethylene epoxide and to form a few other products. The vibrational mode at 400 cm^{-1} appears to be a viable reaction coordinate of the oxametallacycle ring closure to form ethylene epoxide. It is remarkable that the DFT calculation of this reaction coordinate revealed that the activation energy required to form the oxametallacycle is comparable to the activation energies reported in steady-state ethylene epoxidation processes.

Stegelmann *et al.* have published a microkinetic model of ethylene oxidation over silver that seems to join both concepts.[7] Surface-science experiments on Ag(111) and steady-state kinetics on the same surface deliver the reaction mechanism and the model parameters. The main ideas of the model are based on the following: Surface oxide (nucleophilic oxygen) is acting as sites for ethylene adsorption. Electrophilic oxygen, which reacts to form an oxametallacycle is present on the silver surface as well. The oxametallacycle is a common intermediate for ethylene epoxidation, ethylene combustion, and ethylene oxide combustion. The branching ratio of the oxametallacycle into ethylene oxide and acetaldehyde determines the selectivity of the reaction. The model predicts the heat of adsorption, sticking, temperature programmed desorption and temperature programmed reaction measurements. It reproduces a broad range of transient surface-science experiments and it explains the kinetics of epoxidation, ethylene combustion, and ethylene oxide combustion. This demonstrates that the microkinetic model contains the essential physics and chemistry to explain ethylene oxidation over silver. Anyway, no further experimental proofs of the existence of an oxametallacycle are reported in the literature.

9.2.4 Secondary Reactions

The secondary oxidation of ethylene oxide (3) has also been the subject of numerous investigations, both bulk silver (single crystals and silver powders) and supported silver (Ag/α-Al_2O_3, Ag/γ-Al_2O_3, Ag/SiO_2, Ag/TiO_2, *etc.*) being the samples for these studies.[25,63–67] Thus, Backx *et al.* carried out the adsorption of ethylene oxide on clean Ag(110) at 110 K and found a weakly bound species, which desorbed below room temperature.[63] On the oxygen-precovered surface, they observed that the EO saturation coverage was enhanced. These species, however, were relatively weakly bound as the activation energies for desorption were less than 53 kJ/mol. Later, Grant and Lambert reported that EO adsorbing on clean and Cs-precovered Ag(111) desorbed in a broad peak with a peak maximum ranging between 426 and 500 K without undergoing any surface reactions.[25] It has also been shown that EO can be further oxidized to CO_2 *via* acetaldehyde on a silver surface.[25,64] In addition, some authors have argued that ethylene oxide isomerization to acetaldehyde is the rate-limiting step in its further combustion to CO_2.[65,66]

One of the more detailed and systematic investigations of the secondary reaction of ethylene oxide on various silver catalysts were performed by Bulushev *et al.* using the transient-response technique and infra-red spectroscopy.[67] The authors could identify three routes of transformation of ethylene oxide: (1) its decomposition to ethylene and adsorbed oxygen; (2) ethylene oxide isomerization to acetaldehyde; (3) formation of glycol-like species and their further oxidation.

Ethylene oxide decomposition, which was also reported, proceeds on the silver surface, but the contribution of this route is rather small.[5] This is a possible reason why other authors (*e.g.* ref. 25) have not registered any ethylene formation during ethylene oxide interaction with Ag(lll) surface.

Isomerization of ethylene oxide to acetaldehyde occurs on the surface of both silver and support. Positively charged Ag^+ ions on the silver surface and Al^{3+} or H^+ ions on the alumina surface are the active sites for this reaction. Nevertheless, the Lewis-acid sites are more active in isomerization than Broensted ones. This conclusion is based on higher activity of Ag/α-Al_2O_3 catalysts in isomerization compared to Ag/SiO_2 ones, which are characterized by a considerable amount of acid hydroxyl group, but the absence of Lewis-acid groups.[67] The acetaldehyde formed reacts with the alumina surface to produce acetate species (CH_3COO) that quickly oxidizes to carbon dioxide and water. The step of acetaldehyde transformation to rather stable acetate has been concluded to proceed also on the silver surface *via* interaction with the nucleophilic oxygen.[69] The authors estimated that the contribution of isomerization of ethylene oxide to acetaldehyde followed by oxidation to the overall rate of CO_2 and water formation is decreased by three times when cesium is added to the catalyst. Since acid sites of alumina are responsible for isomerization this behavior is usually explained by neutralization of acid sites by cesium.[68,84]

According to Bulushev *et al.*, the main route of secondary reactions of ethylene oxide on Ag/α-Al_2O_3 catalysts involves the formation of glycol species (–OCH_2CH_2OH), the concentration of which depend on the concentration of basic hydroxyl groups.[67] They have proposed that subsequent oxidation of glycol species to CO_2 and H_2O occurs at the support/silver interface. The suggestion about the determining role of the support/silver interface in oxidation of glycol species is sustained by the experimental fact that the ethylene oxide total oxidation on the catalysts with small silver particles is higher than for the catalysts with large particles. Again, silver particles themselves can produce glycol species *via* interaction of ethylene oxide with the nucleophilic oxygen.

Thus, secondary oxidation of ethylene oxide over Ag/α-Al_2O_3 catalysts is a very complex process that proceeds *via* several routes, with both support and silver particles playing an important role in the reaction steps. This result is not only of fundamental, but also of practical significance. Indeed, promotion of the catalysts with alkali metals such as sodium and cesium will depress the concentration of acid groups, but simultaneously increases the number and strength of the basic OH groups.[67] Therefore, cesium will depress isomerization of ethylene oxide to acetaldehyde, but will increase the secondary oxidation *via* the glycol route. One can assume that namely such dual role played by Cs explains the well-known fact of the strong dependence of the epoxidation activity on the concentration of the cesium loaded. The introduction of cesium in high concentration decreases sharply the rate of ethylene-oxide formation over Ag–Cs/α-Al_2O_3 catalysts.[70,71] This seems to be also a reason why researchers try to find other methods of suppressing the formation of acid OH-groups. For example, Jung *et al.* reported the reduction of hydroxyl groups by silane treatment of Ag/MCM-41 catalysts to suppress isomerization of ethylene oxide to acetaldehyde.[72] Thus, the selectivity for ethylene oxide in the ethylene oxidation is enhanced. Another way to neutralize the undesirable part of the dual effect of cesium is the use of the promoters, which are no base metals as alkalis: for example, rhenium that has been added to the epoxidation catalysts for almost thirty years. We should also take into account that chlorine, which is the main promoter of the epoxidation catalysts, accelerates the isomerization of ethylene oxide to acetaldehyde.[66] Finally, opposite trends in variation of secondary ethylene oxide oxidation (increase) and in ethylene epoxidation (decrease, see later) with reduction of silver particle sizes causes the reduction of selectivity towards ethylene oxide so that the use of the catalysts with small silver particles (<50 nm) becomes unprofitable. And finally, the increase of the rate of the C_2H_4O secondary oxidation on the catalysts with small silver particles together with the decrease of the epoxidation rate (the size effect for the Ag/α-Al_2O_3 catalysts will be discussed later), so that the rates of these reaction become comparable, reduces the selectivity towards ethylene oxide with decreasing the silver particle sizes. Since alumina plays an important role both in isomerization of ethylene oxide to acetaldehyde and in glycol formation, it is evident that development of highly selective catalysts also requires optimization of the support.

9.2.5 Mechanisms of Ethylene Epoxidation over Silver

The results presented in the previous chapter and especially the *in-situ* XPS data allow the suggestion of the following reaction mechanism of the epoxidation reaction over silver, as illustrated in Scheme 9.2. It is obvious that silver is used

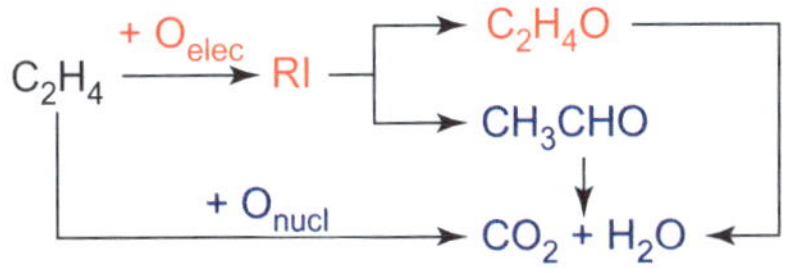

Scheme 9.2 Mechanism of ethylene epoxidation over silver catalysts.

as a catalyst for ethylene epoxidation, since different types of oxygen species are formed on the surface under reaction conditions. At least two different types of oxygen are required for the ethylene epoxidation reaction. One type (nucleophilic oxygen) should create ionic silver in order to form adsorption sites for ethylene.

A second type of oxygen (electrophilic oxygen) is required to form the cyclic ethylene oxide structure. In addition, dissolved oxygen is present in the bulk, which might be transformed into electrophilic oxygen during the reaction. At higher temperatures than the reaction temperature of the ethylene epoxidation further oxygen species are detectable. Ethylene adsorbs on Ag^+ sites that are formed due to the formation of nucleophilic oxygen. When electrophilic oxygen is present, it will interact with the π-system of the adsorbed ethylene to form ethylene epoxide, which will desorb. When there are just further nucleophilic oxygen species available, this oxygen species will activate the C–H bond and CO_2 is formed, the product of the total oxidation. Supply of electrophilic oxygen to adsorbed ethylene occurs *via* migration at kink sites, where electrophilic oxygen is localized, to the sites of ethylene adsorption. At the point of π-complex decomposition, the addition of electrophilic oxygen takes place, resulting in the formation of ethylene epoxide. The mechanism describes the reaction on bulk silver. In the case of supported silver where the reaction occurs near the boundaries of regular and defect regions, electrophilic oxygen is supplied to the reaction zone *via* dynamic processes in the defect phase of silver. Alumina presents a specific area, which is higher by two orders of magnitude compared to silica. Therefore, alumina is used as a support of Ag particles in order to minimize sintering effects during the catalytic reaction. Deactivation of supported catalysts can be prevented by the use of an alumina support.

The addition of chlorine is known to have a positive impact on the reactivity of ethylene over silver. This conclusion was made in a number of papers.[73,92] Thus, according to TPD data, introduction of chlorine on a Ag(110) surface increased the amount of adsorbed ethylene that is accompanied by increasing the heat of ethylene adsorption.[73] This is due to the fact that chlorine, or less efficiently subsurface oxygen, renders the absorbed atomic oxygen on the silver surface more electrophilic by withdrawing electron density from around the adsorbed oxygen species. The presence of negatively charged subsurface oxygen or chlorine generates a positive charge on the uppermost silver layer. This draws negative charge away from adsorbed atomic oxygen, making it more electrophilic. In the absence of a charged upper layer of silver the surface oxygen species carries a stronger negative charge and is less inclined to attack the electron-rich π-system of the adsorbed ethylene molecule. Another impact of the chlorine-induced positive charge on the uppermost silver layer is the possibility for ethylene to produce adsorbed π-complexes without activity in the total oxidation of ethylene.

9.3 Optimization of the EtO Catalysts *via* Nanostructured Materials

9.3.1 Size Effects in Ethylene Epoxidation

Metal particle-size effects are very well known in heterogeneous catalytic processes.[74] They are associated with metal–nonmetal transitions and they are usually observed at particle sizes of tens of angstroms. Many studies have proven that supported silver catalysts show a size effect in ethylene epoxidation.[75–81] In these studies the properties of the particles are the same as for bulk silver if the particle sizes are larger than 100–1000 Å, which is much higher than in other metals. The size effect of supported particles is explained by a change in the structure of the surface, particle morphology, effect of the support, and other factors. Unfortunately, the arguments were vague, and no specific conclusions were drawn about the reason for size effects.

The epoxidation rates of single crystals of Ag(111) and Ag(110) were investigated by Lambert and Campell, respectively.[25,82] Their measurements show that the reaction rate per unit surface area is higher by two orders of magnitude than for supported silver catalysts. Saikowski and Boudart concluded that the reaction is structure insensitive.[83] The variations in the catalytic activity were attributed to the presence of an admixture in dispersed silver or the support, which may segregate on the silver surface during catalyst preparation or during the reaction. Lee *et al.* studied silver supported on various materials under conditions that prevent the segregation of admixtures.[84] In the particle-size range between 150–400 Å the reaction rate increased by a factor of three. The authors explain the rise in activity by the absence of small inactive particles ($d < 200$ Å).

When size effects are discussed, the sintering of silver particles during the reaction has to be considered. Electron-microscopy investigations reveal that silver particles may sinter even at 200–300 °C, which is close to reaction conditions.[85] Obviously, sintering may drastically change the initial distribution of particles, and therefore, a wrong dependence of the reaction rates on the particle size may be found. The sintering can be minimized by the use of a high surface area support like silica.[81]

A detailed investigation on the size effect was published by Bal'zhinimaev *et al.*[55] They synthesized Ag/α–Al_2O_3 (7 m^2/g) and Ag/SiO_2 (158 m^2/g) catalysts with a narrow silver particle-size distribution. The deviation from the average value was at most 30–40%. The total concentration of admixtures in the catalysts was at most 0.01 wt%. The size distribution and particle shape of the silver particles did not change during calcination or ethylene epoxidation, as shown by XRD and electron-microscopy data.[86]

The rates of ethylene epoxidation and total oxidation of ethylene show a strong dependency on the silver particle size, as shown in Figure 9.11. The epoxidation rate increases by more than an order of magnitude when the silver particle size increases to 300–500 Å (Figure 9.11(a)). Further increase in particle size reveals only a small change in the catalytic activity. But when one uses silver powder with $d \approx 20\,000$ Å the reaction rate is several times higher.

The size effect on the total oxidation of ethylene over Ag particles (Figure 9.11(b)) is less pronounced compared to the epoxidation reaction. In the range 300–500 Å the reaction rate increases just several times, but not by an order of magnitude. As a consequence, selectivity towards ethylene oxide is reduced with decreasing Ag particles size (Figure 9.12).

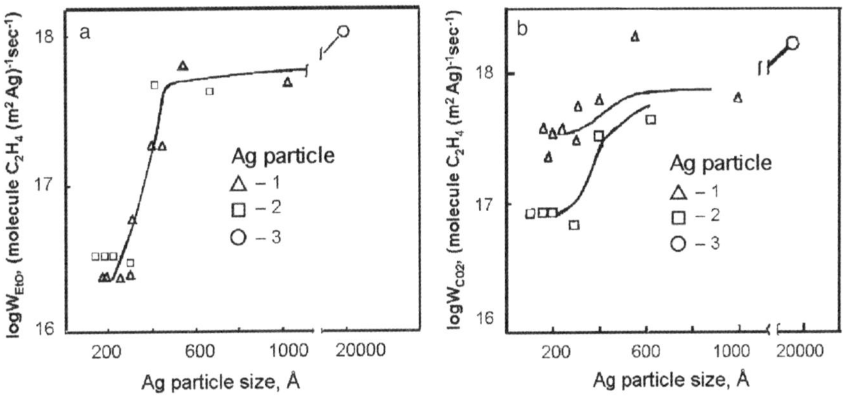

Figure 9.11 (a) Ethylene epoxidation rate and (b) total oxidation rate of ethylene as a function of silver particle size (1) Ag/α-Al_2O_3, (2) Ag/SiO_2, (3) silver powder.[89]

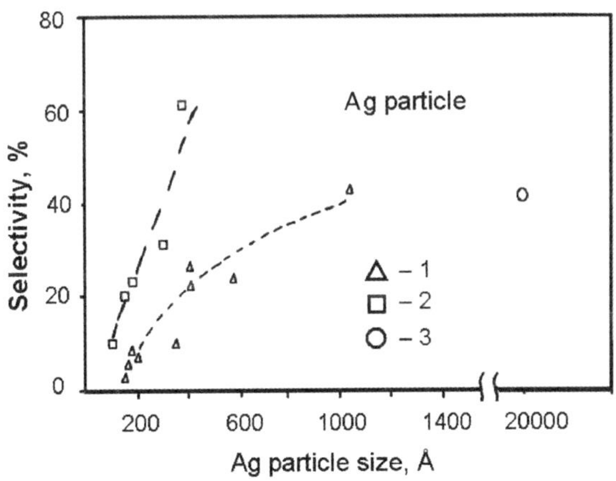

Figure 9.12 Selectivity to ethylene oxide as a function of particle size. (1) Ag/α-Al_2O_3, (2) Ag/SiO_2, (3) silver powder.[89]

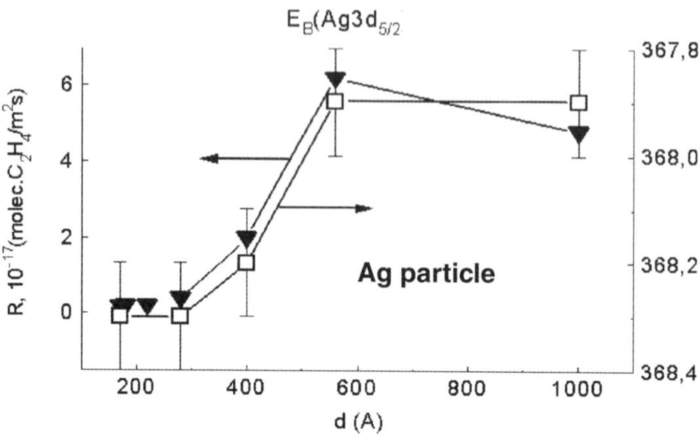

Figure 9.13 Rate of ethylene epoxidation and $Ag3d_{5/2}$ binding energy as a function of Ag particle size.[89]

The CO_2 formation rate shows a strong dependence on the nature of the support for small particles. This effect is not attributed to a change of the intrinsic rate of the complete oxidation. It is related to differences in the acid–base properties of the support, which determines secondary ethylene epoxide transformations, as discussed in the previous subchapter.[67] On basic hydroxyl groups of α-Al_2O_3, ethylene epoxide efficiently transforms into glycol-like compounds, which are further oxidized into CO_2 and H_2O over silver particles. The rate of their oxidation substantially increases as silver becomes more dispersed. As a result, the selectivity to ethylene oxide on silica-gel-supported samples is higher.

The described results allow the conclusion that the structural and electronic properties of silver are substantially changed in the range 300–500 Å.

Variation of the Ag $3d_{5/2}$ binding energy with the silver particle size, which correlates with catalytic properties, seems to confirm this suggestion.[71] Indeed, as seen from Figure 9.13 an increase in the catalytic activity is associated with a decrease in the binding energy of the Ag $3d_{5/2}$ electrons. A lower value $E_b = 367.9$ eV than for bulk silver (368.3 eV) is due to the effect of differential charging of supported silver particles.[71,87] A drastic increase in the binding energy when passing to small particles is due to the appearance of the defect phase in the near-surface layers of silver. The conductivity of this defect phase is lower than that of bulk silver. Note that the change in binding energy is observed only for samples that have been used in the reaction.

The Ag $3d_{5/2}$ binding energy values for the fresh catalysts are close to the binding energy of bulk silver. This points to the fact that the reaction medium favors the formation of defects states of silver, the formation of near-surface oxygen might by one of the possible reason of this phenomenon. Indeed, as shown earlier decrease in silver particle sizes increases the contribution of subsurface oxygen in TPD spectra.[91]

It should be noted that not only electronic properties, but also the structure of silver surface layers in small silver particles can be substantially different from the properties of the bulk. Scanning tunneling microscopy studies by Frenken *et al.* have demonstrated that the Ag(115) surface contains defects already at $T = 78\,°C$, which is lower than the reaction temperature.[88] For smaller particles the surface tension is higher and therefore a majority of the small particles have a rounded shape, whereas larger particles have a cut-stone shape typical of crystals. Electron microscopy investigations have indicated that silver particles in the size range of 200 Å are almost spherical but particles with larger diameter than 500 Å are well cut.[89] The amount of regular surface increases with increasing silver particle size, whereas the portion of defect surfaces decreases.

If this statement is correct, variation of silver particle size will change the relative population of the nucleophilic and electrophilic oxygen. Due to the intense O ls emission from the alumina, this suggestion could not be investigated for Ag/Al_2O_3 samples using XPS. To solve this problem, model Ag/C (C = amorphous graphite) catalysts were prepared and applied to study oxygen adsorption, with the aim of monitoring the nature and distribution of oxygen adspecies as a function of mean particle size.[90,91]

Figure 9.14 shows the O ls difference XP spectra measured after the adsorption of dioxygen (1 mbar for 10 min at 470 K) as a function of silver coverage expressed as *R*, the measured Ag3d to C1s intensity ratio. A weak O(ls) signal was observed to appear on the clean graphite surface after similar treatment, and this was therefore subtracted from the raw data.

These spectra indicate that the relative populations of the nucleophilic and electrophilic oxygen are strongly dependent on silver coverage. Indeed, a single

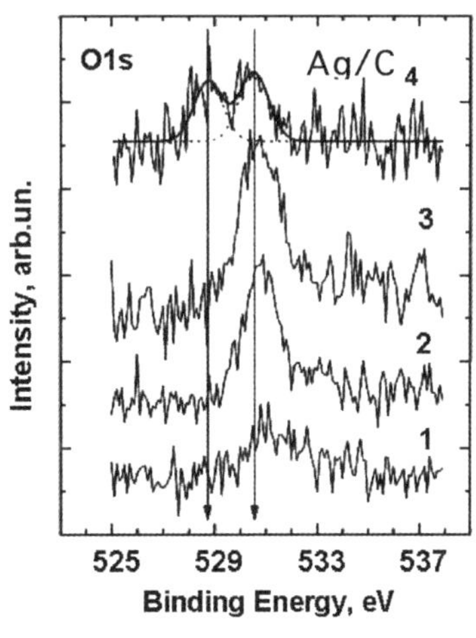

Figure 9.14 O1s spectra measured after oxygen adsorption (100 Pa, 10 min, 420 K) on silver deposited on graphite, for increasing silver coverage (1–4): 1 – R $(= I_{Ag}/I_C) = 0.4$, 2 – $R = 1.5$, 3 – $R = 2.4$, 4 – $R = 13.1$.

O1s feature with a binding energy of 530.5 eV characteristic of the electrophilic oxygen is observed for $R<10$, whereas for larger coverage a second feature develops at a binding energy of 528.4 eV that is due to accumulation of the nucleophilic oxygen. Since Ag atoms on graphite are well known to agglomerate to three-dimensional particles even at room temperatures, it would be more correctly to discuss silver particle size, but not coverage. To recalibrate the XPS intensity ratio to the mean particle size, *ex-situ* STM measurements of Ag particle size were made for two coverages: $R=3$ and 10. The calibration delivers mean particle sizes of about 50 and 100 Å, respectively.

These data allow us to suggest the explanation of the size effect in ethylene epoxidation observed at unusually large sizes of alumina-supported silver particles.[55] Taking into account the following scheme of ethylene epoxidation

$$O_{elec}(ads) + \pi\text{-}C_2H_4(ads)/Ag^+ \rightarrow C_2H_4O \tag{9.7}$$

where Ag^+ is the center for ethylene molecule adsorption created by nucleophilic oxygen, O^{2-}, which disappears on decreasing the Ag particle sizes (Figure 9.14), one can conclude that less-effective adsorption of ethylene in the case of small silver particles will decrease the rate of ethylene epoxidation. In order to check this assumption, the concentration of ethylene π-complexes was determined for different silver catalysts and correlated with the rate of epoxidation (Table 9.3). And indeed, as indicated by Table 9.3 the concentration of π-complexes per unit of silver surface area and the rate of ethylene-oxide formation increases by a similar factor of 10–100 when passing from supported to bulk catalysts. But the rate normalized to the concentration of π-complexes remains the same over the whole range of sizes.

These findings can be summarized as follows: The reaction rate is proportional to the concentration of the π-complexes of ethylene adsorbed on Ag^+ sites. The transition from supported to bulk catalysts is accompanied by an increase in the amount of regular surfaces, which is associated by an increase of the formation rate of ethylene oxide by a factor of 10–100. Ethylene epoxidation on dispersed and on bulk silver occurs most probably *via* the

Table 9.3 Adsorption and catalytic properties of silver catalysts.[89]

Sample	*Ag particle size, Å*	$N_{C2H4} \times 10^{-17}$ *molecule/* $(m^2\ Ag)$	$W \times 10^{-17}$ *molecule* $(C_2H_4\ (m^2\ Ag)^{-1}\ s^{-1})$	$W/NC_2H_4,\ s^{-1}$
Ag(110)	–	–	200	3
Polycrystalline silver foil	–	60	–	–
Silver powder	10 000	10	10	1
Ag/α-Al_2O_3	1000	1	5	5
Ag/α-Al_2O_3	160	≤0.3	0.2	≥1

N_{C2H4}: concentration of ethylene π-complexes
W: rate of ethylene epoxidation

same mechanism. On comparison of particle sizes estimated by TEM and XRD Balzhinimaev concluded that the active sites are localized at the grain boundaries near the interface of regular and defect regions on the silver surface.[89]

9.3.2 Modern Trends in Development of the Catalysts for Ethylene Epoxidation

9.3.2.1 Optimization of Silver Particle Sizes

Taking into account the reduction of the concentration of nucleophilic oxygen, which is active in ethylene total oxidation, or even its full disappearance on small silver particles (Figure 9.14), one can suggest that decreasing the silver particle sizes will increase the selectivity of the reaction towards ethylene oxide. Such dependences were observed by Verykios and coauthors who reported an increase in selectivity from 39 to 48% at 240 °C when the silver particle size was decreased from 1660 to 1280 Å.[92] But a further decrease in size changed the trend, and the selectivity became 26% for a catalyst with the mean size of silver particles at about 550 Å. A similar dependence (Figure 9.12) was observed and was explained by a decrease in epoxidation rate (Figure 9.11(a)) and an increase in the rate of secondary oxidation of ethylene oxide formed on basic hydroxyl groups of α-Al_2O_3 located on the perimeter of small silver particles.[70,89]

This observation means that production of a more-selective catalyst requires optimization of the silver particle sizes around 50–100 nm. A decrease or increase of the size will reduce the selectivity due to an increase in the rate of ethylene oxide secondary oxidation or due to an increase in the rate of the total oxidation of ethylene with the nucleophilic oxygen (Scheme 9.2). As a consequence, the modern Ag/Al_2O_3 catalysts contain silver particles ranged between 50 and 100 nm.

The knowledge about the reaction mechanisms allows to suggest another way to improve the selectivity of the silver catalysts – the application of the support without basic OH-groups. And indeed, replacement of α-alumina with a carbon-originated support, Sibunit, allowed Balzhinimaev to reach a selectivity of about 85–90% on Ag/Sibunit catalysts.[89]

9.3.2.2 Ag–Cu Alloys in Ethylene Epoxidation

Barteau *et al.* have shown that bimetallic catalysts Ag–Cu do have higher selectivity than corresponding monometallic catalysts in the ethylene epoxidation reaction.[93] Computational screening led to formulation of a new Cu/Ag alloy that is more selective than Ag to ethylene epoxide. These predictions were verified by steady-state experiments with monolith-supported catalysts.

In order to gain insight into the reasons of the improved selectivity ambient first-principles calculations and ambient-pressure XPS measurements have been performed recently.[94] The surface energy of the most-stable surface structures on each of the three surface orientations (Ag(111), Ag(100) and Ag(110)) were calculated as a function of the Cu chemical potential at a fixed chemical potential of oxygen $\Delta\mu_O = -0.61$ eV (corresponding to $T = 600$ K and $p_{O2} = 1$ bar). CuO-like structures start to form on the Ag facets with increasing chemical potential of Cu. Thus, the alloy will decompose under these conditions. The results of the calculations were confirmed by XPS measurements. An Ag–Cu bimetallic powder (particle size 100 nm, 2.5% Cu) was measured in a C_2H_4:O_2 = 1:2 mixture at 0.5 mbar. The catalytic activity was measured by mass spectrometry. Cu $2p_{3/2}$ spectra measured in oxygen and in the reaction mixture indicate a CuO-covered surface.

The concentrations of Ag, Cu, O and C were measured in vacuum and under various gas atmospheres, assuming a homogeneous distribution of these elements on the surface. Starting with a Ag/Cu ratio 8 times higher than the nominal loading of 2.5% Cu the Cu concentration even increases more (factor 50) on the surface under oxygen or reaction atmospheres. A significant amount of oxygen was found on the surface, in agreement with the observation of surface-oxide formation. The calculation of the Cu/O ratio indicates more oxygen than the presence of Cu_2O requires. Since the binding energy of the O 1s electrons in Cu_2O is 530.4 eV, it cannot be excluded that oxygen spills over from Cu_2O to Ag to form electrophilic oxygen, which has a similar binding energy.[41]

9.4 The Similarity and Differences in Propylene Epoxidation Compared to Ethylene Epoxidation

Catalysts and reaction conditions that are optimized for ethylene epoxidation do not give comparable selectivities in the direct epoxidation of propylene. The propylene epoxidation over Ag/α-alumina is characterized by typical selectivities of the order of only a few per cent. It was proposed that the very low selectivities in propylene epoxidation are due to the ease with which an allylic hydrogen atom may be stripped from the molecule, a process that would immediately preclude epoxidation and result in combustion.[95] There exist a lot of publications on the modification of Ag catalysts to enhance the propylene oxide selectivity.[96–101] Most of the silver catalysts for propylene epoxidation result in high silver loading combined with heavy promotion.

Lambert *et al.* have shown that Ag/$CaCO_3$ catalysts have a maximum in selectivity towards propylene epoxidation as the amount of potassium is increased.[100] They have shown that there is a correlation between potassium loading and Ag particle size. Particles in the size range between 20 and 40 nm are the most selective catalysts, whereas Ag particles much larger or much smaller are less selective. It was shown that very large Ag particles (>150 nm) supported on α-alumina are very selective in ethylene epoxidation, but they reveal a rather poor selectivity towards propylene epoxidation.[102] In addition, it was found by Bukhtiyarov that decreasing the Ag particle size decreases ethylene epoxidation selectivity in contrast to propylene epoxidation on Ag/$CaCO_3$.[103] These findings suggest that the selectivity of the epoxidation reaction is controlled by primary chemistry in case of ethylene epoxidation, while the secondary chemistry is more important with propylene. Ethylene and propylene epoxidations behave differently towards changes in silver particle size. Potassium-promoted $CaCO_3$ is a good support in propylene epoxidation since it favors the formation of small Ag particles.

Recently Lei *et al.* have published a combined grazing-incidence small-angle X-ray scattering catalytic–DFT study.[104] They have investigated the catalytic activity in the propylene epoxidation reaction of Ag_3 clusters and of Ag nanoparticles (3.5 nm) supported on alumina. They claim that Ag trimers and nanoparticles show high activity at low temperatures and the formation of carbon dioxide was negligible. The trimers start to sinter at about 110 °C. At 50 °C only acrolein and propylene oxide (30%) are formed. At higher temperatures, the amount of acrolein is decreasing and CO_2 is increasing. The nanoparticles show a selectivity of more than 80% to propylene oxide at around 100 °C, whereas no CO_2 was formed. DFT calculations based on the formation of an oxametallacycle support these findings. These results show that the mechanism of propylene is not understood in detail yet.

9.5 Outlook

This review shows that there exist already a lot of facts that might help to describe the reaction mechanism of the ethylene epoxidation on the atomistic level. It is accepted that different atomic oxygen species like nucleophilic, electrophilic and dissolved oxygen are present under reaction conditions. There are indications that the oxygen species might transform into each other, which is very important for the understanding of the reaction mechanism. The dissolved oxygen species O_β seems to play an important role in the reaction, since the formation of this subsurface species creates defects, which are necessary to form electrophilic oxygen. The investigation of the O_β species is hindered by the fact that its O 1s binding energy is very close to that of the electrophilic oxygen. Therefore X-ray emission, which is a more bulk sensitive method, might help in the future to study O_β under reaction conditions.

References

1. C. N. Satterfield, *Heterogeneous Catalysis in Industrial Practice*, Krieger Publishing Company, Malabar, 1996.
2. *Ullmann's Encyclopedia of Industrial Chemistry*, 5th edn, Vol. A11, ed. W. Gerhartz, VCH Verlagsgesellschaft mbH, Weinheim, 1988.
3. *Kirk-Othmer Encyclopedia of Chemical Technology*, 4th edn., Vol. 9, Wiley, USA, 1994, p. 915.
4. H. Takada, *Shokubai*, 1996, **38**, 213.
5. R. A. van Santen and H. P. C. E. Kuipers, *Adv. Catal.*, 1987, **35**, 265.

6. J. G. Serafin, A. C. Liu and S. R. Seyedmonir, *J. Mol. Cat. A: Chemical*, 1998, **131**, 157.
7. C. Stegelmann, N. C. Schiødt, C. T. Campbell and P. Stoltze, *J. Catal.*, 2004, **221**, 630.
8. H. Lynggaard, A. Andreasen, C. Stegelmann and P. Stoltze, *Prog. Surf. Sci*, 2004, **77**, 71.
9. S. Linic and M. A. Barteau, *J. Catal.*, 2003, **214**, 200.
10. P. A. Kilty and W. M. H. Sachtler, *Catal. Rev.-Sci. Eng.*, 1974, **10**, 1.
11. N. W. Kant and W. K. Hall, *J. Catal.*, 1978, **52**, 81.
12. C. T. Campbell and M. T. Paffett, *Surf. Sci.*, 1984, **139**, 396.
13. C. T. Campbell, *J. Catal.*, 1985, **94**, 436.
14. A. I. Boronin, S. V. Koscheev, V. F. Malakhov and G. M. Zhidomirov, *Catal. Lett.*, 1996, **47**, 111.
15. A. I. Boronin, S. V. Koscheev and G. M. Zhidomirov, *J. Electron Spectrosc. Relat. Phenom.*, 1998, **96**, 43.
16. D. Swern, *Organic Reactions*, John Wiley and Sons Inc, New York, NY, 1953, 7.
17. C. T. Campbell and M. T. Paffett, *Surf. Sci.*, 1984, **143**, 517.
18. C. T. Campbell, *Surf. Sci.*, 1985, **157**, 43.
19. C. Backx, C. P. M. de Groot, R. Biloen and W. M. H. Sachtler, *Surf. Sci.*, 1981, **104**, 300.
20. G. K. Boreskov and A. V. Khasin, *Dokl. AN SSSR*, 1984, **274**, 348.
21. A. V. Khasin, *Kinet. Catal.*, 1993, **34**, 42.
22. C. Backx, J. Moolhuysen, P. Geenen and R. A. van Santen, *J. Catal.*, 1981, **72**, 1152.
23. R. A. van Santen and C. P. M. de Groot, *J. Catal.*, 1986, **98**, 530.
24. R. B. Grant and R. M. Lambert, *J. Chem. Soc. Chem. Commun.*, 1983, 662.
25. R. B. Grant and R. M. Lambert, *J. Catal.*, 1985, **92**, 364.
26. V. I. Bukhtiyarov, A. I. Boronin, I. P. Prosvirin and V. I. Savchenko, *J. Catal.*, 1994, **150**, 262.
27. D. W. Park and G. Gau, *J. Catal.*, 1987, **105**, 81.
28. S. Linic and M. A. Barteau, *J. Am. Chem. Soc.*, 2003, **125**, 4034.
29. V. I. Bukhtiyarov, V. V. Kaichev and I. P. Prosvirin, *J. Chem. Phys.*, 1999, **111**, 2169.
30. M. Rocca, L. Savio, L. Vattuone, U. Burghaus, V. Palomba, N. Novelli, F. Buatier de Mongeot and U. Valbusa, *Phys. Rev. B*, 2000, **61**, 213.
31. L. Savio, L. Vattuone, M. Rocca, F. Buatier de Mongeot, G. Comelli, A. Baraldi, S. Lizzit and G. Paolucci, *Surf. Sci.*, 2002, **506**, 213.
32. X. Bao, B. Pettinger, G. Ertl and R. Schlögl, *Berichte der Bunsengesellschaft-PCCP*, 1993, **97**, 322.
33. V. I. Bukhtiyarov, I. P. Prosvirin and R. I. Kvon, *Surf. Sci.*, 1994, **320**, L47.
34. V. I. Bukhtiyarov, V. V. Kaichev, E. A. Podgornov and I. P. Prosvirin, *Catal. Lett.*, 1999, **57**, 233.
35. R. W. Joyner and M. W. Roberts, *Chem. Phys. Lett.*, 1979, **60**, 459.

36. V. I. Bukhtiyarov, A. I. Nizovskii, H. Bluhm, M. Hävecker, E. Kleimenov, A. Knop-Gericke and R. Schlögl, *J. Catal.*, 2006, **238**, 260.
37. A. Knop-Gericke, E. Kleimenov, M. Hävecker, R. Blume, D. Teschner, S. Zafeiratos, R. Schlögl, V. I. Bukhtiyarov, V. V. Kaichev, I. P. Prosvirin, A. I. Nizovskii, H. Bluhm, A. Barinov, P. Dudin and M. Kiskinova, *Adv. Catal.*, 2009, **52**, 213.
38. W. Lindinger, A. Hansel and A. Jordan, *Chem. Soc. Rev.*, 1998, **27**, 347.
39. D. Herein, A. Nagy, H. Schubert, G. Weinberg, E. Kitzelmann and R. Schlögl, *Z. Phys. Chem.-PCCP*, 1996, **197**, 67–96.
40. V. I. Bukhtiyarov, M. Hävecker, V. V. Kaichev, A. Knop-Gericke, R. W. Mayer and R. Schlögl, *Catal. Lett.*, 2001, **74**, 121.
41. V. I. Bukhtiyarov, M. Hävecker, V. V. Kaichev, A. Knop-Gericke, R. W. Mayer and R. Schlögl, *Phys. Rev. B*, 2003, **67**, 235422.
42. J. F. Moulder, W. F. Stickle, P. E. Sobol and K. D. Bomben, in *Handbook of X-ray Photoelectron Spectroscopy*, ed. J. Chastain, Perkin-Elmer, Eden Prairie, 1992.
43. G. Hohlneicher, H. Pulm and H.-J. Freund, *J. Electron Spectrosc. Relat. Phenom.*, 1985, **37**, 209.
44. L. H. Tjeng, M. B. J. Meinders, J. van Elp, G. A. Sawatzky and R. L. Johnson, *Phys. Rev. B.*, 1990, **41**, 3190.
45. P. Ascarelli and G. Moretti, *Surf. Interf. Anal.*, 1985, **7**, 8.
46. L. E. Orgel, *J. Chem. Phys.*, 1958, 4186.
47. E. A. Carter and W. A. Goddard, *J. Catal.*, 1988, **112**, 80.
48. H. Schubert, U. Tegtmeyer, D. Herein, X. Bao, M. Muhler and R. Schlögl, *Catal. Lett.*, 1995, **33**, 305.
49. A. J. Nagy, G. Mestl, D. Herein, G. Weinberg, E. Kitzelmann and R. Schlögl, *J. Catal.*, 1999, **182**, 417.
50. S. N. Trukhan, V. P. Ivanov and B. S. Bal'zhinimaev, *Kinet. Catal.*, 1997, **38**, 617.
51. D. A. Bulushev and B. S. Bal'zhinimaev, *Kinet. Catal.*, 1996, **37**, 149.
52. E. M. Sadovskaya, D. A. Bulushev and B. S. Bal'zhinimaev, *Kinet. Catal.*, 1999, **40**, 61.
53. G. Rovida, F. Pratesi, M. Maglietta and E. Ferroni, *Surf. Sci.*, 1974, **43**, 230.
54. X. Bao, M. Muhler, B. Pettinger, Y. Uchida, G. Lehmpfuhl, R. Schlögl and G. Ertl, *Catal. Lett.*, 1995, **32**, 171.
55. S. N. Goncharova, E. A. Paukshtis and B. S. Bal'zhinimaev, *Appl. Catal.*, 1995, **126**, 67.
56. W. H. Richey, *J. Phys. Chem.*, 1972, **76**, 213.
57. M. Egashira, R. L. Kuczkowskii and N. W. Cant, *J. Catal.*, 1980, **65**, 297.
58. A. L. Larrabee and R. L. Kuczkowskii, *J. Catal.*, 1978, **52**, 72.
59. R. B. Grant and R. M. Lambert, *J. Catal.*, 1985, **92**, 364.
60. R. A. van Santen and C. P. M. de Groot, *J. Catal.*, 1986, **98**, 530.
61. C. Backx, J. Moolhuysen, P. Geenen and R. A. van Santen, *J. Catal.*, 1981, **72**, 364.

62. E. L. Force and A. T. Bell, *J. Catal.*, 1975, **52**, 440.
63. C. Backx, C. P. M. de Groot, P. Biloen and W. M. H. Sachtler, *Surf. Sci.*, 1983, **128**, 81.
64. R. E. Kenson and M. J. Lapkin, *Z. Phys. Chem.*, 1970, **74**, 1493.
65. M. Kanoh, T. Nishimura and A. Ayame, *J. Catal.*, 1979, **57**, 372.
66. S. A. Tan, R. B. Grant and R. M. Lambert, *Appl. Catal.*, 1987, **31**, 159.
67. D. A. Bulushev, E. A. Paukshtis, Yu. N. Nogin and B. S. Balzhinimaev, *Appl. Catal. A*, 1995, **123**, 301.
68. Y. S. Yong, E. M. Kennedy and N. W. Cant, *Appl. Catal. A*, 1991, **76**, 31.
69. A. G. Sault and R. J. Madix, *Surf. Sci.*, 1986, **172**, 276.
70. S. N. Goncharova, E. A. Paukshtis and B. S. Balzhnimaev, *Appl. Catal. A*, 1995, **126**, 67.
71. V. I. Bukhtiyarov, I. P. Prosvirin, R. I. Kvon, S. N. Goncharova and B. S. Bal'zhinimaev, *J. Chem. Soc., Faraday Trans.*, 1997, **93**, 2323.
72. K. H. Jung, K. H. Chung, M. Y. Kim and G. Seo, *Korean J. Chem. Eng.*, 1989, **16**, 396400.
73. C. T. Campbell and M. T. Paffett, *Appl. Surf. Sci.*, 1984, **19**, 28.
74. M. Che and C. O. Bennett, *Adv. Catal.*, 1989, **35**, 55.
75. P. Harriot, *J. Catal.*, 1971, **21**, 56.
76. J. C. Wu and P. Harriot, *J. Catal.*, 1975, **39**, 395.
77. M. Jaroui, B. Moravek, P. C. Gravelle and S. J. Teichner, *J. Chem. Phys.*, 1978, **75**, 1061.
78. M. Jaroui, B. Moravek and S. J. Teichner, *J. Chem. Phys.*, 1978, **75**, 1069.
79. X. E. Verkios, F. P. Stein and R. W. Couglin, *J. Catal.*, 1980, **66**, 368.
80. S. Cheng and A. Clearfield, *J. Catal.*, 1985, **94**, 455.
81. S. R. Seyedmonir, J. K. Plischke, M. A. Vannice and H. W. Young, *J. Catal.*, 1990, **123**, 534.
82. C. T. Campbell, *J. Vac. Sci. Technol. A*, 1984, **2**, 1024.
83. D. J. Saikowski and M. Boudart, *Catal. Rev.-Sci. Eng.*, 1987, **29**, 325.
84. J. K. Lee, X. E. Verykios and R. Pitchai, *Appl. Catal.*, 1989, **50**, 171.
85. E. Ruckenstein and S. H. Lee, *J. Catal*, 1988, **109**, 100.
86. S. V. Tsybuya, G. N. Kryukova and S. N. Goncharova, *J. Catal.*, 1995, **154**, 194.
87. V. I. Bukhtiyarov, I. P. Prosvirin and R. I. Kvon, *J. Electron Spectrosc. Relat. Phenom.*, 1996, **77**, 7.
88. J. W. M. Frenken, R. J. Hames and J. E. Demuth, *J. Vac. Sci. Technol. A*, 1990, **81**, 293.
89. B. S. Balzhinimaev, *Kinetics and Catal.*, 1999, **40**, 879.
90. V. I. Bukhtiyarov, A. F. Carley, L. A. Dollard and M. W. Roberts, *Surf. Sci.*, 1997, **381**, 605.
91. V. I. Bukhtiyarov and V. V. Kaichev, *J. Mol. Catal. A: Chem.*, 2000, **158**, 167.
92. X. E. Verykios, F. P. Stein and R. W. Coughian, *J. Catal.*, 1980, **66**, 368.
93. S. Linic, J. Jankowiak and M. A. Barteau, *J. Catal.*, 2004, **224**, 489.

94. S. Piccinin, S. Zafeiratos, C. Stampfl, T. Hansen, M. Hävecker, D. Teschner, V. I. Bukhtiyarov, F. Girgsdies, A. Knop-Gericke, R. Schlögl and M. Scheffler, *Phys. Rev. Lett.*, 2010, **104**, 035503.
95. M. A. Barteau and R. J. Madix, *J. Am. Chem. Soc.*, 1983, **105**, 344.
96. G. Lu and X. Luo, *Catal. Lett.*, 1999, **58**, 67.
97. J. Lu, M. Luo, H. Lei and C. Li, *Appl. Catal. A*, 2002, **237**, 11.
98. M. Luo, J. Lu and C. Li, *Catal. Lett.*, 2003, **86**, 43.
99. A. Palermo, A. Husain, M. Tikhov and R. Lambert, *J. Catal.*, 2002, **207**, 331.
100. F. Zemicheael, A. Palermo, M. Tikhov and R. Lambert, *Catal. Lett.*, 2002, **80**, 93.
101. G. Jin, G. Lu, Y. Guo, J. Wang and X. Liu, *Catal. Lett.*, 2003, **87**, 249.
102. B. K. Hodnett, *Heterogeneous Catalytic Oxidation*, Wiley, Chichester, 2000, ch.6.
103. V. I. Bukhtiyarov, I. P. Prosvirin, R. I. Kvon, B. S. Bal′zhinimaev and E. A. Podgornov, *Appl. Surf. Sci.*, 1997, **115**, 135.
104. Y. Lei, F. Mehmood, S. Lee, J. Greeley, B. Lee, S. Seifert, R. E. Winans, J. W. Elam, R. J. Meyer, P. C. Redfern, D. Teschner, R. Schlögl, M. J. Pellin, L. A. Curtiss and S. Vajda, *Science*, 2010, **328**, 224.

CHAPTER 10

Ruthenium Active Catalytic States: Oxidation States and Methanol Oxidation Reactions

RAOUL BLUME,[1] MICHAEL HÄVECKER,[1] SPIROS ZAFEIRATOS,[1] DETRE TECHNER,[1] AXEL KNOP-GERICKE,[1] ROBERT SCHLÖGL,[1] LUCA GREGORATTI,[2] ALEXEI BARINOV[2] AND MAYA KISKINOVA*[2]

[1] Fritz-Haber-Institut der Max-Planck-Gesellschaft, Faradayweg 4-6, 14195 Berlin, Germany; [2] Sincrotrone Trieste, AREA Science Park-Basovizza, Trieste-34012, Italy

10.1 Introductory Remarks

Ru-based oxidation catalysis is extremely versatile, with multiple applications ranging from removing CO from the industrial and automobile exhaust gases to important organic syntheses and direct methanol fuel cells for accelerating CO_2 formation.[1,2] The chemistry of industrial Ru catalysts is very complex and the ongoing challenge is to have a total command based on comprehensive understanding of the processes involved down to molecular and atomic scales. To unravel the state of the oxygen that participates in the oxidation process is of particular importance and still there are controversial discussions in this respect. Most often the classical oxygen source considered in modelling the reaction mechanism is the lattice oxygen on nearly stoichiometric surfaces of metal oxides (Mars–van Krevelen mechanism)[3] or adsorbed oxygen on the

RSC Nanoscience & Nanotechnology No. 19
Nanostructured Catalysts: Selective Oxidations
Edited by Christian Hess and Robert Schlögl

Published by the Royal Society of Chemistry, www.rsc.org

metal surfaces. In reality, the catalytically active state is not necessarily the well-defined oxide or adsorption phase, and possible differences in the so-called "oxidation state" of the surface and the bulk have already been considered long ago.[4] In this respect, surface-science research has made serious contributions using as a principle concept exemplary catalytic systems starting from simplified models, expanding towards more complex systems and looking for correlations with the realistic catalytic conditions.[5] The main challenge of the surface-science approach is that the experiments are exclusively carried out under ultrahigh-vacuum (UHV) conditions and often enough the extensive knowledge about important elementary processes on model catalyst surfaces cannot be directly extrapolated to the catalytic processes occurring at ambient pressures. This is due to the fact that the catalytically active structures are not necessarily formed under UHV conditions due to the low chemical potentials of reagents in the gas phase. Along with the thermodynamical and kinetic factors controlling the catalyst composition, restructuring of the catalyst surface induced by the reaction participants has to be considered as well.

A textbook example of the "pressure gap" is CO oxidation over a model Ru(0001) catalyst, one of the most extensively studied systems because of its "simplicity""and technological importance. The anomalously low Ru activity observed in the traditional surface-science experiments carried out at oxygen pressures, P_{O2}, not exceeding 10^{-5} mbar, is attributed to the kinetic limitation for accumulation of oxygen to form the catalytically active surface at low oxygen potentials.[6–9] However, the nature of the catalytically active state even for the simple CO oxidation reaction on the Ru(0001) surface continues to be a very disputable issue.[10–13] In spite of the open questions, all relevant studies agree that P_{O2} plays a crucial role being a prerequisite for formation of the following phases, suggested by different groups as catalytically active in the working catalyst: $\sim(1\times1)$ adsorption phase with 1 monolayer (ML) oxygen,[10] a transient surface oxide (TSO) film (RuO_x, where $1<x<2$) with poorly defined structure and stoichiometry,[13] and a stoichiometric RuO_2.[11,13]

The real-time information about the evolution of chemical state of Ru catalyst during CO oxidation with increasing reaction temperature provided by *in-situ* photoelectron spectroscopy led to the following important conclusions:[13,14] (i) the metallic state of Ru with adsorbed oxygen less than 1 ML cannot catalyze the CO oxidation; (ii) the Ru surface is activated when oxygen is incorporating subsurface at critical coverage ~1 ML, corresponding to the (1×1)-O adsorption phase; and (iii) there is no distinct difference between the catalytic activity of the stochiometric RuO_2 and a few layers thick poorly ordered TSO, RuO_x. The major finding is that below 500 K, when the formation of RuO_2 is kinetically hindered, the RuO_x offers a structure with activity comparable with that of the well-defined RuO_2 phase.[15] This is in fair agreement with the DFT predictions, that the high catalytic activity of RuO_2 is not only due to the weaker CO and O bonding on the oxide surface, compared to metallic Ru, but the reorganization required to achieve the CO–O transition-state configuration plays an important and even decisive role.[16] In the framework of this concept the activation role of the subsurface oxygen is ascribed to the induced deformation of the lattice, which affects the O and CO adsorption configurations and activation energy to the CO–O transition state. Furthermore, the catalytically active region under realistic dynamic reaction conditions can also lie at the boundary between two thermodynamically equilibrium phases.[17] As a result of anisotropic oxide growth or oxide reduction the RuO_x and RuO_2 states can coexist in a wide temperature range and both phases of this "mixed" state are proven to be active towards CO oxidation.[13]

The CO oxidation is a simple single-channel reaction system, which is the final and often undesired step of important oxidation reactions of organic molecules, and according to the recent findings it is not exclusively sensitive to the actual Ru oxidation state once the TSO with a critical amount of accumulated oxygen is formed. However, most of the industrial oxidation processes need catalysts with specific selectivity to particular intermediate products, which supposes catalytically active state different from that favoring full oxidation to CO_2.

In this review, we focus on the catalytically active state of the Ru surface in selective oxidation reactions. We addressed this very important and disputable issue using methanol oxidation as a model system. This is a complex process involving desirable intermediate products, such as formaldehyde, the yield of which turned out to be very sensitive to the Ru oxidation state. The results are a joint research effort using two advanced experimental approaches, described briefly in Section 10.2. The knowledge gained about the evolution of the oxidation states and morphology of the Ru surface under pure oxidation conditions (Section 10.3) provided the basis for identifying the catalytically active states during CH_3OH oxidation under various pressure–temperature conditions (Section 10.4). Section 10.5 concludes with brief summary and outlook.

10.2 Experimental Techniques

One of the best analytical methods for probing the composition and electronic structure of the catalyst's surface, X-ray photoelectron spectroscopy, XPS,[18] has undergone a real renaissance using ultrabright and tunable photon beams at the third generation synchrotron facilities. Maintaining high signal level and short acquisition time, synchrotron-based XPS has opened new opportunities for bridging the pressure gap, performing experiments at pressures up to several mbar or bridging the material gap by pushing the lateral resolution into the submicrometer range. The detailed description of these instruments can be found in many reports[14,19–21] and here we will only briefly describe the potential of the instruments used in the studies under consideration.

The high-pressure XPS instrument, HP-XPS, at the synchrotron radiation facility BESSY in Berlin, designed and constructed in the FHI-MPG,[14] uses the common principal approach with a sample placed in the reaction cell in front of an aperture to the differentially pumped stages of the hemispherical electron analyzer. The setup allows monitoring the evolution of the electron core-level spectra, fingerprints of the chemical state of the catalyst surface under variable close-to-realistic reaction conditions. Simultaneous measurements of the changes in gas-phase composition by mass spectrometry provide information on the reactants consumption and products yield. This allows a direct correlation between the reaction rates (measure of the specific catalytic activity) to the actual chemical status of the catalyst surface. In some cases (partial pressures above ~0.05 Torr) the gas-phase composition (excluding hydrogen) can also be determined from the XPS spectra, provided the core-level binding energy shift between gas phase and surface species is large enough to separate their contributions to the spectra.

The scanning photoelectron microscope (SPEM) operated at the synchrotron facility ELETTRA[22] uses Fresnel zone plate lenses as photon-focusing optics to demagnify the photon beam to a spot with a diameter of less than 0.15 μm. Element specific information about the composition and electronic structure of the substrate surface and adlayer is provided by the photoelectrons emitted from the microprobe, which are collected and energy filtered by a hemispherical (HS) electron analyzer, mounted at a grazing angle with respect to the surface normal that enhances the surface sensitivity. In order to obtain spatially resolved chemical images the analyzer is tuned to collect photoelectrons from a specific atomic core level, while scanning the sample. The contrast of the two-dimensional maps is due to variation of the photoelectron yield within a selected energy window, reflecting the local concentration of the element or its chemical state. This instrument was used to study the morphology and oxidation state of the Ru(0001) surface evolved under different oxidation conditions, which is an important link in understanding the complexity of the studied catalytic oxidation reactions.

10.3 Oxidation States and Morphology of the Ru Catalyst Surface Exposed to O_2 at Different Temperatures

Although the oxidation of Ru(0001) surface has been very extensively studied, only recently was more light shed on disputable issues such as the presence and the role of subsurface oxygen and the actual morphology of the Ru surface under realistic oxidation conditions. According to the pioneering studies the O-rich "active" Ru surface can consist of different phases with microscopic dimensions.[6–8] One of the earliest models in favor of RuO_2 as an active catalytic state suggests that as soon as the oxygen load exceeds the coverage of 1 ML, accommodated by the densest (1×1) adsorption phase, the RuO_2 nuclei are formed and the oxide grows progressively in an autocatalytic manner.[7] According to this nontraditional pathway the main coexisting phases are the inactive (1×1) chemisorbed phase, and the active RuO_2 phase, not excluding the presence of a small amount of subsurface oxygen and oxide buried below the (1×1) adsorption phase. This scenario is based on Ru oxidation carried out at $T > 600$ K and contradicts the classical oxidation mechanism of transition metals, including as an intermediate stage the oxygen incorporation subsurface and formation of a surface oxide as a transient to the formation of stoichiometric bulk oxide.[23–26] According to the theoretical predictions the nucleation and growth of the RuO_2 domains can occur above a critical O coverage of 4 ML.[26] This is in accordance with our extensive SPEM studies on Ru(0001) oxidation at lower temperatures (<500 K), which showed no evidence for the formation of a crystalline RuO_2 phase since the oxygen incorporation is limited to the subsurface region.[27] Here, we would like to stress that at temperatures above ~450 K we never obtained a "pure" (1×1)-O adsorption phase: a small component of incorporated oxygen was always present in the spectra. This is a clear indication that a "critical" coverage of 1 ML is the one that favors oxygen incorporation, occupying different sites between the first and second layer. Apparently, the presence of the incorporated oxygen destabilizes the oxygen at the surface and makes it more reactive, as demonstrated by the CO oxidation experiments.[13]

Using laterally resolved XPS we have overcome one of the obstacles in spectroscopic assignment of the oxidation state of the transition-metal surface due to the fact that the oxide nucleation and growth proceeds at very local scales and several phases may coexist on the so-called "O-rich" Ru surface. The Ru images in Figure 10.1 "visualize" the spatial anisotropy of the initial stages of oxide growth and the dependence of the morphology on the oxidation temperature. This anisotropy was observed in a wide temperature (550–775 K) and oxygen exposure range. The images were measured by tuning the analyzer to the energy window of electrons emitted from the metallic Ru, which has a distinct component in the Ru 3d spectra called bulk Ru, Ru_b. The darker regions in the maps, developing above critical oxygen doses, correspond to attenuated Ru_b intensity, *i.e.* to a more advanced oxidation state, since the Ru_b intensity decreases exponentially with increasing the thickness of the O-containing layer. The intensity profiles taken along the dark O-richer patches allowed us to quantify the thickness of the O-containing phases under varying oxidation conditions.[28] As can be seen in Figure 10.1 the characteristic feature of the 625 K map is that the dark patches, representing the most advanced oxidation stage under these conditions, are decorated by areas that are the brightest, *i.e.* with the lowest oxygen content. These O-depleted regions

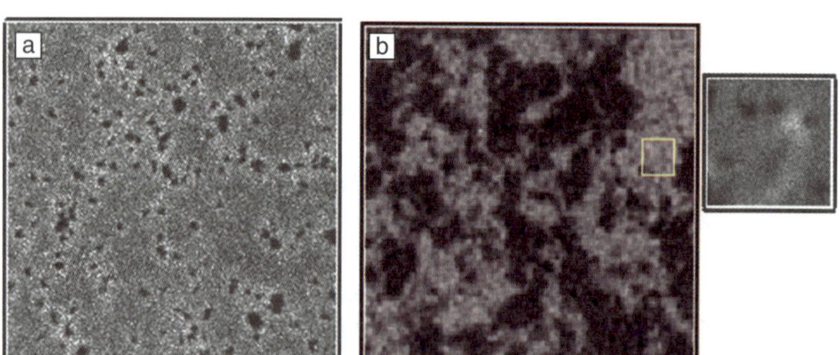

Figure 10.1 64×64 μm^2 Ru 3d$_{5/2}$ images tuned to the Ru_b component illustrating the morphology of the Ru(0001) surface developed after exposure to 10^5 mbar s O_2 at 625 K (a) and 775 K (b). The higher resolution 6×6 μm^2 map in (b) of the bright area, indicated in the large-area image, clearly shows the nucleating and growing RuO_2 islands (darker spots). $P_{O2} = 10^{-3}$ mbar.

evidence the finite mobility of the O atoms from the surrounding areas, which are the oxygen source for "oxide" nucleus growth. Pushing the resolution, we were also able to evidence the initiation of oxide nucleation in the regions that appear brighter in the maps of the advanced oxidation state at 775 K.

The Ru 3d and O 1s spectra measured in the darker and brighter regions revealed the local oxidation state. The corresponding spectral fingerprints, used for elucidation of the "oxidation" state of the Ru catalyst under reaction conditions, are reported in Figure 10.2. The spectra are taken in spots selected from Ru 3d maps, such as those in Figure 10.1. The components, called Ru_{ad}, RuO_x and RuO_2, relevant to the deconvolution of the Ru $3d_{5/2}$ spectra in the reaction experiment, are summarized in Table 10.1. They correspond to the following states: (i) the adsorption phases with ~0.8 and $\geq$1 ML oxygen located exclusively on the surface, where the components reflect double and triple oxygen coordinated Ru surface atoms[29] (ii) the RuO_x ($1<x<2$) transient state of several ML of oxygen incorporated within the top few Ru atomic layers[27] and (iii) the stoichiometric RuO_2 oxide phase. The RuO_2 phase has two components corresponding to cus-Ru and six-fold coordinated Ru atoms in the bulk RuO_2.[30] Another distinct feature of the oxide phase is the broad satellite at $\approx$283 eV between the Ru $3d_{3/2}$ and $3d_{5/2}$ peaks, which we used as unambiguous proof of the development of RuO_2 phase. The relative contribution of the Ru_b component to the Ru $3d_{5/2}$ spectra reflects the exponential decrease of photoelectron emission from the metallic Ru on increasing the thickness of the O-containing phases. Considering the analyzer acceptance angle and photon energy used in the studies under consideration, the extinction of the Ru_b occurred when the RuO_2 becomes thicker than ~10–12 Å (SPEM) and ~12–15 Å (HP-XPS), *i.e.* a probing depth of ~3 and 4 rutile layers, respectively. The O 1s peak of the adsorption and RuO_x phases is at ~530.0 eV, but when the load is higher than 1 ML the O 1s intensity grows and becomes comparable with the O 1s intensity of the nucleating RuO_2 islands. The O 1s spectra from the RuO_2 phase have two components reflecting the emission from the bridge-O atoms at the RuO_2 surface (528.6 eV) and bulk RuO_2 oxygen (529.5 eV).[30] It should also be noted that a well-developed undercoordinated "bridge" oxygen component is present only when the oxide formation is in a very advanced stage and has attained a long-range order. This is observed, for example, in the narrow temperature–pressure range when the growth is in the preferential crystallographic directions.[28]

RuO_x with $\leq$4 ML oxygen is the final oxidation state achieved at 450–550 K. We evidenced the continuous conversion from 1 ML (1×1)-O to 3–4 ML RuO_x, all these stages we call transient surface oxide (TSO). The RuO_x TSO is also present in a wide pressure–temperature range at $T > 550$ K, coexisting with adsorption phases or RuO_2 islands.[27,28] For example, the Ru $3d_{5/2}$ spectra characterizing the dark regions (O-richest state) in the 625 K map, and the bright areas (lower oxygen content) in the 775 map in Figure 10.1 are similar to the spectra in Figure 10.2, dominated by the RuO_x component. Accordingly, the Ru $3d_{5/2}$ spectra from the bright areas in the 625 K map (Figure 10.1) resemble the spectra of the adsorption phases with coverage ~0.7–0.8 ML and less than 0.5 ML for the brightest depleted regions. This indicates that the conversion from (1×1)-O to RuO_x is rather fast at $T > 550$ K and the oxygen supply occurs *via* the low coverage adsorption phases where the sticking coefficient for O_2 from the gas phase is higher. For the advanced oxidation state, represented by the 775 K images in Figure 10.1, the thickness of RuO_2 patches (dark) exceeds our probing depth and the Ru_b can no longer be seen. These regions are characterized by Ru $3d_{5/2}$ spectra as the top ones in Figure 10.2(a).

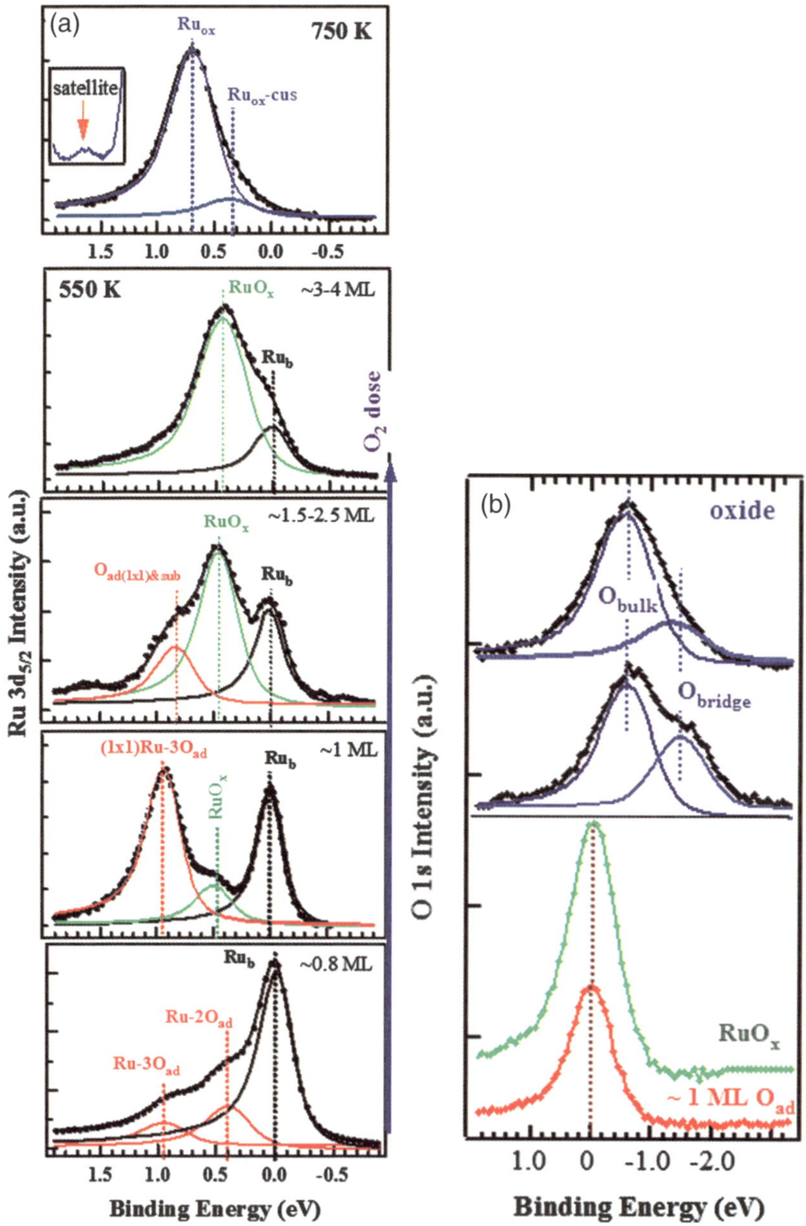

Table 10.1 Ru $3d_{5/2}$ and O 1s binding energies of components, reflecting the oxidation state of Ru catalyst surface.

State	*Component binding energy (shift) in eV*	*Component assignment*	*Ref.*
Ru $3d_{5/2}$			
Metallic Ru	280.1 (0)	Ru_b	29
Adsorption phases	280.48 (0.4)	Ru-2O (0.5 ML)	29
	281.03 (0.93)	Ru-3O (1 ML)	29
RuO_2	280.75 (0.64)	RuO_2-bulk	30
	280.45 (0.35)	Ru-cus	30
	283.0 (2.92)	satellite	30
RuO_x: 1–2 ML O	280.9 (~0.8 eV)	O_{ad}&O_{sub}	27
RuO_x: 2–4 ML O	280.5 ± 0.05 (~0.45)	RuO_x	27
O1s			
	529.5	O in RuO_2-bulk	30
	528.7	"bridge" O	30
	530.0	O_{ad} and RuO_x	1,27

Using a large set of XPS spectroscopy and microspectroscopy results, combined with O_2 thermal desorption data the boundaries in the pressure–temperature (T–P_{O2}) space where the RuO_2 formation is kinetically limited were suggested.[27] In brief, incorporation of oxygen in the subsurface region occurs readily at temperatures above 350 K, but only at temperatures above 550 K does stoichiometric oxide nucleation and growth take place. These results are in general agreement with theoretically predicted temperature–pressure (T–P_{O2}) phase diagrams for thermodynamically stable phases with phase-coexistence regions.[26,31,32]

In realistic oxidation reactions the second reactant usually acts as a reducer (*e.g.* CO, H_2, CH_3OH or other organic molecules) and affects the (T–P_{O2}) space of the oxidation phases, driving the oxidation state away from the equilibrium achieved in a pure O_2 ambient. Consequently, when the reactions have several possible pathways the selectivity and the corresponding catalytically active state can be controlled by varying the reaction temperature and chemical potentials (partial pressures) of the reactants.

Figure 10.2 (a) From top to bottom: deconvoluted Ru $3d_{5/2}$ spectra corresponding to ~0.8 ML adsorbed oxygen, ≥1 ML – (1×1) adsorbed oxygen with some oxygen already incorporated below, RuO_x with total oxygen content of ~2 ML, *i.e.* ≥1 ML incorporated between the first and second Ru layers, RuO_x, with total oxygen content ~4 ML mostly incorporated oxygen located within the top three layers, and RuO_2. The adsorption-related (red), RuO_x (green) and Ru_b (black) components are reported in Table 10.1. For the sake of clarity the binding energy (280.1 eV) of the bulk component, Ru_b, is placed at zero. (b) From top-bottom: O 1s spectra representing the adsorption (red), RuO_x states (green) and RuO_2 phase (blue). The two O 1s spectra for the RuO_2 phase illustrate the effect of imperfect surface structure and/or variations in the surface oxygen vacancies on the relative weight of the O_{bridge} component. The features of the RuO_2 phase appear only for oxidation temperature >550 K.

Our SPEM and HP-XPS experiments showed that heating of thick RuO_2 to high temperatures or exposing it to reducing ambient results in reversible processes, *i.e.* development of lateral anisotropy due to the coexistence of the above-described oxidation states. *In-situ* HP-XPS experiments showed that both the RuO_x and RuO_2 states can be easily reduced (with comparable rate) at $T > 400$ K in excess of CO or CH_3OH in the gas-phase mixture, reaching the same final reduced state, inactive oxygen adsorption phase or almost O-free metallic surface.[13,14] It should be noted that in surface-science experiments, carried out at temperature and pressure conditions far from the realistic ones, no reduction of the $RuO_2(110)$ was observed even in extreme excess of CO, *i.e.* $CO:O_2$ pressure ratios of 10.[33]

10.4 Methanol Oxidation

Selective oxidation of methanol to formaldehyde, CH_2O, being a large-scale industrial catalytic process, is continuing to attract the interest of academic and applied research aiming at further improving the formaldehyde yield. Along with silver, the most commonly industrially used catalyst, there are many reports about the excellent performance of copper, molybdena-, vanadia- and ruthenium-based catalysts as well.[1,34–36] The research efforts focused towards the elucidation of reaction mechanisms, usually consider the following nonexhaustive list of reactions that may be important under industrial reaction conditions.

(1) $CH_3OH \rightarrow CH_2O + H_2$
(2) $CH_3OH \rightarrow CO + 2H_2$
(3) $CH_3OH \rightarrow C + H_2O + H_2$
(4) $CH_3OH + \frac{1}{2}O_2 \rightarrow H_2 + H_2O + CO$
(5) $CH_3OH + \frac{1}{2}O_2 \rightarrow CH_2O + H_2O$
(6) $2CH_3OH + \frac{1}{2}O_2 \rightarrow 2CH_2O + H_2O + H_2$
(7) $CH_3OH + O_2 \rightarrow CO + 2H_2O$
(8) $CH_3OH + 1\frac{1}{2}O_2 \rightarrow CO_2 + 2H_2O$
(9) $CH_2O \rightarrow CO + H_2$
(10) $CH_2O + O \rightarrow HCOOH \rightarrow CO_2 + H_2$
(11) $CO + \frac{1}{2}O_2 \rightarrow CO_2$.

A common intermediate step is the scission of the O–H bond of CH_3OH yielding adsorbed H and methoxy, CH_3O species.[37] In absence of oxygen the reactions paths (1), (2) and (3) are possible, involving further activation of the C–H bonds in CH_3O and also scission of the C–O bond, leading to the products of partial or complete dehydrogenation, CH_2O, CO, H_2, H_2O and C. In the presence of oxygen the pathways relevant to the present study are the partial (4)–(7) or full (8) oxidation reactions. Apparently, the CH_2O production will be favored if the formation rate of the HCOO intermediate to CO_2 is low. For RuO_2 catalysts the rate-limiting step to formaldehyde formation is supposed to be C–H activation in the methoxy species through redox cycles using the lattice O atoms.[36] The steady-state concentrations of these adsorbed species are low because the following dehydrogenation steps are fast and accordingly the lifetime of all further intermediates is much shorter.

Regarding the nature of the catalytically active state favoring CH_2O production there is no clear consensus. For example, the remarkable activity for CH_2O production at rather low temperatures (300–400 K) on supported RuO_2 catalysts is attributed to the fast redox cycles, which preserve the active RuO_2 state not allowing the formation of inactive Ru metal clusters.[36] Somewhat contradictory are the results based on temperature-programmed desorption, where a significant CH_2O desorption was monitored only from partially reduced polycrystalline RuO_2.[38]

The main issue addressed by the studies on CH_3OH oxidation reported here is the role of nonequilibrium RuO_x TSO states in tuning the catalyst selectivity. The controversial assignment of the catalytically active phase has been overcome using the HP-XPS station, which allowed us to follow the evolution of the topmost catalyst layers caused by the coupling of catalyst structure and composition to the temperature and chemical potential of the gas-phase reactants, designated as "chemical dynamics". The dominance of chemical dynamics has been confirmed by the observed minor impact of the initial surface structure and order on attained steady catalytically active states, comparing the results obtained with Ru(0001), Ru(1010) and polycrystalline Ru samples. All results relevant to this report were almost identical for the three samples: the same catalytically active steady states were attained with small differences mainly due to the variation of onset temperature of oxidation by 20–30 K when starting from a metallic Ru precatalyst state.[39]

The experiments were carried out at different temperatures and partial pressures of O_2 and CH_3OH in a wide pressure range, starting from a metallic Ru or RuO_2 precatalyst state. As reported below sensible formaldehyde production was monitored only at the highest pressures (10^{-1} mbar range) and the high selectivity conditions were obtained for a very narrow P_{O2}/P_{CH3OH} ratio and $T > 450$ K. Fortunately, as described in the previous section, all three "candidate" catalytically active states to be taken into consideration, namely metallic Ru with adsorbed oxygen, the RuO_x TSO with ($1 < x < 2$) and RuO_2, have characteristic Ru 3d and O 1s spectral features and their presence can be easily evidenced. Thus, combining *in-situ* XPS measurements with simultaneous mass-spectrometric monitoring of the gas-phase products we identified the Ru oxidation states evolving under operation conditions (10^{-6}–10^{-1} mbar range varying the T–P_{O2}–P_{CH3OH} space) and we established their correspondence to the catalytic performance.

The first surprising result of these investigations is the observed "pressure gap" in the CH_3OH selective oxidation to CH_2O, which cannot simply be related to the kinetically limited activation of the metallic Ru by the low oxygen potentials, as in the case of CO oxidation.[40] In order to exclude this kinetic constraint we explored the "pressure gap" in the selectivity towards CH_2O formation starting with a RuO_2 precatalyst. The experiments were carried out in the $10^{-6} - 10^{-1}$ mbar range, at the same P_{O2}/P_{CH3OH} ratio of ~0.7, which was found to sustain the CH_2O production at pressures higher than 10^{-2} mbar.

For the 10^{-6}–10^{-2} mbar pressure range the only significant reaction in the 470–720 K range was the partial reduction of RuO_2 to RuO_x, as illustrated by the Ru $3d_{5/2}$ spectra in Figure 10.3(a). The attained RuO_x state was catalytically inactive at these low O_2 and CH_3OH potentials. The catalyst can be activated in the 10^{-4} range if it is reoxidized to RuO_2, which takes place at temperatures above 600 K by increasing the P_{O2}/P_{CH3OH} ratio, as illustrated by the Ru $3d_{5/2}$ spectra in Figure 10.3(b). However, as expected the steady RuO_2 state under these O-rich conditions catalyzes the undesired total oxidation reaction to $H_2O + CO_2$, as shown in Figure 10.3(c), with a very low CH_2O yield only at 720 K.

In brief, the results obtained in the pressure range 10^{-6}–10^{-2} mbar varying temperature, P_{O2} and P_{CH3OH}, showed the dominance of the products of the

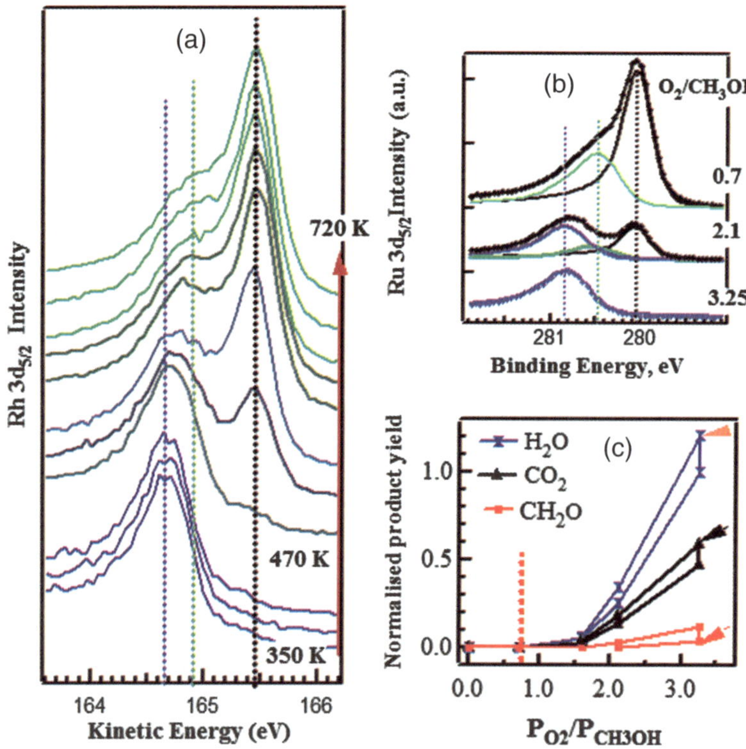

Figure 10.3 (a) Evolution of the Ru $3d_{5/2}$ spectra, as monitored *in-situ* during increasing of the reaction temperature from 350 to 720 K. The dotted lines indicate the RuO_2, RuO_x and Ru_b components. (b) Deconvoluted Ru $3d_{5/2}$ spectra, measured at 720 K for three P_{O2}/P_{CH3OH} pressure ratios, representing O-rich, O-optimum and O-poor reaction conditions. (c) Normalized mass-spectra intensity of reaction products as a function of P_{O2}/P_{CH3OH} ratio, corresponding to the evolution of the Ru $3d_{5/2}$ spectra. The arrows indicate the change in the products yield after decreasing the temperature from 720 to 600 K. The dashed line indicates the O-optimum conditions favoring a selective CH_2O production in the 10^{-1} mbar range. $P_{O2} + P_{CH3OH} = 3.4 \times 10^{-4}$ mbar.

two nonselective reaction channels, full oxidation to $CO_2 + H_2O$ or partial oxidation to $CO + H_2O + H_2$. The RuO_2 state exhibits high activity for nonselective full oxidation to CO_2 and is preserved at higher P_{O2}/P_{CH3OH} pressure ratios, as illustrated by Figure 10.3. The traces of CH_2O are attributed to desorption of a small fraction of adsorbed CH_2O before its conversion to HCOO or dehydrogenation to CO, which occurs only at rather high temperatures, confirmed by the drop of CH_2O production reducing the temperature from 720 to 600 K (Figure 10.3(c)). Lowering the P_{O2}/P_{CH3OH} ratio affects the stability of the RuO_2 precatalyst structure and depending on the actual pressure ratio RuO_2 undergoes partial reduction to a RuO_x surface oxide (see Figure 10.3(a)) or full reduction to metallic Ru when $P_{O2}/P_{CH3OH} \leq 1$.[40] The metallic Ru state also changes the product yield, catalyzing effectively only the nonselective partial oxidation to $CO + H_2O + H_2$ in the whole pressure range under consideration.[40]

The most extensive studies of CH_3OH oxidation were carried out in the $2\text{–}4 \times 10^{-1}$ mbar range beyond the "selectivity" pressure gap, starting from metallic Ru or RuO_2 precatalyst and increasing the reaction temperature stepwise from 350 to 650 K to exploit the nature of the catalytically active states.[39] By varying the P_{O2}/P_{CH3OH} partial pressure ratio from 0.4 (O-poor) to 1.4 (O-rich), we revealed that very small changes in the Ru oxidation state can have a strong impact on the catalyst selectivity, favoring full or partial oxidation pathways. For O-poor mixture ($P_{O2}/P_{CH3OH} = 0.4$) the least desired partial oxidation to $H_2 + H_2O + CO$ is the dominant reaction with onset above ~450–470 K. Low $CH_2O + H_2O$ yield is monitored above 550 K, whereas the CO_2 yield remains negligible even at 650 K. The steady-state Ru $3d_{5/2}$ and O 1s spectra evidenced that this reaction is catalyzed by metallic Ru. The evolution of Ru $3d_{5/2}$ spectra showed that the metallic Ru precatalyst undergoes insignificant changes reaching steady state with <0.5 ML adsorbed oxygen, whereas the RuO_2 precatalyst undergoes rigorous reduction in the temperature range 430–470 K to the same steady "catalytically active" metallic state with adsorbed oxygen. Since these results are not relevant to the specific topic of this chapter more details can be found in ref. 39.

The intriguing result of the CH_3OH oxidation studies is that even under low-pressure conditions, selected to attain the same catalytically active chemical state as at high pressures, not necessarily the same selectivity can be obtained. This concerns, in particular, the transient surface oxide state RuO_x, which becomes active only in the 10^{-1} mbar range, as clearly evidenced by comparing the results reported in Figure 10.3 and Figure 10.4, obtained for the same P_{O2}/P_{CH3OH} ~0.7 in the 10^{-4} mbar and 10^{-1} pressure range, respectively.

As noted above, the desired partial oxidation reaction $CH_3OH + \frac{1}{2}O_2 \rightarrow CH_2O + H_2O$ becomes dominant for P_{O2}/P_{CH3OH} ~0.7. The evolution of the Ru $3d_{5/2}$ spectra demonstrated that both precatalysts undergo conversion to the same catalytically active steady state with the characteristic features of the transient RuO_x state, confirmed also by the O 1s spectra peaked at 530 eV. In order to attain this state the RuO_2 undergoes partial reduction, whereas the metallic Ru has to be enriched with incorporated oxygen, which readily occurs

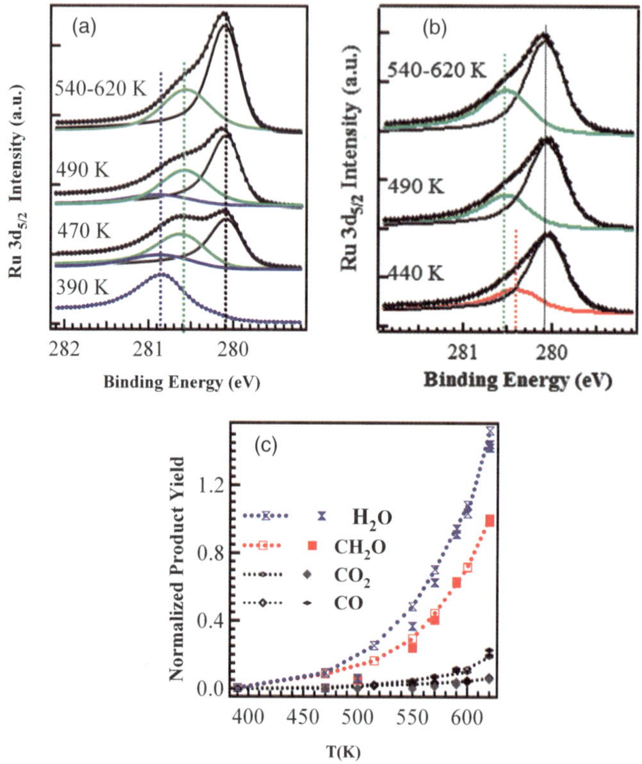

Figure 10.4 Deconvoluted Ru $3d_{5/2}$ spectra illustrating the evolution of the oxidation state with increasing reaction temperature for RuO_2 (a) and metallic (b) precatalysts. The dotted lines indicate the positions of the components RuO_2 (blue), RuO_x(green), Ru_b (black), and Ru bonded to adsorbed CH_3OH-related adspecies (red). (c) Yield of reaction products as a function of temperature starting from metallic (full markers) and oxidized (dashed line) precatalyst states. The yields are normalized to the CH_2O yield at 650 K. Reaction conditions: total pressure: 2.4×10^{-1} mbar, $P_{O2}/P_{CH3OH}=0.67$.

above 450 K. The fits of the Ru $3d_{5/2}$ spectra in Figure 10.4(a) clearly points towards a patchy surface of coexisting RuO_2 and RuO_x at temperatures below ~500 K. Above 500 K the RuO_2 component is practically absent and according to the weight of the RuO_x-related component the catalytically active state is a thin few-layer TSO. Starting from a metallic Ru precatalyst state higher temperatures are required for oxygen incorporation and formation of RuO_x (470–550 K compared to 450–510 K for partial reduction of RuO_2), which accounts for ~30° higher reaction onset temperature. Apparently, the catalytically active RuO_x is a natural oxidation (reduction) transient, as manifested by the evolution of the chemical state of the Ru and RuO_2 precatalysts in Figures 10.4(a) and (b). This steady state is maintained by the actual P_{CH3OH}/P_{O2} pressure ratio in the temperature range 550–620 K and for the reaction rate

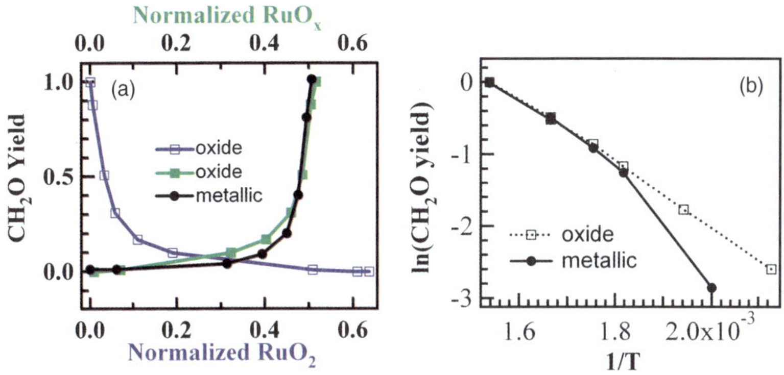

Figure 10.5 (a) Relative CH_2O yield as a function of the weight of the RuO_x component, developed starting from oxide or metallic precatalyst. The intensity of the RuO_x and RuO_2 components was normalized against the intensity of the Ru_b component in the Ru $3d_{3/5}$ spectra measured for metallic Ru at 320 K before the reaction. (b) Arrhenius plot of CH_2O production, starting from metallic Ru and oxide precatalyst. Reaction conditions: total pressure: 2.4×10^{-1} mbar, $P_{O2}/P_{CH3OH} = 0.67$.

becomes independent on the precatalyst nature, as illustrated in Figure 10.4(c). Above 550 K CO, the product of the parallel $CH_3OH + O_2 \rightarrow CO + 2H_2O$ reaction and some traces of CO_2 appear as well.

The correlation between RuO_x formation and high CH_2O production rate is demonstrated by the plots in Figure 10.5(a), showing the dependence of the formaldehyde yield on the development of catalytically active TSO state with optimal oxygen content. The development of the same catalytically active state independently of the precatalyst chemical state is further confirmed by the overlap of the Arrhenius plots in Figure 10.5(b) for reaction temperatures above 550 K. The deviation from a straight line at lower temperatures observed for the Ru precatalyst reflects the activation barrier for the oxygen incorporation and formation of the active RuO_x state.

The dynamics of the transient surface oxide RuO_x is very sensitive to the reactants mixture ratio, which controls the x value. In particular, x also reflects the relative concentration of the steady-state subsurface and surface oxygen, which can be expected to have a strong impact on the reaction pathway. This is confirmed by the results obtained under O-rich conditions, $P_{O2}/P_{CH3OH} = 1.4$, favoring full oxidation to CO_2. The most surprising observation is that the corresponding steady catalytically active state also has the characteristic RuO_x spectral features, *i.e.* the oxide precatalyst undergoes partial reduction with the onset of the reaction, reaching the steady state at ~550 K, as illustrated in Figure 10.6(a). Correspondingly, the plots in Figure 10.6(b) confirm that the reaction $CH_3OH + 1\frac{1}{2}O_2 \rightarrow 2H_2O + CO_2$ dominates, whereas the $CH_2O + H_2O$ production is strongly suppressed.

The reported results clearly demonstrate that the desired partial oxidation pathway to formaldehyde is catalyzed only when the reduction–oxidation

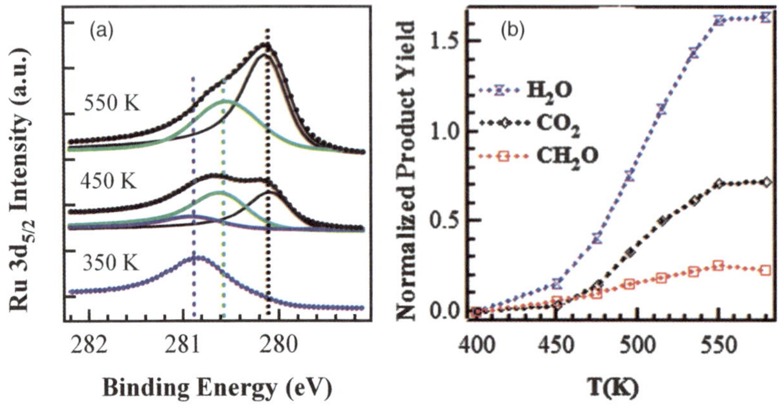

Figure 10.6 (a) Deconvoluted Ru $3d_{5/2}$ spectra, illustrating the evolution of the RuO_2 precatalysts with increasing reaction temperature. The dotted lines indicate the positions of the components RuO_2 (blue), RuO_x (green), Ru_b (black). Reaction conditions: total pressure: 2.4×10^{-1} mbar, $P_{O2}/P_{CH3OH} = 1.4$ (b) Yield of reaction products as a function of temperature starting from metallic (full markers) and oxidized (dashed line) precatalyst states. For the sake of comparison the yields are normalized to the CH_2O yield at 650 K, measured for $P_{O2}/P_{CH3OH} = 0.67$ (see Figure 10.4(c)).

equilibrium maintained by the reactant potentials stabilizes a specific RuO_x state, which evolves independently of the initial Ru precatalyst state (oxide or metal). The dynamics of the catalyst oxidation state illustrated in Figures 10.4 and 10.6 shows how small variations in the stoichiometry of the RuO_x transient result in dramatic changes in the catalyst selectivity. Indeed, both partial oxidation to CH_2O and full oxidation to CO_2 are catalyzed by RuO_x transient states, but these states contain different amounts of oxygen. This difference was quantified by comparing the $O1s/Ru3d_{5/2}$ and RuO_x/Ru_b intensity ratios, corresponding to the two RuO_x TSO states. They both increased from 0.1 to 0.12 ($O1s/Ru3d_{5/2}$ intensity ratio) and from 0.46 to 0.56 (RuO_x/Ru_b intensity ratios), changing the conditions from selective CH_2O production to full oxidation.[39] This supposes that there is critical amount of incorporated and surface oxygen, which selectively activates the catalyst surface to provide the most favorable energetic of interlinked processes of adsorption, desorption, surface diffusion and surface reactions for the desired reaction channel. The observed "pressure gap" in the RuO_x activity to CH_2O formation may be ascribed to kinetic limitations at low reactant potentials to get the optimal coverage of reacting species with specific adsorption configurations, favoring the immediate desorption of CH_2O to prevent further oxidation.

10.5 Concluding Remarks

The observed dynamics of Ru oxidation state and the related catalytic activity and selectivity have demonstrated the compositional complexity of catalytically active states, which are nonstochiometric transient surface oxides, RuO_x, lying in the boundary region between two thermodynamically stable phases. These nonequilibrium transient oxides evolve under the specific reaction conditions and reach their steady catalytically active state independently of the initial Ru precatalyst chemical state. The dramatic changes in the selectivity, correlated to relatively small differences in the oxygen content of the RuO_x states, demonstrate the compositional complexity of the catalyst surfaces under dynamic reaction conditions and the important role of pressure and temperature in fine tuning the catalyst performance. This findings are not unique to the Ru catalysts. Catalytically active state with incorporated subsurface oxygen, which can be described as a disordered surface oxide film is reported to sustain the selective oxidation of CH_3OH to formaldehyde on Cu catalysts as well.[41,42]

The findings reported here are of general validity for understanding the disparity observed between the high catalytic activity of metal and metal-oxide catalysts under technically relevant ambient pressures and the lack of reactions under UHV conditions. They also evidence the interconnection between the pressure and material gap, *i.e.* by varying the chemical potential of the reactants the catalyst structure and composition dynamically change. Such changes may occur on very local scales resulting in structural and/or compositional heterogeneity and may depend on structural features ranging from atomic to micrometer scales.

The results in this work apply well to real Ru catalyst systems, which are nanoparticles forming amorphous oxide with poorly defined stoichiometry.[43,44] The active role of subsurface oxygen and formation of surface oxide have also been considered in explaining the catalytic behavior of Ru electrodes in fuel cells.[45] These states are comparable to the transient RuO_x state rather than with the well-structured $RuO_2(110)$ surface, which represents a limiting case of a well-ordered model structure.

The observed dynamics of the Ru oxidation state and related to that catalytic activity and selectivity varying the methanol and oxygen chemical potentials in the mbar pressure regime should also occur under technically relevant ambient pressures, but not necessarily at the same temperatures and partial pressure ratios of the reactants observed in the present model system.

Acknowledgements

M. Kiskinova thanks the AvH Foundation for the Award to pursue research in FHI-Berlin in 2004-2005 and the financial support under EU Contract No. NMP3-CT-2003-505670 (NANO2). The critical reading of the manuscript by Tevfik Onur Mentes is acknowledged.

References

1. I. W. C. E. Arends, T. Kodama and R. A. Sheldon, *Oxidation Using Ruthenium Catalysts, in Topics in Organometallic Chemistry*, ed. C. Bruneau and P. H. Dixneuf, Springer, Berlin, 2004, Vol. 11, p. 277–320.

2. F. Maillard, G.-Q. Lu, A. Wieckowski and U. Stimming, *J. Phys. Chem. B*, 2005, **109**, 16230.
3. P. Mars and D. W. van Krevelen, *Chem. Eng. Sci.*, 1954, **3**, 41.
4. J. A. Labinger and K. C. Ott, *Catal. Lett.*, 1990, **4**, 245.
5. G. Ertl, *Angew. Chem. Intl. Ed.*, 2008, **47**, 3524.
6. H. Over, Y. D. Kim, A. P. Seitsonen, E. Lundgren, M. Schmid, P. Varga, A. Morgante and G. Ertl, *Science*, 2000, **287**, 1474.
7. H. Over and M. Muhler, *Prog. Surf. Sci.*, 2003, **72**, 3 and references therein.
8. A. Böttcher and H. Niehus, *J. Chem. Phys.*, 1999, **110**, 3186; *Phys. Rev. B* 1999, **60**, 14396.
9. M. S. Chen, Y. Cai, Z. Yan, K. K. Gath, S. Axnanada and D. W. Goodman, *Surf. Sci.*, 2007, **601**, 5659.
10. D. W. Goodman, C. H. F. Peden and M. S. Chen, *Surf. Sci.*, 2007, **601**, L124; ibid 5663.
11. J. Assman, V. Narkhede, N. A. Breuer, M. Muhler, A. P. Seitsonen, M. Knapp, D. Crihan and G. Mellau, *J. Phys.: Condens. Matter.*, 2008, **20**, 184017.
12. H. Over, M. Muhler and A. P. Seutsonen, *Surf. Sci.*, 2007, **601**, 5659.
13. R. Blume, M. Hävecker, S. Zafeiratos, D. Teschner, E. Kleimenov, A. Knop-Gericke, R. Schlögl, A. Barinov, P. Dudin and M. Kiskinova, *J. Catal.*, 2006, **239**, 354.
14. A. Knop-Gericke, E. Kleimenov, M. Hävecker, R. Blume, D. Teschner, S. Zafeiratos, R. Schlögl, V. I. Bukhtiyarov, V. V. Kaichev, I. P. Prosvirin, A. I. Nizovskii, H. Bluhm, A. Barinov, P. Dudin and M. Kiskinova, *Advances in Catalysis*, 2009, **52**, 213, Academic Press, Burlington.
15. R. Blume, H. Niehus, H. Conrad, A. Böttcher, L. Aballe, L. Gregoratti, A. Barinov and M. Kiskinova, *J. Phys. Chem. B*, 2005, **109**, 14052, and references therein.
16. X.-Q. Gong, Z.-P. Liu, R. Raval and P. Hu, *J. Am. Chem. Soc.*, 2004, **126**, 8.
17. K. Reuter and M. Scheffler, *Phys. Rev. Lett.*, 2003, **90**, 46103; *Phys. Rev. B*, 2003, **60**, 45407.
18. K. Siegbahn, *J. Electron Spect. Relat. Phenom.*, 1990, **51**, 11.
19. D. F. Ogletree, H. Bluhm, G. Lebedev, C. S. Fadley, Z. Hussain and M. Salmeron, *Rev. Sci. Instrum.*, 2002, **73**, 3872.
20. Günther, B. Kaulich, L. Gregoratti and M. Kiskinova, *Progr. Surf. Sci.*, 2002, **70**, 187.
21. A. Barinov, P. Dudin, L. Gregoratti, A. Locatelli, T. O. Mentes, M. A. Nino and M. Kiskinova, *Nucl. Instrum. Methods Phys. Res. A.*, 2009, **601**, 195.
22. http://www.elettra.eu/experiments/beamlines/esca/index.html.
23. E. Lundgren, A. Mikkelsen, J. N. Andersen, G. Kresse, M. Schmid and P. Varga, *J. Phys.: Condens. Matter.*, 2006, **18**, R481.
24. C. I. Calistle, T. Fujimoto, W. S. Sim and D. A. King, *Surf. Sci.*, 2000, **470**, 15.
25. M. Todorova, W. X. Li, M. V. Ganduglia-Pirovano, C. Stampfl K. Reuter and M. Scheffler, *Phys. Rev. Lett.*, 2002, **89**, 96103.
26. K. Reuter, *Nanocatalysis: Principles, Methods, Case Studies*, ed. I. Heiz, H. Hakkinen, U. Landman, Springer-Verlag, Berlin, 2005.

27. R. Blume, H. Niehus, H. Conrad, A. Böttcher, L. Aballe, L. Gregoratti, A. Barinov and M. Kiskinova, *J. Phys. Chem. B.*, 2005, **109**, 14052 and references therein.
28. A. Böttcher, U. Starke, H. Conrad, R. Blume, L. Gregoriatti, B. Kaulich, A. Barinov and M. Kiskinova, *J. Chem. Phys.*, 2002, **117**, 8104.
29. S. Lizzit, A. Baraldi, A. Groso, K. Reuter, M. V. Ganduglia-Pirovano, C. Stampfl, M. Scheffler, M. Stichler, C. Keller, W. Würth and D. Menzel, *Phys. Rev. B*, 2001, **205**, 419.
30. H. Over, A. P. Seitsonen, E. Lindgren, M. Wiklund and J. N. Andersen, *Chem. Phys. Lett.*, 2001, **342**, 467.
31. C. Stampfl, *Catal. Today*, 2005, **105**, 17.
32. A. Soon, M. Todorova, B. Delley and C. Stampfl, *Phys. Rev. B*, 2006, **73**, 165424.
33. J. Wang, V. Y. Fan, K. Jacobi and G. Ertl, *J. Phys. Chem. B*, 2002, **106**, 3422.
34. H. R. Gerberich and G. C. Seaman, *"Formaldehyde" in Kirk-Othmer Encicrlopedia of Chemical Technology*, 4th edn, 1994, Vol. 11, Wiley, p. 929.
35. G. Busca, *Catal. Today*, 1996, **27**, 457.
36. H. Liu and E. Iglesia, *J. Chem Phys. B*, 2005, **109**, 2155.
37. M. Mavrikakis and M. Barteau, *J. Mol. Cat A: Chemical.*, 1998, **131**, 135 and references therein.
38. H. Madhavaram, H. Idriss, S. Wendt, Y. D. Kim, M. Knapp, H. Over, J. Amann, E. Löffer and M. Muhler, *J. Catal.*, 2001, **202**, 296.
39. R. Blume, M. Havecker, S. Zafeiratos, D. Teschner, E. Kleimenov, A. Knop-Gericke, R. Schlögl A. Barinov, S. Lizzit P. Dudin and M. Kiskinova, *Phys. Chem. Chem. Phys.*, 2007, **27**, 3648.
40. R. Blume, M. Hävecker, S. Zafeiratos, D. Teschner, A. Knop-Gericke, R. Schlögl, P. Dudin, A. Barinov and M. Kiskinova, *Catal. Today*, 2007, **124**, 71.
41. H. Bluhm, M. Hävecker, A. Knop-Gericke, E. Kleimenov, R. Schlögl, D. Teschner, V. I. Bukhtiyarov, D. F. Ogletree and M. Salmeron, *J. Phys. Chem. B*, 2004, **108**, 14340.
42. L. Zhou, S. Gunther, D. Moszynski and R. Imbihl, *J. Catal.*, 2005, **235**, 359.
43. W. Vogel and N. Alonso-Vante, *J. Catal.*, 2005, **232**, 395.
44. B.-Z. Zhan, M. A. White, T.-K. Sham, J. A. Pincock, R. J. Doucet, K. V. Ramana Rao, K. N. Robertson and T. S. Cameron, *J. Am. Chem. Soc.*, 2003, **125**, 2195.
45. H. Schulenberg, M. Hilgendorff, I. Dorbandt, J. Radnik, P. Bogdanoff, C. Fiechter, M. Bron and H. Tributsch, *J. Power Sources.*, 2006, **155**, 47 and references therein.

CHAPTER 11

Styrene Synthesis: In-Situ *Model Catalysis Studies on Ethylbenzene Dehydrogenation*

WOLFGANG RANKE

Fritz Haber Institute of the Max Planck Society, Faradayweg 4-6, DE-14195 Berlin, Germany

11.1 Introduction

Model catalytic studies take advantage of the advanced methods of surface science for surface preparation and characterization that are able to yield detailed information on the chemical composition, the geometrical surface structure (including defects), the electronic surface structure (distribution of electronic states at and near the surface) and the resulting chemical properties (adsorption, conversion, desorption).

However, surface science is in general the regime of low pressures, small and mostly single-crystalline samples, small amounts of adsorbed species and controlled modification of surface composition and structure,[1] while catalysis is the regime of high pressures, temperatures and coverages, dispersed, polycrystalline or even porous samples with ill-defined surfaces of large area that may undergo fast and dramatic changes during the catalytic reaction.[2]

The difficulties in the transferability of surface-science results to real catalysis conditions and *vice versa* have been expressed in terms of a number of "gaps" between both fields of research.

RSC Nanoscience & Nanotechnology No. 19
Nanostructured Catalysts: Selective Oxidations
Edited by Christian Hess and Robert Schlögl

Published by the Royal Society of Chemistry, www.rsc.org

The idea is not new to overcome the pressure p (and temperature T and coverage Θ) gap (p–T–Θ gap) by combination of a conventional UHV chamber with a high-pressure cell.[3] Different types of arrangements have been proposed (some of them are discussed in ref. 4). Preparation and characterization of a catalytically interesting surface is performed in UHV using the methods of surface science and after that the sample is transferred under vacuum into a high-pressure cell where catalytic conversion measurements can be performed at pressures far enough away from UHV to yield information relevant to real catalysis conditions.

Often, the catalyst changes composition or structure under the conditions of real catalysis (sometimes termed a materials gap). Even on seemingly simple metallic systems like Cu[5] and Ag,[6] *in-situ* measurements showed that activity is only generated under the influence of the reactants at high pressure (see Chapter 9). And most of the relevant catalysts are even more complex: Oxides, usually promoted, doped or impregnated, zeolites, not to speak of enzymes in the regime of biochemistry.[2] Also, most reactants are complex. It is rather the rule that the working catalyst differs from the state before the start of the reaction or after quenching from reaction conditions. Quenching is a problem in itself. The application of *in-situ* analysis methods is unavoidable in order to shine light on such processes. A number of such *in-situ* methods has been developed during the last decade.[7,8] If a material gap exists, the existence of a reaction gap has to be suspected: Do the reactants take part in the change of the catalyst composition? How important are stoichiometric reactions?

Real catalysts are either highly dispersed on a support or have a high inner surface formed by pores. Transport properties for matter and heat are often decisive but often difficult to separate from the properties of the elementary catalytic reaction. Model catalysts can be prepared undispersed and free of pores. The catalytic reaction can therefore be studied without transport limitations. The influence of polycrystallinity, dispersion and pores can then be introduced step by step in a controlled way.[9]

11.1.1 Nonoxidative Dehydrogenation (DH) of Ethylbenzene

Industrially, most of ST is produced using potassium-promoted iron-oxide catalysts.[10] With more than 20 miot per year worldwide,[11] this reaction belongs to the 10 most important catalyzed organic reactions. The reaction (Figure 11.1(a)) is endothermic ($\Delta H = 125\,\text{kJ/mol}$) and has to be run at fairly high temperature, typically 870 K. The pressure is 1 bar but since steam in excess (molar ratio EB:H_2O between 1:6 and 1:10) is added to the reactant mixture, the partial pressure of EB is correspondingly lower. The steam is considered to transport heat, to counteract coking by the coal gasification reaction and to act as a diluent so that the equilibrium is shifted towards the product side.

The catalyst is complex. Since the divalent Fe^{2+} and the trivalent Fe^{3+} state are similarly stable, there exist several stable or metastable oxide phases.[12,13] Addition of K adds several stable or metastable ternary KFe_xO_y phases to the zoo of possible catalyst phases.[14] The reaction atmosphere is reducing due to the H_2 liberated in the reaction. This influences the catalyst phase during reaction. X-ray diffraction studies of a fresh K-promoted catalyst by Lee[15] showed that it contains hematite α-Fe_2O_3 and $K_2Fe_{22}O_{34}$, while after use the main component is reduced magnetite (Fe_3O_4). Hirano[16–18] identified potassium ferrite K_2FeO_2 at the surface of the active catalyst and assumed that it was an ingredient of the active state. This was confirmed by extensive *ex-situ* and *in-situ* studies by Muhler *et al.*[19,20] The active state was characterized by the presence of $KFeO_2$ and $K_2Fe_{22}O_{34}$. Also, some potassium carbonate K_2CO_3 was identified. The long-term deactivation was observed to be due to reduction of Fe^{3+} and removal of potassium from the active surface by spatial disintegration resulting in KOH in the bulk of the pellets or grains and Fe_3O_4 at the surface. The added steam was shown to counteract deactivation by carbonaceous deposits by the coal gasification reaction.

Since reduction of the catalyst material is easy, it may well be that the catalytic process includes oxidation of the liberated hydrogen by lattice oxygen to form water. In the steady state, the formed oxygen vacancies have to be reoxidized in a subsequent step. Such a cycle is often called the Mars–van Krevelen (MvK) mechanism.[21] The only oxygen-containing species in the feed is water. Energetically, it makes no difference if hydrogen desorbs directly as H_2 or if the lattice is invoked: formation of H_2O with lattice oxygen followed by reoxidation of the oxygen vacancy by dissociating H_2O from the feed and liberating H_2 in this way. Kinetically, however, a MvK mechanism may be advantageous because reoxidation may be timely and (if diffusion is possible) spatially separated from the oxygen removal from the surface. The catalytic process would then also be oxidative with the oxidant being water. As will be shown, a MvK mechanism is in fact very likely.

An understanding of the fundamental processes could only be achieved by starting projects capable of splitting up the complex reaction into more elementary processes on well-characterized model catalyst surfaces and by applying microkinetic modeling. Such studies were performed at the Fritz-Haber-Institute (FHI) of the Max-Planck-Gesellschaft in Berlin, Germany (W. Weiss, W. Ranke, R. Schlögl). Surface-science studies were combined with conversion measurements in an *in-situ* microflow reactor. The bridge to technical application was completed by microkinetic modeling at the Institute of Chemical Process Engineering of the University of Stuttgart, Germany (A. Schüle, G. Kolios, U. Nieken).

Also, the reactant ethylbenzene (EB) C_8H_{10} seems complex. However, the aromatic benzene ring is a fairly stable and well-defined entity, as is the ethyl group where the dehydrogenation has to occur. Deuterium exchange at the α-position of the ethyl group of EB (Figure 11.1) is much faster than at the β-position.[22] Therefore dehydrogenation will start at the α-position and the structure of the active site should fit to interact at this position.

A major problem is the formation of carbonaceous deposits (here shortly termed "coke") as a result of side reactions, leading to further or even full dehydrogenation of adsorbates. The coke covers the active surface and

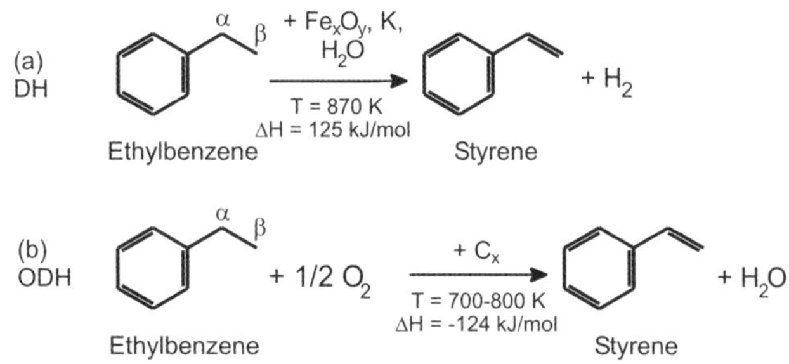

Figure 11.1 (a) Nonoxidative dehydrogenation(DH) and (b) oxidative dehydrogenation (ODH) of ethylbenzene to styrene.

deactivates it. The added steam is believed to counteract coking by the coal gasification reaction. Coke is not necessarily inactive and may be used as catalyst as well, especially in oxidative dehydrogenation.[23]

Hematite, α-Fe_2O_3, is an electrical insulator that makes the application of surface-science techniques using charged particles difficult. Therefore, thin FeO, Fe_3O_4 and Fe_2O_3 films of single-crystalline quality were grown epitaxially on single-crystalline Pt supports. Films about 10–20 nm thick still have sufficient electric conductivity. Thin films have further advantages: The film can easily be renewed if destroyed or contaminated, it consists of a limited and known amount of bulk material, which is an advantage if reactions with bulk material are suspected. Further, the time-consuming procedure of removal of bulk contaminants segregating at the surface is avoided.

11.1.2 Oxidative Dehydrogenation (ODH) of Ethylbenzene

The oxidative dehydrogenation (ODH) process of ethylbenzene using oxygen as a means of oxidation (Figure 11.1(b)) is so far not applied industrially. ODH is attractive because the reaction is exothermic and proceeds at lower temperatures (down to 650 K). It is not equilibrium-limited since the hydrogen is removed by reaction with oxygen. Finally, the high heat consumption for the steam added in the DH process is avoided and the separation of steam from the product is unnecessary.

As reviewed by Lisovskii and Aharoni[23] and by Vrieland and Menon,[24] oxides of metals with multiple valency like Mo, Sb, Sn, Fe, V, Bi–Mo, and Co–Mo were suggested as catalysts. But also insulators with acid–base properties (alumina, phosphates, zeolites) appeared active. It turned out, however, that rather the carbonaceous products were catalytically active. They are formed on them, catalyzed by the substrate which in this way becomes the support of the ODH catalyst.

The catalytically active carbons are characterized by the existence of oxygen-containing functional groups, the favored candidate being quinone-like groups.[25,26] Since burning of carbon is related to defects, fairly well-ordered graphene-like surfaces are important. On the other hand, enough defects carrying active functional groups have to exist. Therefore, much work has been concentrated on the test of different carbon materials.

Presently, the investigations concentrate on three main questions: Which form of carbon is most active? Which functional groups are necessary and responsible for catalytic activity? How can active carbon material be brought into a suitable form for technical application?

The most active research groups just now are centered at Berlin[27–30] and at Porto.[31,32] But other groups are also active in the field.[33,34] Compared to DH on iron-oxide-based catalysts, much less is known on the elementary processes of ODH on carbon-based materials. Here, ODH will therefore not be reviewed further.

11.2 Dehydrogenation of Ethylbenzene: Model Catalysis Studies

11.2.1 Experimental

11.2.1.1 Instrumentation

For the model catalysis investigation, three ultrahigh-vacuum (UHV) systems were used, all combined with a high-pressure chamber and a sample transfer system. The high-pressure chambers are separated from the main chamber by a gate valve and also serve as a load-lock. One of the systems (Figure 11.2)

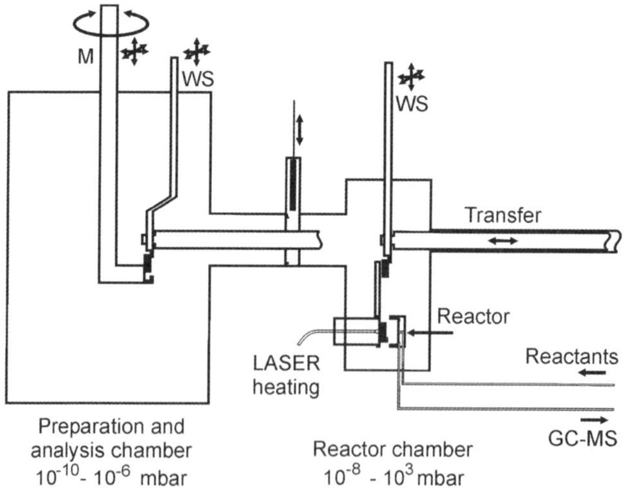

Figure 11.2 Experimental setup, schematic consisting of the preparation and analysis chamber working at ultrahigh vacuum and the reactor chamber, working at pressures up to 1 bar. The sample on its support (Figure 11.3) is moved by a magnetically coupled transfer rod. The transfer between the rod and manipulator M or reactor is accomplished by wobble sticks WS.

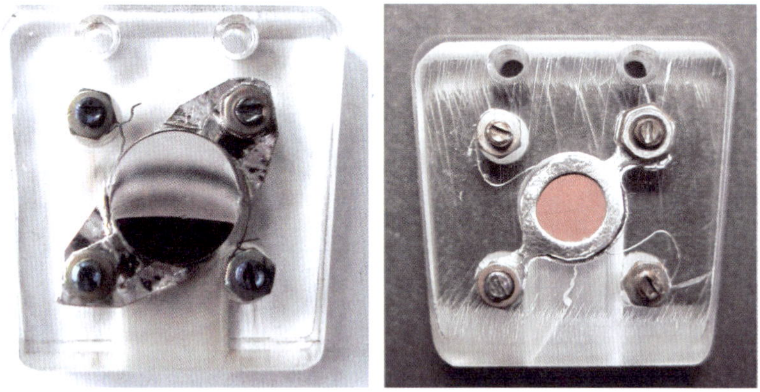

a b

Figure 11.3 Samples on sapphire support. The screws are contacted from behind to the manipulator or reactor when inserted. Two screws are used to fix the sample, the others for a K-type thermocouple. The support has a hole for heating by electron bombardment or laser irradiation from behind. (a) Pt sample for Fe_xO_y or KFe_xO_y thin film samples. (b) pressed Fe_2O_3 powder sample.

contains a microflow reactor integrated into the high-pressure chamber that will be described in more detail in Section 11.2.3.2. All systems contain facilities for preparation (heating, cooling, sputter gun, evaporators, gas-inlet systems) and an optics for low-energy electron diffraction (LEED) for controlling the phase of the thin films. One system is equipped for thermal desorption mass spectroscopy (TDS) and Auger electron spectroscopy (AES). The sample can be transferred into a further chamber with a photoelectron emission microscope (PEEM) and an ion scattering spectrometer (ISS). The second system contains a scanning tunneling microscopy (STM) system and the third one a spectrometer for X-ray and ultraviolet photoelectron spectroscopy (XPS, UPS) AES and a high-resolution LEED system.

Samples are mounted on sapphire transfer supports as shown in Figure 11.3. Sapphire is chemically inert, mechanically stable and has unique thermal properties. At low temperature, its heat conductivity is high and allows cooling, while it is a thermal insulator at higher temperature, which is favorable for annealing.

NEXAFS measurements were performed in a cooperation project at the former synchrotron storage ring BESSY-1.[35]

11.2.1.2 Preparation

Figure 11.4 shows the preparation of single-crystalline Fe_xO_y films by stepwise evaporation and oxidation of iron on a clean single-crystalline Pt(111) substrate, as reviewed in ref. 36. Also, Pt(001),[37] Ru(0001),[38,39] and Cu(111)[40] substrates can be used. When an iron layer is exposed to 10^{-6} mbar O_2 at 870–1000 K, it oxidizes and forms an oxygen-terminated layer with the

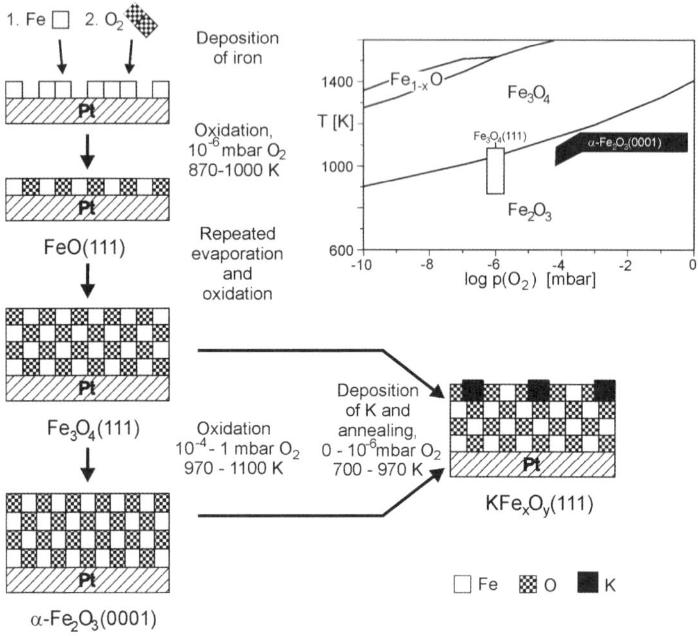

Figure 11.4 Preparation of Fe_xO_y and KFe_xO_y films. Included is the Fe–O phase diagram.[13]

characteristics of FeO(111), 1–4 Fe–O bilayers thick, depending on temperature and substrate. Upon further iron deposition and oxidation at 870 K, Fe_3O_4(111) islands grow on and into the FeO layer that eventually coalesce into a closed film. Postannealing at 1000 K improves the long-range order of Fe_3O_4(111). Films used as model catalysts were grown to a thickness of 5–20 nm. According to the iron-oxygen phase diagram in Figure 11.4[13] Fe_2O_3 is the thermodynamically stable phase for the growth conditions. However, its formation is kinetically hindered and oxygen pressures between 10^{-4} and 1 mbar have to be applied to convert the Fe_3O_4 film into α-Fe_2O_3.

K-promoted films are prepared by K deposition on Fe_3O_4 or Fe_2O_3 films, followed by annealing in vacuum or at $p(O_2) = 10^{-6}$ mbar.[14,41] The most-stable K–Fe–O phases are $KFeO_2$ (K-ferrite) and $K_xFe_{22}O_{34}$ with x = 2 (K–β-ferrite) or $x = 4$ (K–β″-ferrite). $K_xFe_{22}O_{34}$ has a hexagonal layer structure consisting of a sequence of "spinel-blocks" like Fe_3O_4, separated by –Fe–$K_{x/2}$O–Fe layers. Lower K contents ($x < 2$) are formed either by increasing the thickness of the spinel blocks, thus continuously approaching the pure Fe_3O_4, or by reducing the occupation of the potassium sublayers in analogy to the structure of corresponding alumina clay minerals. Because of the Fe_3O_4-like spinel blocks the lattice constant of the $K_xFe_{22}O_{34}$(0001) and Fe_3O_4(111) surfaces are almost identical, resulting in epitaxial growth. The tetragonal $KFeO_2$ structure does not fit with the lattice of Fe_3O_4 and the thin films have no long-range order. In its bulk, the K, Fe and O atoms are fairly homogeneously distributed.

11.2.2 Catalyst Characterization

11.2.2.1 Surface Structure

Figure 11.5 presents LEED patterns, STM images and structure models of well-ordered films. The hexagonal oxygen layers are very similar in all cases. The O–O distances vary only between 0.290 nm in Fe_2O_3, 0.296 nm in Fe_3O_4 and $K_xFe_{22}O_{34}$ and 0.304 nm in bulk FeO (thin layers slightly expanded). The strong differences in the LEED patterns and the STM images result therefore from the different arrangement of the Fe atoms between and on top of the oxygen layers. The thin FeO(111) layers form characteristic satellite patterns around the integral order spots in LEED and Moiré patterns in the STM image. Both result from the low film thickness (the shown patterns are for a film only one Fe–O bilayer thick) and the misfit between the Pt(111) and the FeO(111) lattice. The surface is oxygen terminated.[36,42] However, for the used parameters, oxygen is invisible for STM and the bright dots reflect the positions of the Fe atoms underneath.[43]

Also in the STM image of $Fe_3O_4(111)$, the bright spots correspond to the position of Fe atoms. The spots are quite clear on $Fe_3O_4(111)$ because Fe forms the top layer.[44] The STM image of α-$Fe_2O_3(0001)$ is again formed by Fe-derived spots but the contrast is lower than on Fe_3O_4. The reason is that the surface most likely is oxygen-terminated or possibly even hydroxylated.[45]

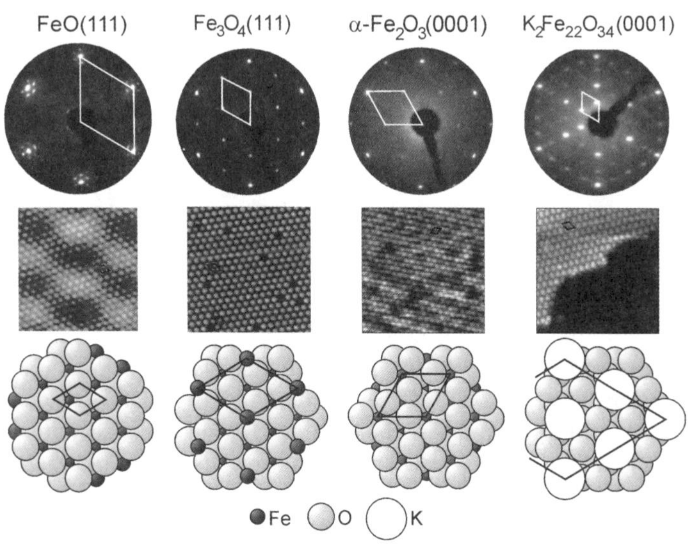

Figure 11.5 LEED patterns, STM images and structure models of ordered binary Fe_xO_y films and the long-range ordered ternary $K_xFe_{22}O_{34}(0001)$ film ($x \approx 0.67$). The structure models for the binary films result from structure analysis, that of $K_xFe_{22}O_{34}$ is hypothetical but it reflects the surface composition as deduced from XPS measurements.[41]

The preparation of the Fe_2O_3 surface is difficult and many complex surface structures exist.[46]

Different ternary K–Fe–O compounds form, depending on the amount of K and the annealing temperature.[47] Quantitative XPS studies[41] showed that K diffuses already at 300 K into the bulk. Near the surface it forms KOH by reacting with water from the residual gas but even faster when water is admitted.

In the range 400–600 K, KOH desorption competes with decomposition and K incorporation in the form of $KFeO_2$, related with water desorption.[48] STM imaging of $KFeO_2$ is difficult and only possible on a very thin (2 nm) film because the compound is an insulator. The film is rough and atomic resolution is not achieved.[14] When annealing to 800–900 K, the potassium content decreases further. The presence of water speeds this process up, showing that it proceeds *via* formation and desorption of KOH. A $KFeO_2$ layer of decreasing thickness remains at the surface, while the underlying film converts to $K_xFe_{22}O_{34}$. At 970 K, the $KFeO_2$ has disappeared completely, leaving a $K_xFe_{22}O_{34}$ layer. XPS results[41] suggest that it is terminated by a Fe–K–O layer while the bulk contains less potassium ($x \approx 0.67$).

11.2.2.2 Adsorption and Interaction

11.2.2.2.1 Ethylbenzene and Styrene. The nature of the adsorbed species (molecular or dissociated) after exposure to EB[49] and ST[50] was studied by UPS, their orientation by NEXAFS.[35] The energetic and kinetic data for adsorption and desorption were measured using TDS[51] and isosteric methods.[50] Figure 11.6 summarizes the knowledge for EB and ST and correlates it to structural data of the substrate.

The FeO film is O-terminated and the first Fe layer is relatively deep below the top O-layer (–0.068 nm[42] or –0.062 nm[43]). According to the best-fit structure, also Fe_2O_3 is O-terminated[45] with the first Fe layer only –0.046 nm below the surface O-layer. In contrast, Fe_3O_4 is Fe-terminated with the Fe atoms +0.038 nm above the O-layer.[44] Figure 11.6(b) demonstrates that the position of the reactive Fe atoms is decisive for the interaction with adsorbates and their orientation.

According to UPS, all adsorbate states of EB and ST at room temperature and below are molecular (for decomposition behavior see Section 11.2.3.1). NEXAFS measurements on FeO revealed that EB and ST adsorb strongly tilted, corresponding to a physisorbed state. Most likely adsorbate–adsorbate interaction prevails resulting in island formation and since interaction occurs *via* the π-orbitals of the benzene rings, the molecules get tilted. In contrast, both EB and ST chemisorb almost flat on Fe_3O_4 and Fe_2O_3 (γ-state) and get tilted near saturation, when adsorbate–adsorbate interaction is unavoidable. Chemisorption is followed by adsorption of a second, physisorbed layer. Due to Fe termination, chemisorption on Fe_3O_4 is stronger than on Fe_2O_3. And it is stronger for ST because its conjugated π-system extends over the ethyl group if adsorbed in a planar configuration.

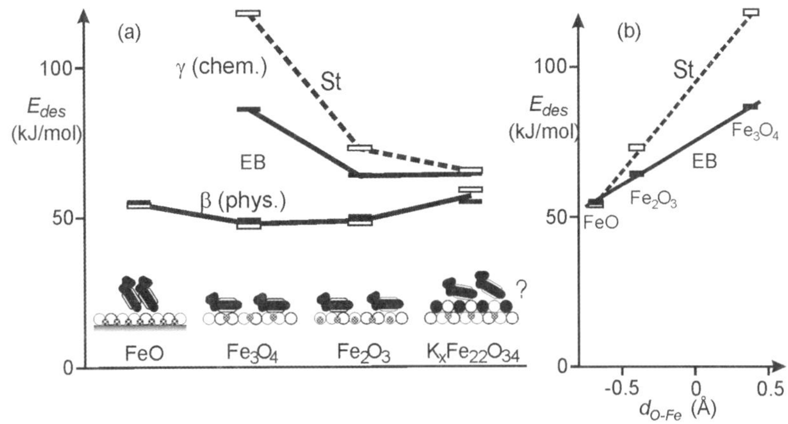

Figure 11.6 Compilation of energetic and structural results for adsorption of EB (filled rectangles) and ST (open rectangles) on different substrate films. (a) Desorption energies from TDS [52] for chemisorbed (γ) and physisorbed (β) species and schematic adsorbate arrangement at low coverages of the initially adsorbing species (β on FeO, γ on the others) from NEXAFS measurements.[35] Shown is adsorbed EB, the arrangement for ST is similar. The arrangement on $K_xFe_{22}O_{34}$ is hypothetical. (b) Dependence of the desorption energy for the initially adsorbing species on the position of the first iron layer relative to the first oxygen layer.

From UPS measurements of EB on Fe_3O_4 in adsorption–desorption equilibrium it was deduced that saturation of the chemisorbed γ-state corresponds to about 83% of a close-packed adlayer (Van der Waals area of EB $\sim 0.5\,nm^2$). This relates to the concentration of Fe surface sites with two sites per EB molecule. In contrast, the physisorbed first β-layer on FeO saturates at a full monolayer without correlation of the adsorbate with the surface structure.[49]

Chemisorption is thus dominated by the interaction between the polarizable π-system of the adsorbate representing a soft base[53] and the acidic Fe ions. This is highlighted by Figure 11.6(b). It shows the dependence of E_{des} for the initially adsorbing species (β for FeO, γ for Fe_2O_3 and Fe_3O_4) on the position of the Fe layer with respect to the top O layer. E_{des} is highest on Fe_3O_4 where Fe atoms form the top layer. On Fe_2O_3, the Fe-layer is below the O-layer but the molecules can obviously still "feel" it and arrange in a planar chemisorbed state. Although the Fe-layer is not located much deeper on FeO, the interaction appears just too weak to compete with the Van der Waals interaction between the adsorbates.

There exists no LEED structure analysis for the $K_xFe_{22}O_{34}$ film but quantitative XPS[41] suggests that it is terminated by a Fe–K–O layer similar to such layers in the bulk of $K_2Fe_{22}O_{34}$. The first Fe atoms would then be located below the O atoms and an interaction with the adsorbate is expected to be fairly weak. The existence of a chemisorbed state with a similar binding energy as on Fe_2O_3 suggests similar binding situations. The difference of the binding energies for

EB and ST has disappeared. However, nothing is known about the influence of surface K ions on adsorption.

11.2.2.2.2 Water. The interaction of water with single-crystalline iron-oxide films was studied using thermodesorption,[48,54,55] photoelectron spectroscopy,[50,54] infrared absorption spectroscopy[56] and also theoretically.[57] On the oxygen-terminated FeO film, adsorption is molecular and weak. A unique case is the Fe-terminated Fe_3O_4 surface. Chemisorption is observed but although it is fairly weak and completely reversible, it is dissociative.[54,56,57] Two different OH groups form, one on the surface Fe atoms and one by interaction of the remaining H atoms with one kind of surface oxygen atoms. This adsorption saturates when all surface Fe is occupied. Further water adsorbs by bridging the two OH groups, resulting in an unusual kind of water dimers that resemble a H_3O^+–OH^-–surface complex.[57] The capability of surface Fe and O sites to dissociate water is interesting for the styrene process. It may represent the decisive step in reoxidation of reduced surface sites by water from the feed in a MvK mechanism. The desorption of H_2 that has to accompany such a process was also observed (see next section). Yet, details of the mechanism of reoxidation and H_2 desorption are not known.

The interaction of water with Fe_2O_3 is less clear. Thermodesorption[48,55] showed the existence of chemisorption and physisorption similar to Fe_3O_4, but further data do not exist. The main reason is that this surface is more difficult to prepare.

The interaction of water with K-promoted surfaces is strongly related to the kinetics of formation, decomposition and removal of KOH as described in Section 11.2.2.1.

11.2.2.2.3 Hydrogen. The interaction of iron-oxide surfaces with molecular H_2 is weak and the H–H bond cannot be broken. During dehydrogenation of EB, however, hydrogen is supposed to appear in atomic form and to form hydroxyl groups at the surface. Since the catalyst is reduced during use, the hydroxyl groups recombine at least partly to form water that desorbs, leaving oxygen vacancies behind. This would eventually result in complete reduction to metallic iron (see also Section 11.2.3.3 and Figure 11.11) if no steam were added. Water is able to preserve the Fe_3O_4 state. According to a thermodynamic study[13] it is in fact not the water directly that is responsible but the oxygen partial pressure according to the dissociation equilibrium of water at reaction temperature (870 K).

$$H_2O \leftrightarrow H_2 + \tfrac{1}{2}O_2 \tag{11.1}$$

Since very small concentrations of O_2 appear to be sufficient, it is possible that even O_2 contaminations or O_2 dissolved in the process water may play a role in technical catalysis.

In order to study the oxide reduction by hydrogen (H_2O desorption) and the possibility of nonreductive removal of hydrogen (H_2 desorption), systematic studies using a source of atomic hydrogen H_{at} were performed.[58–60] XPS and LEED studies on Fe_2O_3 [60] showed that the surface and bulk composition and surface structure were already modified during exposure to H_{at} around 300 K. Only an exposure equivalent to about one monolayer (1 L, Langmuir unit) was necessary to saturate a Fe_2O_3 film with OH groups. According to XPS, the film still consisted of Fe_2O_3, but higher exposures resulted in an increasing amount of Fe_3O_4. After 1000 L H_{at}, the film was completely reduced to Fe_3O_4. The Fe_3O_4 domains were even large enough and oriented to be observable in LEED, which showed that even near 300 K, oxygen or, more likely, iron is mobile over distances corresponding to the film thickness (about 10 nm) to allow rearrangement.

After 1 L H_{at}, TDS showed only desorption of H_2O, indicating that Fe_2O_3 is easily reduced. After 1000 L, where the film consisted almost completely of Fe_3O_4, H_2 desorption also occurred. Reduction of this phase is more difficult and H_2 desorption can compete with reductive H_2O desorption. Interestingly, the surface remained partly terminated by Fe_2O_3 domains even after heavy bulk reduction. This termination thus appears quite stable. Since Fe_2O_3 is the catalytically most active phase (see Section 11.2.3.3), this may contribute to catalytic activity even if the bulk is strongly reduced.

Also, Fe_3O_4 is quickly saturated with OH groups and can be reduced further by atomic H_{at} at 355 K.[59] Reduction occurs in water-desorption bursts, probably starting at defects formed by H_2O desorption. Such an autocatalytic mechanism was clearly identified on a very thin FeO film (one Fe–O bilayer thick). However, the energetics in this case is influenced by the interface to the Pt substrate. The water-desorption bursts on Fe_3O_4 were always accompanied by desorption of H_2.

In summary, the adsorption–desorption studies using H_{at} showed that Fe_2O_3 is easily reduced to Fe_3O_4. Initially, no H_2 but only H_2O desorption occurs. Fe_3O_4 is less easily reduced and, along with H_2O desorption, always direct desorption of H_2 is observed. In catalytic dehydrogenation of EB on Fe_2O_3 the removal of hydrogen is therefore either stoichiometric (reduction) or follows a Mars–van Krevelen mechanism, while both MvK and direct H_2 desorption are possible on Fe_3O_4.

11.2.3 Catalytic Characterization

11.2.3.1 Low- and Medium-Pressure Reactivity Measurements

The first reactivity measurements were performed on unpromoted model catalysts using mass-spectrometric analysis in the main UHV chamber using a mixture EB:H_2O = 1:5 at a total pressure of 3.5×10^{-6} mbar.[61] The degree of surface order was qualitatively deduced from the initial LEED patterns. It turned out that well-ordered Fe_2O_3 hematite samples were quite inactive, while conversion over poorly ordered samples was clearly observable. After several

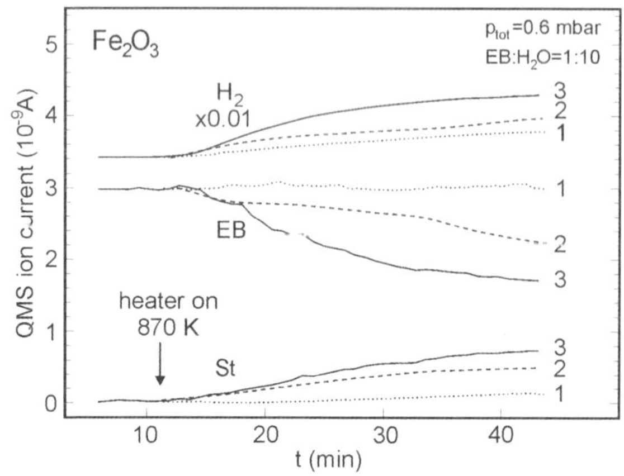

Figure 11.7 Mass spectrometric analysis of a batch reactor experiment at intermediate pressure conditions for three Fe_2O_3 model catalysts with differing surface quality. (1) well ordered, (2) intermediate order, (3) poorly ordered (from ref. 63).

such cycles, deactivation occurred on a poorly ordered film and the LEED pattern showed weak spots from Fe_3O_4, indicating reduction. Also, coking occurred according to photoelectron emission microscopy (PEEM). Magnetite Fe_3O_4 samples showed no measurable activity.

In a first attempt to bridge the pressure gap, batch-reactor experiments were performed in the reactor chamber in Figure 11.2 (at that time without a reactor). The mass spectrometer of the main UHV chamber was applied for analysis, using a bypass line with a dosing valve between the chambers. 10 min after admitting the gas mixture ($EB{:}H_2O = 1{:}10$, total pressure 0.6 mbar), sample heating was started, reaching the reaction temperature after some minutes. Figure 11.7 shows evolution of styrene and hydrogen and simultaneous ethylbenzene consumption. Again, activity increases with the degree of initial disorder. After about 30 min, the most-active film was deactivated and completely coked, as shown by PEEM and AES. The coke could be removed by a mild oxidation cycle, after which LEED showed a mixture of the patterns of Fe_2O_3 and Fe_3O_4, confirming catalyst reduction during the experiment. As in the low-pressure experiments, Fe_3O_4 showed no measurable activity.

Since the amount of formed coke followed the activity of the different Fe_2O_3 films, it was concluded that the product molecule ST is mainly responsible for coking.

UPS[50] on FeO and Fe_3O_4 confirm this tendency. Below 400 K, adsorption of EB on FeO and Fe_3O_4 is molecular and reversible. But if a ST layer condensed on FeO at low temperature is heated off, an irreversibly adsorbed residue with spectral features of polystyrene remains on the surface. Its amount increases upon further condensation–desorption cycles. This shows that polymerization is quite effective as a reaction channel open for ST but not for EB. Adsorption on Fe_3O_4 is much stronger (Figure 11.6). Also here, ST causes irreversible adsorption but the adsorbate is more strongly decomposed and does not look like polystyrene. In a ST atmosphere, this species accumulates continuously.

A similar behavior was observed on Pt.[62] There, decomposition proceeds further the higher the temperature is and beyond 700 K, only a graphitic coke layer is left. It is thus evident that strong interaction with the substrate results in strong coking.

Under real catalysis conditions it is likely that decomposition of both EB and ST are side reactions. But the UHV studies have shown that coke formation by ST is by far more effective because oligomers can be formed. Oligomers as larger entities are less easily desorbed so that decomposition is effective. Clearly the product molecule ST is mainly responsible for coking.

After the low- and medium-pressure conversion experiments it was clear that model catalyst surfaces as small as 0.5 cm^2 were sufficient to perform meaningful experiments. As in earlier experiments, Fe_2O_3 proved to be clearly more active than Fe_3O_4 and deactivation of Fe_2O_3 is connected with reduction and coking. The catalyst phase is decisive for good performance but a well-ordered surface is inactive. Although an influence of defects was often assumed, the proof of their importance was new.

11.2.3.2 In-Situ *Microflow Reactor*

In order to study the time dependence of conversion and deactivation at still higher pressures, a stagnation-point microflow reactor as shown in Figure 11.8 was constructed.[64] Analysis of sample order and composition was possible after interruption and sample transfer into the UHV chamber.

The reactor was constructed on a 70-mm od flange with feedthroughs for the thermocouple and for two fiber rods for coupling in laser radiation for sample heating. Two flexible capillaries for gas inlet and outlet are connected to the reactor cap. For insertion of the sample on its sapphire support, the reactor cap is retracted (Figure 11.8(a)). After retraction of the wobble stick, the cap is closed (Figure 11.8(b)) and the arrangement represents a reactor with a total volume of only about 4 ml. This determines the response time. For typical flow conditions ($p = 1$ bar, 25 ml min^{-1}), it is less than 30 s. Due to the small distance between sample and cap (typically 1 mm), the gas volume in contact with the sample is much smaller ($\sim$0.05 ml) and the contact time is only about 0.12 s.

The sample is heated from behind by radiation from two diode laser stacks with a total power of 100 W, which is necessary to reach a reaction temperature of 870 K under flow conditions. The sample is always the warmest spot, 100–200 K warmer than any part in the environment.

Product analysis was achieved with a gas chromatograph equipped with an ion-trap mass detector (GC-MS), optimized for separation and analysis of EB, ST, toluene and benzene but not suitable for smaller hydrocarbons, CO_2, CO and H_2. GC analysis is cyclic every three minutes.

Since surface characterization cannot be performed during reaction, the procedures of sample transfer, gas admission and heating before reaction as

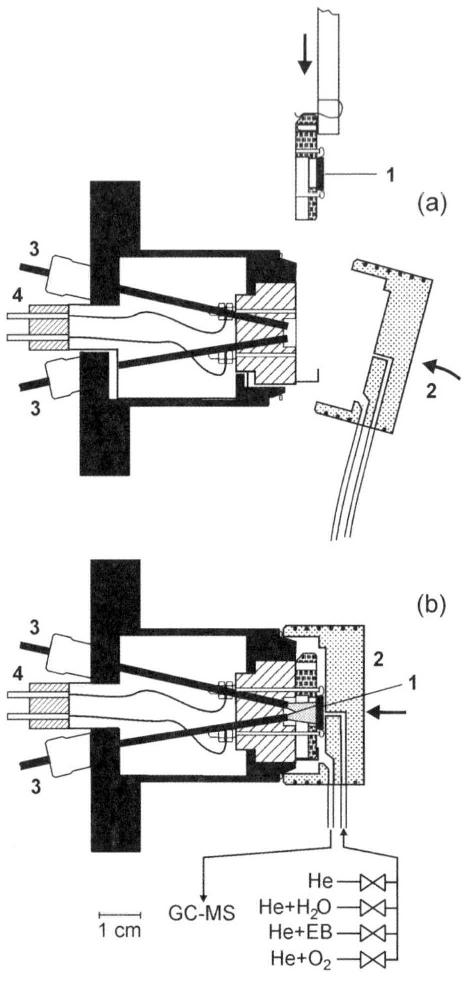

Figure 11.8 Stagnation-point microflow reactor for model catalysis at high pressure. 1: sample on sapphire support; 2: reactor cap; 3: fiber rods for coupling in laser irradiation; 4: thermocouple feedthrough. (a) during insertion of the sample on a wobble stick, reactor cap withdrawn. (b) reactor cap closed, gas admission and analysis lines schematically shown.

well as quenching and evacuation after reaction are of decisive importance.[65] After transfer from the UHV chamber, the reactor cap is preheated and outgassed under vacuum at 670 K, then the chamber is backfilled with N_2, a flow of pure He (K-promoted samples) or He with admixture of H_2O or H_2O and O_2 (Fe_3O_4, Fe_2O_3) is admitted to the reactor and the sample is heated to the reaction temperature (870 K). This takes about 20 min. Then the flow is switched to the reaction conditions (Table 11.1) and the GC measurement started. After the measurements, the reactive gases EB, H_2O and O_2 are switched off

Table 11.1 Partial pressures and molar ratios of reactive gases in the gas feed for the used standard reaction conditions. The rest of the working pressure of 1 bar is He. The standard reaction temperature is 870 K, the standard total flow 25 ml min^{-1}.[65]

Reaction conditions	*p (mbar)*	*Molar ratios*
Normal	$p(EB) = 3.3$	$EB:H_2O$
	$p(H_2O) = 33$	1:10
Reductive	$p(EB) = 3.3$	–
Oxidative	$p(EB) = 3.3$	$EB:H_2O:O_2$
	$p(H_2O) = 33$	1:10:0.5
	$p(O_2) = 1.7$	

and the sample is cooled in a flow of pure He to 600 K within less than 60 s. After reaching 500 K, the reactor chamber is evacuated and the sample transferred for postreaction analysis.

Two checks were performed to make sure that the surfaces are not changed by these procedures:[65] (i) An Fe_2O_3 sample that was exposed to a heat-up and cool-down cycle without reaction turned out to be clean and showed the characteristic LEED pattern. (ii) Reactivity measurement was interrupted, the sample cooled down, transferred and characterized, transferred back and the reaction restarted. It showed the same catalytic activity as before interruption.

AES and LEED analysis after reaction[65] showed always that the samples were disordered and covered by coke to an extent increasing with time on-stream. In order to analyze the substrate phase, the coke was removed by one or several mild thermal programmed oxidation (TPO) cycles[63] under conditions not affecting the oxide phase.[13]

The used standard reaction conditions are listed in Table 11.1.

11.2.3.3 Unpromoted Fe_3O_4, Fe_2O_3

Figure 11.9(a) shows the dependence of the ST production rate on the time on-stream for a magnetite (Fe_3O_4) sample, using the normal reaction conditions ($EB:H_2O = 1:10$).[65] The initial rate is low and decays further to a final value that is reached after about 70 min. Postreaction analysis at B shows that the final state consists of Fe_3O_4, covered with coke. The decay is thus due to coking alone.

For hematite (Fe_2O_3) (Figure 11.9(b)),[65,66] the initial rate is clearly higher and decays to the same final value as observed for Fe_3O_4. The experiment was repeated several times with freshly prepared samples and stopped at the positions D, E and F. Postreaction analysis showed that the substrate was only moderately coked at D. One standard TPO cycle was enough to remove it. The LEED pattern of Fe_2O_3 was recognizable, but the spots were diffuse, indicating disorder. At E, coking was strong and 3 TPO cycles were necessary to remove it. After that, both the patterns of Fe_2O_3 and Fe_3O_4 appeared, indicating that

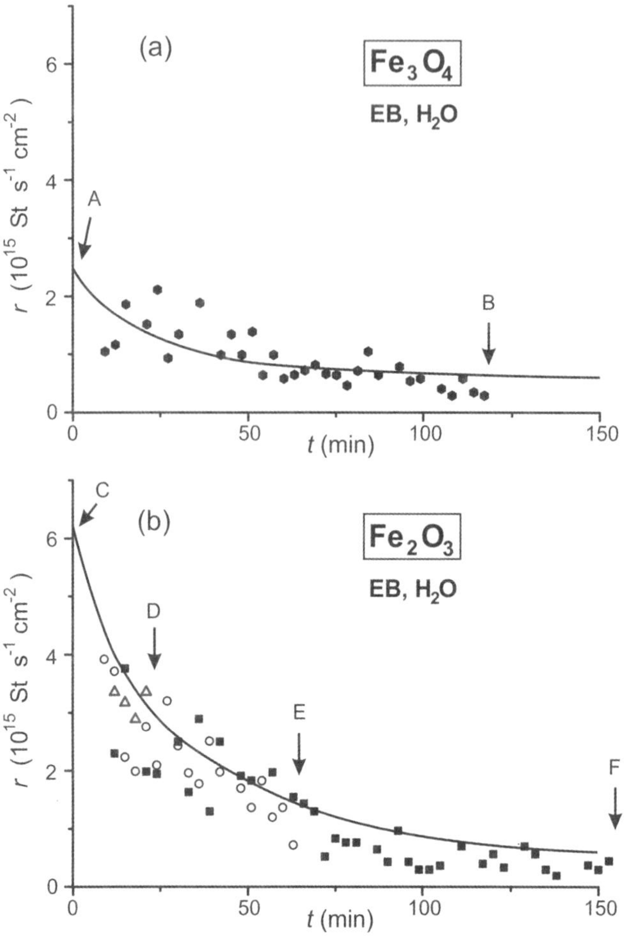

Figure 11.9 (a) ST production rate over Fe_3O_4 under normal conditions (see Table 11.1); postreaction analysis at B. (b) ST conversion over Fe_2O_3 under normal conditions, three measurements, finished at D, E, F for post-reaction analysis. Solid lines are from microkinetic modeling (see Section 11.2.4).

the substrate was partially reduced. At F, coking had not increased but LEED after 3 TPO cycles showed complete reduction to Fe_3O_4.

Comparison of the different initial activities of Fe_3O_4 and Fe_2O_3 proves immediately that reduction is also a reason for deactivation, in addition to coking. Since coking is caused mainly by the product ST, the higher initial production rate over Fe_2O_3 is expected to cause faster coking than on Fe_3O_4. The contributions of both can therefore not be separated by simple difference formation of the production rate curves of Figures 11.9(a) and (b) but modeling is needed.

Since oxygen is liberated during reduction of Fe_2O_3, the dehydrogenation in this period (approximately until E in Figure 11.9(b)) might not be catalytic according to the equation

$$C_8H_{10} \rightarrow C_8H_8 + H_2 \qquad (11.2)$$

but stoichiometric, involving reduction of the substrate:

$$C_8H_{10} + 3Fe_2O_3 \rightarrow C_8H_8 + 2Fe_3O_4 + H_2O \qquad (11.3)$$

In our case, the amount of catalyst material was limited and known so that the amount of liberated oxygen could be calculated (assuming complete reduction to Fe_3O_4) and compared to the amount of ST (which equals the amount of H_2) produced during this period. It turns out that the amount of ST exceeds the amount of available oxygen by at least a factor of 700. The reaction on Fe_2O_3 is thus essentially catalytic but may be stoichiometric at the very beginning.

As outlined in Section 11.1.1, water could take an active part in a Mars–van Krevelen type of reaction. The time dependence of production rate was therefore also measured without water (Figure 11.10, lower curve).[65,66] The initial rate is very similar to that with water, which makes it unlikely that water is involved on clean Fe_2O_3. At first glance, the deactivation behavior seems similar. But interruption at K and surface analysis revealed heavy coking and strong oxygen depletion and even after several TPO cycles, no LEED pattern

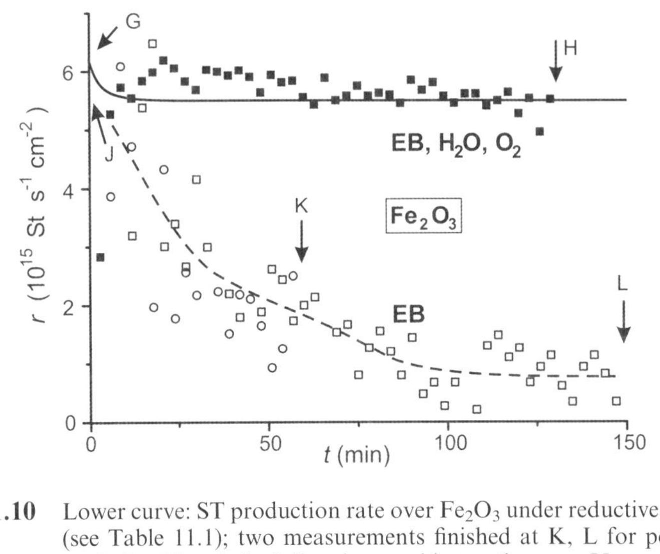

Figure 11.10 Lower curve: ST production rate over Fe_2O_3 under reductive conditions (see Table 11.1); two measurements finished at K, L for postreaction analysis. The dashed line is a guide to the eye. Upper curve: ST conversion over Fe_2O_3 under oxidative conditions (see Table 11.1). Postreaction analysis at H. The solid line is from microkinetic modeling (see Section 11.2.4).

reappeared. At L, AES showed only carbon. After many TPO cycles, a weak pattern of the clean Pt(111) substrate reappeared. Oxidation reproduced Fe_2O_3, proving that Fe had not been lost. It may have formed islands or alloyed with the Pt substrate. Iron had thus been completely reduced to Fe^0. The final production rate is characteristic for carbon as in Figures 11.9(a) and (b). This proves clearly that, apart from the very beginning, the reaction on partially reduced Fe_2O_3 and on Fe_3O_4 follows a MvK mechanism with water acting as reoxidation agent.

It was tried to prevent reduction and coking by addition of oxygen to the feed[65,66] (Figure 11.10, upper curve). The production rate stabilizes at a value even higher than the initial rates without O_2 (lower curve and Figure 11.9(b)). Assuming that the mechanism of catalytic dehydrogenation with and without O_2 is the same, this suggests that the first measured points without O_2 already represent a partly deactivated state. Postreaction analysis at H reveals only slight coking and LEED after only one TPO cycle shows the pattern of unreduced hematite with increased disorder.

These observations agree with experimental equilibrium determinations for the Fe–O_2–H_2O system[67] and calculations for the low pressures applied here.[13] They show that Fe_2O_3 is reduced at 870 K in the presence of H_2. Without H_2O, this proceeds to metallic Fe^0, while the presence of H_2O limits reduction to Fe_3O_4. The hydrogen produced during EB dehydrogenation is thus responsible for the observed reduction. It also confirms that the role of H_2O is not only to remove coke but also to limit reduction beyond the magnetite phase. Hematite reduction can be prevented by addition of oxygen.

Experimentally, the steady-state production rate after a 50-min reaction period decreases when the oxygen partial pressure is reduced. The optimal value as used in Figure 11.10 (upper curve) corresponds to the oxygen necessary for oxidation of the hydrogen originating from the EB dehydrogenation and from coking reactions as well as for oxidation of the coke itself.[65] For higher oxygen flow rates total oxidation becomes dominant.

The reduction of the hematite to magnetite observed under normal conditions agrees with observations on technical catalysts.[10,68] However, the high initial rate associated with clean hematite was never observed before because it was hidden in the start-up and equilibration periods. It is therefore likely that most investigations on technical catalysts were performed on partly or completely deactivated material.

Therefore, the time dependence of conversion was also tested on pressed Fe_2O_3 powder pellets (BET surface 2.2 $m^2 g^{-1}$, mean grain size 300 nm, mean pore diameter 240 nm).[69,70] They were pressed so that they could be used in the same sample support (Figure 11.3(b)). Compared to a single-crystalline Fe_2O_3 film, the initial production rate was about 2.5 times higher and deactivation was slower. However, the activity did not at all scale with the BET surface (powder: 2200 cm^2, single crystal 0.5 cm^2). It rather scaled with the roughness-related increase of the geometric surface area. Since diffusion of H_2O and EB even at room temperature is fast, blocking or deactivation of the inner pore surface by coke formation by the product molecule ST must be responsible. An initial activity decrease due to pore deactivation was not seen in the experiment. But it may be fast and masked by switch-on and equilibration effects.

As on single-crystalline films, addition of oxygen (see Table 11.1) could prevent long-term deactivation but did not increase the initial production rate. It could thus not prevent pore deactivation. The probability is fairly high that ST molecules formed in a pore are readsorbed, form oligomers and are dehydrogenated. This is in line with an estimation of coke-formation efficiency on the single-crystalline sample where about 10% of the EB was converted into ST and about 5% into coke.[65] This means that even on a flat surface only about 2/3 of the formed ST was able to escape the coking process.

In the low- and medium-pressure experiments, there was clear evidence that defective Fe_2O_3 surfaces are catalytically more active than well-ordered ones. In the high-pressure experiments, no indications for a defect-dependent initial activity was observed but postreaction analysis showed always that disorder had formed during reaction. Formation of disorder is not unexpected because of the observed reduction that implies out-diffusion of oxygen and nucleation of magnetite. Also, when O_2 is added so that no net reduction occurs, surface reduction–oxidation processes may occur dynamically. At low and medium pressures, formation of defects is slow enough and the influence of initial surface disorder is clearly visible. At high pressure defect formation may be so fast due to the more violent reaction conditions that it has happened within the first minutes and may at most influence the first GC data point after gas admission. These first data points are sometimes lower but this is more likely due to a temporary temperature drop upon admission of EB. The importance of defects for the mechanism of the catalytic reaction can thus only be deduced from low- and medium-pressure experiments.

Deactivation reduces the production rate considerably but the final activity is not zero. It is equal, irrespective of the degree of oxidation of the final surface. Thus, Auger analysis showed that the final surface in Figure 11.9(a) contained oxygen, while that in Figure 11.10 (lower curve) was essentially free of oxygen. Therefore the final activity can be ascribed to coke. Certain forms of carbon are active catalysts in oxidative dehydrogenation of EB.[71,72] Possibly, the coke formed here contains the necessary active centers.

11.2.3.4 K-promoted KFe_xO_y

Figure 11.11 shows representative production-rate measurements for K-promoted model catalysts with different K content.[65] A relative measure of the K content is the I_K/I_{Fe} Auger intensity ratio. The $K_xFe_{22}O_{34}$ film used in Figure 11.11(a) had $I_K/I_{Fe} \approx 2.3$, which is not far from the value of the 2×2 surface ($I_K/I_{Fe} \approx 2.7$, $x \approx 0.67$, see Section 11.2.2.1). The initial production rate is near to that of the unpromoted film (also shown for comparison in Figure 11.11(a)) but the deactivation is much slower. Postreaction analysis showed only weak coking (carbon to iron Auger intensity ratio $I_C/I_{Fe} = 0.5$) and a moderate reduction of the K content to $I_K/I_{Fe} \approx 1.8$.

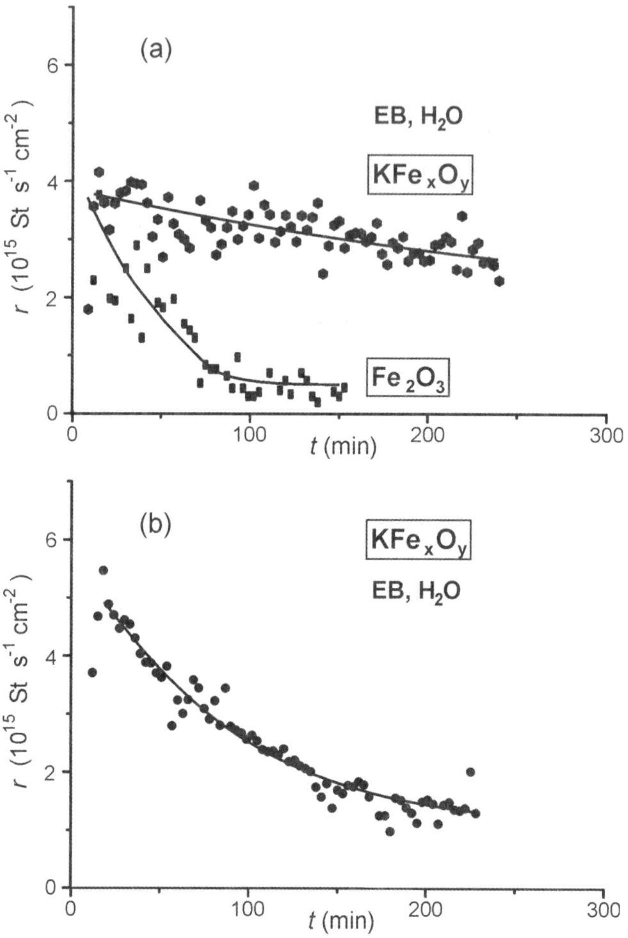

Figure 11.11 (a) ST production rate over a K-promoted catalyst with intermediate K content (upper curve) and unpromoted Fe_2O_3 (lower curve, same as Figure 11.9(b)) under normal conditions (see Table 11.1). (b) ST production rate after "regeneration" of the K-promoted catalyst from (a) for 15 min with only H_2O and no EB in the feed, resulting in a low K content.

In the industrial process, deactivated catalysts are regenerated by running the reactor with only steam in the feed ("steaming"). This treatment was simulated on the model catalyst used in Figure 11.11(a) by running it for 15 min without EB but with the same flux of water as before. Then, the EB flux was admitted again and the rate measured. The result is presented in Figure 11.11(b). The initial rate is higher than before but the decay is faster. Postreaction analysis showed a strongly reduced K content ($I_K/I_{Fe} \approx 0.6$) and strong coking ($I_C/I_{Fe} = 4.4$). Control measurements revealed that the strong K-reduction had occurred during the "regeneration" in water and not during the second production-rate measurement.

That water causes K depletion is understandable. If KOH is formed at the surface, this may evaporate eventually because its vapor pressure at the reaction temperature is about 0.1 mbar. That the depletion in the presence of EB is much slower means that carbon is necessary to avoid excessive K depletion. A coke layer may directly protect the substrate from K removal but in this case also the production rate would suffer. A result of K promotion is just that the equilibrium amount of coke is lowered. A more likely possibility is the formation of carbonate. It was proposed that K_2CO_3 formation is part of a cyclic process by which coke is removed. Oxidic K reacts with water to form KOH that – competing with desorption – reacts with carbon to form K_2CO_3. The cycle is closed by thermal decomposition of carbonate into CO_2, which goes off and K oxides that remain.[15,73]

Figure 11.12 summarizes the dependence of the initial production rate r_{in} and of the time constant of deactivation τ_{de} on the K content in terms of I_K/I_{Fe}.[65] Also, the initial rates for the unpromoted samples (Fe_2O_3, Fe_3O_4, Fe_2O_3 with O_2 in the feed) are included. The initial rate decreases and the time constant for deactivation increases with growing K content. Obviously, K prevents catalyst deactivation by reduction and coking but at the same time excess K reduces or blocks active surface sites. For $I_K/I_{Fe} > 2.7$, the surface is increasingly covered by $KFeO_2$.[14,41] The further decrease of the rate in this range proves that $KFeO_2$ is not the catalytically active phase. However, representing a K reservoir, it may be responsible for slow deactivation. For practical applications, slow deactivation may be the decisive property.

For low K content, the initial rate in Figure 11.12 tends towards the rate of Fe_2O_3 with O_2 in the feed, while the rate of Fe_2O_3 without O_2 is lower

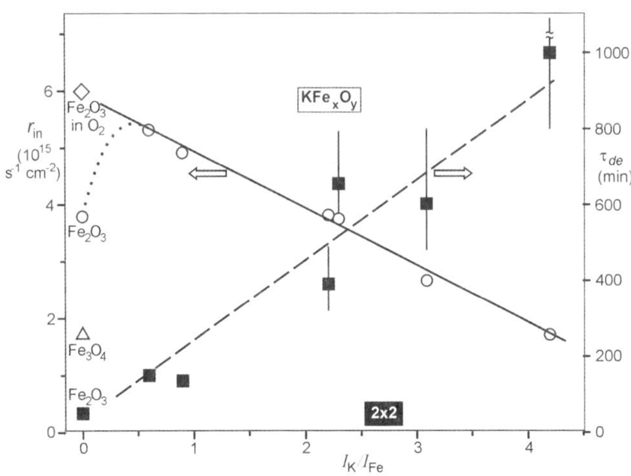

Figure 11.12 Initial ST production rate r_{in} and time constant for deactivation τ_{de} for samples with different initial K content in terms of the Auger peak height ratio I_K/I_{Fe}. The composition where the ordered (2×2) structure is formed is indicated.

(dotted curve). This confirms that Fe_2O_3 is partly deactivated before the first data point of the rate curve (Figure 11.9(b)) could be taken.

That potassium also prevents reduction of the catalyst was checked by using a catalyst with moderate K content ($I_K/I_{Fe} = 1.2$) for a production-rate measurement with EB but without water in the feed, *i.e.* under the same conditions as for the unpromoted film in Figure 11.10 (lower curve). After 45 min, the surface composition was checked by AES, revealing an essentially unchanged oxygen content on the promoted catalyst.

11.2.4 Microkinetic Modeling

After having proved that the role of the promoter potassium is to slow down catalyst deactivation by reduction and coking but does not actively take part in the catalytic reaction itself, microkinetic modeling could be restricted to the production-rate measurements on the unpromoted samples reviewed in Section 11.2.3.3. Since both Fe_2O_3 and Fe_3O_4 and even coke display catalytic activity and since both reduction of Fe_2O_3 to Fe_3O_4 and coking occurs during the course of the measurement, the number of reactions to be considered is large. Figure 11.13 displays them in graphical form. The partially reduced iron oxide surface is simulated as a coverage of Fe_3O_4 domains in a Fe_2O_3 matrix. The contribution of adsorption, desorption and surface reactions on both oxide phases are weighted according to their respective coverages.

The gigantic number of 26 single-step reactions with altogether 30 parameters had to be considered. But the procedure was divided in a clever way into three steps with only part of the total reaction scheme active in each of them. Further, frequency factors and adsorption energies for water, EB and ST on the three surface phases (Fe_2O_3, Fe_3O_4 and coke) had been determined experimentally as described in Section 11.2.2.2, so that 18 of the 30 parameters were known.

The inputs for modeling are the time dependence of the production rate under normal (Figure 11.9) and oxidative conditions (Figure 11.10, upper curve) and the information on composition and surface phases determined by surface-science methods before the production-rate measurement and after

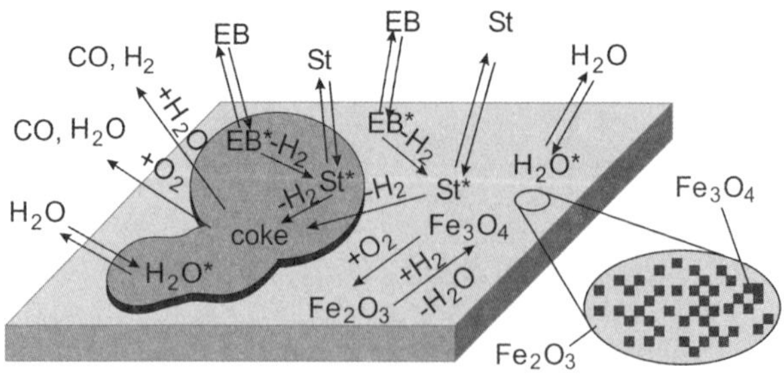

Figure 11.13 Mechanistic model of ethylbenzene dehydrogenation over iron oxide including all reactions considered in microkinetic modeling.

stopping it after different times on-stream. The letters A–L in Figures 11.9 and 11.10 indicate the positions where surface analysis was performed.

In a first step, the nonoxidative production rate and change of surface phase on Fe_3O_4 was modeled. Since no Fe_2O_3 phase existed and no oxygen was admitted, the decrease of production rate was ascribed to coking only. 5 parameters were unknown and could be fitted. In a second step, the non-oxidative production rate on Fe_2O_3 was fitted that includes the influence of oxide reduction to Fe_3O_4. A further 4 parameters could be fitted. Finally, the oxidative production rate was fitted resulting in the determination of the remaining 3 parameters. The solid lines in Figures 11.9 and 11.10 represent the result of fitting. The detailed discussion including sensitivity analysis and time dependence of phase composition can be found in refs. 70 and 74.

A crucial test of the results of modeling is presented in Figure 11.14. After oxidative dehydrogenation over Fe_2O_3 for more than 2 h, the oxygen was switched off and after 30 min switched on again. The observed decrease and re-establishment of the production rate could be modeled excellently using the parameters determined in the fitting procedure described above.

Also the rate measurements on a Fe_2O_3 pressed powder sample under normal and oxidative conditions was modeled, including the effects of pore diffusion and deactivation of the inner pore surface. Measurement and simulation agree remarkably concerning initial and final activity, while the dynamics of deactivation could not be fitted equally well.[70] Reasons may be that the experimental basis for pressed powder results was limited and that the model needs refinement. The value of the modeling is mainly to show that it is promising to extend it to porous materials.

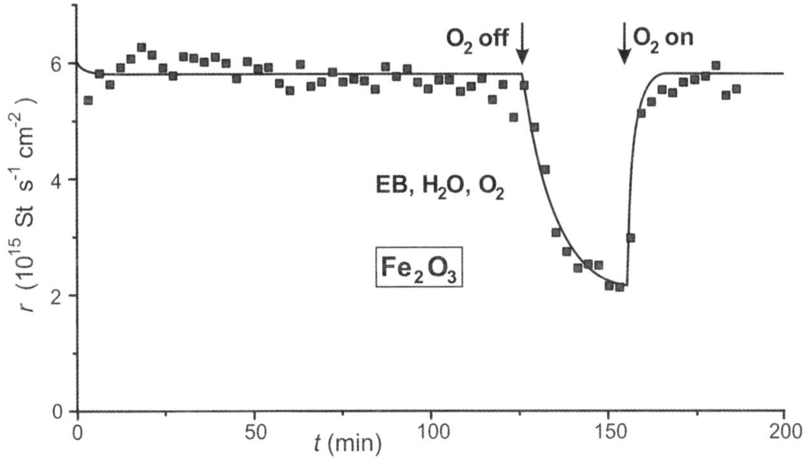

Figure 11.14 ST production rate over Fe_2O_3 under oxidative conditions (see Table 11.1). O_2 was switched off and on again at the indicated positions. Solid line: Rate calculated using the parameters deduced from microkinetic fitting of Figures 11.9 and 11.10.

11.3 Summary and Outlook

An in-depth study on a complex model catalysis system is reviewed. The model reaction is the dehydrogenation of ethylbenzene to styrene in the presence of steam over unpromoted and K-promoted iron oxides. The study consists of a systematic and detailed analysis of the structure, composition and adsorption properties of 10–20-nm thick epitaxially grown model catalyst films of the composition Fe_xO_y and KFe_xO_y, followed by *in-situ* reactivity experiments at different pressures of the reactive gases: 3.5×10^{-6} mbar, 0.6 mbar and 36 mbar, always at the reaction temperature of 870 K and always with EB:H_2O ratios similar to real catalysis. The bridge over the pressure gap thus has several supports and is quite safe. The studies were supplemented by microkinetic modeling using physicochemically meaningful parameters. Measurements and modeling are extended to unpromoted porous material to bridge the gap towards technical application.

In order to be catalytically active, the surface of unpromoted Fe_2O_3 has to contain defects. At sufficiently high pressures, defects are quickly created during reaction. Deactivation is accompanied by reduction to Fe_3O_4 and by coking. This agrees with powder catalyst studies but it had not been realized before that the initial activity was much higher and reached the level for promoted catalysts. By addition of low concentrations of oxygen, reduction and coking can be avoided and the high production rate level maintained. This may open a new reaction route using oxygen instead of promotion with K. It is an engineering problem to implement it.

On the pore-free model catalysts, the observed production rates could easily be referred to the active geometric surface. Production rates are in the range of 1 to 10×10^{15} styrene molecules per cm^2 and s. The corresponding turnover frequencies (TOF) are of the order of 10 per site and s or even higher if the concentration of active defects is low. The catalytic activity is thus quite high. The limitations in technical catalysis are the accessibility of the catalyst surface and matter and energy transport.

The initial activity on unpromoted Fe_2O_3 is equal to or even higher than on promoted films. But promotion slows down deactivation. It prevents coking or accelerates its removal and it prevents catalyst reduction. But the catalytically active sites appear to be the same and to be related to Fe^{3+}. This seems to confirm an old proposal for polycrystalline catalysts[16,75] based on apparent activation energies. However, in these cases the activity of the unpromoted catalyst was an order of magnitude lower than for the promoted catalyst. According to the results reviewed here, the unpromoted catalyst in the former studies[16,75] must have been deactivated, *i.e.* reduced and coked. The model catalysis experiments showed indeed that the activity of the deactivated catalyst is about an order of magnitude lower but not zero. Most likely it is related to the catalytic activity of carbonaceous deposits (coke).

The importance of disorder on promoted catalysts could not be studied but the similarity of the initial activities of promoted and unpromoted model catalysts suggests the same mechanism.

Addition of oxygen does not change the initial activity on Fe_2O_3, only its deactivation behavior. Also, oxygen obviously does not influence the catalytic mechanism, which is thus not oxidative in the sense that oxygen directly takes part in the catalytic reaction. It rather balances the effect of catalyst reduction by released hydrogen and it oxidizes deposited coke. It stabilizes the catalyst in the fully oxidized state Fe_2O_3, while water can only stabilize Fe_3O_4. If not even water is present, full reduction to metallic iron Fe^0 occurs. The role of water (or oxygen) as reoxidation agents proves that a Mars–van Krevelen mechanism is active on unpromoted catalysts. Whether it accounts for the complete reaction is not sure since hydrogen-desorption studies showed that H_2 desorption (without substrate reduction) is also possible. It is not clear if a MvK mechanism is active also on promoted catalysts since potassium effectively slows down reduction and complete reduction was not observed.

Measurements on unpromoted Fe_2O_3 powder samples showed that the inner pore surface hardly contributes to the production rate. Most likely it is quickly deactivated by polymerization of ST followed by coking. Corresponding results for promoted powder samples do not exist but it is likely that also in technical catalysis the inner pore surface is fairly inactive so that the pellet bulk serves mainly as K reservoir.

The active site requires Fe^{3+}. Although also containing Fe^{3+}, Fe_3O_4 is not equally active. It could be that a pure Fe^{3+} environment is necessary as present in Fe_2O_3, $K_2Fe_{22}O_{34}$ and $KFeO_2$. But since $KFeO_2$ is obviously not more active than Fe_3O_4, it is more likely that the different binding energies of EB and ST on the different model catalysts are decisive. Due to iron in the top layer of Fe_3O_4, both EB and ST are bound so strongly that they block the surface. In contrast, the interaction with Fe_2O_3 and $K_xFe_{22}O_{34}$ is sufficient to bind the EB molecule long enough to the surface to enable dehydrogenation but neither EB nor ST so strongly that they block the surface.

A model for the catalytic cycle has been proposed[51,55] that uses the observations that the right adsorption strength and – at least on unpromoted Fe_2O_3 – defects are necessary for high production rates. A mobile adsorption state *via* the π-system of the benzene ring is responsible for holding the molecules long enough on the surface. When meeting a defect site exposing basic oxygen that attracts the H atoms, the ethyl group may be dehydrogenated. This reduces two Fe^{3+} ions in the vicinity to Fe^{2+}, which explains why Fe^{3+} is necessary and why iron with its variable valency is essential. The formed styrene desorbs. The hydrogen desorbs either directly as H_2 or in the form of H_2O by consuming a substrate O atom that is later replaced by dissociation of water or by reaction with O_2 from the feed (MvK mechanism).

It is likely that a K carbonate forms on the promoted catalyst but its structure is unknown. Possibly, the local structure contains O above Fe^{3+} as in the case of unpromoted Fe_2O_3 while surface K carries a carbonate group. This would agree with the observation that surface K reduces the activity by reducing the concentration of active sites.

In previous studies, $KFeO_2$ was observed to exist on the active promoted catalyst and to be essential for high and long-term activity. It was therefore considered to represent the catalytically active phase,[18–20] while K–β-ferrite $K_2Fe_{22}O_{34}$ as bulk phase was proposed to represent a K reservoir for the re-establishment of $KFeO_2$. The model catalyst studies[14,41] confirm that $KFeO_2$ forms when a layer of K deposited on Fe_3O_4 (or Fe_2O_4) is annealed at 700 K. Even at 870 K, a thin $KFeO_2$ layer is left over a possibly K-deficient K–β-ferrite. However, such a $KFeO_2$-containing sample is much less active than samples with low K content. Most likely, the roles of active and reservoir phase are inverted which intuitively seems plausible: K–β-ferrite is the active phase, while $KFeO_2$ as K reservoir re-establishes stoichiometric K–β-ferrite under the conditions of continuous slow K-removal. The existence of $KFeO_2$ would thus "buffer" the K–β-ferrite phase against K depletion by H_2O.

Initial activity is high but deactivation is fast when only little K is deposited or when K is removed by steaming. Most likely, no complete Fe–O-K surface layer can form then, which according to the observations needs at least a bulk stoichiometry corresponding to $K_xFe_{22}O_{34}$ ($x \geq 0.67$).

In the steady state, the model catalysts and surely also real catalysts are always totally (unpromoted samples) or partially (K-promoted samples) covered by carbonaceous deposits. It has been demonstrated for unpromoted catalysts that this coke is also catalytically active. On the promoted catalysts, the catalytic role of coke is not well established. K is important for coke removal but some coke is always present. Oxidative dehydrogenation studies of EB over carbon catalysts have shown that graphite, nanofilaments and nano-onions are highly active,[71,72] while soot is not.[76] It is not known what kind of carbon deposits are formed in the absence and in the presence of K, and even less what their activity in presence of H_2O or O_2 is.

For a number of mechanistic details, it was important that the model catalysts consisted of a limited amount of material in the form of thin films. It could easily be shown that the oxygen liberated during reduction of Fe_2O_3 is not responsible for the main dehydrogenation reaction path of ethylbenzene in a stoichiometric reaction but that the main reaction is catalytic.

Also, the process of K depletion could only be observed and assigned so clearly because the films contained a limited amount of K. At reaction temperature, K diffusion over distances of the film thickness is fast. Its distribution is always near to equilibrium and leads to K accumulation in the topmost layer. As long as the bulk concentration is high enough to maintain this surface composition, deactivation is slow. Lower K concentrations even increase the initial activity, but this is quickly compensated by a faster deactivation. Not unexpectedly, the presence of pure water causes fast K depletion by formation of KOH that evaporates. A surprising observation was that this depletion is much slower when ethylbenzene is admitted. Obviously, a surface compound is formed in the presence of EB and its reaction products (ST, CO_2, coke) that attenuates K removal. A candidate is K_2CO_3 that is also likely to be involved in the coke removal process.

Since potassium does not actively take part in the catalytic reaction, microkinetic modeling could be restricted to unpromoted samples. Microkinetic means that modeling was performed using physicochemically meaningful parameters like adsorption energies and frequency factors for desorption (determined independently) that together determine adsorption-desorption kinetics. Also, the experimentally determined catalyst composition (oxidation state, coke coverage) before, during and after reaction was used. The activity of three phases (Fe_2O_3, Fe_3O_4 and coke) in the presence of up to four gases (EB, St, H_2O, O_2) had to be considered and the kinetics of phase conversion $Fe_2O_3 \leftrightarrow Fe_3O_4$ and of coke formation had to be modeled. For these reactions, parameters were determined by fitting the experimentally determined deactivation kinetics.

This enormous task was tackled in a clever way by dividing it into three steps with increasing complexity. This is probably the first successful modeling of a complex system including phase transformations. Finally, even the activity of a porous sample was modeled, which needed the inclusion of pore diffusion. It was possible to model the experimental observation that pores contribute only weakly to the production rate because of quick coking. The feasibility of modeling a practically relevant sample was demonstrated.

The presented study demonstrates that the model-catalysis approach using single crystalline thin films of complex composition is possible and yields essential contributions for a deeper understanding of catalysis. But it demonstrates also that this has a long way to go. And it works only for systems that can be prepared as thin films and that do not have too high vapor pressures. Improvements of technological processes on the basis of model-catalytic results are feasible.

References

1. G. Ertl and J. Küppers, *Low Energy Electrons and Surface Chemistry*, 2nd edn, VCH, Weinheim, 1985; G. A. Somorjai, *Surface Chemistry and Catalysis*, John Wiley & Sons, Inc., New York, 1994.
2. G. Ertl, J. Knözinger and J. Weitkamp (ed.), *Handbook of Heterogeneous Catalysis*, Wiley-VCH, Weinheim, 1997.
3. D. W. Blakely, E. I. Kozak, B. A. Sexton and G. A. Somorjai, *J. Vac. Sci. Technol.*, 1976, **13**, 1091.
4. W. Reichl, G. Rosina, G. Rupprechter, C. Zimmermann and K. Hayek, *Rev. Sci. Instrum.*, 2000, **71**, 1495.
5. A. Knop-Gericke, M. Hävecker, T. Schedel-Niedrig and R. Schlögl, *Top. Catal.*, 2001, **15**, 27.
6. V. I. Bukhtiyarov, M. Hävecker, V. V. Kaichev, A. Knop-Gericke, R. W. Mayer and R. Schlögl, *Phys. Rev. B*, 2003, **67**, 235422.
7. J. Haw, *In-Situ Spectroscopy in Heterogeneous Catalysis*, Wiley-VCH, Weinheim, 2002.

8. B. M. Weckhuysen, *In-situ Spectroscopy of Catalysts*, American Scientific Publishers, Stephenson Ranch, 2004.
9. H.-J. Freund, N. Ernst, M. Bäumer, G. Rupprechter, J. Libuda, H. Kuhlenbeck, T. Risse, W. Drachsel, K. Al-Shamery and H. Hamann, *Surface Chemistry and Catalysis*, ed. A. F. Carley, P. R. Davis, G. J. Hutchings, M. S. Spencer, Kluwer Academic/Plenum Publishers, New York, 2002.
10. K. Kochloefl, in *Handbook of Heterogeneous Catalysis*, ed. G. Ertl, J. Knözinger, J. Weitkamp, Vol. 5, VCH Weinheim, 1997, p. 2151.
11. www.styreneforum.org.
12. R. M. Cornell and U. Schwertmann, *The Iron-oxides*, VCH, Weinheim, 1996.
13. G. Ketteler, W. Weiss, W. Ranke and R. Schlögl, *Phys. Chem. Chem. Phys.*, 2001, **3**, 1114.
14. G. Ketteler, W. Ranke and R. Schlögl, *J. Catal.*, 2002, **212**, 104.
15. E. H. Lee, *Catal. Rev.*, 1973, **8**, 285.
16. T. Hirano, *Appl. Catal.*, 1986, **26**, 65.
17. T. Hirano, *Appl. Catal.*, 1986, **26**, 81.
18. T. Hirano, *Appl. Catal.*, 1986, **28**, 119.
19. M. Muhler, J. Schütze, M. Wesemann, T. Rayment, A. Dent, R. Schlögl and G. Ertl, *J. Catal.*, 1990, **126**, 339.
20. M. Muhler, R. Schlögl and G. Ertl, *J. Catal.*, 1992, **138**, 413.
21. G. Centi, in *Catalysis from A to Z*, ed. B. Cornils, W. A. Herrmann, R. Schlögl and C.-H. Wong, Wiley-VCH, Weinheim, 2000.
22. H. Miura, H. Kawai and T. Taguchi, *Studies Surf. Sci. Catal.*, 1994, **90**, 461.
23. A. E. Lisovskii and C. Aharoni, *Catal. Rev. Sci. Eng.*, 1994, **36**, 25.
24. G. E. Vrieland and P. G. Menon, *Appl. Catal.*, 1991, **31**, 1.
25. Y. Iwasawa, H. Nobe and S. Ogasawara, *J. Catal.*, 1973, **31**, 444.
26. G. Mestl, N. I. Maksimova, N. Keller, V. V. Roddatis and R. Schlögl, *Angew. Chem. Int. Ed.*, 2001, **40**, 2066.
27. J. Zhang, D. S. Su, A. Zhang, D. Wang, R. Schlögl and C. Hébert, *Angew. Chem. Int. Ed.*, 2007, **46**, 7319.
28. J. J. Delgado, D. S. Su, G. Rebmann, N. Keller, A. Gajovic and R. Schlögl, *J. Catal.*, 2006, **244**, 126.
29. J. J. Delgado, R. Vieira, R. Rebmann, D. S. Su, N. Keller, M. J. Ledoux and R. Schlögl, *Carbon*, 2006, **44**, 809.
30. D. S. Su, N. Maksimova, J. J. Delgado, N. Keller, G. Mestl, M. J. Ledoux and R. Schlögl, *Catal. Today*, 2005, **102**, 110.
31. M. F. R. Pereira, J. L. Figueiredo, J. J. M. Orfao, P. Serp, P. Kalck and Y. Kihn, *Carbon*, 2004, **42**, 2897.
32. M. F. R. Pereira, J. J. M. Orfao and J. L. Figueiredo, *Carbon*, 2002, **40**, 2393.
33. P. Li, T. J. Zhao, J. H. Zhou, Z. J. Sui, Y. C. Dai and W. K. Yuan, *Carbon*, 2005, **43**, 2701.
34. D. B. Tagiyev, G. O. Gasimov and M. I. Rustamov, *Catal. Today*, 2005, **102**, 197.

35. Y. Joseph, M. Wühn, A. Niklewski, W. Ranke, W. Weiss, C. Wöll and R. Schlögl, *Phys. Chem. Chem. Phys.*, 2002, **2**, 5314.
36. W. Weiss and W. Ranke, *Prog. Surf. Sci.*, 2002, **70**, 1.
37. M. Ritter, H. Over and W. Weiss, *Surf. Sci.*, 1997, **371**, 245.
38. G. Ketteler and W. Ranke, *Mater. Res. Soc. Symp. Proc.*, 2003, **775**, p 6.1.1.
39. G. Ketteler and W. Ranke, *J. Phys. Chem. B.*, 2003, **107**, 4320.
40. A. N. Koveshnikov, R. H. Madjoe, J. Karunamuni, R. L. Stockbauer and R. L. Kurtz, *J. Appl. Phys.*, 2002, **87**, 5929.
41. Y. Joseph, G. Ketteler, C. Kuhrs, W. Ranke, W. Weiss and R. Schlögl, *R. Phys. Chem. Chem. Phys.*, 2001, **3**, 4141.
42. C. S. Fadley, M. A. Van Hove, Z. Hussain and A. P. Kaduwela, *J. Electron Spectrosc. Relat. Phenom.*, 1995, **75**, 273.
43. Sh. K. Shaikhutdinov, R. Meyer, D. Lahav, M. Bäumer, T. Klüner and H.-J. Freund, *Phys. Rev. Lett.*, 2003, **91**, 076102.
44. M. Ritter and W. Weiss, *Surf. Sci.*, 1999, **432**, 81.
45. G. Ketteler, W. Weiss and W. Ranke, *Surf. Rev. Lett.*, 2001, **8**, 661.
46. Sh. K. Shaikhutdinov and W. Weiss, *Surf. Sci.*, 1999, **432**, L627.
47. Sh. K. Shaikhutdinov, W. Weiss and R. Schlögl, *Appl. Suf. Sci.*, 2000, **161**, 497.
48. C. Kuhrs, *PhD thesis*, Freie Universität Berlin, 2000.
49. W. Ranke and W. Weiss, *Surf. Sci.*, 1998, **414**, 236.
50. W. Ranke and Y. Joseph, *Phys. Chem. Chem. Phys.*, 2002, **4**, 2483.
51. C. Kuhrs, Y. Arita, W. Weiss, W. Ranke and R. Schlögl, *Top. Catal.*, 2001, **14**, 111.
52. S. K. Shaikhutdinov, Y. Joseph, C. Kuhrs, W. Ranke and W. Weiss, *Faraday Discuss.*, 1999, **114**, 363.
53. R. I. Masel, *Principles of Adsorption and Reaction on Solid Surfaces*, John Wiley & Sons, New York, 1996.
54. Y. Joseph, W. Ranke and W. Weiss, *J. Phys. Chem. B*, 2000, **104**, 3224.
55. W. Weiss and R. Schlögl, *Top. Catal.*, 2000, **13**, 75.
56. U. Leist, W. Ranke and K. Al-Shamery, *Phys. Chem. Chem. Phys.*, 2003, **5**, 2435.
57. M. E. Grillo, M. W. Finnis and W. Ranke, *Phys. Rev. B*, 2008, **77**, 075407.
58. W. Huang, W. Ranke and R. Schlögl, *J. Phys. Chem. B*, 2005, **109**, 9202.
59. W. Huang and W. Ranke, *Surf. Sci.*, 2006, **600**, 793.
60. W. Huang, W. Ranke and R. Schlögl, *J. Phys. Chem. B*, 2007, **111**, 2198.
61. D. Zscherpel, W. Weiss and R. Schlögl, *Surf. Sci.*, 1997, **382**, 326.
62. W. Ranke and W. Weiss, *Surf. Sci.*, 2000, **465**, 317.
63. W. Weiss, D. Zscherpel and R. Schlögl, *Catal. Lett.*, 1998, **52**, 215.
64. C. Kuhrs, M. Swoboda and W. Weiss, *Top. Catal.*, 2001, **15**, 13.
65. O. Shekhah, W. Ranke and R. Schlögl, *J. Catal.*, 2004, **225**, 56.
66. O. Shekhah, W. Ranke, A. Schüle, G. Kolios and R. Schlögl, *Angew. Chem. Int. Ed.*, 2003, **42**, 5760.
67. A. Muan, *Am. J. Sci.*, 1958, **256**, 171.
68. G. R. Meima and P. G. Menon, *Appl. Catal. A*, 2001, **212**, 239.

69. W. Ranke, A. Schüle, O. Shekhah, G. Kolios, U. Nieken, R. Schlögl, *Book of extended abstracts*, 10th Int. Symp. Catal. Deactivation, Berlin, 05–08 Jan. 2006, DECHEMA e.V., Frankfurt am Main, 2005.
70. A. Schüle, U. Nieken, O. Shekhah, W. Ranke, R. Schlögl and G. Kolios, *Phys. Chem. Chem. Phys.*, 2007, **9**, 3619.
71. G. Mestl, N. I. Maksimova, N. Keller, V. V. Roddatis and R. Schlögl, *Angew. Chem. Int. Ed.*, 2001, **40**, 2066.
72. N. Keller, N. I. Maksimova, V. V. Roddatis, M. Schur, G. Mestl, Y. V. Butenko, V. L. Kuznetsov and R. Schlögl, *Angew. Chem. Int. Ed.*, 2002, **41**, 1885.
73. K. Coulter, D. W. Goodman and R. G. Moore, *Catal. Lett.*, 1995, **31**, 1.
74. A. Schüle, O. Shekhah, W. Ranke, R. Schlögl and G. Kolios, *J. Catal.*, 2005, **231**, 172.
75. W. P. Addiego, C. A. Estrada, D. W. Goodman, M. P. Rosynek and R. G. Windham, *J. Catal.*, 1994, **146**, 407.
76. N. I. Maksimova, V. V. Roddatis, G. Mestl, M. Ledoux and R. Schlögl, *Eurasian Chem. Tech. J.*, 2000, **2**, 231.

Section III

Supported Vanadium-Oxide Systems and Mechanistic Studies

CHAPTER 12

Highly Dispersed Vanadium-Oxide Catalysts

CHRISTIAN HESS

Technische Universität Darmstadt, Eduard-Zintl-Institut für Anorganische und Physikalische Chemie, Petersenstr. 20, 64287 Darmstadt, Germany

12.1 Introduction

Supported vanadium-oxide catalysts play a tremendous role in the commercial production of important chemicals *via* selective oxidation (*e.g.* phthalic anhydride, SO_3) and the removal of hazardous pollutants by selective catalytic reduction (*e.g.* NO_x emissions from power plans).[1–3] Commonly used support materials are titania, alumina and silica. For example, titania-supported vanadium oxide is employed for the selective oxidation of *o*-xylene to phthalic anhydride,[4] whereas commercial DeNOx catalysts are based on titania-supported mixed vanadium and tungsten oxide systems.[5] Other examples of commercial catalysts based on supported vanadium oxide include silica-supported vanadium-oxide systems that serve as catalyst for SO_2 oxidation to SO_3 as well as alumina- or silica-supported mixed vanadium and molybdenum oxides, which are used for the selective oxidation of benzene to maleic anhydride.[1]

Technically, mixed-metal oxide systems may also be considered as supported systems if there are significant differences between surface and bulk regarding structure and composition. In addition, the transformation of the catalyst precursor to the working catalyst may be considered to involve the generation of a (supported) active phase driven by the chemical potential of the reactants.

RSC Nanoscience & Nanotechnology No. 19
Nanostructured Catalysts: Selective Oxidations
Edited by Christian Hess and Robert Schlögl

Published by the Royal Society of Chemistry, www.rsc.org

Thus, as vanadium is an essential component of many efficient (bulk) selective oxidation catalysts (*e.g.* VPO, MoVTeNbO) a likely scenario may be based on the formation of supported vanadium oxide as active phase (see Chapter 15).

A detailed understanding of the mode of operation of supported vanadium-oxide catalysts requires profound knowledge of the molecular structure of the active site. However, despite numerous studies even the structure of the surface vanadium-oxide species without being exposed to reaction conditions is a matter of debate. Recently, the discussion has focused in particular on the dehydrated state of vanadium oxide supported by alumina and silica.[6–9] Important results of this discussion are (i) the reassignment of a vibrational feature at $\sim 950\,cm^{-1}$ (previously assigned to V–O–V and used as an indicator for polymeric species) to an interface stretching mode,[7] (ii) the spectroscopic evidence for V–O–V in the case of silica-SBA-15-supported vanadium-oxide systems implying the presence of associated vanadium-oxide species (in contrast to the previously accepted picture of exclusively monomeric species).[10,11] These efforts are justified by the fact that the dehydrated vanadium-oxide structure serves as a well-defined starting point for *in-situ* spectroscopic studies.

Knowledge of the active-site structure is the key to a profound understanding of support effects in supported vanadium-oxide catalysts. A strong influence of the support material on the catalytic performance has been reported, *e.g.* for the selective oxidation of methanol and the oxidative dehydrogenation of propane.[12–20] The origin of this behavior may either be attributed to intrinsic support properties possibly resulting in specific interactions with the supported phase (metal-oxide support effect) or differences in the vanadium-oxide structure(s) present on the different support materials, or even a combination of both effects. The isoelectric points of common support materials such as silica, titania or alumina cover a broad pH range leading to very different pH values near the surface. As the vanadium-oxide structure depends on pH (Pourbaix diagram[21]), a variation of the surface vanadium-oxide precursor is expected upon impregnation on different support materials. Thus, new approaches for the preparation of (stable) supported vanadium-oxide catalysts need to be developed that allow for control of the structure of the deposited vanadium-oxide species. One strategy pursued is to prepare molecular compounds of supported vanadium-oxide systems, *e.g.* vanadium-substituted polyhedral oligomeric silsesquioxane and vanadium calixarene complexes.[22–24] Alternatively, organometallic precursor species have been grafted onto oxide support materials followed by thermal activation.[25] Recently, we have developed a synthesis scheme based on grafting/ion exchange using silica SBA-15 as support material.[8,26] As the reactivity of hydroxyl groups depends on the intrinsic support properties,[27] they are first replaced by ammonium groups as reactive centers. Subsequently, using ion exchange the vanadium precursor is introduced into the functionalized porous silica matrix. All approaches outlined above have been successfully applied to the formation of silica-supported vanadium-oxide catalysts. However, future experiments will show whether their transfer to other support materials results in structurally equivalent vanadium-oxide species at the surface.

Dispersion is an important property of supported catalysts. Changes in the dispersion of metal-oxide species are often linked with structural changes. For example, it has been reported that in the oxidative dehydrogenation (ODH) of propane a decrease in the dispersion of silica-supported vanadium-oxide species (by formation of crystalline V_2O_5 at higher loadings) leads to a drop in propene selectivity.[28] The dispersion may also be influenced by products or reactants. Water as one of the products is omnipresent in selective oxidation reactions and often even added to the feed to improve the catalyst performance. Under reaction conditions hydrothermal processes may lead to polymerization/depolymerization of an initial vanadium-oxide condensate. A similar scenario may be envisioned in the presence of methanol. In fact, the effect of water adsorption (followed by calcination) as well as the exposure to methanol oxidation reaction conditions has been studied indirectly by monitoring the presence of crystalline V_2O_5 using visible Raman spectroscopy.[29,30] However, only a few studies have directly addressed the changes in the dispersion of supported vanadium-oxide systems in the presence of water or methanol.[31]

Based on the above brief discussion the following fundamental questions regarding supported vanadium-oxide catalysts for selective oxidation reactions may be raised:

- What is the molecular structure of the active site?
- Is the active-site structure influenced by the presence of reaction products such as water or by the support material?
- Do reactants or products have an influence on the vanadium-oxide dispersion?
- Can the catalytic performance be improved by functionalization of the support material?

To tackle these problems catalyst systems with variable but controlled chemical complexity are prepared. Focus is put on vanadium oxides as the active phase due to their crucial role in many selective oxidation catalysts. In addition, accelerated progress is expected to be facilitated by the large number of studies on vanadium-oxide catalysts. Tailored catalytic materials are prepared by controlled synthesis on the basis of spectroscopic characterization, while the activation and specific function are studied by *in-situ* spectroscopic methods. The active phase is supported on a high-surface area mesoporous support material, which allows for (i) a higher dispersion of the active surface species (ii) a better spectroscopic traceability due to the higher concentration of species as compared to most conventional support materials. Besides the questions mentioned above, supported systems may also open the possibility to gain insight into the role of deep lattice oxygen in selective oxidation reactions and to contribute to the question of the active sites in bulk systems if the reactivity behavior can be successfully mimicked (see Chapter 15).

In this contribution, our current state of knowledge regarding the above questions is summarized extending previous discussions of the outlined approach.[6,8] In the first part, after a brief literature survey on the synthesis of (silica-supported) highly dispersed vanadium-oxide catalysts, a detailed description of the grafting/ion-exchange approach is given. A synthesis mechanism is proposed based on results from spectroscopic characterization and thermal analysis. In the second part, the influence of water on the structure and dispersion of supported vanadium-oxide catalysts is discussed. To this end, particular attention will be given to recent findings regarding the structure of silica-supported vanadium oxide as well as the application of a new multiple *in-situ* experimental setup. In the final part, the reactivity behavior of supported vanadium oxides in methanol selective oxidation and propane oxidative dehydrogenation is reviewed. In each case, a literature survey is followed by a description of recent results from our own work.

12.2 Synthesis of Highly Dispersed Vanadium Oxide

For the preparation of supported catalysts with a high dispersion of transition-metal oxides the use of high-surface area support materials is essential. Nanostructured materials offer high surface areas. This is illustrated by a simple calculation for amorphous SiO_2 materials by using a typical density value of 2.2 g/cm^3, which shows that spherical particles with radius $r = 1$ nm exhibit a specific surface area of ~ 1400 m^2/g.[32] A similar increase in specific surface area results if holes are drilled into a solid, thus creating a porous matrix. In fact, mesoporous silica MCM and SBA as representatives of such porous materials, exhibit internal surface areas of typically >800 m^2/g. Other high surface area materials available for the preparation of supported catalysts include alumina, titania, zeolites, alumosilicates and active carbon.[33] However, in contrast to many other high surface area materials, MCM-41 and SBA-15 also possess a well-defined structure. As the hydrothermal stability of SBA-15 exceeds that of the thinner-walled MCM-41 materials,[34] SBA-15 appears as the silica support material of choice for preparation of highly dispersed vanadium-oxide catalysts.

As shown in Figure 12.1 SBA-15 is based on uniform channels (left image) with a pore diameter of typically 7 nm and a narrow pore-size distribution (FWHM: 0.35 nm), the channels are hexagonally arranged (right image).[35,36] The pore diameter can easily be tuned by variation of the synthesis conditions.[37] While SBA-15 as compared to other mesoporous silica materials exhibits a good hydrothermal stability *per se*,[34] it can be further improved by the presence of internal porous plugs or heteroatoms.[38] Recently, we discovered that a significant increase in hydrothermal stability can also be achieved by simple postsynthesis surface modification, thus avoiding pore blockage or heteroatom incorporation (see Section 12.5.2).[39]

Supported metal-oxide catalysts with a high dispersion of metal-oxide species have been prepared by a variety of synthesis procedures, which include impregnation, ion exchange, grafting, coprecipitation and thermal spreading.[40] Regarding silica systems, highly dispersed vanadium-oxide species have been obtained for example by incipient wetness impregnation using high surface area

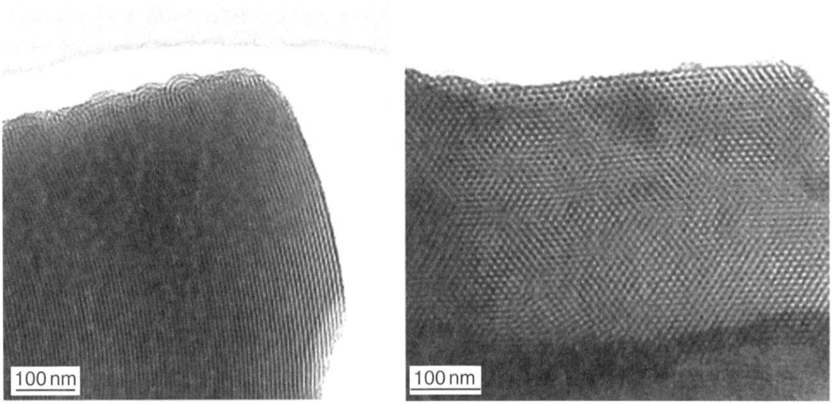

Figure 12.1 TEM images of mesoporous silica SBA-15 in a (left) cross-sectional view and (right) top view.

silica materials (Cabosil EH-5) and vanadyl triisopropoxide.[41] Aiming at uniform vanadium-oxide species, single-source tri(alkoxy)siloxy precursor compounds were grafted onto the surface of mesoporous silica SBA-15 followed by thermal decomposition of the precursor.[42,43] Alternatively, MCM-48 and SBA-15-supported vanadium-oxide systems have been prepared by the molecular designed dispersion (MDD) technique using $VO(acac)_2$.[44,45] A high dispersion of vanadium-oxide species is also accessible by atomic layer deposition (ALD) by sequentially exposing the silica surface to volatilized triisopropoxide and oxygen.[46] As an alternative to the above synthesis approaches based on post-treatment of silica, direct synthesis has been used to prepare, for example, vanadium-containing MCM-41 materials.[47]

For preparation of silica-supported catalysts with a high dispersion of vanadium oxide a grafting/ion-exchange approach (described in detail below) has proved very powerful due to its simplicity.[8] In a typical synthesis using SBA-15 as support material the maximum experimental dispersion of vanadium is achieved at $\sim$7.2 wt% V (2.3 V/nm^2), which is comparable to values previously achieved by atomic-layer deposition or multiple incipient wetness impregnation.[41,46] Thus, this approach allows for a very homogeneous distribution of vanadium oxide within the pores of the mesoporous matrix in a single wet-chemical synthesis run. Importantly, at these loadings the mesoporous structure of SBA-15 is fully preserved, as evidenced by TEM and also corroborated by integral methods such as nitrogen adsorption–desorption and XRD.[26]

Figure 12.2 summarizes the reaction steps involved in the grafting/ion-exchange approach. It consists of (i) functionalization of the silica surface by grafting of 3-aminopropyltrimethoxysilane (APTMS), and subsequent treatment with aqueous HCl solution resulting in the formation of the corresponding ammonium salt (APTMS-SBA-15), (ii) anion exchange of a suitable vanadium precursor, *e.g.* decavanadate ($V_{10}O_{28}^{6-}$) from water soluble butylammonium decavanadate, leading to the formation of a decavanadate-containing

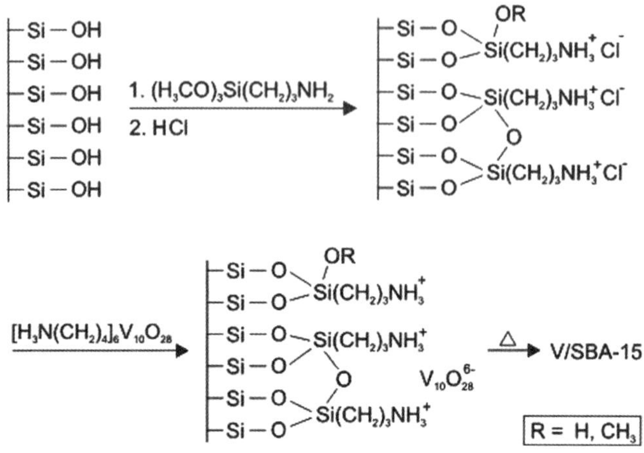

Figure 12.2 Reaction mechanism for the controlled synthesis of silica-SBA-15-supported vanadium oxide.[8]

SBA-15 matrix (deca-SBA-15), and (iii) thermal decomposition of the precursor material yielding the final supported vanadium-oxide catalyst (V/SBA-15).[48]

Details regarding the reaction mechanism and the structure of the reaction intermediates were obtained on the basis of spectroscopic characterization and thermal analysis. In the following, a few key results will be discussed. Details can be found elsewhere.[48] Figure 12.3 shows a series of low-frequency Raman spectra obtained during the synthesis of 2.7 wt% V/SBA-15 using 532-nm laser excitation. A sample containing 7.2 wt% V, which corresponds to the maximum vanadium loading using the above procedure without observation of crystalline V_2O_5, gave comparable results. The spectrum of blank SBA-15 (spectrum a) shows Raman features around 485, 600 and 810 cm^{-1}, which are attributed to cyclic tetrasiloxane rings,[49] cyclic trisiloxane rings and symmetrical Si–O–Si stretching, respectively.[50–52] The Raman band around 977 cm^{-1} is assigned to the Si–OH stretching vibration of surface hydroxyl groups.[50–52] Functionalization of the surface of SBA-15 with APTMS followed by treatment in aqueous HCl results in the appearance of new major bands at 940 and 1040 cm^{-1} (spectrum b), which are assigned to CH_2 rocking and C–C stretch vibrations of the propylammonium group, respectively.[53,54] Incorporation of decavanadate into the channels of the functionalized mesoporous matrix leads to the formation of deca-SBA-15 (spectrum d), which is characterized by Raman bands at 596, 939, 991 and 1040 cm^{-1}, respectively. Qualitatively, spectrum d can be considered as a superposition of spectra b (APTMS-SBA-15), and c (butylammonium decavanadate), which strongly suggests that decavanadate is incorporated intact into the porous matrix. After calcination of deca-SBA-15 bands at 520, 656, 707, 998 and 1021 cm^{-1}, respectively, were observed. They can be assigned to a hydrated vanadium oxide gel due to the close resemblance with the Raman bands observed for

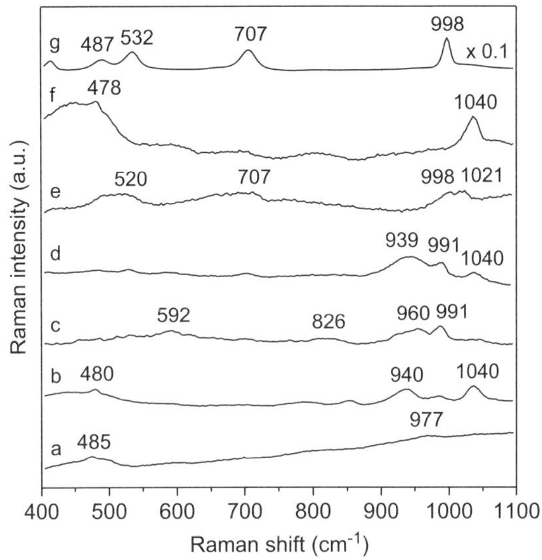

Figure 12.3 Room-temperature Raman spectra of (a) SBA-15, (b) APTMS-SBA-15, (c) butylammonium decavanadate, (d) deca-SBA-15, (e) 2.7 wt% V/SBA-15 under ambient conditions, (f) dehydrated 2.7 wt% V/SBA-15, and (g) crystalline V_2O_5. Spectra are offset for clarity.[48]

xerogels of type $V_2O_5 \cdot nH_2O$.[55,56] Dehydration leads to a completely different spectrum (spectrum f), which is dominated by an asymmetric band around $1040\,cm^{-1}$. Its main contribution is characteristic of the V=O stretching vibration of vanadium in a tetrahedral coordination (see Section 12.3.2).[57–60] For dehydrated 2.7 wt% V/SBA-15 (7.2 wt% V/SBA-15), no Raman bands due to (micro)crystalline V_2O_5 (spectrum g) were observed, demonstrating the highly dispersed state of the anchored vanadium-oxide species.

To gain insight into the calcination conditions necessary for complete removal of the organic framework thermal analysis was employed. The left panel of Figure 12.4 shows the evolution of the weight loss and the differential scanning calorimetry (DSC) traces during thermal decomposition of deca-SBA-15 (2.7 wt% V) in 20% oxygen in argon at a heating rate of 5 K/min. The right panel displays the corresponding signals from mass spectrometry (MS) for $m/z = 18$, $m/z = 27$, $m/z = 28$, $m/z = 30$ and $m/z = 44$. They are assigned to water, C_2H_3, CO/N_2, NO/N_2O and CO_2/N_2O, respectively. The C_2H_3 fragment represents a large number of C_xH_y species detected during the decomposition process, which show a similar intensity evolution. The main sharp weight loss observed at 85 °C is correlated with a strong endothermic DSC signal and the appearance of a signal at $m/z = 18$ indicating the evaporation of water from the silica matrix. The DSC measurements indicate exothermic decomposition steps around 240, 290 and 500 °C, respectively. Based on the MS signals they can be assigned to decomposition and combustion processes of the grafted ammoniumpropyl moieties. In summary, the results from thermal analysis strongly

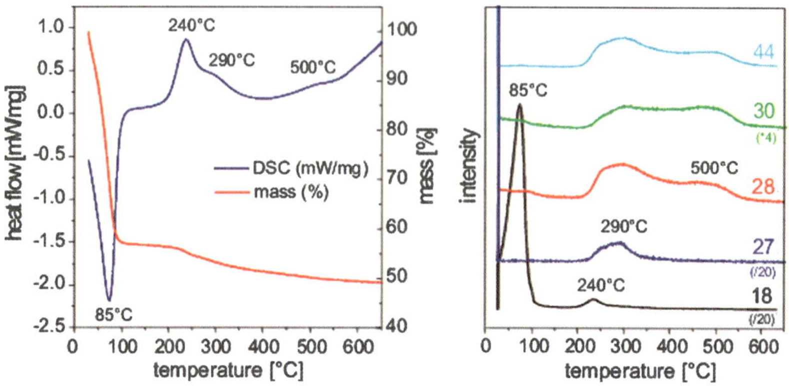

Figure 12.4 Left panel: Evolution of the differential scanning calorimetric trace and weight loss during decomposition of deca-SBA-15 in 20% oxygen in argon at a heating rate of 5 K/min. Right panel: Corresponding mass signals offset for clarity.

suggest that the conditions employed for calcination of the precursor, *i.e.* treatment for 6 h at 550 °C, lead to decomposition and removal of the organic framework constituents, in good agreement with results obtained from Raman and IR spectroscopy.[48]

12.3 Influence of Water on Structure and Dispersion

12.3.1 Structure

It is known from the literature that the catalytic performance of supported vanadium-oxide catalysts crucially depends on the structure of the surface vanadium-oxide species.[12–20] To understand the relationship between structural and catalytic properties of dispersed vanadium oxide various techniques including IR,[58,59,61,62] Raman,[13,20,22,59] UV-Vis,[58,59,63–65] ^{51}V NMR,[22] and X-ray absorption spectroscopy[62,66] have been used. However, despite numerous studies the origin of this behavior is still not well understood as it may either be attributed to intrinsic support properties or differences in the vanadium-oxide structure(s) present on the different support materials, or even a combination of both effects. To this end, Figure 12.5 shows Raman data for a variety of supported vanadium-oxide catalysts containing dispersed vanadium-oxide species.[20] All catalysts were prepared by exactly the same procedure using, *e.g.*, identical precursor concentrations *etc.* Obviously, comparison of the vanadyl stretch vibrations of the dispersed vanadium-oxide species (1015–1045 cm^{-1}) reveals significant differences regarding the position and shape of the Raman bands, indicating the presence of different vanadium-oxide structures on the different support materials. It therefore remains an important future task to develop a more detailed understanding of the Raman spectra in the vanadyl region to unravel these structural differences.

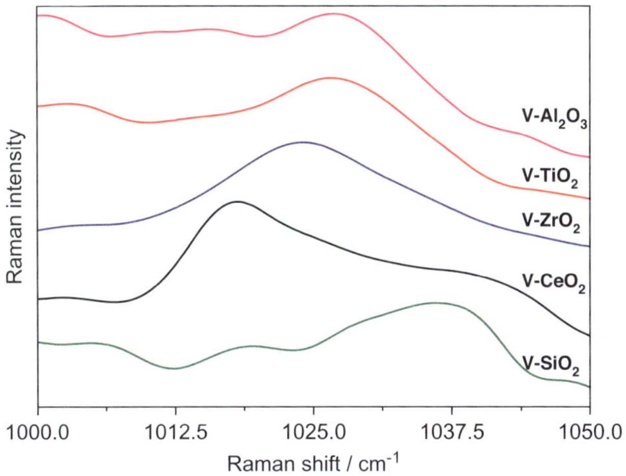

Figure 12.5 Raman spectra of dehydrated V–Al_2O_3, V–TiO_2, V–CeO_2, V–SiO_2 and V–ZrO_2. Spectra are offset for clarity.[20]

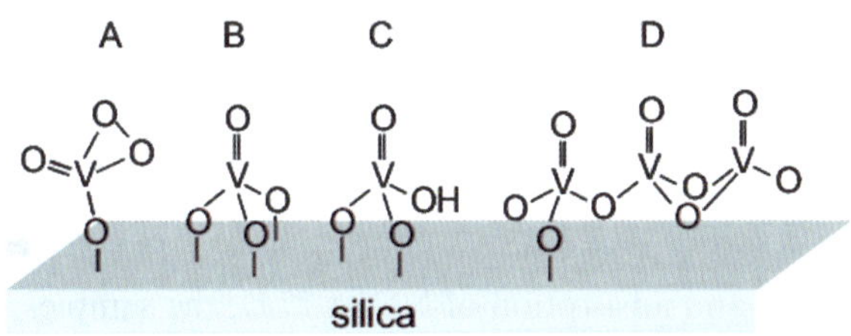

Figure 12.6 Structural motifs for silica-supported vanadium-oxide species.

In particular, the silica system is the subject of an ongoing debate regarding the detailed structure of the dispersed vanadium oxide.[7,10,67] Depending on the type of preparation, the maximum density of this dispersed vanadium-oxide species corresponds to 0.5–2.3 V/nm.[2,26,41,46] Figure 12.6 shows structural motifs of silica-supported vanadium-oxide species proposed in the literature. While it is now generally accepted that in the dehydrated state vanadium is present in a distorted tetrahedral coordination with one short vanadyl bond, there is mainly controversy about (i) the question whether monomeric species, dimeric/oligomeric species or a mixture of both types of species are present and (ii) the details of the structure of the surface vanadium-oxide species. A common structural motif of all nonmonomeric species is a V–O–V fragment. Thus, question (i) may be rephrased: Is it possible to detect V–O–V fragments and if yes, can the amount of V–O–V containing vanadium-oxide species be quantified? It should be mentioned that from the analytical point of view answering the first part of the question represents a demanding task on its own partly due to the low concentration of vanadium-oxide species. Also, at this stage, there is no experimental basis for a further distinction of the polymeric species (D) regarding size or structure (single *versus* double oxygen bridges). To this end it should be mentioned that Schraml-Marth *et al.* have suggested the presence of vanadium-oxide ribbons with vanadium ions located in the center of square pyramids besides tetrahedrally coordinated vanadium.[59]

Mainly structures of type (A), (B), and (C) have been proposed in the literature with a strong preference of structure (B). As has been pointed out by Keller *et al.*, compounds of type (A) are expected to be thermally unstable.[67] Also, they should give rise to a peroxide CT band in the UV-Vis spectrum, which has not been observed. Structures B and C differ by the number of Si–O–V bonds. In principal, also a structure with only one Si–O–V linkage is conceivable. The presence of V–OH has been revealed using IR spectroscopy.[59,68] However, quantification is not possible as the molar extinction coefficient of the V–OH band (at $\sim$3660 cm^{-1}) is unknown.

In the following, recent results on the structure of (dehydrated) highly dispersed silica-SBA-15-supported vanadium-oxide species will be discussed. In particular, it will shown that the current literature picture of exclusively monomeric species is incomplete and that a significant fraction of the surface vanadium oxide is present as associated species containing V–O–V (dimers/oligomers). Similar results were obtained for samples containing low (2.7 wt% V, 0.7 V/nm^2) and high (10.8 wt% V, 4.7 V/nm^2) loadings of dispersed vanadium-oxide species. This suggests that within the range studied the structure of the surface vanadium-oxide species is independent of vanadium loading.

The first indication of the presence of associated vanadium-oxide species came from transmission IR experiments using NO as probe molecule.[11] After reduction at 400 °C and adsorption of NO at 77 K, the samples were heated to room temperature and exposed to oxygen resulting in nitrate formation. Based on the literature,[69–71] the formed nitrate bands can be assigned to bridging and bidentate nitrates. Importantly, formation of bridging nitrates implies the presence of V–O–V fragments and thus associated vanadium-oxide species.

The presence of associated vanadium-oxide species in these V/SBA-15 catalysts has recently been confirmed by X-ray absorption spectroscopy using soft X-rays (NEXAFS, O K-edge)[10,72] and hard X-rays (EXAFS, V K-edge).[73] As an example, Figure 12.7 depicts a detailed view of the onset of the O-edge NEXAFS spectrum together with theoretical partial spectra obtained by DFT calculations on the basis of a $V_2Si_6O_{14}H_6$ model cluster. Details of the calculations can be found elsewhere.[10] The partial spectra represent the contribution from differently coordinated oxygen, *i.e.* vanadyl oxygen (V=O), oxygen in V–O–V and oxygen in V–O–Si, respectively. While the comparison between experiment and theory clearly shows that that oxygen in V–O–V bridges needs to be included to account for the observed O K-edge intensity pattern, a full assignment of all experimental features will make it necessary to include other clusters besides $V_2Si_6O_{14}H_6$ into the calculations. In full agreement with the above results, EXAFS analysis shows that the surface vanadium-oxide species form a regular arrangement of neighboring V_2O_7 units (dimers).[73]

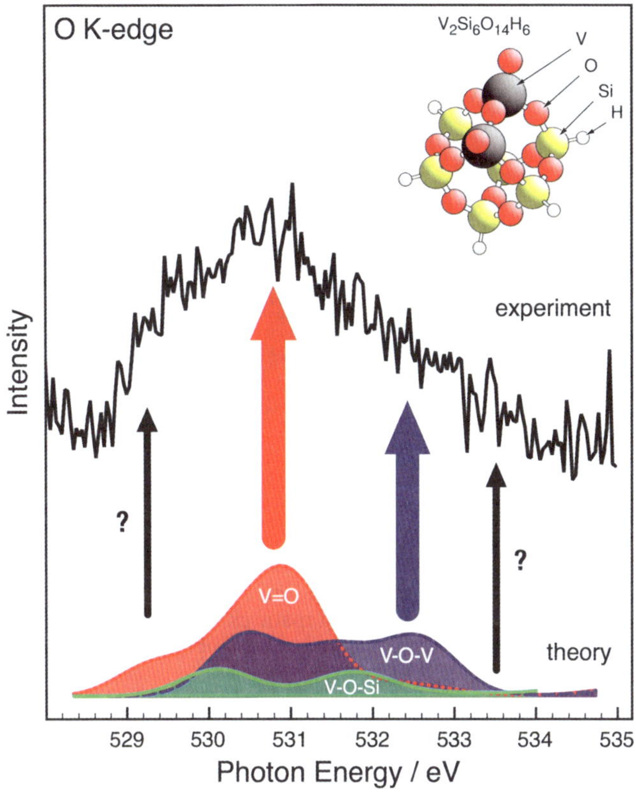

Figure 12.7 Comparison of the experimental O K-edge NEXAFS spectrum of 10.8 wt%V/SBA-15 (top panel) with theoretically obtained partial spectra of a $V_2Si_6O_{14}H_6$ dimer cluster representing the contribution of different oxygen coordinations (lower panel).[72]

12.3.2 Correlation of Changes in Structure and Dispersion

Beside the structure, the dispersion of supported vanadium oxide plays an important role in determining the catalytic performance as agglomeration of highly dispersed species may result in a loss of activity due to a reduced number of available active sites. Previously, the dispersion of supported vanadium oxide has been determined on the basis of an oxygen chemisorption experiment.[60] After sample reduction the oxygen uptake is measured, which allows for determination of the fraction of accessible surface vanadium atoms. A potential source of error may be the incomplete reduction of the sample. Recently, information on the dispersion of supported vanadium-oxide systems has also been obtained by XPS analysis.[74–76] To this end, the experimentally determined vanadium to support cation intensity ratio is converted into a V:Si atomic ratio by using atomic sensitivity factors.[76]

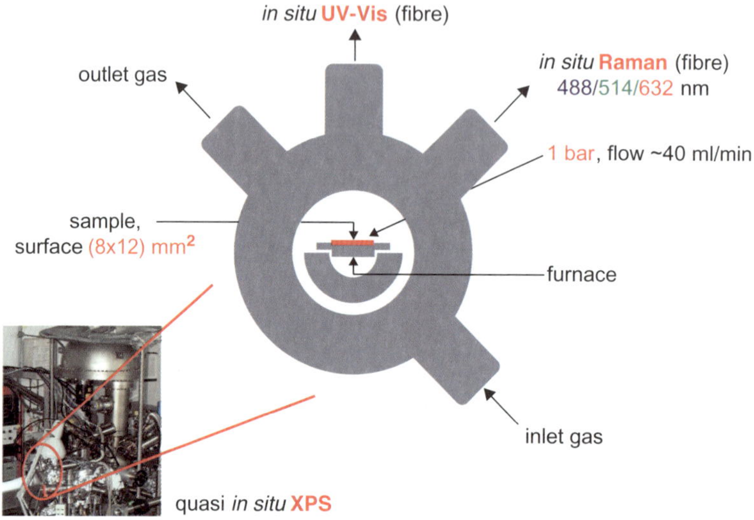

Figure 12.8 Cross-sectional scheme of the flow cell of the modified XPS setup illustrating the combined *in-situ* Raman and UV-Vis capabilities. The inset shows a picture of the XPS apparatus. The catalyst sample is transferred between reaction cell and analysis chamber without exposure to air thus allowing for quasi-*in-situ* XPS analysis.

For a detailed analysis of the dispersion of supported vanadium oxides (*i.e.* without prior exposure to air) as well as its correlation with structure a multiple *in-situ* spectroscopic experiment was set up, which combines XPS with optical methods (Raman, UV-Vis) thus avoiding any interference of sample-cell effects. As shown in Figure 12.8, an evacuation cell of the XPS apparatus was modified to allow for continuous gas flow coupled with analysis of the outgoing gas mixture by mass spectrometry (reaction cell) as well as Raman and UV-Vis fiber probes to be inserted into the cell (*in-situ* cell). Visible Raman spectra can be measured using 488, 514 or 632 nm excitation. For combined optical and XPS experiments, the sample was transferred between reaction cell and analysis chamber without exposure to air (quasi-*in-situ*). Experimental details can be found elsewhere.[31]

Recent results from multiple *in-situ* spectroscopic experiments on the dehydration of a 3.3 wt% V/SBA-15 catalyst are summarized in Figure 12.9.[31] As discussed in the following, comparison of the spectra obtained for the hydrated state and a dehydrated state allows for a direct correlation of changes in the structure and dispersion. The hydrated sample did not show any noticeable spectroscopic changes for several months, which suggests that it was present in its (fully) hydrated equilibrated state under ambient conditions. The dehydrated state studied was prepared by treatment in 20% O_2/He flow at 300 °C for 30 min. While UV-Vis and Raman spectroscopy provide structural information, XPS is used to gain insight into changes of the dispersion (see Chapter 13). First, the structural changes associated with the transition from

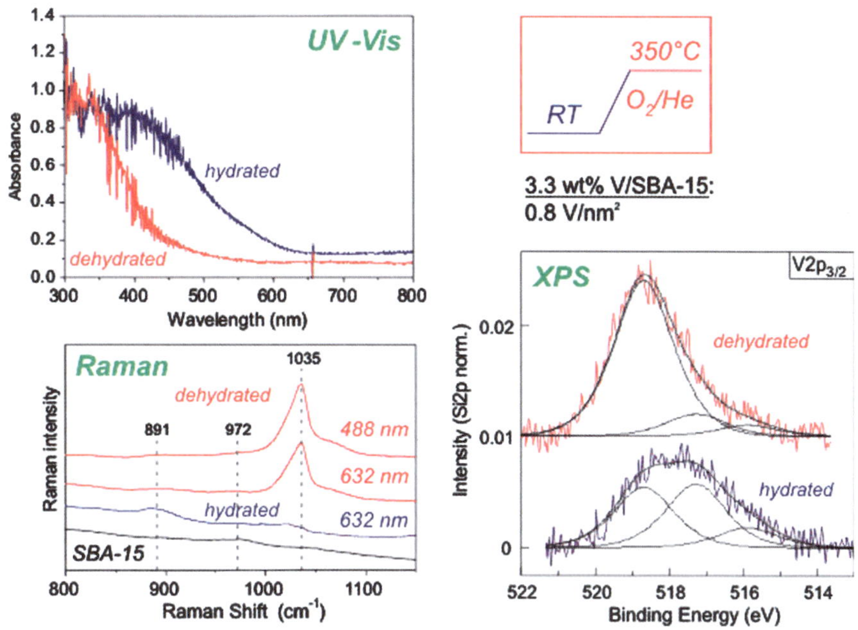

Figure 12.9 Multiwavelength excitation visible Raman, UV-Vis and V2p3/2 XP spectroscopic results for 3.3 wt% V/SBA-15 in its hydrated (blue) and dehydrated (red) state at 300 °C. Raman spectra were collected using either 488-nm or 632-nm laser excitation. The XP spectra are shown together with the results of a least-square fitting analysis. Raman and XP spectra are offset for clarity.[8]

hydrated to dehydrated vanadium oxide are discussed. The broad oxygen → vanadium CT band extending up to ~650 nm and the close resemblance of the Raman spectrum with that of $V_2O_5 \cdot 1.2H_2O$ strongly suggest that hydrated V/SBA-15 consists of layers of polymerized vanadium oxide with vanadium in pseudooctahedral coordination and water molecules coordinated to about every second vanadium ion. The small offset in the UV-Vis spectrum indicates the presence of d–d transitions of reduced vanadium oxide.

After treatment in 20% O_2/He flow at 300 °C for 30 min, the UV-Vis spectrum shows mainly absorption below 400 nm, which is characteristic of vanadium in a tetrahedral coordination by comparison with reference compounds. Raman spectra within (488 nm) and outside (632 nm) the range of the absorption band extension did not show a noticeable difference, indicating no significant contribution from other structurally different vanadium-oxide species.[9] The Raman spectra show bands around 891, 972 and 1035 cm^{-1}, respectively. The latter band exhibits a complex line shape, which is asymmetric towards the low-frequency side and has a shoulder extending up to ~1100 cm^{-1}. As discussed previously,[31] the smaller features at 891 and 972 cm^{-1} are assigned to the out-of phase Si–O–V interface mode and the

Si–OH stretching mode of surface hydroxyl, respectively. The band around 1035 cm^{-1} is commonly accepted to originate from the vanadyl stretch vibration of tetrahedrally coordinated vanadium. The shoulder around 1068 cm^{-1} can be explained by contributions from the in-phase Si–O–V interface mode as well as Si-O^- and Si(-O^-)$_2$ functionalities. Please note the redshift with respect to the room-temperature positions as the shown spectra were recorded at 300 °C. Despite these assignments it is obvious that more Raman studies will be necessary to give a detailed account of all features within the region of the vanadyl band, *e.g.* the asymmetry of the band.

XP spectra of the hydrated and dehydrated state are shown in the right panel of Figure 12.9 together with the result of a fitting analysis. Changes in the V2$p_{3/2}$ emission upon dehydration can be described on the basis of three contributions located at 515.9, 517.3, and 518.7 eV, which are attributed to V^{3+}, V^{5+} and highly dispersed vanadium oxide, respectively. Dehydration leads to an intensity shift of the V2$p_{3/2}$ emission to higher binding energies. This behavior implies a significant increase in the relative amount of the highly dispersed species at 518.7 eV from 42% to 82%. A more detailed discussion of the XPS data can be found elsewhere.[31]

In summary, based on XPS analysis dehydration of the catalyst leads to a strong decrease in cluster size, *i.e.* an increase in the dispersion of the catalyst. The corresponding structural changes consist of a transformation of the surface vanadium-oxide species from layers of polymerized vanadium oxide with vanadium in pseudo-octahedral coordination and water molecules coordinated to about every second vanadium into small aggregates of linked VO_4 units (dimers/oligomers) with tetrahedrally coordinated vanadium.

12.4 Selective Oxidation of Methanol

12.4.1 Reactivity

Methanol-selective oxidation to formaldehyde is an important probe reaction, as the selectivity pattern depends on the surface concentration of redox, basic and acidic sites.[12,77–80] Redox sites give rise to the formation of formaldehyde (FA), methyl formate, and dimethoxy methane. Acidic and basic sites produce dimethyl ether and carbon oxides. In Table 12.1, results of a literature survey on the catalytic performance of supported vanadium-oxide catalysts in the selective oxidation of methanol to formaldehyde are summarized. Although a direct comparison of the data sets is difficult due to the different catalytic conditions and sample preparation routes, it is obvious that some of the results are contradictory.[81] While Deo and Wachs[12] reported titania- (anatase) supported vanadium oxide to be very selective towards FA, Roozeboom *et al.*[13,14] found the titania system to be the least selective, in agreement with results from Kijenski *et al.*[15] Thus more fundamental work on well-characterized samples will be necessary to unravel the influence of the support material on catalytic performance.

Table 12.1 Reactivity of supported vanadium-oxide catalysts in methanol oxidation to formaldehyde. Catalysts were prepared by incipient wetness impregnation (iwi) and ion exchange (ion-ex.).

Support	*Prep.*	V *(mmol/g)*	V/*nm*^{2}	T *(°C)*	$CH_3OH/O_2/He$ *(ml/min)*	m *(g)*	X*(MeOH)* *(%)*	S*(FA)* *(%)*	Y*(FA)* *(%)*	*Ref.*
Al_2O_3 (Harshaw)	iwi	0.33	1.10[a]	230	6/13/81		<20	2	<0.4	12
Al_2O_3 (Harshaw)	iwi	2.2	7.36[a]	230	6/13/81		<20	46	<9.2	12
Nb_2O_5 (NPC)	iwi	0.11	1.20[a]	230	6/13/81		<20	91	<18.2	12
SiO_2 (Cab-O-Sil)	iwi	0.33	0.66[a]	230	6/13/81		<20	45	<9	12
TiO_2 (Degussa)	iwi	0.11	1.32[a]	230	6/13/81		<20	>90	(18)	12
TiO_2 (Degussa)	iwi	0.33	3.97[a]	230	6/13/81		<20	98	<19.6	12
ZrO_2 (Degussa)	iwi	0.06	0.85[a]	230	6/13/81		<20	54	<10.8	12
ZrO_2 (Degussa)	iwi	0.33	5.1[a]	230	6/13/81		<20	92	<18.4	12
Al_2O_3 (Degussa)	ion-ex.	1.33	10.0	230	1.6/9.6/28.8	0.2	100	28	28	13,14
TiO_2 (Degussa)	ion-ex.	0.33	4.6	230	1.6/9.6/28.8	0.2	100	0	0	13,14
CeO_2	ion-ex.	0.58	14.4	230	1.6/9.6/28.8	0.2	100	10	10	13,14
ZrO_2	ion-ex.	0.47	5.1	230	1.6/9.6/28.8	0.2	100	0	0	13,14
V_2O_5 (Baker)				230	1.6/9.6/28.8	0.2	22	70	15.4	13,14
SiO_2[b]	iwi	0.32	0.43–0.47[c]	285	24/58/216[e]	1		93		15
Al_2O_3[b]	iwi	0.34	1.0	250/265	24/58/216[e]	1		7		15
MgO[b]	iwi	0.22	3.4	275	24/58/216[e]	1		72		15
TiO_2 (Degussa)	iwi	0.05	0.91	270	24/58/216[e]	1		0		15
V_2O_5				260	24/58/216[e]	1		76		15

[a]Calculated on the basis of specific surface area values of the bare support.
[b]Alumina, silica and magnesia were prepared by hydrolysis of aluminum isopropoxide, ethyl orthosilicate and $Mg(NO_3)_2 \cdot 6H_20$, respectively.[15]
[c]The value for singly impregnated catalyst is expected to be within the given range. Boundary values are based on the surface areas obtained for a doubly impregnated catalyst and bare SiO_2, respectively.[15]

Regarding the use of silica-supported vanadium-oxide catalysts in methanol selective oxidation a more consistent picture is available. Table 12.2 summarizes literature data on the catalytic performance of vanadium oxide supported on various conventional and mesoporous silica materials. While silica-supported vanadium oxide is very selective towards FA, a rather low activity in observed as compared to catalysts based on other support materials such as titania.[12,77] Recently, silica MCM-48 supported vanadium-oxide catalysts were tested in methanol oxidation at 400 °C and compared to that of (amorphous) silica-supported vanadium oxide (see Table 12.2).[83] Interestingly, for the same vanadium loading the MCM-48-based systems gave significantly higher FA yields as the materials based on conventional silica. This behavior may be explained by the higher surface area offered by mesoporous support materials allowing for a higher dispersion of the surface V_xO_y species. In general, for increasing vanadium loading an increase in methanol conversion and FA yield is observed. On the other hand, the presence of crystalline V_2O_5 leads to a decrease in activity and FA yield due to the loss of accessible active sites. Based on these findings a better catalytic performance may be expected for vanadium-oxide systems that contain a high density of dispersed vanadium-oxide species. As a first step in that direction, the SBA-15 supported highly dispersed vanadium oxide (V/SBA-15) systems described in Section 12.2 were tested for the selective oxidation of methanol to FA.[30] The results in Table 12.2 demonstrate the excellent catalytic performance of the V/SBA-15 catalysts. In fact, at 400 °C and at a methanol conversion of 87% a FA yield of 72% is obtained, which represents the highest FA yield reported in the literature for silica-supported vanadium-oxide catalysts (see Table 12.2).

12.4.2 Spectroscopic Characterization

To gain insight into changes of the structure and dispersion of the supported vanadium oxide in the course of the reaction the multiple spectroscopic setup described in Section 12.3.2 was employed (see Figure 12.8). In the following, Raman and XPS results will be discussed, which were obtained for a SBA-15-supported vanadium-oxide catalyst ($0.7\,V/nm^2$) before and after methanol oxidation reaction at 350 °C for 1 h.

Figure 12.10 depicts Raman spectra recorded at room temperature under ambient conditions using 514-nm laser excitation before (a) and after (b) reaction as well as after subsequent treatment in O_2/N_2 at 350 °C for 1 h (c). Spectrum a is characterized by bands at 507, 688, and $1020\,cm^{-1}$, which are attributed to hydrated vanadium oxide forming a $V_2O_5\,nH_2O$ gel-like structure as discussed in Section 12.2. Exposure to reaction conditions gives rise to dramatic spectral changes. Spectrum b recorded after reaction shows Raman bands at 497, 671, 928, 1033, 1067, 2832, 2862, 2938, 2962, and $2994\,cm^{-1}$. The bands in the high-frequency region are readily assigned to C–H stretch vibrations of methoxy species bound to vanadium oxide (2832 and $2938\,cm^{-1}$) and the silica support (2862, 2962, and $2994\,cm^{-1}$), respectively.[85] The presence of

Table 12.2 Reactivity of silica-supported vanadium-oxide catalysts in methanol oxidation to formaldehyde. Catalysts were prepared by incipient wetness impregnation (iwi), ion exchange (ion-ex.), molecular designed dispersion (MDD), and sol-gel synthesis (sol-gel).

Support	*Prep.*	V *(mmol/g)*	V/*nm*^{2}	T *(°C)*	$CH_3OH/O_2/He$ *(ml/min)*	m *(g)*	X*(MeOH) (%)*	S*(FA) (%)*	Y*(FA) (%)*	*Ref.*
Cab–O–Sil	iwi	0.11	0.22[a]	230	6/13/81		<20	72	<14.4	12
Cab–O–Sil	iwi	0.33	0.66[a]	230	6/13/81		<20	45	<9	12
SBA–15	ion-ex.	0.45	0.6	230	3/7/90	0.2	3.2	96.3	3.1	30
MCM–48 mod.[b]	MDD	1.2	0.67	300	3.2/6.4/70.4	0.2	48	21	10.1	82
MCM–48	MDD	1.4	0.75	300	3.2/6.4/70.4	0.2	24	60	14.4	83
SBA–15	ion-ex.	1.4	2.3	300	3/7/90	0.2	7.8	95.8	7.5	30
SiO_2[c]	iwi	0.32	0.43–0.47[d]	310	24/58/216[e]	1		90		15
SiO_2[c]	iwi	0.32	0.43–0.47[d]	345	24/58/216[e]	1		84		15
SBA–15	ion-ex.	0.45	0.6	350	3/7/90	0.2	15.7	94	14.8	87
MCM–48 mod.[b]	MDD	1.2	0.67	350	3.2/6.4/70.4	0.2	80	44	35.2	82
MCM–48	MDD	1.4	0.75	350	3.2/6.4/70.4	0.2	38	85	32.3	83
SBA–15	ion-ex.	1.4	2.3	350	3/7/90	0.2	39	93.3	36.4	87
Cab–O–Sil	iwi	0.11	0.17	350	3/6.5/40.5	$15–45\times10^{-3}$	9.1	84.6	7.7	84
Cogel	sol-gel	0.11	0.17	350	3/6.5/40.5	$15–45\times10^{-3}$	13.1	87.0	11.4	84
V–silicalite	sol-gel	0.03	0.17	350	3/6.5/40.5	$15–45\times10^{-3}$	12.9	87.0	11.2	84
MCM–48	MDD	0.2	0.1	400	3.2/6.4/70.4	0.2	31	98.1	30.4	83
MCM–48	MDD	0.4	0.2	400	3.2/6.4/70.4	0.2	60.9	94.7	57.7	83
MCM–48	MDD	0.7	0.31	400	3.2/6.4/70.4	0.2	75	92.8	69.6	83
Kieselgel 60	MDD	0.2	0.4	400	3.2/6.4/70.4	0.2	21	79.3	16.7	83
SBA–15	ion-ex.	0.45	0.6	400	3/7/90	0.2	49	91.3	44.7	30
MCM–48 mod.[b]	MDD	1.2	0.67	400	3.2/6.4/70.4	0.2	84	58	48.7	82
MCM–48	MDD	1.4	0.75	400	3.2/6.4/70.4	0.2	55	85	46.8	82
Kieselgel 60	MDD	0.4	0.94	400	3.2/6.4/70.4	0.2	45	88	39.6	83
SBA–15	ion-ex.	1.4	2.3	400	3/7/90	0.2	86.7	83.2	72.1	30

[a]Calculated on the basis of specific surface area values of the bare support.
[b]Silica MCM-48 was modified *via* reaction with dimethyldichlorosilane. Subsequently, $VO(acac)_2$ was grafted on the silylated surface. Finally, the material was calcined at 450 °C.
[c]Silica was prepared by hydrolysis of ethyl orthosilicate.[15]
[d]Value for singly impregnated catalyst is expected to be within the given range. Boundary values are based on the surface areas obtained for a doubly impregnated catalyst and bare SiO_2, respectively.[15]
[e]In this study N_2 was used instead of He.

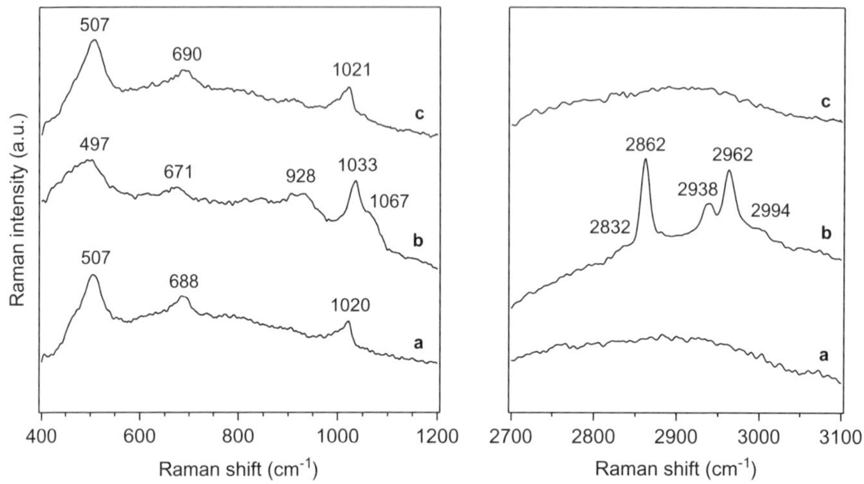

Figure 12.10 Raman spectra of 2.7 wt% V/SBA-15 (a) before methanol oxidation (b) after methanol oxidation at 350°C for 1 h, and (c) after subsequent treatment in O_2/N_2 at 350 °C for 1 h. All spectra were taken under ambient conditions and at room temperature. They are offset for clarity.

methoxy also explains the band at 671 cm^{-1} and the shoulder at 1067 cm^{-1}, which are attributed to V–O and C–O stretch vibrations of a vanadium bound methoxy, respectively.[41,86] As expected, all methoxy-related features disappear upon heating in oxygen (see spectrum c). After reaction, the vanadyl band is observed at 1033 cm^{-1}, as compared to its position at 1039 cm^{-1} in the dehydrated state (see Figure 12.3). This redshift is consistent with the calculated shift for the V=O stretch vibration of tetrahedrally coordinated vanadium upon addition of methoxy.[86] However, as water is a reaction product and as the Raman experiments were performed under ambient conditions, the exact position of the vanadyl band can also be influenced by the presence of water. To this end, the position of the vanadyl band (1020 cm^{-1}) in the (fully) hydrated state may serve as a reference (see spectrum a). In a previous study, for a partially hydrated vanadium-oxide sample a blueshift of the V=O stretch vibration to 1035 cm^{-1} as well as the appearance of an additional band at ~915 cm^{-1} was observed.[41] Therefore, the spectrum of the supported vanadium-oxide catalyst obtained after reaction (spectrum c) can be explained by adsorbed methoxy and the presence of water either from the reaction or from exposure to ambient. According to a recent theoretical study the rate-limiting step in the selective oxidation of methanol is the methyl hydrogen transfer from adsorbed methoxy to vanadyl oxygen.[86] After desorption of the product formaldehyde the catalytic cycle is completed by release of water and reoxidation of the reduced vanadium sites.

The above Raman results show that exposure to reaction conditions leads to the formation of methoxy species on both the vanadium-oxide species and the silica support. For the V/SBA-15 sample (0.7 V/nm^2) discussed here the changes associated with the surface vanadium-oxide species are fully reversible, as shown by the treatment in oxygen. However, for a V/SBA-15 catalyst with the maximum loading of dispersed vanadium oxide achieving a grafting/ion-exchange (2.3 V/nm^2), extensive reaction at 400–460 °C resulted in agglomeration of part of the dispersed vanadium oxides species into crystalline V_2O_5.[30] The formation of methoxy species under reaction conditions is confirmed by XPS.[87] To this end, samples were transferred from the reaction cell to the analysis chamber without exposure to air (quasi-*in-situ* XPS). As discussed elsewhere in detail,[87] based on C1s analysis the overall amount of methoxy bound to vanadium can be estimated as 74%. Interestingly, besides methoxy, a comparable amount of carbonaceous species is formed during the reaction. This result clearly underlines the importance of understanding the role of carbonaceous deposits in future studies on selective oxidation reactions.

Changes in the dispersion of the V/SBA-15 catalyst (0.7 V/nm^2) after exposure to the reaction conditions were studied using quasi-*in-situ* XPS. Figure 12.11 depicts the V2p$_{3/2}$ emission before reaction, *i.e.* in a hydrated state (a), after treatment in an O_2/N_2 mixture at 350 °C leading to dehydration (b), and after methanol oxidation reaction at 350 °C for 1 h (c). As discussed in detail in Section 12.3.2, dehydration leads to a V2p$_{3/2}$ intensity shift to higher binding energies as a result of a significant increase in the dispersion of the supported vanadium oxide (see spectrum b). Spectrum c shows that after exposure to reaction conditions a backshift in intensity to lower binding energies is observed. These results suggest that in the presence of the reaction

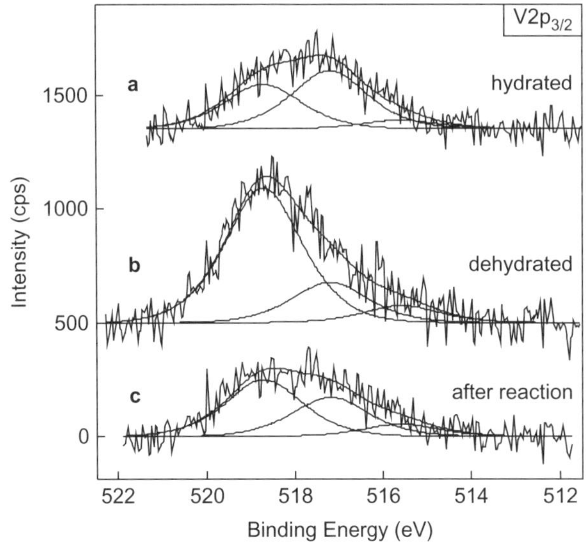

Figure 12.11 X-ray photoelectron V2p$_{3/2}$ spectra of 2.7 wt% V/SBA-15 (a) in its hydrated state, (b) in its dehydrated state, and (c) after methanol oxidation at 350 °C for 1 h together with least-square fits to the data. The spectra are offset for clarity.

mixture part of the supported vanadium-oxide agglomerates resulting in a decrease of the dispersion. This interpretation is corroborated by the values of the V:Si atomic ratio before reaction (0.0239), after dehydration (0.0294) and after reaction (0.0277). The larger values in case of the hydrated and the after-reaction sample can be explained by the larger size of the vanadium-oxide aggregates exceeding those of the characteristic escape depth of the V2p electrons. In summary, the above discussion demonstrates the benefits of an experimental approach that combines complementary techniques such as Raman spectroscopy and XPS. While Raman spectroscopy can provide valuable information on the structure of the catalyst as well as the adsorbed species, XPS allows for the detection of changes of the catalyst composition and catalyst dispersion during reaction.

12.5 Oxidative Dehydrogenation of Propane

12.5.1 Reactivity

The oxidative dehydrogenation (ODH) of propane over supported vanadium oxide using Al_2O_3,[88–90] SiO_2,[28,91,92] TiO_2[93,94] or ZrO_2[95] as support material has been the focus of a large number of studies. Reactivity studies as a function of the support material by Arena *et al.* revealed the following trend regarding the selectivity at low propane conversions (3.0–5.1%) and 500 °C: $V/Al_2O_3 > V/SiO_2 > V/ZrO_2 > V/MgO > V/TiO_2$.[16,17] Surprisingly, under similar experimental conditions (1.5–6.9% propane conversion, 500 °C), Lemonidou *et al.* observed a different order: $V/TiO_2 > V/Al_2O_3 > V/MgO > V/ZrO_2$.[18]

Figure 12.12 shows recent results by Dinse *et al.* on the influence of the support material on the performance of supported vanadium-oxide catalysts.[20]

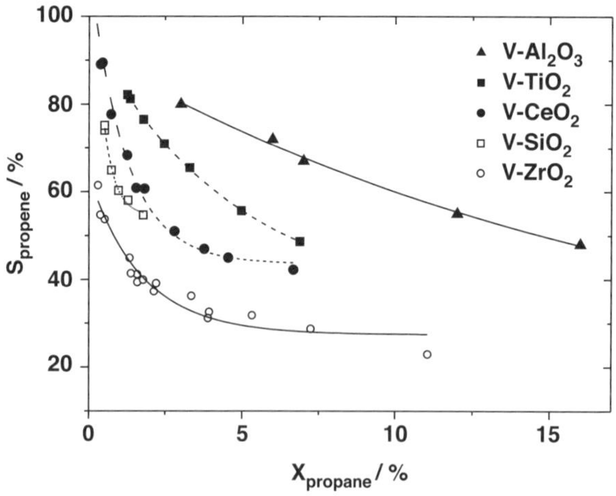

Figure 12.12 Selectivity–conversion trajectories for $V–Al_2O_3$, $V–TiO_2$, $V–CeO_2$, $V–SiO_2$ and $V–ZrO_2$ at 400 °C. $C_3H_8/O_2/N_2 = 29.1/14.5/56.4$ at a total gas flow of 60 ml min^{-1}.[20]

At 400 °C and 2% propane conversion again a different trend with respect to propene selectivity has been observed: $V/Al_2O_3 > V/TiO_2 > V/CeO_2 > V/SiO_2 > V/ZrO_2$. A similar behavior has also been observed previously by Khodadov *et al.* at 333 °C based on propene selectivities extrapolated to zero conversion: $V/Al_2O_3 > V/TiO_2 > V/SiO_2 > V/ZrO_2 > V/HfO_2$.[19] Obviously, further studies will be necessary to resolve the observed differences and to clarify the role of the support material. To this end, significant progress will largely depend on the availability of (stable) samples based on different support materials but containing active vanadium-oxide species with the same structure. Presently, supported vanadium-oxide catalysts are prepared mostly by application of the same synthetic procedure irrespective of the specific properties of the support material. As a result, structurally different vanadium-oxide surface species are obtained (see Figure 12.5).

Recently, highly dispersed vanadium-oxide systems on the basis of mesoporous silica have been tested in the ODH of propane.[11,20,28,91,96–103] As an example, a 2.8 wt% V/SBA-15 catalyst with a vanadium density of 0.69 V/nm^2 gave a higher propene yield as the corresponding catalyst based on conventional silica, 2.8 wt% V/SiO_2 (1.49 V/nm^2).[102] Recently, a slightly higher catalytic activity of 2.7 wt% V/MCM-41 (0.37 V/nm^2) as compared to 2.9 wt% V/SiO_2 (1.38 V/nm^2) has been observed.[91] The improved catalytic performance has been attributed to the higher dispersion of the vanadium-oxide species and, in the case of the SBA-15-based systems, the lower surface acidity as compared to conventional silica-supported vanadium-oxide catalysts.

12.5.2 Influence of Surface Functionalization

As discussed in the previous section, vanadium-oxide catalysts based on mesoporous silica are promising catalysts for propane ODH. However, besides their high costs an important issue regarding industrial applications is the poor mechanical and hydrothermal stability of the mesoporous support materials. To this end, Herbert *et al.* have recently shown that surface functionalization of mesoporous silica SBA-15 increases the resistance towards mechanical, thermal and hydrothermal damage and, as a result, improves the catalytic performance.[39]

Figure 12.13 shows catalytic results obtained for propane ODH at 500 °C using 2.7 wt% V/SBA-15 catalysts (0.7 V/nm^2). The tested catalysts were prepared by incipient wetness impregnation and by the grafting/ion-exchange procedure (see Section 12.2), which involves surface functionalization in the first step. Both systems were tested before and after pressure treatment at 752 MPa.[39] Before pressure treatment, the functionalized and the incipient wetness impregnated sample show a comparable catalytic performance. However, after pressure treatment, a beneficial effect of surface functionalization on catalytic performance is observed (see Figure 12.13). As an example, at a conversion of 3–4%, the functionalized sample shows still a selectivity of 64%, whereas that of the incipient wetness impregnated sample has dropped to 58%.

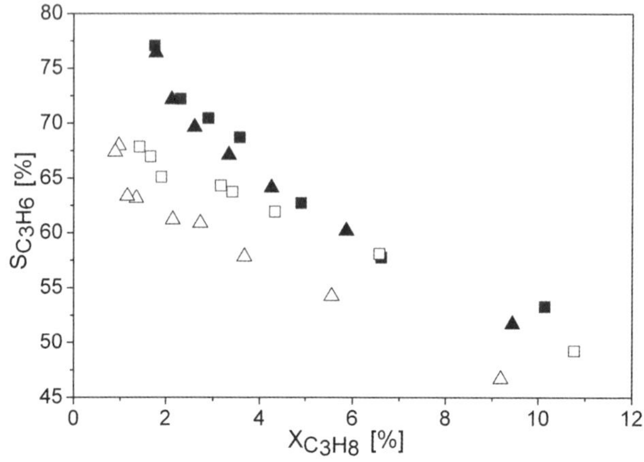

Figure 12.13 Selectivity towards propene before pressure treatment of 2.7 wt% V/SBA-15 (■) multistep, (▲) incipient wetness impregnation and after pressure treatment at 752 MPa with BET area correction (□) multistep, (△) incipient wetness impregnation.[39]

12.6 Conclusions and Outlook

The above discussion clearly demonstrates the potential of the approach outlined based on catalyst systems with controlled chemical complexity. In fact, our recent studies on silica-SBA-15-supported vanadium-oxide catalysts have revealed new insight into the properties of supported vanadium-oxide systems. In this contribution, emphasis was put on the questions raised in the introduction. The results presented show that the current picture of the molecular structure of silica-supported vanadium oxide in its dehydrated state is incomplete as besides monomeric VO_4 species also associated vanadium-oxide species (dimmers/oligomers) are present. While preliminary experiments did not give any indication of a change in structure upon exposure to reaction conditions systematic *in-situ* experiments as a function of the gas-phase chemical potential will be necessary to unravel all details of the structural dynamics. To this end, an important aspect is the influence of water. The above results reveal that at room temperature the presence of water has a dramatic effect on the structure and dispersion of silica-supported vanadium oxide. However, the role of water under reaction conditions has not been studied in detail. Interestingly, during methanol oxidation a decrease in dispersion, *i.e.* an agglomeration of surface vanadium oxide is observed, which in the case of harsh reaction conditions (elevated temperature, high conversion) may even lead to the formation of crystalline V_2O_5. Finally, it was shown that surface functionalization can improve the resistance of mesoporous silica SBA-15 towards mechanical and hydrothermal damage thus reducing the impact of this obvious disadvantage of mesoporous materials regarding technical applications. With the new insight gained for the silica system, we are now in an excellent starting position to elucidate the influence of the support material on the properties of the supported vanadium-oxide species.

Acknowledgement

Most of the results discussed above were obtained during my time at the Department of Inorganic Chemistry at the Fritz-Haber-Institut of the Max Planck-Gesellschaft. In particular, I would like to thank my coworkers Dr. Rita Herbert, Jörg Thielemann and Ute Wild. The support from Gisela Lorenz, Norbert Pfänder and Gisela Weinberg is gratefully acknowledged. Also, I thank the members of the SFB 546 for fruitful discussions. Finally, I would like to express my sincere appreciation to Prof. Robert Schlögl for very fruitful discussions, his interest in the progress of the work as well as his continuous support. Financial support by the Max-Plank-Gesellschaft and the Deutsche Forschungsgemeinschaft (Emmy Noether, SFB 546) is gratefully acknowledged.

References and Notes

1. *Handbook of Heterogeneous Catalysis*, ed. G. Ertl, H. Knözinger and J. Weitkamp, Wiley-VCH, Weinheim, 1997.
2. *Catalyst Handbook*, ed. M. V. Twigg, Manson Publishing, London, 1996.
3. *Handbook of Commercial Catalysts: Heterogeneous Catalysts*, H. F. Rase, CRC Press, Boca Raton, 2000.
4. G. Busca, L. Lietti, G. Ramos and F. Berti, *Appl. Catal. B*, 1998, **18**, 1.
5. G. Centi, *Appl. Catal. A*, 1996, **147**, 267.
6. C. Hess, Habilitationsschrift, Humboldt-Universität zu Berlin, 2007.
7. N. Magg, B. Immaraporn, J. B. Giorgi, T. Schroeder, M. Bäumer, J. Döbler, Z. Wu, E. Kondratenko, M. Cherian, M. Baerns, P. C. Stair, J. Sauer and H.-J. Freund, *J. Catal.*, 2004, **226**, 88.
8. C. Hess, *ChemPhysChem*, 2009, **10**, 319.
9. Z. Wu, H.-S. Kim, P. C. Stair, S. Rugmini and S. D. Jackson, *J. Phys. Chem. B*, 2005, **109**, 2793.
10. M. Cavalleri, K. Hermann, A. Knop-Gericke, M. Hävecker, R. Herbert, C. Hess, A. Oestereich, J. Döbler and R. Schlögl, *J. Catal.*, 2009, **262**, 215.
11. T. V. Venkov, C. Hess and F. C. Jentoft, *Langmuir*, 2007, **23**, 1768.
12. G. Deo and I. E. Wachs, *J. Catal.*, 1994, **146**, 323.
13. F. Roozeboom, M. C. Mittelmeljer-Hazeleger, J. A. Mouljin, J. Medema, V. H. J. d. Beer and P. J. Gellings, *J. Phys. Chem.*, 1980, **84**, 2783.
14. F. Roozeboom, P. D. Cordingley and P. J. Gellings, *J. Catal.*, 1981, **68**, 464.
15. J. Kijenski, A. Baiker, M. Glinski, P. Dollemeire and A. Wokaun, *J. Catal.*, 1986, **101**, 1.
16. F. Arena, F. Frusteri and A. Parmaliana, *Catal. Lett.*, 1999, **60**, 59.

17. G. Martra, F. Arena, S. Clouccia, F. Frusteri and A. Parmaliana, *Catal. Today*, 2000, **63**, 197.
18. A. A. Lemonidou, L. Nalbandian and I. A. Vasalos, *Catal. Today*, 2000, **61**, 333.
19. A. Khodakov, B. Olthof, A. T. Bell and E. Iglesia, *J. Catal.*, 1999, **181**, 205.
20. A. Dinse, B. Frank, C. Hess, D. Habel and R. Schomäcker, *J. Mol. Catal. A*, 2008, **289**, 28.
21. D. L. Kepert, *The Early Transition Elements*, Academic Press, London, 1972.
22. N. Das, H. Eckert, H. Hu, I. E. Wachs, J. F. Walzer and F. J. Feher, *J. Phys. Chem.*, 1993, **97**, 8240.
23. E. Hoppe, C. Limberg, B. Ziemer and C. Mugge, *J. Mol. Catal. A*, 2006, **251**, 34.
24. E. Hoppe, C. Limberg and B. Ziemer, *Inorg. Chem.*, 2006, **45**, 8308.
25. K. L. Fujdala and T. D. Tilley, *J. Catal.*, 2003, **216**, 265.
26. C. Hess, J. D. Hoefelmeyer and T. D. Tilley, *J. Phys. Chem. B*, 2004, **108**, 9703.
27. S. Grasser, C. Haessner, K. Kohler, F. Lefebvre and J. M. Basset, *Phys. Chem. Chem. Phys.*, 2003, **5**, 1906.
28. E. V. Kondratenko, M. Cherian, M. Baerns, D. Su, R. Schlögl, X. Wang and I. E. Wachs, *J. Catal.*, 2005, **234**, 131.
29. S. Xie, E. Iglesia and A. T. Bell, *Langmuir*, 2000, **16**, 7162.
30. C. Hess, I. J. Drake, J. D. Hoefelmeyer, T. D. Tilley and A. T. Bell, *Catal. Lett.*, 2005, **105**, 1.
31. C. Hess, *J. Catal.*, 2007, **248**, 120.
32. F. Schüth, *Angew. Chem.*, 2003, **42**, 3604.
33. *Preparation of Solid Catalysts*, ed. G. Ertl, H. Knözinger and J. Weitkamp, Wiley-VCH, Weinheim, 1997.
34. K. Cassiers, T. Linssen, M. Matthieu, M. Benjelloun, K. Schrijnemakers, P. v. d. Voort, P. Cool and E. F. Vansant, *Chem. Mater.*, 2002, **14**, 2317.
35. D. Y. Zhao, J. Feng, Q. Huo, N. Melosh, G. H. Fredrickson, B. F. Chmelka and G. D. Stucky, *Science*, 1998, **279**, 549.
36. Note that inside the walls of SBA-15 a second porosity (microporosity) has been observed that disappears in samples prepared at higher temperatures.
37. A. Katiyar, S. Yadav, P. G. Smirniotis and N. G. Pinto, *J. Chromatogr. A*, 2006, **1122**, 13.
38. B. Celer, M. Kruk, Y. Zuzek and M. Jaroniek, *J. Mater. Chem.*, 2006, **16**, 2824.
39. R. Herbert, D. Wang, R. Schomäcker, R. Schlögl and C. Hess, *ChemPhysChem*, 2009, **10**, 2320.
40. M. S. Wong, in *Metal Oxides: Chemistry and Applications*, ed. J. L. G. Fierro, Taylor and Francis, Boca Raton, 2006, p. 31.
41. X. Gao, S. R. Bare, B. Weckhuysen and I. E. Wachs, *J. Phys. Chem. B*, 1998, **102**, 10842.

42. K. L. Fujdala and T. D. Tilley, *J. Catal.*, 2003, **216**, 265.
43. R. Rulkens, J. L. Male, K. W. Terry, B. Olthof, A. Khadakov, A. T. Bell, E. Iglesia and T. D. Tilley, *Chem. Mater.*, 1999, **11**, 2966.
44. P. v. d. Voort, M. Morey, G. D. Stucky, M. Mathieu and E. F. Vansant, *J. Phys. Chem. B*, 1998, **102**, 585.
45. Y. Segura, P. Cool, P. Kustrowski, L. Chmielarz, R. Dziembaj and E. F. Vansant, *J. Phys. Chem. B*, 2005, **109**, 12071.
46. J. Keranen, C. Guimon, E. Iiskola, A. Auroux and L. Niinisto, *J. Phys. Chem. B*, 2003, **107**, 10773.
47. D. Wei, H. Wang, X. Feng, W.-T. Chueh, P. Ravikovitch, M. Lyubovsky, C. Li, T. Takeguchi and G. L. Haller, *J. Phys. Chem. B*, 1999, **103**, 2113.
48. C. Hess, U. Wild and R. Schlögl, *Microp. Mesop. Mater.*, 2006, **95**, 339.
49. C. J. Brinker, R. Kirkpatrick, D. R. Tallant, B. C. Bunker and B. Montez, *J. Non-Cryst. Solids*, 1988, **99**, 418.
50. D. R. Tallant, B. C. Bunker, C. J. Brinker and C. A. Balfe, *Mater. Res. Soc. Symp. Proc.*, 1986, **73**, 261.
51. R. H. Stolen and G. E. Walrafen, *J. Chem. Phys.*, 1976, **64**, 2623.
52. B. C. Brinker, D. R. Tallant, E. P. Roth and C. S. Ashley, *Mater. Res. Soc. Symp. Proc.*, 1986, **61**, 387.
53. D. Lin-Vien, N. B. Colthup, W. G. Fateley and J. G. Grasselli, *The Handbook of Infrared and Raman Characteristic Frequencies of Organic Molecules*, Academic Press, Boston, 1991.
54. H. Okabayashi, K. Taga, K. Miyagai, T. Uehara, T. Yoshida and E. Nishio, *J. Phys. Chem.*, 1991, **95**, 7932.
55. L. Abello, E. Husson, Y. Repelin and G. Lucazeau, *J. Solid State Chem.*, 1985, **56**, 379.
56. Y. Repelin, E. Husson, L. Abello and G. Lucazeau, *Spectrochem. Acta*, 1985, **41A**, 993.
57. M. Banares, X. Gao, J. L. G. Fierro and I. E. Wachs, *Stud. Surf. Sci. Catal.*, 1997, **110**, 295.
58. B. Jonson, B. Rebenstrof, R. Larsson and S. L. Andersson, *J. Chem. Soc., Faraday Trans.*, 1988, **84**, 1897.
59. M. Schraml-Marth, A. Wokaun, M. Pohl and H. L. Krauss, *J. Chem. Soc., Faraday Trans.*, 1991, **87**, 2635.
60. S. T. Oyama, G. T. Went, K. B. Lewis, A. T. Bell and G. A. Somorjai, *J. Phys. Chem.*, 1989, **93**, 6786.
61. M. Anpo, M. Sunamoto and M. Che, *J. Phys. Chem.*, 1989, **93**, 1187.
62. K. Inumaru, T. Okuhara, M. Misono, N. Matsubayashi, H. Shimada and A. Nishijima, *J. Chem. Soc., Faraday Trans.*, 1992, **88**, 625.
63. W. Hanke, R. Bienert and H.-G. Jerschkewitz, *Z. Anorg. Allg. Chem.*, 1975, **414**, 109.
64. B. Horvath, J. Geyer and H. L. Krauss, *Z. Anorg. Allg. Chem.*, 1976, **426**, 141.
65. F. Arena, F. Frusteri, G. Martra, S. Coluccia and A. Parmaliana, *J. Chem. Soc., Faraday Trans.*, 1997, **93**, 3849.
66. T. Tanaka, H. Yamashita, R. Tsuchitani, T. Funabiki and S. Yoshida, *J. Chem. Soc. Faraday Trans.*, 1988, **84**, 2987.

67. D. E. Keller, T. Visser, F. Soulimani, D. C. Koningsberger and B. M. Weckhuysen, *Vibr. Spectrosc.*, 2007, **43**, 140.
68. L. D. Nguyen, S. Loridant, H. Launay, A. Pigamo, J. L. Dubois and J. M. M. Millet, *J. Catal.*, 2006, **237**, 38.
69. K. Hadjiivanov, *Catal. Rev.-Sci. Eng.*, 2000, **42**, 71.
70. T. Dines, C. Rochester and A. Ward, *J. Chem. Soc. Faraday Trans.*, 1991, **87**, 1617.
71. H. Miyata, S. Konishi, T. Ohno and F. Hatayama, *J. Chem. Soc. Faraday Trans.*, 1995, **91**, 1557.
72. M. Hävecker, M. Cavalleri, R. Herbert, R. Follath, A. Knop-Gericke, C. Hess, K. Hermann and R. Schlögl, *Phys. Status Solidi B*, 2009, **264**, 1459.
73. A. Walter, C. Hess, R. Herbert and T. Ressler, *Chem. Central J.*, 2010, **4**.
74. K. Inumaru, M. Misono and T. Okuhara, *Appl. Catal. A*, 1997, **149**, 133.
75. G. C. Bond, J. P. Zurita and S. Flamerz, *Appl. Catal.*, 1986, **27**, 353.
76. C. Hess, G. Tzolova-Müller and R. Herbert, *J. Phys. Chem. C*, 2007, **111**, 9471.
77. G. C. Bond and S. F. Tahir, *Appl. Catal. A*, 1991, **71**, 1.
78. G. Deo, I. E. Wachs and J. Haber, *Crit. Rev. Surf. Chem.*, 1994, **4**, 141.
79. M. Baltes, P. v. d. Voort, O. Collart and E. F. Vansant, *J. Porous Mater.*, 1998, **5**, 357.
80. P. v. d. Voort, M. Baltes, M. G. White and E. F. Vansant, *Interf. Sci.*, 1997, **5**, 209.
81. P. Forzatti, E. Tronconi, A. S. Elmi and G. Busca, *Appl. Catal. A*, 1997, **157**, 387.
82. P. v. d. Voort, M. Baltes and E. F. Vansant, *Catal. Today*, 2001, **68**, 119.
83. M. Baltes, K. Cassiers, P. V. d. Voort, B. M. Weckhuysen, R. A. Schoonheydt and E. F. Vansant, *J. Catal.*, 2001, **197**, 160.
84. C.-B. Wang, G. Deo and I. E. Wachs, *J. Catal.*, 1998, **178**, 640.
85. J.-M. Jehng, H. Hu, X. Gao and I. E. Wachs, *Catal. Today*, 1996, **28**, 335.
86. J. Döbler, M. Pritzsche and J. Sauer, *J. Am. Chem. Soc.*, 2005, **127**, 10861.
87. C. Hess, *Surf. Sci.*, 2006, **600**, 3695.
88. M. D. Argyle, K. Chen, A. T. Bell and E. Iglesia, *J. Catal.*, 2002, **208**, 139.
89. F. Klose, T. Wolff, H. Lorenz, A. Seidel-Morgenstern, Y. Suchorski, M. Piorkowska and H. Weiss, *J. Catal.*, 2007, **247**, 176.
90. O. Schwarz, B. Frank, C. Hess and R. Schomäcker, *Catal. Commun.*, 2008, **9**, 229.
91. O. Ovsitser, M. Cherian and E. V. Kondratenko, *J. Phys. Chem. C*, 2007, **111**, 8594.
92. A. Dinse, S. Khennache, B. Frank, C. Hess, R. Herbert, S. Wrabetz, R. Schlögl and R. Schomäcker, *J. Mol. Catal. A*, 2009, **307**, 43.
93. K. Routary, K. R. S. K. Reddy and G. Deo, *Appl. Catal. A*, 2004, **265**, 103.
94. D. Habel, J. B. Stelzer, E. Feike, C. Schröder, A. Hösch, C. Hess, A. Knop-Gericke, J. Caro and H. Schubert, *J. Eur. Ceram. Soc.*, 2006, **26**, 3287.

95. K. Chen, A. Khodakov, J. Yang, A. T. Bell and E. Iglesia, *J. Catal.*, 1999, **186**, 325.
96. J. M. Lopez Nieto, *Top. Catal.*, 2001, **15**, 189.
97. O. V. Buyevskaya, A. Brückner, E. V. Kondratenko, D. Wolf and M. Baerns, *Catal. Today*, 2001, **67**, 369.
98. B. Solsona, T. Blasco, J. M. L. Nieto, M. L. Pena, F. Rey and A. Vidal-Moya, *J. Catal.*, 2001, **203**, 443.
99. Y. Wang, Q. Zhang, Y. Ohishi, T. Shishido and K. Takehira, *Catal. Lett.*, 2001, **72**, 215.
100. Q. Zhang, Y. Wang, Y. Ohishi, T. Shishido and K. Takehira, *J. Catal.*, 2001, **202**, 308.
101. J. Santamaria-Gonzalez, J. Luque-Zambrana, J. Merida-Robles, P. Maireles-Torres, E. Rodriguez-Castellon and A. Jimenez-Lopez, *A. Catal. Lett.*, 2000, **68**, 67.
102. Y.-M. Liu, Y. Cao, N. Yi, W.-L. Feng, W.-L. Dai, S.-R. Yan, H.-Y. He and K.-N. Fan, *J. Catal.*, 2004, **224**, 417.
103. Y.-M. Liu, W.-L. Feng, T.-C. Li, H.-Y. He, W.-L. Dai, W. Huang, Y. Cao and K.-N. Fan, *J. Catal.*, 2006, **239**, 125.

CHAPTER 13

Surface-Science Models with Chemical Complexity

CHRISTIAN HESS

Technische Universität Darmstadt, Eduard-Zintl-Institut für Anorganische und Physikalische Chemie, Petersenstr. 20, 64287 Darmstadt, Germany

13.1 Introduction

The ultimate goal of catalyst characterization is a molecular-level understanding of the catalytically active surface. Despite extensive research activities over the last decades, there is only little understanding regarding the mode of operation of metal-oxide catalysts. An essential component of many effective selective oxidation catalysts is vanadium. Prominent examples are VPO catalysts used commercially for maleic anhydride production from butane (see Chapter 6),[1] and MoVTeNb oxides that constitute today's most efficient catalysts for propane-selective oxidation to acrylic acid (see Chapter 4).[2] In general, very little has been reported concerning the origin of their high catalytic performance. To gain insight into the mode of operation of catalysts, in the past, a variety of model catalysts ranging from single crystals to impregnated planar systems have been employed.[3] In the case of metal-oxide catalysts for selective oxidation reactions the reactivity is strongly linked to the presence of defects and thus suitable model systems need to be developed.[4,5] Regarding vanadium oxide, *e.g.* (partially) reduced single crystals,[6] nanoparticles,[7] supported planar models,[8–10] or molecular compounds such as vanadium calixarene complexes have been employed.[11] Recently, we have shown that silica-SBA-15-supported vanadium-oxide catalysts represent valuable model systems

RSC Nanoscience & Nanotechnology No. 19
Nanostructured Catalysts: Selective Oxidations
Edited by Christian Hess and Robert Schlögl

Published by the Royal Society of Chemistry, www.rsc.org

(see Chapter 12),[12,13] which exhibit high selectivity in a number of selective oxidation reactions such as methanol oxidation to formaldehyde or oxidative dehydrogenation of propane.[12–15] Thereby, the silica support can be considered as an inert diluent, which stabilizes the nanostructured surface vanadium oxide in its highly dispersed state during reaction, preventing it from sintering.

As part of our strategy for understanding the properties of silica-supported vanadium oxide in detail, we intend to bridge the gap between powder and well-defined thin-film and cluster model systems prepared by physical methods. Previous studies have shown that chemically prepared planar systems can represent realistic models for industrial catalysts.[3,16–18] Importantly, in contrast to their real counterparts, charging effects are largely removed. Therefore, surface-analytical tools such as X-ray photoelectron spectroscopy (XPS), secondary ion mass spectroscopy (SIMS) or ion scattering can readily be applied. A versatile tool to prepare such planar metal-oxide model systems is spin-coating impregnation developed by Kuipers *et al.* and van Hardeveld *et al.*[19–21] Based on the model proposed by van Hardeveld the amount of material deposited during spin coating can be predicted quantitatively.[21] Thus, the spin-coating impregnation seems like a promising way to prepare silica-supported vanadium-oxide model systems with a high degree of control over the vanadium loading. Previously, silica-supported vanadium-oxide model catalysts have been prepared by vapor deposition and grafting methods and characterized by AFM and XPS.[9,10] These samples contained a considerable amount of crystalline V_2O_5 leading to a mixture of structurally different vanadium-oxide species. However, to approach the ultimate goal of establishing a correlation between structure and activity, the controlled preparation of structurally homogeneous vanadium-oxide model systems is an important prerequisite.

In the following the use of spin-coating impregnation to prepare surface-science models with chemical complexity is illustrated for silica-supported vanadium oxide. By comparison with silica-SBA-15-supported vanadium-oxide powder catalysts (see Chapter 12), it is shown that the properties of these planar model systems as studied here by XPS strongly resemble those of their powder analogs both in the hydrated and dehydrated state.

13.2 Surface-Science Models with Chemical Complexity

13.2.1 Preparation

The most common approach to realistic surface-science models is based on three steps. First, a thin oxide film such as silica or alumina is deposited on an inert, electrically conducting substrate (silicon, aluminum, gold). The use of thin oxide layers largely prevents charging effects even in the case of insulating oxide supports. In the second step, the surface is rehydroxylated. The presence of hydroxyl groups is essential to anchor the active phase to the support. Finally, the active phase is deposited by spin-coating impregnation.

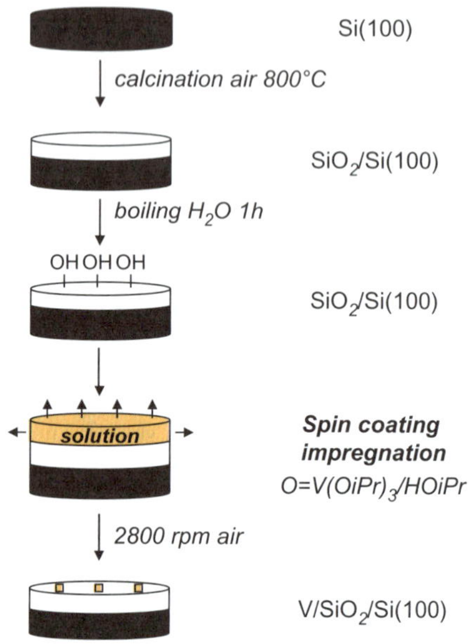

Figure 13.1 Preparation of $V/SiO_2/Si(100)$ surface-science models by spin-coating impregnation.

This procedure is illustrated in Figure 13.1 for silica-supported vanadium-oxide systems, denoted as $V/SiO_2/Si(100)$, and described in detail in the following.

Prior to use, the Si(100) wafers (p-type doped with boron, Silchem) were cleaned by successive treatment in an ultrasonic bath using chloroform, acetone, ethanol and bidistilled water. Then the wafers were calcined in air at 800 °C to form an amorphous silica overlayer. The calcined samples were rehydroxylated in boiling water for 1 h and finally dried in air overnight. Spin coating was performed at 20 °C in air at 2800 rpm using a commercial spin coater (Laurell Technol., model WS-400A-6TFM). As vanadium precursor, 2-propanol solutions of vanadium isopropoxide (Alfa Aesar, 95-99%) were used. Prior to spinning 1 ml of solution was squirted through a 0.45 μm filter onto the center of the wafer piece. After impregnation, the samples were calcined in air at 550 °C for 2 h. The vanadium loading was determined based on the model by van Hardeveld *et al.*[21]

13.2.2 Characterization

The model by van Hardeveld *et al.* allows for a quantitative prediction of the amount of active phase material deposited.[21] The validity of this model has been tested for silica-supported molybdenum oxide ($Mo/SiO_2/Si(100)$) by comparing the theoretically predicted values for the amount of deposited molybdenum with those determined experimentally by Rutherford backscattering.[21] The accuracy of the predicted values was found to be within 20% of the deposited material.

As similar conditions were applied here using a 2-propanol solution compared to those studied previously for ethanol and butanol solutions, the evaporation process can be expected to be determined by mass transfer in the vapor phase.[21] The amount of deposited material is then calculated according to $L_v = 4.87 (t_{evap})^{-[0.5]} C_0$. L_v, t_{evap} and C_0 are the metal loading in atoms/nm^2, the evaporation time in seconds and the solute concentration used for spin coating, respectively. The evaporation time can be determined experimentally by the disappearance of the Newton rings. For a 2-propanol solution a value of 4.5 s is obtained under the conditions applied here (air, 2800 rpm). The resulting vanadium loadings of the prepared $V/SiO_2/Si$ samples are shown in Table 13.1.

XPS analysis is a well-suited tool to characterize thin silica-supported model systems due to its sensitivity. Besides, in comparison with silica-based powder catalysts charging effects are largely reduced (to ~0.5 eV) and the resolution is better (see Figure 13.6). As an example, Figure 13.2 depicts spectra of the O1s

Table 13.1 Atomic ratios (%) of the detected elements of $SiO_2/Si(100)$ and $V/SiO_2/Si(100)$ as obtained from XPS.

	Loading (V/nm^2)	*O1s*	*$V2p_{3/2}$*	*C1s*	*Si2p*	*V:Si*
$SiO_2/Si(100)$	0	66.7	0	4.2	29.1	0
0.7 $V/SiO_2/Si(100)$	0.68	66.0	0.4	2.0	31.6	0.013
1.4 $V/SiO_2/Si(100)$	1.37	65.2	0.8	5.7	28.3	0.032
2.7 $V/SiO_2/Si(100)$	2.73	64.7	1.3	6.0	28.0	0.048
4.8 $V/SiO_2/Si(100)$	4.8	64.1	1.8	6.0	28.1	0.064
7.3 $V/SiO_2/Si(100)$	7.25	62.9	2.4	7.3	27.4	0.088

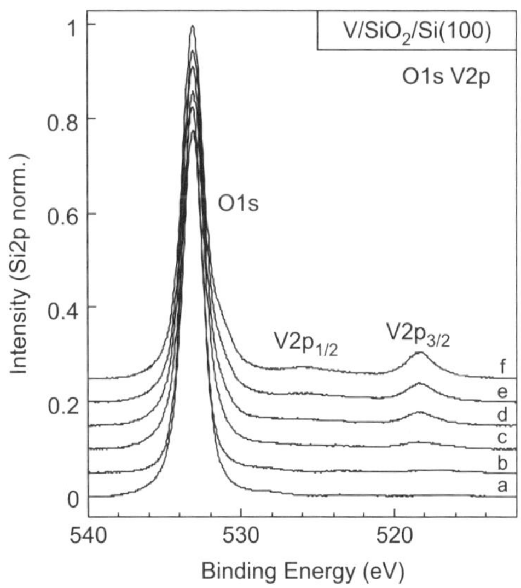

Figure 13.2 O1s and V2p XP spectra of $V/SiO_2/Si(100)$ samples after spin-coating impregnation. The V loading is (a) 0, (b) 0.7, (c) 1.4, (d) 2.7, (e) 4.8 and (f) 7.3 V/nm^2. The spectra are offset for clarity.

V2p region of $V/SiO_2/Si$ samples as a function of vanadium loading. The results will be discussed in detail below. The experimental conditions and data-analysis routines correspond to those applied previously.[22] Intensities were normalized to the Si2p peak area (Si2p norm.).

13.3 Studies on Silica-Supported Vanadium Oxide

13.3.1 XPS Results on Planar Systems

The atomic ratios of the detected elements, as obtained from the XPS analysis, are summarized in Table 13.1. For all $V/SiO_2/Si(100)$ samples only oxygen, vanadium, carbon and silicon were detected. Based on the XPS results no significant increase in the amount of carbon was observed for the supported vanadia samples as compared to the bare support after calcination in air at 550 °C, which suggests that the spin-coating impregnation introduces only negligible amounts of carbon impurities. Figure 13.2 shows the O1s V2p region of a series of $V/SiO_2/Si(100)$ samples with loadings between 0.7 and 7.3 V/nm^2 after satellite subtraction. As reference, the spectrum of the bare $SiO_2/Si(100)$ support is shown on the bottom (spectrum a). Besides the strong O1s and $V2p_{3/2}$ lines, the weaker $V2p_{1/2}$ emission is observed at higher loadings. With increasing vanadium loading, a shoulder at around 530.5 eV develops at the low-energy side of the O1s emission of the silica support at 532.8 eV. Binding energies around 530 eV are typical for oxygen in binary vanadium oxides.[23] Therefore, it is assigned to oxygen of the deposited vanadium oxide.

Figure 13.3 displays the corresponding $V2p_{3/2}$ spectra after subtraction of a Shirley background. The spectrum of the bare support shows structure in the

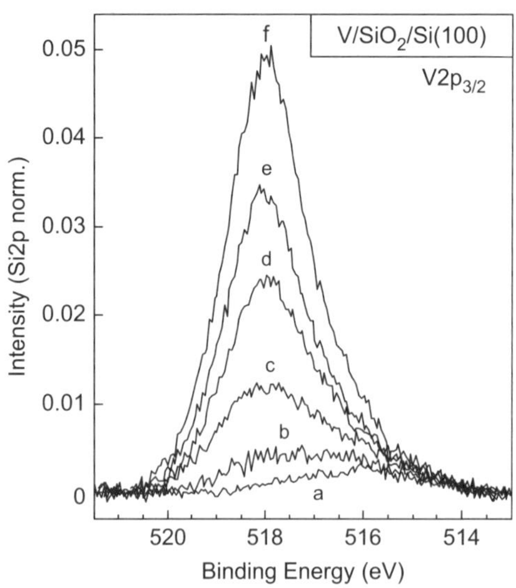

Figure 13.3 XP $V2p_{3/2}$ spectra of $V/SiO_2/Si(100)$ samples after spin-coating impregnation. The V loading is (a) 0, (b) 0.7, (c) 1.4, (d) 2.7, (e) 4.8 and (f) 7.3 V/nm^2.

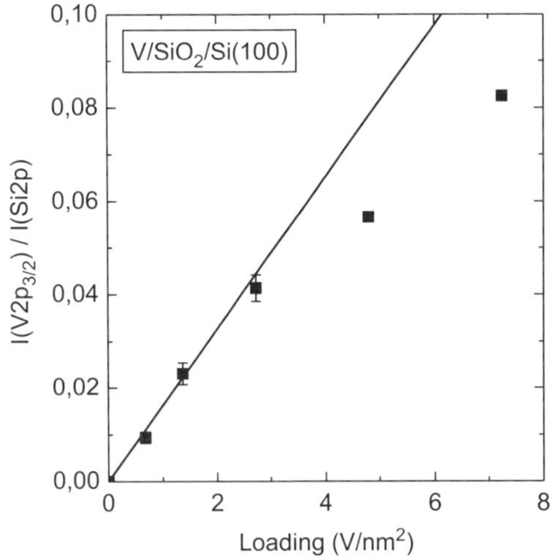

Figure 13.4 Integrated background subtracted XP $V2p_{3/2}$ intensities of $V/SiO_2/Si(100)$ samples after spin-coating impregnation as a function of V loading.

$V2p_{3/2}$ region due to inadequate removal of the $K\alpha_5$ satellite of the nearby O1s line. Hence, for further analysis the support spectrum was used for background subtraction. The resulting Si2p normalized $V2p_{3/2}$ intensities (integrated areas) as a function of vanadium loading are plotted in Figure 13.4. A linear correlation between XP intensity and vanadium loading is observed up to a vanadium loading of about 2.7 V/nm^2. For comparison, silica-supported vanadium-oxide powder catalysts have shown maximum dispersions within 2.3–2.6 V/nm^2 for samples prepared by ion exchange (2.3 V/nm^2),[12] multiple incipient wetness impregnation (2.6 V/nm^2),[24] or gas-phase deposition (2.3 V/nm^2)[25] of vanadium isopropoxide. For higher loadings of vanadium on silica SBA-15 (>2.3 V/nm^2), the formation of crystalline V_2O_5 has been observed by visible Raman spectroscopy.[12] Regarding the interpretation of this behavior it should be mentioned that for XPS analysis spin-coated vanadium-oxide samples in their "as is", *i.e.* hydrated, state were used, as the preparation was performed under ambient conditions. Previous visible Raman studies have shown that the spectra of the hydrated state of silica-supported vanadium-oxide powder samples show a strong resemblance to those of $V_2O_5\ nH_2O$ gels forming water-containing vanadium-oxide aggregates.[26] Furthermore, there is no indication of significant structural changes of the hydrated vanadium-oxide species within 0.7–4.7 V/nm^2. Thus, it is tempting to conclude that also for the planar model system the decrease in the vanadium XP signal at higher loadings is caused by the decreased dispersion of V_2O_5 (micro)crystallites as compared to the hydrated vanadium-oxide aggregates and therefore reduced XPS visibility of V_2O_5 crystallites. While visible Raman spectroscopy does not have sufficient sensitivity for vanadium oxide in such planar model systems, conclusive evidence on this matter may be expected from future experiments using UV Raman spectroscopy, which has been successfully employed to study $Mo/SiO_2/Si$ systems.[27]

13.3.2 Comparison with Powder Catalysts

13.3.2.1 Loading Dependence

The XPS results for the planar $V/SiO_2/Si$ samples were compared to those of silica-SBA-15-supported vanadium-oxide powder catalysts (see Chapter 12).[12,13] First, the results of the physical characterization of the studied powder samples will be briefly described. Details are given elsewhere.[26] Table 13.2 summarizes the results of the N_2 physisorption analysis of bare SBA-15 and SBA-15-supported vanadium-oxide samples with loadings between 0.7 and 4.7 V/nm^2.[26]

As evidenced by XRD, the hexagonal structure of SBA-15 is preserved in the presence of vanadium oxide. Deposition of vanadium oxide leads to a decrease in the surface area, pore radius and pore volume, which suggests that the vanadium-oxide species are located inside the pores of SBA-15, coating the inner walls of the mesoporous matrix. For vanadium loadings higher than 7.2 wt% V, Raman spectroscopy has shown the presence of 3-dimensional V_2O_5 crystallites. Therefore, the drop in surface area observed for the 10.8 wt% V/SBA-15 sample is most probably caused by blocking of some of the channels.

Table 13.3 displays the atomic ratios of the detected elements, as obtained from the XPS analysis as well as the corresponding V:Si ratios. For all

Table 13.2 BET characteristics of blank SBA-15 and the SBA-15-supported vanadium oxides.

	V *(wt%)*	V *(nm^2)*	V *(mmol/g)*	S_{BET} *(m^2/g)*	d_p *(nm)*	V_p *(mL/g)*
SBA-15	0.0	0.0	0.0	897	7.0	1.1
2.7 wt% V/SBA-15	2.7	0.7	0.53	445	6.7	0.5
5.4 wt% V/SBA-15	5.4	1.4	1.05	440	6.6	0.5
10.8 wt% V/SBA-15	10.8	4.7	2.12	273	5.5	0.3

Table 13.3 Atomic ratios (%) of the detected elements of blank SBA-15 and the silica-SBA-15-supported vanadium oxides as obtained from XPS.

	O1s	*V2p$_{3/2}$*	*C1s*	*Si2p*	*V/Si*
SBA-15	69.4		0.6	30.0	
2.7 wt% V/SBA-15	70.1	0.7	0.7	28.5	0.025
5.4 wt% V/SBA-15	68.2	1.5	1.3	29.0	0.052
10.8 wt% V/SBA-15	70.5	2.4	0.9	26.2	0.092

V/SBA-15 samples only oxygen, vanadium, carbon and silicon were detected. There is no significant increase in the amount of carbon for the SBA-15-supported vanadium-oxide samples compared to the bare SBA-15 support. These results indicate that during the calcination step essentially all organic residues, which were introduced during the functionalization of SBA-15, were removed.

Figures 13.5 and 13.6 compare the XP O1s V2p region of planar (bottom) and powder samples (top), respectively. For each system, a series of spectra is depicted corresponding to vanadium loadings of 0.7, 1.4, and 4.8 V/nm^2, respectively. The O1s V2p region shows three emissions due to O1s, V2p$_{1/2}$ and V2p$_{3/2}$. The O1s spectra of the planar and powder system at a loading of 1.4 V/nm^2 are displayed in the left panel of Figure 13.6. The corresponding linewidth values (FWHM) are 1.65 and 2.0 eV. For comparison, the Si2p emissions (not shown) of the same planar and powder samples exhibit a linewidth (FWHM) of 1.8 and 2.2 eV, respectively. These results can either be explained by charging or inhomogeneous broadening of the silica support of the powder sample.[26,28]

The V2p$_{3/2}$ region of the planar and the powder model samples are displayed in more detail in Figure 13.7. The close resemblance of the corresponding V2p$_{3/2}$ emissions regarding position and width is evident. Also, there is no apparent difference in the shape of the spectra with increasing loading. The right panel of Figure 13.6 displays the V2p$_{3/2}$ emission of the planar and powder system at a loading of 1.4 V/nm^2. The corresponding linewidth values (FWHM) of the planar (2.5 eV) and powder (2.6 eV) system compare favorably

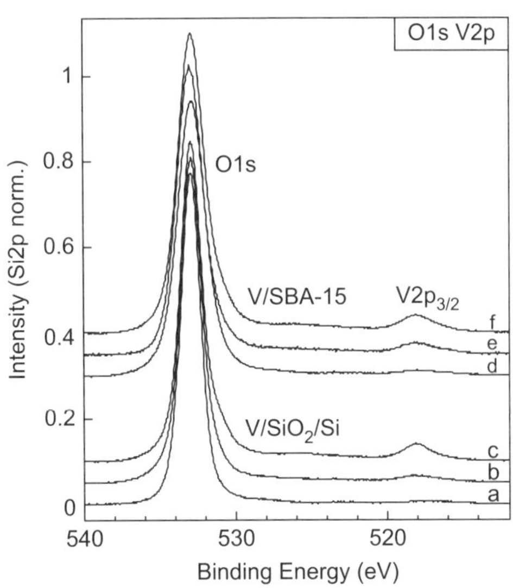

Figure 13.5 O1s and V2p XP spectra of V/SiO$_2$/Si(100) samples with a loading of (a) 0.7, (b) 1.4, (c) 4.8 V/nm^2, and V/SBA-15 samples with a loading of (d) 0.7, (e) 1.4, (f) 4.7 V/nm^2. The spectra of the V/SBA-15 samples are offset for clarity.

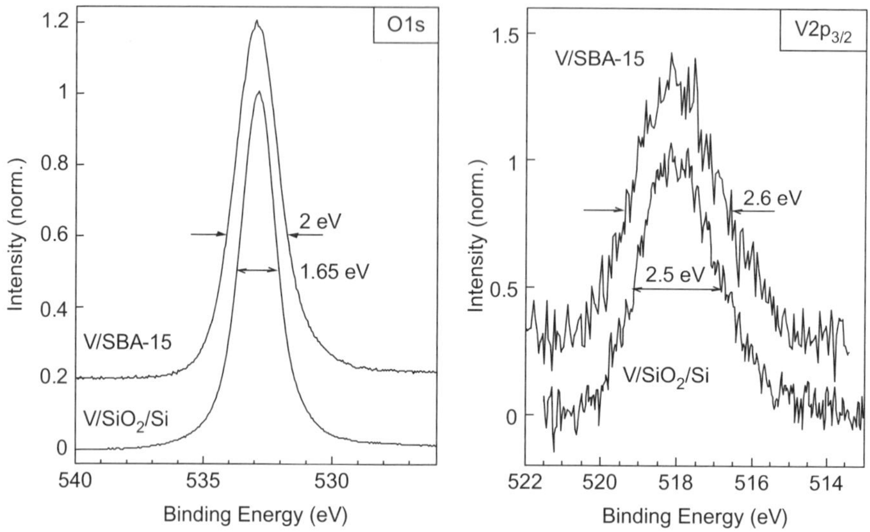

Figure 13.6 Left panel: Normalized XP O1s spectra of $V/SiO_2/Si(100)$ (bottom) and V/SBA-15 (top) at a loading of 1.4 V/nm^2. Right panel: Normalized XP $V2p_{3/2}$ spectra of $V/SiO_2/Si(100)$ (bottom) and V/SBA-15 (top) at a loading of 1.4 V/nm^2. Spectra are offset for clarity.

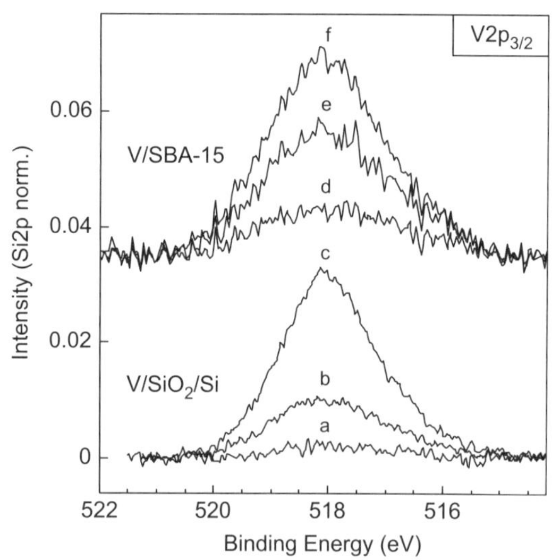

Figure 13.7 XP $V2p_{3/2}$ spectra of $V/SiO_2/Si(100)$ samples with a loading of (a) 0.7, (b) 1.4, (c) 4.8 V/nm^2, and V/SBA-15 samples with a loading of (d) 0.7, (e) 1.4, (f) 4.7 V/nm^2. The spectra of the V/SBA-15 samples are offset for clarity.

with those observed for well-ordered vanadium-oxide films containing vanadium in the same oxidation states, indicating a high degree of homogeneity of the deposited vanadium oxide for both systems studied here.[29]

A quantitative description of the $V2p_{3/2}$ emissions is given below. In contrast to the O1s and Si2p results no significant differences in the $V2p_{3/2}$ linewidths of the planar and the powder sample are observed. These observations suggest that the mechanisms influencing the powder support are too small to be detectable in the case of the deposited vanadium-oxide samples. At the highest vanadium loading shown in Figure 13.7 (4.8 V/nm^2) besides highly dispersed vanadium-oxide species a small amount of V_2O_5 is present, as evidenced by previous visible Raman experiments on the powder 10.8 wt% V/SBA-15 sample.[26] Therefore, one can conclude that the $V2p_{3/2}$ emissions of the powder samples strongly resemble those of their planar counterparts up to the point of maximum loading of highly dispersed species beyond which in case of the powder sample V_2O_5 formation is observed.

13.3.2.2 Dehydrated State

Spectra under dehydrated conditions were recorded at room temperature after sample treatment in 20% O_2/He (50 ml/min) at 300 °C for 60 min and subsequent transfer to the analysis chamber without exposure to air (quasi-*in-situ*). Figure 13.8 compares normalized $V2p_{3/2}$ spectra of the planar and the powder model system at a vanadium loading of 1.4 V/nm^2. Spectra a and b correspond to the hydrated, and spectra c and d to the dehydrated state of the 1.4 V/SiO_2/Si(100) (5.4 wt% V/SBA-15) sample. No noticeable change of the XP spectra

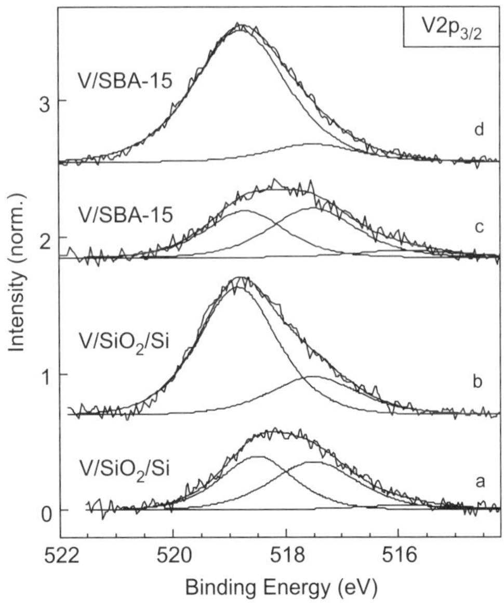

Figure 13.8 Normalized XP $V2p_{3/2}$ spectra of (a) 1.4 V/SiO_2/Si(100), (b) dehydrated 1.4 V/SiO_2/Si(100), (c) 5.4 wt% V/SBA-15 and (d) dehydrated 5.4 wt% V/SBA-15 together with least-square fits to the data. The spectra are offset for clarity. For details see text.

was observed after several months of sample storage in air. This suggests that the hydrated samples represent a fully equilibrated state of silica-supported vanadium oxide in air at room temperature. The planar and the powder model sample show very similar $V2p_{3/2}$ emissions both in their hydrated as well as their dehydrated state. Most striking, for both samples, dehydration leads to a significant increase in total intensity and a substantial redistribution of spectral weight to higher binding energies (BE).

Quantitative analysis was performed using three fit functions. The results from the peak-fitting analysis are summarized in Table 13.4. Dehydration leads to an intensity increase of the band around 518.8 eV from 47% (43%) to 76% (89%) for 1.4 V/SiO_2/Si(100) (5.4 wt% V/SBA-15). Similarly, the band centered at 517.5 eV shows an intensity decrease from 49% (49%) to 11% (25%). In their hydrated state, both samples also show a small contribution around 515.9 eV. The bands at 515.9 eV and 517.5 eV can be assigned to V^{3+} and V^{5+}, respectively, by comparison with literature data on $V2p_{3/2}$ BE for well-ordered vanadium-oxide films and binary vanadium oxides.[23,29] As can be seen from Table 13.4, for the dehydrated samples no V^{3+} was observed. Therefore, both samples are fully oxidized during the treatment in oxygen flow at elevated temperatures leading to dehydration. Please note that the widths of the three fit functions used for quantitative analysis (see Table 13.4) are lower than those used to describe the well-ordered vanadium-oxide films,[29] confirming the homogeneity of the model systems, in particular the powder samples. As EPR experiments on dehydrated 2.7 wt% V/SBA-15 showed no V^{4+} signal and those on hydrated 2.7 wt% V/SBA-15 gave a V^{4+} signal corresponding to only 0.3% of the total amount of V, the quantitative analysis was restricted to V^{5+} and V^{3+} species.

Evidently, the $V2p_{3/2}$ contribution around 518.8 eV can not solely be described by a change in oxidation state. However, Wertheim has shown that final-state effects can result in size-dependent BE shifts for small conducting particles on insulating substrates as a result of the Coulomb interaction between the photoelectron and the positively charged particle.[30] Such a situation applies to small vanadium oxide particles with V ions in oxidation state 3+ and vanadyl groups at the surface, which are known to be formed

Table 13.4 Results of the $V2p_{3/2}$ analysis of SiO_2/Si(100) and SBA-15-supported vanadium oxides with a loading of 1.4 V/nm^2.

	position (eV)	*width (eV)*	*%*
1.4 V/SiO_2/Si(100)	518.5	1.7	46.5
(hydr)	517.5	2.0	48.5
	515.9	2.2	5.0
1.4 V/SiO_2/Si(100)	518.8	1.8	75.5
(dehydr)	517.5	2.0	24.5
5.4 wt% V/SBA-15	518.7	1.8	43.0
(hydr)	517.5	2.0	49.2
	515.9	2.2	7.8
5.4 wt% V/SBA-15	518.8	2.1	88.7
(dehydr)	517.5	2.0	11.3

under reduced-pressure conditions.[31] Therefore, our XPS results suggest that dehydration of the two model systems is accompanied by a decrease in cluster size, *i.e.* an increase in the dispersion of the catalyst.[26,32] Based on the position of the high BE contribution in both cases dehydration leads to surface vanadia species of comparable cluster size, emphasizing the close resemblance of the powder and the planar vanadium-oxide model system. Previous studies on silica-SBA-15-supported vanadium oxides using Raman and UV-Vis spectroscopy have shown that the dehydration process is also associated with changes in the molecular structure of the surface vanadium-oxide species, which transforms from a hydrated vanadium oxide xerogel with vanadium in pseudo-octahedral coordination to smaller ensembles of dehydrated vanadium oxide with vanadium in tetrahedral coordination.[26,32]

13.4 Conclusions

Spin-coating impregnation allows for the controlled preparation of planar models of silica-supported vanadium-oxide catalysts. These chemically prepared surface-science model systems have the potential to establish a close link between well-studied thin-film and cluster systems prepared by physical methods on the one hand and powder catalyst systems on the other. Here, we demonstrate their strong resemblance with recently developed silica-SBA-15-supported powder catalysts with respect to dispersion properties and electronic structure both in their hydrated and dehydrated state. Our results underline that the materials gap between chemically prepared surface-science and powder model systems can be successfully bridged. It will be a future challenge to develop analytical tools to study the catalytic performance of such planar models in selective oxidation reactions.

Acknowledgement

The experiments were performed in the Department of Inorganic Chemistry at the Fritz-Haber-Institut of the Max-Planck-Gesellschaft. The author would like to thank Rita Herbert for preparation of some of the powder samples, Gisela Lorentz for performing the BET measurements, Oksana Storcheva for performing the EPR experiments and Ute Wild for technical assistance with the XPS setup. The author is grateful to Robert Schlögl for fruitful discussions and continuous support of the project. The work was supported by the Deutsche Forschungsgemeinschaft (SFB 546, Emmy Noether fellowship).

References

1. G. Centi, F. Cavani and F. Trifiro, *Selective Oxidation by Heterogeneous Catalysis*, Kluwer Academic/Plenum Publishers, New York, 2000.
2. M. M. Lin, *Appl. Catal. A*, 2001, **207**, 1.

3. P. J. L. Gunter, J. W. Niemantsverdriet, F. H. Ribeiro and G. A. Somorjai, *Catal. Rev. Sci. Eng.*, 1997, **39**, 77.
4. R. Schlögl and S. B. Abd Hamid, *Angew. Chem. Int. Ed.*, 2004, **34**, 1628.
5. R. Schlögl, A. Knop-Gericke, M. Hävecker, U. Wild, D. Frickel, T. Ressler, R. E. Jentoft, J. Wienold, G. Mestl, A. Blume, O. Timpe and Y. Uchida, *Top. Catal.*, 2001, **15**, 219.
6. H.-J. Freund, *Top. Catal.*, 2008, **48**, 137.
7. M. Hävecker, N. Pinna, K. Weiß, H. Sack-Kongehl, R. E. Jentoft, D. Wang, M. Swoboda, U. Wild, M. Niederberger, J. Urban, D. S. Su and R. Schlögl, *J. Catal.*, 2005, **236**, 221.
8. J. Nickl, R. Schlögl, A. Baiker, H. Knözinger and G. Ertl, *Catal. Lett.*, 1989, **3**, 379.
9. J. Nickl, Ch. Schild, A. Baiker, M. Hund and A. Wokaun, *Fresenius J. Anal. Chem.*, 1993, **346**, 79.
10. Ch. Schild, J. Engweiler, J. Nickl, A. Baiker, M. Hund, M. Kilo and A. Wokaun, *Catal. Lett.*, 1994, **25**, 179.
11. C. Limberg, *Eur. J. Inorg. Chem.*, 2007, 3303.
12. C. Hess, *ChemPhysChem*, 2009, **10**, 319.
13. C. Hess, *Surf. Sci.*, 2006, **600**, 3695.
14. C. Hess, I. J. Drake, J. D. Hoefelmeyer, T. D. Tilley and A. T. Bell, *Catal. Lett.*, 2005, 1.
15. A. Dinse, S. Khennache, B. Frank, C. Hess, R. Herbert, S. Wrabetz, R. Schlögl and R. Schomäcker, *J. Mol. Catal. A*, 2009, **307**, 43.
16. J. W. Niemantsverdriet, A. F. P. Engelen, A. M. de Jong, W. Wieldraaijer and G. J. Kramer, *Appl. Surf. Sci.*, 1999, **144–145**, 366.
17. J. C. Muijsers, T. Weber, R. M. Vanhardeveld, H. W. Zandbergen and J. W. Niemantsverdriet, *J. Catal.*, 1995, **157**, 698.
18. P. C. Thüne, C. P. J. Verhagen, M. J. G. van der Boer and J. W. Niemantsverdriet, *J. Phys. Chem. B*, 1997, **101**, 8559.
19. E. W. Kuipers, C. Laszlo and W. Wieldraaijer, *Catal. Lett.*, 1993, **17**, 71.
20. E. W. Kuipers, C. Doornkamp, W. Wieldraaijer and R. E. van der Berg, *Chem. Mater.*, 1993, **5**, 1367.
21. R. M. van Hardeveld, P. L. J. Gunter, L. J. van IJzendoorn, W. Wieldraaijer, E. W. Kuipers and J. W. Niemantsverdriet, *Appl. Surf. Sci.*, 1995, **84**, 339.
22. C. Hess, U. Wild and R. Schlögl, *Microp. Mesop. Mater.*, 2006, **95**, 339.
23. G. A. Sawatzky and D. Post, *Phys. Rev. B*, 1979, **20**, 1546.
24. X. Gao, S. R. Bare, B. Weckhuysen and I. E. Wachs, *J. Phys. Chem. B*, 1998, **102**, 10842.
25. J. Keranen, C. Guimon, E. Iiskola, A. Auroux and L. Niinisto, *J. Phys. Chem. B*, 2003, **107**, 10773.
26. C. Hess, R. Herbert and G. Tzolova-Müller, *J. Phys. Chem. C*, 2007, **111**, 9471.
27. B. L. Mojet, L. Coulier, J. van Grondelle, J. W. Niemandsverdriet and R. A. van Santen, *Catal. Lett.*, 2004, **95**, 1.

28. C. Hess and R. Schlögl, *Chem. Phys. Lett.*, 2006, **432**, 139.
29. A.-C. Dupuis, M. Abu Haija, B. Richter, H. Kuhlenbeck and H.-J. Freund, *Surf. Sci.*, 2003, **539**, 99.
30. G. K. Wertheim, *Z. Physik. B*, 1987, **66**, 53.
31. S. Guimond, M. Abu Haija, S. Kaya, J. Lu, J. Weissenrieder, S. Shaikhutdinov, H. Kuhlenbeck, H.-J. Freund, J. Döbler and J. Sauer, *Top. Catal.*, 2006, **38**, 117.
32. C. Hess, *J. Catal.*, 2007, **248**, 120.

CHAPTER 14

Mechanistic Aspects of Short Contact Time Oxidative Functionalization of Propane and Ethane from Temporal Analysis of Products

EVGENII V. KONDRATENKO

Leibniz Insitute for Catalysis at the University of Rostock, Albert-Einstein-Str. 29a, D-18059 Rostock, Germany

14.1 Introduction

Ethylene and propene are the key building blocks of the petrochemical industry. The world demand for these olefins increases continuously. Currently, their industrial production is based on crude-oil technologies. Natural gas, reserves of which are estimated to exceed those of crude oil, may be considered now and, particularly, in the future as a preferable raw material replacing crude oil. Ethane, propane and butane are constituents (up to 20%) of natural gas. Therefore, their selective oxidative functionalization has become a permanent challenge during the last decade. Numerous catalytic systems have been evaluated for oxidative conversion of ethane and propane. The relevant results are summarized in refs. 1 and 2. However, this is still a challenging area. The main challenge is to minimize the formation of carbon oxides.

RSC Nanoscience & Nanotechnology No. 19
Nanostructured Catalysts: Selective Oxidations
Edited by Christian Hess and Robert Schlögl

Published by the Royal Society of Chemistry, www.rsc.org

Since the pioneering work of Huff and Schmidt,[3] the autothermal oxidative dehydrogenation of ethane (ODE) and propane (ODP) to olefins at short contact times (millisecond range) established a new promising technology for olefin production.[4–9] Catalytic performance ($S(C_2H_4) = 87\%$ at $X(C_2H_6) = 77\%$[10]) of Pt-coated monolith reactors in the ODE reaction is comparable with conventional steam cracking. Baerns and associates have reported similar ODE performance over oxide-based catalytic materials.[11,12] For the ODP reaction at short contact times, ethylene, but not propene, was found as the main olefin. Kondratenko and Ovsitser[13] have recently introduced a novel process scheme and a catalyst for selective short-contact time conversion of C_2H_6 to C_2H_4 and H_2 with simultaneous near to 100% removal of N_2O in a broad range of concentrations (0.15 to 40 vol.%). The suggested process scheme has potential for N_2O-abatement technologies in the production of adipic and/or nitric acids.

Concerning the mechanism of the short contact time ODE and ODP reactions, it was suggested that both heterogeneous and homogeneous reactions play an essential role in ethane and propane transformations.[5,14,15] However, a detailed mechanistic scheme describing the interplay between the above concepts is missing. This is due to the fact that the potential of most common spectroscopic techniques applied for mechanistic and kinetic investigations of oxidative dehydrogenation of light alkanes at short contact times and high temperatures is very limited because the concentration of surface and gas intermediates is usually very low in order to be properly detected. This is why the thorough mechanistic and microkinetic analysis is among the few tools that can be used for investigation of these processes. Generally, transient techniques provide more mechanistic information than steady-state experiments, because they permit study of the individual reaction steps. Therefore, they are very helpful for unraveling complex reaction networks in terms of near to elementary reaction steps. Transient experiments in the temporal analysis of products (TAP) reactor are very suitable for mechanistic analysis of short contact time alkane dehydrogenation due to the following reasons: (i) TAP experiments are performed at millisecond contact times and with submillisecond time resolution; (ii) isothermal conditions are ensured; (iii) gas-phase reactions are minimized as compared to heterogeneous ones.

Based on the above backgrounds, steady-state ambient pressure catalytic tests, microkinetic and mechanistic investigations in the TAP reactor were combined to elucidate the reaction network of short contact time ODE and ODP reactions over two families of catalysts, *i.e.* materials based on nonreducible (rare-earth and alkaline-earth) and reducible (vanadia-based) metal oxides. The catalysts were chosen taking into account the fact that they differ in their ability for generation of active surface oxygen species from gas-phase O_2. It is well known that this catalyst property influences the catalytic performance.[1,16] This work was aimed at further elucidation of the factors determining catalytic performance, which are required for rational catalyst or process design. To this end, microkinetic evaluation of O_2 activation was performed. The developed microkinetics was used for correlation with catalytic performance of reducible and nonreducible metal oxides in short contact time alkane dehydrogenation. It was shown that the TAP-derived microkinetics of O_2 activation is not only a tool for deriving mechanistic insights into O_2 activation but also provides useful guidelines for catalyst design.

14.2 Catalytic Performance of Nonreducible and Reducible Metal Oxides

Catalytic materials based on oxides of rare-earth and alkaline-earth metals lead to ignition of C_2H_6/O_2 reaction mixtures producing enhanced heat generation.[11,12] Similar catalytic performance has been reported for catalytic materials based on noble metals (Pt and Rh).[4–9] This reaction heat resulted in an increase in temperature (T_{max}) up to *ca.* 1200 K in the catalyst bed and the catalytic reaction proceeded without external heating. When C_2H_6 reacted with N_2O over $Na_{0.009}CaO$ catalyst at millisecond contact time, the catalyst temperature increased above 1100 K too.[13] This is due to the fact that the heat production *via* the chemical reaction (selective and nonselective oxidation of hydrocarbons and N_2O decomposition) is faster than the heat removal from the reactor. Since the reactor geometry and its material influence the heat transport in the reactor and therefore the resulting temperature gradients, all catalytic tests were performed in the same reactor. The reactor used in this study is schematically shown in Figure 14.1.

Ignition of alkane/oxygen mixtures takes place above a certain furnace temperature, which is called the ignition temperature ($T_{ignition}$). The ignition temperature strongly depends on the nature of catalytic material and varies from 613 to 983 K (Table 14.1). Besides, the studied catalytic materials also showed pronounced differences in their catalytic performance. A maximal

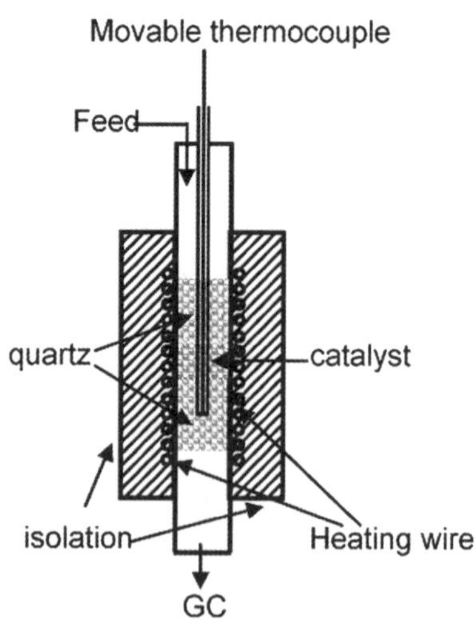

Figure 14.1 Simplified scheme of catalytic reactor for short contact time alkane oxidation.

Table 14.1 Catalytic performance of different materials in the ODE and ODP reactions at short contact times: (a) $C_2H_6/O_2/N_2 = 45/18/37$, (b) $C_2H_6/O_2/N_2 = 40/20/40$, and (c) $C_3H_8/O_2/N_2 = 40/20/40$.

Oxidative dehydrogenation of ethane					
Catalysts	$T_{ignition}/K$	T_{max}/K	$X(C_2H_6)/\%$	$Y(C_2H_4)/\%$	*Ref.*
$SmSr_{0.2}O_x$[a]	723	1061	43	30	18
$Sr_{1.0}La_{1.0}Nd_{1.0}O_x$[b]	613	1137	62	45	12
$Na_{0.001}CaO_x$[b]	643	1078	53	36	present
$SmNa_{0.028}P_{0.014}O_x$[a]	603	1023	43	28	18
$SmCa_5O_x$[a]	863	1102	46	31	18
$Na_{0.009}CaO_x$[b]	873	1200	78	56	present
$Na_{0.065}CaO_x$[b]	Slow reaction, no ignition below 973 K				present

Oxidative dehydrogenation of propane						
Catalysts	$T_{ignition}/K$	T_{max}/K	$X(C_3H_8)/\%$	$Y(C_3H_6)/\%$	$Y(C_2H_4)/\%$	*Ref.*
$SmSr_{0.2}O_x$[c]	500	831	30.3	5.2	8.9	11
$SmNa_{0.0028}P_{0.0014}O_x$[c]	539	833	30.5	5.8	9.2	11
$LaSr_{0.02}O_x$[c]	692	933	39.7	7.7	13.5	11
$VO_x(5\%)/Al_2O_3$[c]	623	931	23	11.3	0.46	present

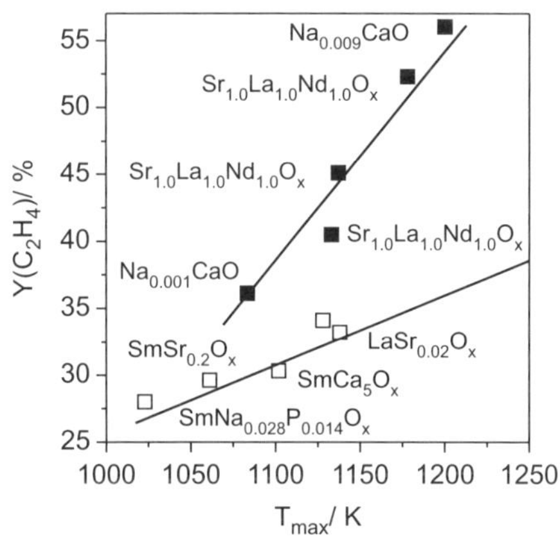

Figure 14.2 Ethylene yield as a function of temperature measured in the catalyst bed. Each symbol represents a catalyst. $C_2H_6/O_2/N_2 = 45/18/37$ (open symbols), $C_2H_6/O_2/N_2 = 40/20/40$ (closed symbols).

ethylene yield of 56% was achieved in the ODE reaction. As shown in Figure 14.2, high temperatures in the bed of catalyst particles are important for achieving high ethylene yields. The role of temperature in enhanced ethylene production is discussed in Section 14.4 taking into account the microkinetics on O_2 activation (Section 14.3.1) and reaction of adsorbed oxygen species with gas-phase ethane (Section 14.3.3).

Catalytic materials based on oxides of rare-earth and alkaline-earth oxides also ignited a C_3H_8/O_2 reaction mixture. As for the ODE reaction, the ignition temperature is a function of catalyst composition (Table 14.1) and is in the range from 500 to 700 K. However, the selectivity towards propene in the ODP reaction was considerably lower than the ethylene selectivity in the ODE reaction. Besides, ethylene but not propene was the main olefin over non-reducible metal oxides (Table 14.1). In contrast to these catalytic materials, propene was the main olefin over a VO_x(5 wt.%)/γ-Al_2O_3 catalyst. The latter is a redox-type catalyst. As demonstrated in ref. 17, the ODP reaction occurs over this catalyst *via* a Mars–van Krevelen sequence. It is important to highlight that the difference in the ratio of C_3H_6/C_2H_4 in the ODP reaction over reducible and nonreducible metal oxides cannot be exclusively explained by thermal cracking of propane/propene over the latter materials, since the reaction temperature in the catalyst bed was very similar for all the catalysts studied (Table 14.1). Therefore, the observed product distribution should at least partially be related to catalytic heterogeneous reactions.

Summarizing the above discussion two important results should be especially emphasized: (i) ignition temperature in the ODE and ODP reactions is strongly influenced by the catalyst composition, and (ii) catalytic materials based on reducible and nonreducible metal oxides differ in the ratio of C_3H_6/C_2H_4 in the ODP reaction, which is highest over the reducible metal oxides. The next section is aimed at the elucidation of possible fundamental reasons for the observed phenomena.

14.3 Mechanistic Analysis of Heterogeneous Reaction Pathways

14.3.1 Microkinetics of Oxygen Activation

It is well accepted for various heterogeneous oxidation reactions[16,19] that oxygen activation, *i.e.* generation of active oxygen species from gas-phase O_2 on the catalyst surface, is an essential reaction step determining the catalyst activity and selectivity. In order to derive mechanistic insights from the role of O_2 activation in the short contact time oxidation of ethane and propane, a detailed microkinetic analysis of O_2 activation on the basis of elementary reaction steps including surface intermediates was performed. To this end, transient responses of O_2 obtained after O_2 pulsing in the TAP reactor over rare-earth, alkaline-earth oxide materials and VO_x(5 wt.%)/γ-Al_2O_3 were kinetically evaluated using the approach described in refs. 20,21. According to the theory of the TAP experiments,[22] transient responses contain information both on chemical (adsorption, desorption and reaction) and transport processes occurring inside the TAP reactor. Figure 14.3 exemplifies oxygen transient responses over selected catalytic materials, when O_2 was pulsed over the catalysts at 773 K. The responses are normalized for better comparison of their shapes. It is clear that the transient responses in Figure 14.3 differ strongly in their shapes. This result proves that gas-phase O_2 reacts differently with the catalytic materials studied.

The microkinetic models in the present study were developed on the basis of experimental results, *i.e.* each step considered in the models was supported by experimental observation or DFT calculations. In this way the physical

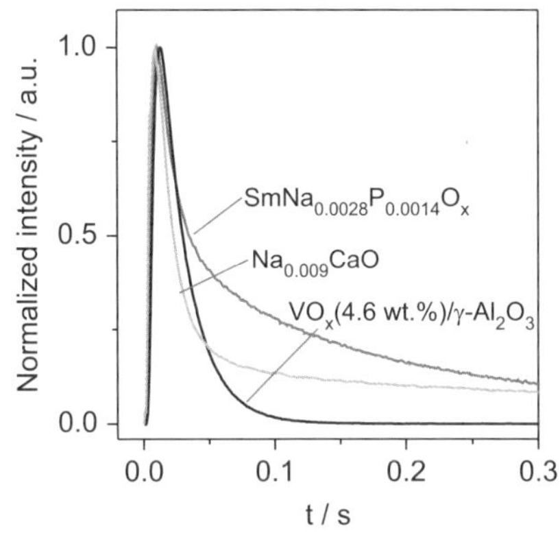

Figure 14.3 Transient oxygen responses upon pulsing an $O_2/Ne = 1/1$ mixture over various catalysts at 773 K. O_2 pulse size is *ca.* 2×10^{14} molecules.

Table 14.2 Microkinetic models for O_2 activation evaluated in this study.

Model	*Elementary reaction steps*	
1	$O_2 + * \rightarrow *\text{-}O_2$	(1.1)
	$*\text{-}O_2 \rightarrow * + O_2$	(1.2)
2	$O_2 + 2* \rightarrow 2*\text{-}O$	(2.1)
	$2*\text{-}O \rightarrow 2* + O_2$	(2.2)
3	$O_2 + * \rightarrow *\text{-}O_2$	(3.1)
	$*\text{-}O_2 \rightarrow * + O_2$	(3.2)
	$*\text{-}O_2 + * \rightarrow 2*\text{-}O$	(3.3)
	$2*\text{-}O \rightarrow * + *\text{-}O_2$	(3.4)
4	$O_2 + 2* \rightarrow 2*\text{-}O$	(4.1)

significance, which is the main criterion for the adequacy of the model,[23] was assured. The models for O_2 activation are summarized in Table 14.2. It has to be stressed that our modeling approach provides no insights into the structure and/or oxidation state of the active sites or the charge of oxygen species adsorbed on these sites.

No adequate description of oxygen transients was achieved when using models 1 and 2 in Table 14.2. Thus, no set of kinetic parameters was found to fit the measured responses in this case. This is valid for all the catalysts studied. O_2 transient responses over nonreducible metal oxides were perfectly described assuming model 3 in Table 14.2. This model considers that gas-phase O_2 adsorbed reversibly and dissociatively *via* a surface biatomic oxygen precursor. The presence of mono- and biatomic oxygen species was experimentally confirmed by the measurements of contact potential difference.[24] In contrast to nonreducible metal oxides, oxygen activation over VO_x(5 wt.%)/γ-Al_2O_3 occurs irreversibly and dissociatively on two active surface sites, yielding lattice oxygen (model 4 in Table 14.2). This observation is in agreement with experimental data on isotopic exchange of oxygen[25] and kinetic modeling of ODP.[26]

The derived kinetic parameters of O_2 activation over both types of catalytic materials were used for calculating surface coverage by adsorbed oxygen species. As a result, the catalyst ability for oxygen activation was identified as an important catalyst function for igniting C_2H_6–O_2 and C_3H_8–O_2 reaction mixtures. The microkinetics of O_2 activation predicted that the ignition occurred at temperatures at which high coverage by active oxygen species is achieved. In order to prove this conclusion experimentally, oxygen isotopic exchange and hydrocarbon activation by surface oxygen species over selected catalytic materials were studied in the TAP reactor. The results are presented and discussed in the following section.

14.3.2 Reactivity of Adsorbed Oxygen Species for Oxygen Isotopic Exchange

For studying oxygen isotopic exchange, catalytic materials were pretreated by $^{16}O_2$ pulses followed by $^{18}O_2$ pulsing. The time interval between the pulses of $^{16}O_2$ and $^{18}O_2$ was *ca.* 20 s. In these experiments, the reactivity of oxygen species, which were formed during the $^{16}O_2$ pretreatment and stayed on the catalyst surface, is proven. $^{16}O^{18}O$ and $^{16}O_2$ were observed as products of pure heterogeneous oxygen isotopic exchange upon $^{18}O_2$ pulsing over nonreducible metal oxides. This result supports that O_2 adsorption over these metal oxides is dissociative and reversible. In order to quantify the catalyst ability for the oxygen isotopic exchange, the degree of conversion of $^{18}O_2$ was calculated from $^{18}O_2$ pulse experiments at different temperatures; the higher the degree of $^{18}O_2$ conversion, the higher the reactivity of adsorbed oxygen species or the higher their concentration. Figure 14.4 demonstrates the influence of temperature and catalyst composition on the degree of $^{18}O_2$ conversion.

It is clearly seen that the catalytic materials differ strongly in their performance for oxygen isotopic exchange. $Sr_{0.2}SmO_x$, $Na_{0.001}CaO$ and $Sr_{1.0}La_{1.0}Nd_{1.0}O_x$ are very active, starting to exchange gas-phase O_2 from 673 K,

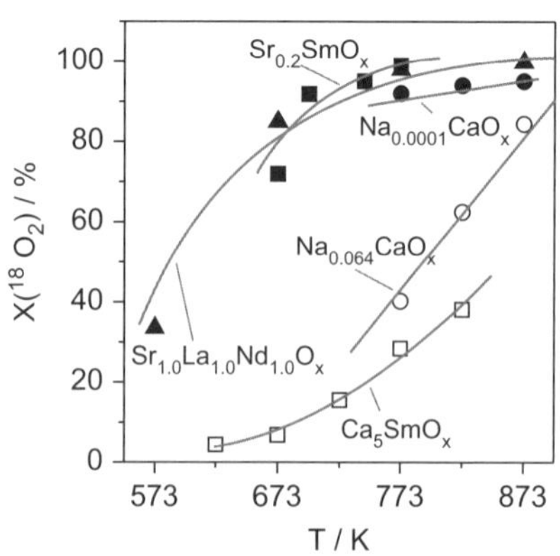

Figure 14.4 Influence of temperature on the degree of $^{18}O_2$ conversion upon pulsing $^{18}O_2$ over various oxide catalysts having been pretreated by $^{16}O_2$.

while $Na_{0.064}CaO_x$ and Ca_5SmO_x show significant activity above 800 K. It is important to emphasize that the order of catalytic activity towards oxygen isotopic exchange correlates well with the ignition performance of oxide materials in the ODE and ODP reactions in Table 14.1. Catalysts that are active for oxygen isotopic exchange under vacuum conditions are also active for ignition of C_2H_6/O_2 and C_3H_8/O_2 mixtures. Thus, this experimental observation supports the conclusion from microkinetic analysis of O_2 activation on the importance of adsorbed oxygen species for the ignition performance. However, for high catalytic activity in alkane oxidation, the catalyst ability for oxygen activation is not the only important property. The catalyst must also be able to supply adsorbed oxygen for the reaction with hydrocarbon, *i.e.* to activate C–H bond. This aspect is discussed in the next section.

14.3.3 Oxygen-Assisted Heterogeneous Activation of C–H Bond in C_2H_6 and C_3H_8

The reactivity of adsorbed oxygen species for alkane oxidation was tested by sequential pulsing of $O_2/Xe = 1/1$ and $C_2H_6(C_3H_8)/Ne = 1/1$ mixtures with different (from 0 to 1 s) time delays (Δt) between the mixtures. Such operation enables investigation of whether adsorbed oxygen species formed in the O_2 pulse participate in hydrocarbon activation. By contrast, it can be seen if surface intermediates, which are formed in the alkane pulse and are retained on the catalyst surface, can be oxidized in the subsequent O_2 pulse. In should be stressed that reaction products were observed in the alkane pulse only, *i.e.* no carbon-containing surface intermediates are stabilized under vacuum conditions during the alkane pulse.

Figure 14.5(a) exemplifies the transient responses resulting from sequential pulsing of oxygen ($O_2/Xe = 1/1$) and ethane ($C_2H_6/Ne = 1/1$) at $\Delta t = 0.2$ s over

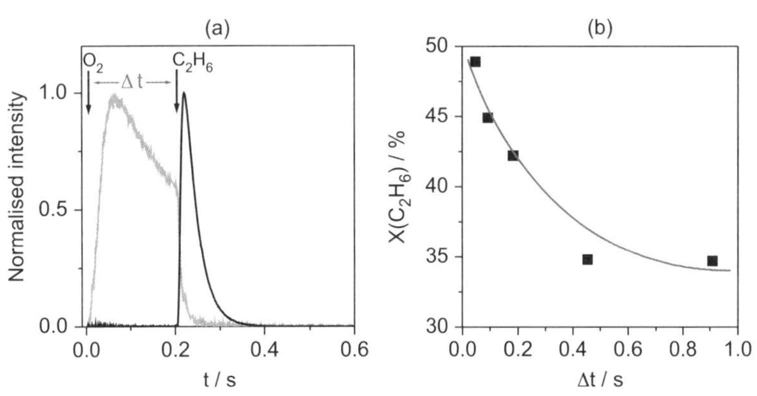

Figure 14.5 (a) Transient responses of sequential pulsing $O_2/X = 1/1$ and $C_2H_6/Ne = 1/1$ mixtures with Δt of 0.2 s over $Na_{0.0001}CaO$ at 873 K. Pulse sizes of O_2 and $C_2H_6 \sim 5 \times 10^{14}$ molecules. (b) $X(C_2H_6)$ over $Na_{0.0001}CaO$ at 873 K *versus* time delay between O_2 and C_2H_6 pulses. Pulse sizes of O_2 and $C_2H_6 \sim 5 \times 10^{14}$ molecules.

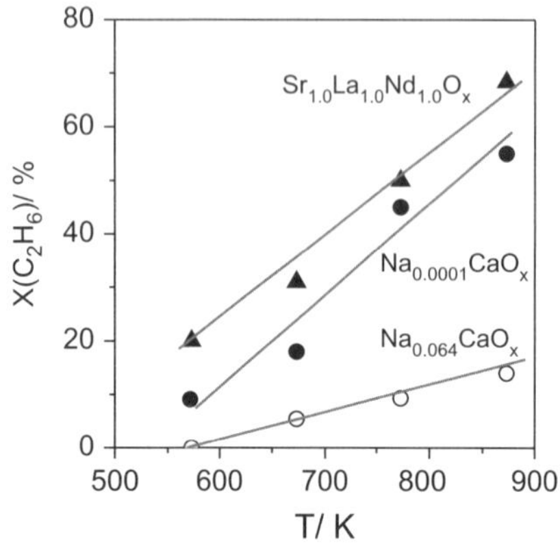

Figure 14.6 $X(C_2H_6)$ over various catalytic materials *versus* temperature. Sequential pulsing oxygen (O_2/Xe = 1/1) and ethane (C_2H_6/Ne = 1/1) with Δt of 0.2 s.

$Na_{0.0001}CaO$ at 873 K. From the oxygen transient response it is clear that the concentration of gas-phase O_2 sharply decreases when ethane enters the reactor ($t = 0.2$ s in Figure 14.5). This sharp decrease in the oxygen signal indicates that adsorbed oxygen species formed in the O_2 pulse rapidly react with ethane. As shown in Figure 14.5(b), the degree of ethane conversion decreases upon increasing the time delay between the oxygen (O_2/Xe = 1/1) and ethane (C_2H_6/Ne = 1/1) pulses. This decrease is due to the fact that the concentration of adsorbed oxygen species decreases due to O_2 desorption, when the time delay between O_2 and C_2H_6 pulses increases. The lower the coverage by adsorbed oxygen species, the lower the degree of ethane conversion.

Figure 14.6 compares ethane conversions over different catalysts and temperatures, when oxygen (O_2/Xe = 1/1) and ethane (C_2H_6/Ne = 1/1) were sequentially pulsed with Δt of 0.2 s. $Na_{0.0001}CaO$ and $Sr_{1.0}La_{1.0}Nd_{1.0}O_x$ are very active starting to oxidize ethane at 573 K, while $Na_{0.064}CaO_x$ shows significant activity only above 873 K. In contrast to the two former catalysts, the latter one is also not active for short-contact time ethane oxidation below 873 K at ambient pressure. Thus, these results together with the microkinetics of O_2 activation (Section 14.3.1) and the results of oxygen isotopic exchange (Section 14.3.3) highlight the catalyst ability for generation of active oxygen species as an important prerequirement for ignition performance of rare-earth and alkali-earth oxide materials in the short contact time ODE and ODP reactions.

14.4 Interplay Between Heterogeneous and Homogeneous Reaction Pathways

14.4.1 Oxidative Dehydrogenation of Ethane

Taking into account the above discussion, the mechanistic concept of short contact time oxidative dehydrogenation of ethane can be summarized as follows. As shown in Figure 14.5, oxygen species are required to break C–H bond in ethane. Since nonreducible oxide catalysts without adsorbed oxygen species are not active for ethane activation, active adsorbed oxygen species should be generated on the catalyst surface from gas-phase O_2. According to the microkinetics of O_2 activation (Section 14.3.1), these adsorbed oxygen species are of biatomic and monoatomic nature. The biatomic oxygen species are responsible for total heterogeneous oxidation, while monoatomic oxygen species catalyze selective dehydrogenation of ethane to ethylene.[27]

With respect to ethane activation and ethylene formation both heterogeneous and homogeneous reaction steps should be taken into account in agreement with the previously reported results of ethane oxidation of Pt-based catalytic materials.[5,14,15] The short contact time ODE reaction is heterogeneously initiated by oxygen-assisted ethane activation. This reaction leads to C_2H_4 and CO_x. Due to the high exothermicity of the last reaction, the operating temperature inside the catalyst bed dramatically rises. If the reactor temperature is high enough, gas-phase thermal dehydrogenation of ethane to ethylene and hydrogen takes place. Since this reaction pathway is endothermic, its contribution to ethylene production increases with temperature. Figure 14.2 nicely demonstrates that independent of the catalyst composition, the higher the temperature inside the catalyst bed, the higher the yield of ethylene.

Thus, summarizing the above discussion it is concluded that both heterogeneous and homogeneous reactions play a key role in the short contact time conversion of ethane to ethylene over nonreducible oxide catalysts. Heterogeneous processes are responsible for generation of reactive oxygen species, which are responsible for heterogeneous ethane oxidation to C_2H_4, and CO_x. Additionally, selective ethylene formation occurs homogeneously *via* thermal dehydrogenation of nonreacted ethane.

14.4.2 Oxidative Dehydrogenation of Propane

Pure homogeneous reactions cannot exclusively explain the differences in the performance of reducible and nonreducible metal oxides in the ODP reaction. This statement is based on two experimental observations: (i) the ratio of C_3H_6/C_2H_4 is high (*ca.* 25) over reducible, but low (*ca.* 0.6) over nonreducible metal oxides and (ii) reaction temperature in the catalyst bed was very similar for all the catalytic systems in Table 14.1. Therefore, the observed product distribution in the ODP reaction with short contact times should also be related to heterogeneous reaction pathways. In order to derive mechanistic insights into the role of the nature of oxygen species in product formation in the ODP reaction, possible transformations of final and intermediate products of propane oxidation were analyzed by single pulsing of C_3H_8, and C_3H_6 as well as by sequential pulsing of O_2 and C_3H_8 over selected reducible and nonreducible metal-oxide catalysts in the TAP reactor. It has to be stressed that the transient experiments were performed in the Knudsen diffusion regime, where homogeneous reactions are minimized as compared to the heterogeneous ones.

Therefore, the results discussed below are due to heterogeneous reactions. For VO_x(5 wt.%)/γ-Al_2O_3, propene, formaldehyde and carbon oxides were detected at the reactor outlet, when propane was pulsed in the absence of gas-phase oxygen. Formaldehyde and carbon oxides were also observed upon propene pulsing. Ethylene and methane were not observed under transient vacuum conditions. This result agrees well with the catalytic performance of the VO_x(5 wt.%)/γ-Al_2O_3 material under ambient-pressure steady-state conditions in Table 14.1. Since gas-phase O_2 was not present in C_3H_8 and C_3H_6 pulse experiments, all the above reaction products were formed *via* C_3H_8 or C_3H_6 oxidation with lattice oxygen of VO_x species. This is also valid for VO_x/γ-Al_2O_3 materials with higher or lower vanadium loading.[17]

Insights into the sequence of product formation from C_3H_8 and C_3H_6 can be obtained from the analysis of the shape and the order of appearance of feed components and reaction products. Figure 14.7 exemplifies typical height-normalized transient responses of the observed carbon-containing products along with the response of unreacted propane/propene. The normalization was performed for proper comparison of the shapes of the transient responses. Each transient response has a unique mean residence time, temporal maximum and decay. When C_3H_8 is pulsed over the VO_x(5 wt.%)/γ-Al_2O_3 material, its transient response appears after the shortest time interval and has the fastest decay. The transient response of C_3H_6 is similar to that of C_3H_8 but it is slightly shifted to longer times. Formaldehyde was observed with a short delay after the transient of C_3H_6. The transient responses of CO and CO_2 are significantly shifted to longer times compared to those of the previous gas-phase reactants. Moreover, they have the broadest shape. According to ref. 22 the order of appearance of the reaction products in Figure 14.7(a) can be interpreted in the following manner. Since the transient response of C_3H_6 follows directly that of C_3H_8, it is concluded that the initial step of the ODP reaction is the formation

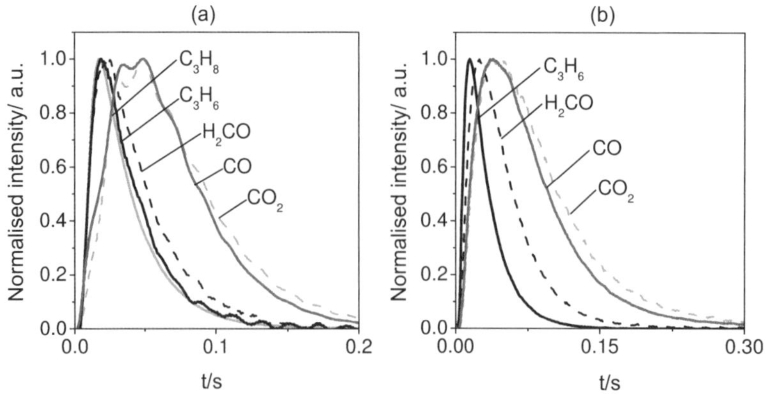

Figure 14.7 Transient responses to pulsing C_3H_8 (a) and C_3H_6 (b) over VO_x(5 wt.%)/γ-Al_2O_3 at 848 K.

of propene *via* breaking of two C–H bonds in a propane molecule as reported previously.[28] The primarily formed propene reacts further at the C–C bond yielding formaldehyde followed by its further oxidation to CO and CO_2. A similar idea was discussed by Bielanski and Haber for alkenes oxidation over V_2O_5.[16] However, the latter authors suggested that electrophilic (O_2^-, O_2^{2-} or O^-) but not lattice oxygen species participates in breaking of C–C bond in the alkene molecule. The present results evidence that lattice oxygen of VO_x species is also active for C–C bond scission.

In contrast to the vanadium-based catalyst, adsorbed oxygen species participate in propane activation over nonreducible metal oxides. This conclusion is based on the discussion below. Figure 14.8 exemplifies height-normalized transient responses of sequential pulsing O_2 and C_3H_8 over $SmNa_{0.028}P_{0.014}O_x$ with a time delay of 0.1 s between them.[11] The O_2 transient response clearly shows that the concentration of gas-phase oxygen sharply decreases when propane enters the reactor ($t = 0.1$ s in Figure 14.8). This sharp decrease in the oxygen signal indicates that adsorbed oxygen species rapidly react with propane.

Another important observation from the sequential pulsing of O_2 and C_3H_8 is the formation of C_2H_4 and CH_4 in the C_3H_8 pulse. Since these transient experiments were performed under conditions, where gas-phase reactions are suppressed, ethylene and methane formation occurs heterogeneously. Such reactions can proceed *via* the addition of various electrophilic oxygen species (O^-, O_2^-) to a double C=C bond in C_3H_6, which is a primary product of C_3H_8 oxidation by adsorbed oxygen species. In contrast to electrophilic oxygen species, nucleophilic lattice oxygen of VO_x species in VO_x(5 wt.%)/γ-Al_2O_3 breaks single C–C bond but with formation of H_2CO, which is further oxidized to CO_x.

Thus, the nature of oxygen species determines the product distribution in propane oxidation at short contact times under conditions where gas-phase

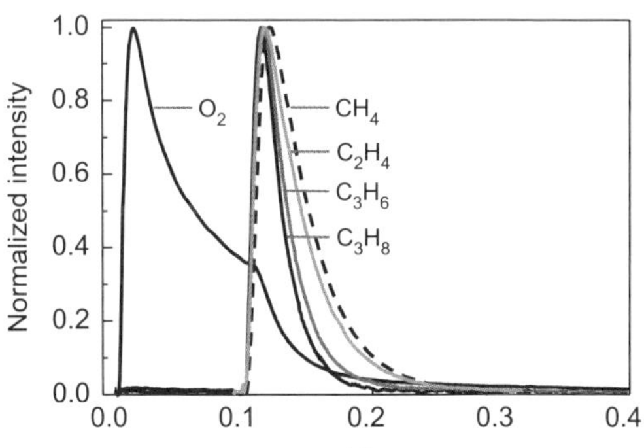

Figure 14.8 Transient responses of sequential pulsing of O_2 (O_2/Xe = 1/1) and C_3H_8 (C_3H_8/Ne = 1/1) with $\Delta t = 0.1$ s over $SmNa_{0.028}P_{0.014}O_x$ at 723 K. From ref. 11.

reactions are strongly inhibited. In order to derive further insights into the role of the nature of oxygen species and/or their surface density in the ODP reaction under conditions including gas-phase reactions, possible transformations of final and intermediate products of propane oxidation were analyzed[29] taking into account the model suggested previously for heterogeneous–homogeneous transformations of light alkanes.[30] If propene radicals are formed as the primary intermediates, ethylene forms *via* cracking of propene radicals (eqn (14.1)), which proceeds concurrently with H-atom elimination leading to propene (eqn (14.2)):

$$C_3H_7^{\bullet} \rightarrow C_2H_4 + CH_3^{\bullet} \tag{14.1}$$

$$C_3H_7^{\bullet} \rightarrow C_3H_6 + H^{\bullet} \tag{14.2}$$

According to the existing kinetic data for the gas-phase reactions,[31–33] the above reaction pathways similarly increase with temperature. Over rare-earth/alkaline-earth oxide catalysts, however, ethylene is formed in higher concentration than propene. This is explained by the fact that propene total oxidation is going to proceed much faster than that of ethylene. The low concentration of ethylene over VO_x(5 wt.%)/γ-Al_2O_3 as compared to rare-earth oxides in Table 14.1 is due to the following reasons. As a rule, catalytic materials contain active oxygen species with high affinity to H atoms, which determine their ability to activate C–H bonds. Since the C–H bond in C_3H_7 radicals is far less strong than in propane (~35 *vs.* 95 kcal/mol, respectively), H-atom abstraction from them yielding propene can proceed with very low activation energy. In other words, at a sufficiently high density of oxidizing states (DOS) this reaction is much more efficient than reactions of C_3H_7 radicals in the gas phase. In this particular case, olefin distribution ($C_{3=}/C_{2=}$ ratio) is much more strongly affected by DOS rather than their strength. As a result, higher DOS on vanadium-containing catalyst(s) causes higher propene-to-ethylene ratios compared to those obtained over rare-earth oxide-based catalysts under comparable reaction conditions.

14.5 Conclusions

The described transient kinetic approach enabled determination of the role of the nature and surface coverage of oxygen species in short contact time oxidative dehydrogenation of ethane and propane over two families of catalytic materials (oxides of reducible and nonreducible metals). An important catalyst function of both catalytic systems for igniting of O_2/C_2H_6 and O_2/C_3H_8 reaction mixtures was identified, *i.e.* their ability for generation of active surface oxygen species from gas-phase O_2. This comprised reversible and dissociative activation *via* a molecular surface precursor over nonreducible metal oxides. As microkinetics of O_2 activation predicts, the ignition of C_3H_8/O_2 and C_2H_6/O_2 mixtures occurs at a temperature at which certain coverage by adsorbed oxygen species is achieved. Thus, for the design of active catalysts for the ODE or ODP reactions at short contact times, the catalyst ability for O_2 activation should be increased. This can be achieved by creating anion vacancies in the lattice of alkali-earth or rare-earth metal oxides upon doping with lower valence nonreducible metal salts.

Over vanadia-based catalytic materials oxygen activation occurred irreversibly and dissociatively on two active surface sites yielding lattice oxygen. In contrast to adsorbed oxygen species (electrophilic), nucleophilic lattice oxygen species of VO_x sites are less active for breaking C–C bonds of the propane molecule to yield C_2H_4 and CH_4. The nature of oxygen species determines the ratio of C_3H_6/C_2H_4 in the ODP reaction over reducible and nonreducible metal oxides. Moreover, kinetic modeling of heterogeneous and homogeneous reaction pathways predicts that a high density of oxidizing sites is also essential for further heterogeneous dehydrogenation of propene radicals. This helps to suppress the concurrent cracking reaction channel. This knowledge may eventually be used for further improvement of catalytic materials and process operation.

References

1. O. V. Buyevskaya and M. Baerns, *Catalysis*, 2002, **16**, 155.
2. M. Baerns, G. Grubert, E. V. Kondratenko, D. Linke and U. Rodemerk, *Oil Gas-Eur. Mag.*, 2003, **1**, 36.
3. M. Huff and L. D. Schmidt, *J. Phys. Chem.*, 1993, **97**, 11815.
4. M. C. Huff and L. D. Schmidt, *AIChE J.*, 1996, **42**, 3484.
5. A. S. Bodke, D. A. Olschki, L. D. Schmidt and E. Ranzi, *Science*, 1999, **285**, 712.
6. D. W. Flick and M. C. Huff, *J. Catal.*, 1998, **178**, 315.
7. R. Lodeng, O. A. Lindvag, S. Kvisle, H. Reier-Nielsen and A. Holmen, *Stud. Surf. Sci. Catal.*, 1998, **119**, 641.
8. A. Beretta, P. Forzatti and E. Ranzi, *J. Catal.*, 1999, **184**, 469.
9. M. Fathi, R. Lødeng, E. S. Nilsen, B. Silberova and A. Holmen, *Catal. Today*, 2001, **64**, 113.
10. K. Venkataraman, J. M. Redenius and L. D. Schmidt, *Chem. Eng. Sci.*, 2002, **57**, 2335.
11. O. V. Buyevskaya and M. Baerns, *Catal. Today*, 1998, **42**, 315.
12. S. A. R. Mulla, O. V. Buyevskaya and M. Baerns, *J. Catal.*, 2001, **197**, 43.
13. E. V. Kondratenko and O. Ovsitser, *Angew. Chem. Int. Ed.*, 2008, **47**, 3227.
14. R. Burch and E. M. Crabb, *Appl. Catal. A*, 1999, **97**, 49.
15. M. C. Huff, I. P. Androulakis, J. H. Sinfelt and S. C. Reyes, *J. Catal.*, 2000, **191**, 46.
16. A. Bielanski, J. Haber, *Oxygen in Catalysis*, Marcel Dekker, New York, 1991.
17. E. V. Kondratenko, N. Steinfeldt and M. Baerns, *Phys. Chem. Chem. Phys.*, 2006, **8**, 1624.

18. E. V. Kondratenko, S. A. R. Mulla, O. V. Buyevskaya, M. Baerns, *5th International Congress on Oxidation Catalysis*, Sapporo, Japan, 2005, p. 342.
19. G. I. Panov, A. K. Uriarte, M. K. Rodkin and V. I. Sobolev, *Catal. Today*, 1998, **41**, 365.
20. M. Rothaemel and M. Baerns, *Ind. Eng. Chem. Res.*, 1996, **35**, 1556.
21. M. Soick, D. Wolf and M. Baerns, *Chem. Eng. Sci.*, 2000, **55**, 2875.
22. J. T. Gleaves, G. S. Yablonsky, P. Phanawadee and Y. Schuurman, *Appl. Catal. A*, 1997, **160**, 55.
23. D. M. BatesD. G. Watts, *Nonlinear Regression Analysis and its Application*, John Wiley & Sons, New York, 1988.
24. E. V. Kondratenko, D. Wolf and M. Baerns, *Catal. Lett.*, 1999, **58**, 217.
25. K. Chen, A. Khodakov, J. Yang, A. T. Bell and E. Iglesia, *J. Catal.*, 1999, **186**, 325–333.
26. D. B. Creaser and B. Andersson, *Appl. Catal. A*, 1996, **141**, 131.
27. E. V. Kondratenko, O. V. Buyevskaya, M. Soick and M. Baerns, *Catal. Lett.*, 1999, **63**, 153–159.
28. K. Chen, E. Iglesia and A. T. Bell, *J. Catal.*, 2000, **192**, 197–203.
29. E. V. Kondratenko and M. Yu. Sinev, *Appl. Catal. A*, 2007, **325**, 353.
30. M. Y. Sinev, V. S. Arutyunov, *5th International Congress on Oxidation Catalysis*, Sapporo, Japan, 2005, p. 164.
31. J. Warnatz, Rate coefficients in the C/H/O system., in *Combustion Chemistry*, ed. W. C. Gardiner, Springer-Verlag, Berlin, 1984, pp. 197–277.
32. W. Tsang and R.F. Hampson, *J. Phys. Chem. Ref. Data*, 1986, **15**, 1087.
33. R. Atkinson, D. L. Baulch, R. A. Cox, R. F. Hampson, J. A. Kerr, M. J. Rossi and J. Troe, *J. Phys. Chem. Ref. Data.*, 1997, **26**, 521.

CHAPTER 15

Characteristics of Selective Oxidation Reactions

ROBERT SCHLÖGL*[1] AND CHRISTIAN HESS[2]

[1] Fritz-Haber-Institut der Max-Planck-Gesellschaft, Abteilung Anorganische Chemie, Faradayweg 4-6, 14195 Berlin, Germany; [2] Technische Universität Darmstadt, Eduard-Zintl-Institut für Anorganische und Physikalische Chemie, Petersenstr. 20, 64287 Darmstadt, Germany

15.1 Introduction

Heterogeneously catalyzed selective oxidation reactions have been the subject of various books[1–7] and reviews.[8–13] Despite the high industrial relevance and environmental impact the current development of selective oxidation catalysts is largely based on phenomenological concepts (principles) rather than a profound understanding of their mode of operation. Further empirical optimization of selective oxidation catalysts on the atomic level (additives, promoters) will increase the chemical complexity leading to added catalyst heterogeneity, which becomes difficult to control. With respect to rational development (design) of selective oxidation catalysts it would be highly desirable to replace chemical by structural complexity. However, for this approach an improved level of scientific understanding is desperately needed.

In this chapter the current state of knowledge on selective oxidation reactions as discussed in the above chapters combined with previous findings is taken as a basis to develop a *consistent picture of the working principle* of selective oxidation catalysts. In this context, the widely accepted separation of substrate redox chemistry (alkane oxidation) and catalyst redox chemistry (oxygen

RSC Nanoscience & Nanotechnology No. 19
Nanostructured Catalysts: Selective Oxidations
Edited by Christian Hess and Robert Schlögl

Published by the Royal Society of Chemistry, www.rsc.org

reduction) referred to as Mars–van Krevelen (MvK)-type reaction mechanism will be revisited. This involves a clear definition of the term "lattice oxygen".

In the following discussion we will examine the functions of active sites in a redox catalyst. We will assume for simplicity a catalyst containing supported active sites[14–17] of the V_xO_y type. The formal processes that are necessary for converting the chosen model molecule, propane, into its olefin propene and into its oxidation product acrylic acid can be described by a set of chemical equations that are useful to discuss the current status of concepts in selective oxidation reactions. These equations are not elementary steps in the sense that they describe the processes changing the reaction coordinate of the substrate (the sum of all bond distances). These equations rather give the minimum of molecular transformations in correct stoichiometric relations for atoms and formal charges that are needed to achieve the desired reaction.

15.2 Activation and Reoxidation

We understand today that the reaction of propane activation is a sequence of consecutive reactions initiated by the rate-determining step of C–H activation of propane. This is a common result from kinetic analysis[18–20] and of quantum-chemical calculations[21,22] (see Chapter 2). Experimentally, for propane oxidative dehydrogenation over supported vanadium oxide a kinetic isotope effect was found for hydrogen *versus* deuterium in the central position.[18] Hence, the first step in propane activation involves the abstraction of methylene hydrogen:

$$\mathrm{R{-}H} + \mathrm{H^+} \rightarrow \mathrm{R^+} + \mathrm{H_2} \quad \text{(activation by acidic function)} \tag{15.1}$$

$$\mathrm{R{-}H} + \mathrm{cat{=}O} \rightarrow \mathrm{R^-} + \mathrm{cat^+{-}OH} \quad \text{(activation by nucleophilic oxygen)} \tag{15.2}$$

For a discussion of the reactivity behavior the electronic properties of the propane molecule need to be considered. A powerful model to describe chemical reactivity is frontier molecular orbital (FMO) theory.[23] Within FMO theory the important aspect is the localization of the highest occupied and lowest unoccupied molecular orbital (HOMO and LUMO). Thus, assuming FMO theory to give a valid picture on the reactivity behavior, for propane activation to proceed according to reaction (15.1) a HOMO with a high electron density in the methylene position is necessary. On the other hand, reaction (15.2) requires localization of the LUMO at the methylene-hydrogen positions. Figure 15.1 shows the HOMO and LUMO of propane. It is evident that only the LUMO has electron density in the methylene position and therefore allows for nucleophilic attack of oxygen.

Figure 15.2 shows a diagram of the families of reactions forming together the process of propane activation. In block (A) of Figure 15.2 we see the options for activating propane. The common catalytic one occurs through nucleophilic oxygen[25] present on an oxide catalyst surface. Halogenated catalysts[26] will strongly support this reaction through the stronger nucleophilicity of the

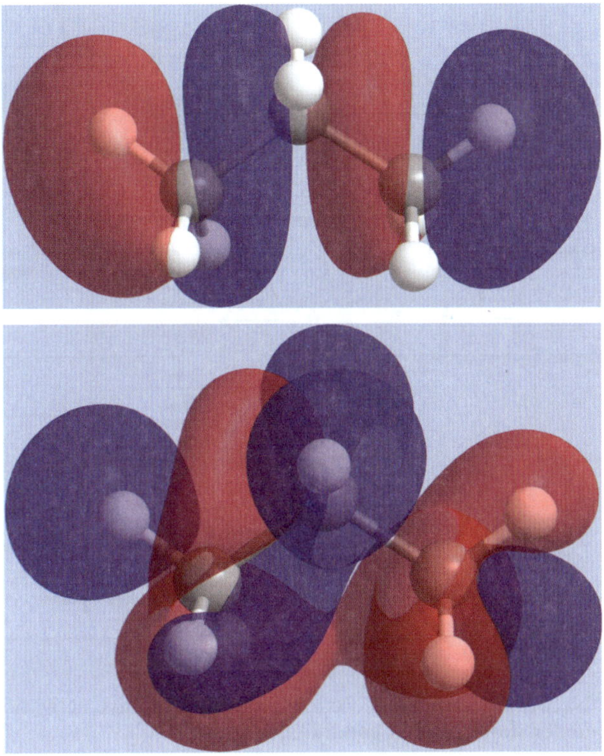

Figure 15.1 Top: Highest occupied molecular orbital (HOMO) of propane. Bottom: Lowest unoccupied molecular orbital (LUMO) of propane.[24]

halogen allowing to reduce the reaction temperature. At high temperature[27–29] (ca. 873 K) the first intermediate radical C_3H_7* can also form in homogenous pyrolytic chemistry. At high temperatures the radical can undergo the reactions indicated in block (B) of Figure 15.2. The chemical potential of oxygen available will affect the reaction pathways together with many influences of the transport of heat and mass in the reactor. The result can either be soot (low potential) or CO_x (high potential, combustion). Only under special conditions will a large fraction of propene leave the reactor, as a reactive intermediate with respect to the two stable end products, soot and CO_x.

The preferred heterogeneous catalytic reaction will occur as shown in block (C) through the formation of an alkoxide intermediate that can either desorb as propene or stay on the surface and become more deeply oxidized. The oxidation state of the sorption site of the alkoxide will critically decide over the lifetime of the adsorbate; the local geometry and aggregation of the active site as well as the surface coverage with activated oxygen will control the oxidation state of the active site. In this block a critical selection occurs of the reaction sequence and of the selectivity of the overall process irrespective of the fact that the overall rate-determining step occurs already in block (A). The desired

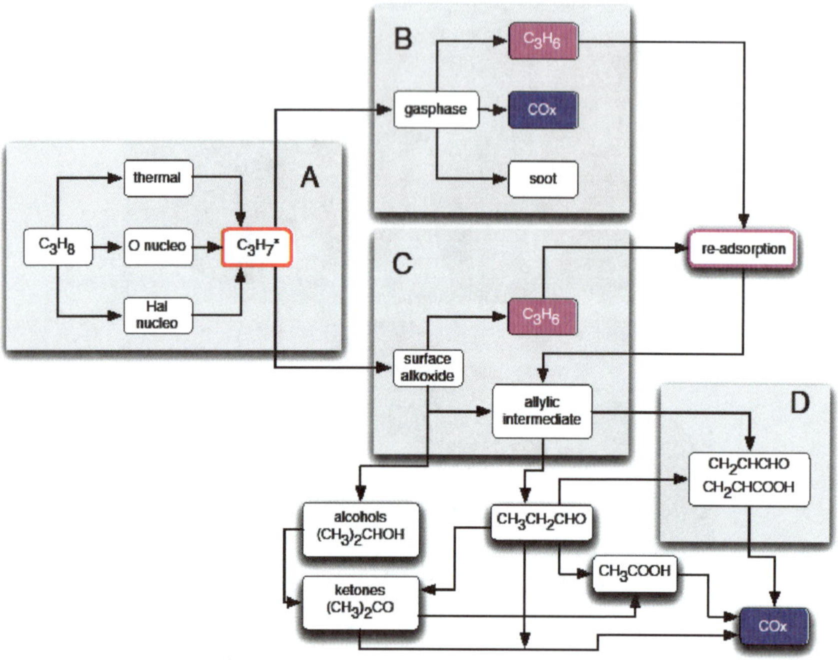

Figure 15.2 Network of processes occurring during oxidative activation of propane under wide range of conditions.

deeper oxidation products acrolein[30–36] and acrylic acid[29,37–60] occur through the sequential oxidation by electrophilic oxygen,[61–64] as indicated in block (D) of Figure 15.2, whereas the undesired deeper oxidation products all lead to CO_2. The selection variables for pathways to deep oxidation will be again the chemical potential of oxygen regulating the surface coverage of active oxygen and the stability of the oxidation state of the active site. In addition, for these reactions the chemical potential of water,[64–79] being either cofed or occurring as reaction product, will also affect the selectivity through multiple roles. The possibility of the initial product to readsorb on the catalyst creates a feedback loop in the sequence of reactions by enhancing the surface coverage with activated propane beyond the value given by the primary splitting of reaction pathways in block (B). The rate of readsorption is controlled by the properties of the catalyst surface, by the feed composition and in addition, by the transport properties of the catalyst in a given reactor. The analysis shows clearly that the selective oxidation of an alkane is a sequence of consecutive and parallel reactions that is controlled by a system of feedback loops operated by the chemical potential of the gas phase. The potential is not constant in a reactor but exhibits gradients and fluctuations. All this acts upon the solid-state chemistry of the oxide catalyst. It is thus no surprise that the observation of a product stream at the outlet of a reactor allows only limited conclusion about the sequence of events that caused this outlet composition. Most controversial debates in the literature can be resolved under this consideration; the necessary reconstruction on the sides of material characterization and on catalytic testing is, however, hampered by the incomplete documentation of facts. The present discussion excludes these levels of complexity and focuses on the redox reaction steps to elucidate the nature and critical properties of active sites for selective oxidation.

The critical role of the surface coverage of activated oxygen for the selectivity in propane activation is illustrated in Figure 15.3. At low enough coverage the deep oxidation routes are not open for the activated intermediate of propane and we observe a high initial selectivity for propene. At a critical surface coverage all deeper oxidation reactions can set in and lead to combustion. The selection of useful oxygenates (block (D) in Figure 15.2) is largely done through kinetics by carefully dosing activated oxygen and by taking advantage of the relative thermodynamic stability of carboxylic acids with respect to all other chemical structures shown in Figure 15.2. The isolation kinetics will be substantially affected by the presence of water (or synonymously by the acid–base properties of the catalyst[64,80–84]) supporting (acidic) or inhibiting (basic) desorption of carboxylic acids.

In the following discussion we will investigate what relation exists between catalyst properties defined during its synthesis and the action of the resulting active sites. As a model catalyst we use a collection of V=O sites without considering their exact chemical structure nor their support. The discussion here is no substitute nor in contradiction to quantum-chemical

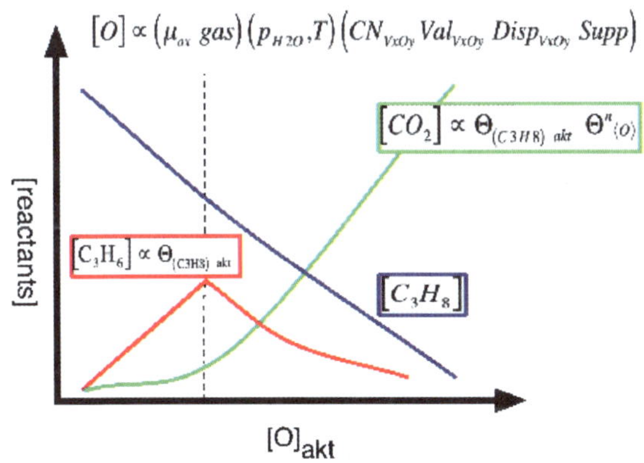

Figure 15.3 Schematic representation of the abundance of selected species during propane activation. The concentration of activated oxygen depends on reaction parameters and on several properties of the catalyst: coordination number of the active vanadia species (CN), valence or oxidation state of the active species (Val), the dispersion or reducibility of the active species (Disp) and on the electronic properties of the support or surrounding of the active site (Supp) representing a whole set of structural properties not individually specified. Other variables controlling the performance such as water and coadsorbates are not shown for clarity.

descriptions[21,22,85] of the elementary step reactions but allows derivation of a set of boundary conditions for the description of selective oxidation reactions based on chemical insights. These conditions may augment the standard model of selective oxidation as described in the "concepts of selective oxidation" put forward[86–90] by R. K. Grasselli (see Chapter 5). We will use the notation of individual redox reactions (one equation for oxidation and one for reduction) and apply correct stoichiometric factors.

For oxidative dehydrogenation eqn (15.3) describes the alkane oxidation.

$$C_3H_8 + 2OH^- \rightarrow C_3H_6 + 2H_2O + 2e^- \quad (15.3)$$

$$V^{+5}{=}(O) + e^- + H_2O \rightarrow V^{+4}{-}(OH) + OH^- \quad (15.4)$$

The necessary oxidation agent is the catalyst. In its simplest formation this would be a single-site metal species as formulated in eqn (15.4). It is worth mentioning that this formulation deviates from catalytic mechanisms over metals where the reactants first are adsorbed, then activated and only then will react with each other. In these formulations the catalyst provides adsorption sites and redox equivalents (electrons). In the present case, the active site provides in addition to sites and redox equivalents ("hole states") the reactant-activated oxygen that was generated in a reaction step decoupled from the conversion of the organic molecule. This difference led to the custom of calling such reactions of "Mars-van-Krevelen" (MvK) type,[15,91–94] defining processes that provide reactive oxygen from the active-site ensemble. The authors note, however, that for other types of heterogeneous catalysis the chemisorption of a reactive species is usually not related to a "lattice" species. The notion that oxides in selective oxidation are fundamentally different catalysts from other systems as they provide some of their structural constituents (oxygen ions) also for the reaction is not justified in general. In particular, the constituting oxygen ions can be used for catalytic conversion in the absence of gas-phase oxygen[95] being, however, a stoichiometric rather than a catalytic process occurring also during oxide reduction to the metal. The special operation of a partial reduction of the catalyst followed by a space-time separated reoxidation[96–98] ("riser mode") is a boundary case[99] rarely occurring in science and practice.

Before resuming the detailed discussion on chemical equations we would like to illustrate the above picture of V=O active sites performing a complete catalytic cycle during propane oxidative dehydrogenation (ODH). It is a common assumption in the literature that the active sites provide two redox equivalents. Accordingly, for the reduction one may formulate

$$V^{+5}{=}(O) + 2e^- + H_2O \rightarrow V^{+3} + 2OH^- \quad (15.5)$$

A recent theoretical study[21] addresses the activation of propane over a supported vanadia catalyst containing exclusively isolated (monomeric) sites. Unfortunately, reoxidation of the reduced catalyst is not described. In any case, reoxidation involves an oxygen molecule. On the basis of monomeric sites the formation of peroxo species and its decomposition is the only scenario for further reaction. Recently, a peroxo species on a V^{3+} site of a vanadia model system has been identified.[100] Decomposition of peroxo would then require one oxygen atom to be transferred to the product, leading to (undesirable) oxidation. As a result, for each product molecule formed one oxygen atom would need to be transferred to a substrate or due to its higher reactivity much more likely to a product molecule to reoxidize a V^{3+} catalytic site. Assuming complete product oxidation to CO_2 and H_2O, which corresponds to the best-case scenario of the sacrificial path, the selectivity would still drop to 8/9. However, experimental results on propane ODH over supported vanadia catalysts indicate 100% selectivity to propene at low conversions.[19]

Focusing on the role of the vanadia active sites first it is desirable to consider a support that serves as an "inert" platform for the deposit. To this end, silica seems to be the material of choice considering its weak interaction with vanadia and its irreducibility. As described in detail in Chapter 12, studies on silica-SBA-15-supported vanadium-oxide catalysts show that besides VO_4 monomers a significant amount of vanadium-oxide surface species containing V–O–V bridges (dimeric/oligomeric) is present.[17,101,102] Such catalysts show activity in propane ODH[102] and methanol oxidation to formaldehyde.[103,104] In passing, it is mentioned that recent studies suggest the presence of M–O–M bridges also in the case of (dispersed) molybdenum oxide catalysts.[105]

Interestingly, a dimer represents the smallest active site that can entertain a full catalytic cycle including product formation and reoxidation without the need for auxiliary (sacrificial) compounds. This is illustrated in Figure 15.4 for propane ODH over a dimeric vanadia site. As a profound theoretical treatment of propane activation over such sites is not available in the literature so far, the sequence of reaction steps in Figure 15.4, while based on the previous DFT results, is still of purely hypothetical nature with respect to the detailed structures. Nevertheless, it allows for the discussion of important characteristics of selective oxidation reactions.

Under ambient conditions silica-supported vanadia strongly resembles a water containing V_2O_5 gel (xerogel) $V_2O_5 \cdot 1.2H_2O$, which consists of polymerized pyramidal VO_5 units similarly as in V_2O_5.[66,102,106] Catalyst dehydration at the reaction temperature leads to the formation of highly dispersed V^{5+} species with tetrahedrally coordinated vanadium. In the course of one catalytic cycle over a dimeric site the four electrons accumulated during ODH of two propane molecules are used for activation of gas-phase oxygen. It is easily seen that such active sites (dimers, trimers, tetramers) fulfill the condition of site isolation, *i.e.* the limited availability of active oxygen species to avoid overoxidation. From the above example, one can deduce the following details regarding active-site requirements: (i) A sufficient number of metal ions to intermediately store the electrons accumulated during one transformation and necessary for reoxidation by gas-phase oxygen, (ii) a sufficient number of oxygen ions to store the substrate protons as hydroxy prior to final water formation. The activated substrate needs to be attached to the catalyst surface prior to the second H abstraction leading to product formation *e.g.* by an

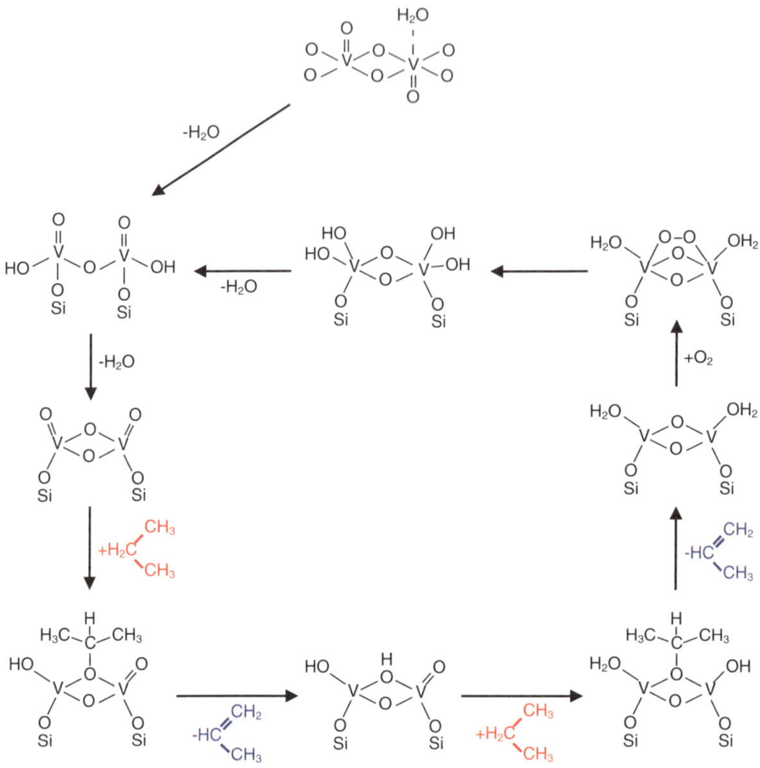

Figure 15.4 Hypothetical reaction steps in the ODH of propane over silica-supported vanadium-oxide dimers.

alkoxide intermediate (see Figure 15.4). Such species have been observed spectroscopically during selective oxidation reactions[107] and have been proposed in theoretical studies on oxidative dehydrogenation of propane.[21]

In the following, the question of the number of redox equivalents provided per vanadium site (+III *versus* +IV) will be resumed. While the assumption of +III formation is common in the literature recent EPR results on supported vanadia catalysts after propane ODH suggest the presence of +IV.[108] Therefore, in the following discussion it is assumed that metal sites should change their redox state only by one formal unit (one redox equivalent per metal site). A relatively small energy barrier will result, as no spin-pairing contributions occurs. This principle is also observed in biological redox systems such as in photosystem II where 4 manganese centers change their oxidation states between +II and +III. The relevance of spin-pairing barriers in selective oxidation was questioned[109] on the basis of theoretical calculations. The result that reoxidation with and without spin pairing was in both cases faster than the initial reduction is consistent with the statement that it is not rate determining for the selected case of methanol oxidation. Given the multiple role of activated oxygen that is different when bonded to one or two metal centers, as discussed below, a deeper analysis of the possible relevance of spin-pairing effect in oxidation catalysis is desirable.

No free protons are allowed to occur in the reaction formalism. This requires the participation of water molecules as products of acid–base reactions. The formalism accounts for the simultaneous participation of redox and acid-base processes[110–112] in hydrocarbon transformations. The critical need in selective oxidation catalysis for the reaction product water or for added water or the beneficial effect of unavoidable water from sacrificial total oxidation find in this way a natural explanation. This does not exclude the action of multiple other effects[66,69] of water as hydration agent, as moderator of the redox chemical potential of the feed and as cocatalyst for solid-state dynamical processes[65,71,113–117] of the catalytic material. Its negative effects as destabilizing agent, as hydrothermal corrosion agent and as site blocker are also not denied by this formulation. The equations allow, however, to formulate the balance of mass in the overall reaction.

The stoichiometrically correct sum of eqns (15.3) and (15.4) is given in eqn (15.6). It indicates that two single sites are required for converting one alkane molecule.

$$C_3H_8 + 2V{=}(O) \rightarrow C_3H_6 + 2V{-}(OH) \tag{15.6}$$

Equation (15.6) symbolizes the dual functions of the active sites as an inorganic oxidant and as a Brønsted base. This is in accord with literature suggestions[93,94,118–120] about the mode of operation without defining mechanistic aspects such as the number of steps[21] in the C–H bond activation.

The stoichiometric reaction (15.6) describes the formation of the desired product. To qualify for a catalytic process this reaction must be complemented by a process regenerating the initial chemical structure of the active site described in eqn (15.4). This occurs through the transfer of the accepted electrons to oxygen.

$$V^{+4}{-}(OH) + OH^- \rightarrow V^{+5}{=}(O) + e^- + H_2O \tag{15.7}$$

$$O_2 + 2H_2O + 4e^- \rightarrow 4OH^- \tag{15.8}$$

$$4V^{+4}{-}(OH) + O_2 \rightarrow 4V^{+5}{=}(O) + 2H_2O \tag{15.9}$$

The combination of the sum of eqns (15.6) and (15.9) into eqn (15.10) highlights the catalytic nature of the process and gives the correct reaction stoichiometry. This formulation avoids the frequently found fractional reactants such as "$0.5O_2$". The reactions (15.4) and (15.7) describing reduction and reoxidation of the active sites cancel each other in the sum equation, as required for a catalytic process. This is indicated by the bracket symbol for the active sites in eqn (15.10).

$$O_2 + 2C_3H_8 + \langle 4V{=}O \rangle \rightarrow 2C_3H_6 + 2H_2O + \langle 4V{=}O \rangle \tag{15.10}$$

The sequence of events in which the participation of lattice oxygen (a MvK process) is assumed, is not expressed in this equation. This does not deny that in the absence of gas-phase oxygen[121] lattice oxygen may execute reaction (15.6). Without regeneration[81] in the "riser mode"[97,99,122] the reaction of propane will stop as soon as all mobile lattice oxygen is consumed and/or the surface structure of the catalysts collapses. It is also not excluded that immediate regeneration according to eqns (15.7)–(15.9) and deeper reduction according to the lattice oxygen mode can coexist at low oxygen potentials (high concentrations of organics) that are below the bulk decomposition potential of the oxide catalyst (see below). Such a situation would lead to phase segregation by reaction-induced disproportionation of the oxide as soon as a nucleus of the limiting phases can form. For surface vanadia species this is quite a real pathway of deactivation, as can be seen from TPR experiments of such systems where the loss of surface lattice oxygen and the formation of vanadium suboxides[123,124] such as V_6O_{13}, V_2O_4 or V_2O_3 occur with little energetic separation.

The sequence for lattice oxygen regeneration in analogy to that with cofed oxygen described in eqns (15.7)–(15.9) reads as:

$$2V^{+4}-(OH) + O_L^{2-} \rightarrow 2V^{+5}=(O) + 2e^- + H_2O \tag{15.11}$$

$$O_2 + 4e^- \rightarrow 2O^{2-} \tag{15.12}$$

$$4V^{+4}-(OH) + 2O_L^{2-} + O_2 \rightarrow 4V^{+5}=(O) + 2O^{2-} + 2H_2O \tag{15.13}$$

The equations symbolize that in this catalyst mode electrons have to flow between different locations and that the conversion of oxide anions at the surface to lattice oxygen may be a diffusion-limited process. The reactions (15.11)–(15.13) involve the participation of a semiconducting solid; they separate locally the regeneration of an active site in alkane processing from an active site for oxygen reduction that is not specified in the equations. It is noted that such a modus requires careful control of the charge-flow kinetics as this will determine the abundance of activated oxygen (eqn (15.12)) and hence the selectivity of the hydrocarbon conversion. The principle[96,120,125,126] of kinetic control of the abundance of reactants needs to be observed. The vanadia–titania[127,128] system and the vanadia–ceria[129,130] system are examples where a collaboration is found between "active sites" and "support sites" that in the view of the authors are inadequate discriminations. The designation of a "heterodimer" structure seems more adequate for this situation of "symbiosis"[125,131] of species. In the modus confining the reaction only to one type of active sites (reactions (15.7)–(15.9) with immediate recovery) this control parameter is absent.

Equation (15.10) describes a fundamental difficulty in oxidation reactions using dioxygen as reactant: the necessity to involve 4 sites of single-electron transfer calls for a cooperation of redox centers. Assuming that all reaction processes occur in adsorbed states, the notion of independent single-site[21] oxidation catalysts is not likely, as a minimum of 2 electron-pair transfers is required to close the catalytic cycle with dioxygen. Rather than controlling the electron exchange between 4 independent sites it seems more plausible to ascribe high catalytic efficiency to oligomeric sites such as a pair of dimer sites[132] or even larger units[133–138] joined through a common bonding system. The concept of kinetic control of the abundance of activated reactants (see Figure 15.3) limits the size of such common electronic systems to dimensions where only few dioxygen molecules are activated at a time. The conceptual alternative that both the redox site and a coadsorbed organic substrate may transfer 4 electrons from one redox site to an oxygen molecule is not excluded but seems an unlikely alternative for a selective oxidation reaction where intermediate peroxo-species[85,139] would be detrimental for the integrity of the organic substrate.

At this point we consider the chemistry of the active sites. These sites do not occur in free form but are part of an active phase. This phase is either supported or forms the termination layer of a bulk oxide catalyst. For the following deliberation these two cases resulting from quite different material chemistries are not distinguished as for the concept of the active phase its materialization is of secondary relevance. We assume that the active sites come into existence as single hydrated ions that for vanadium would be vanadyl species. These species are either impregnated onto a carrier surface[133,140–142] or occur during the activation period of a bulk catalyst material.[44,116,143–148]

As a side comment it is noted that here a fundamental difference exists between the synthesis of high-performance catalysts and model systems prepared under ultrahigh-vacuum (UHV) conditions. The high-performance systems are obtained in the presence of water either as solvent or as reactant. The UHV model systems can be prepared free of water[129,149–151] and thus will not unfold the structural complexity arising from condensation processes of hydrated monomers in sequential oxolation leading to a manifold of metal–oxygen polyhedra in polyoxometallates[71,152–159] of early transition-metal ions. For this reason it is little surprise that UHV-derived systems present the closest possible packing of metal polyhedra compatible with their oxidation state[160,161] that is given by the oxygen chemical potential during synthesis.

From these model systems we know that the surface free-energy minima of the dense phases V_2O_5 and V_2O_3 are identical. They both consist of a vanadyl (V^{+5})-terminated layer[151,161] irrespective of the oxidation state of the bulk system. This layer is chemically inert and requires severe treatment by, *e.g.*, hydrogen atoms to become defective and thus reactive. This is in excellent agreement with theoretical studies[149,161,162] and after all justifies the approach that theoretical models of reactive sites use V_xO_y species terminated with OH groups. It is assumed that such a description[149,161,162] developed for model systems is also valid for active sites on high-performance catalysts. In the present context we see no structural difference between model systems exhibiting defects in their perfect termination and active-site ensembles found on high-performance catalysts.

Single redox sites, advocated as highly relevant forms[163–166] of active sites, are unlikely to be abundant on high-performance systems from chemical arguments respecting the well-developed knowledge of vanadium species in aqueous solution. Under most conditions of practical impregnation chemistry, the vanadium source is an isopolyoxometallate[167–170] species, usually the decavanadate $(V_{10}O_{28})^{-6}$. This species forms a spherical cluster[168] consisting of 10 edge-sharing octahedra. Six terminal single-bonded oxygen species provide reactivity, whereby two each form a plane of coordination. As a result, depending on the stability of the support, a salified form of the oligomer that is successively protonated with falling pH exists during impregnation as indicated by eqn (15.14). This form prevails at pH ranges between 6 and about 2.

$$[V_{10}O_{28}]^{6-} + 2MOH \rightarrow \{[V_{10}O_{26}](MO)_2\}^{4-} + 2OH^- \quad (15.14)$$

At lower pH values that occur inevitably upon drying and activation, the decavanadate decomposes and tends to form either bulk hydrated vanadium oxide, being chemically probably isopolyvanadates of ill-defined structure or oligomeric species of the metavanadate series. This decomposition is controlled by the presence of free coordination sites of the support (MOH). Their availability regulates the partition of the reaction sequence between isopolyvanadate formation[167,171] and spreading[76] as metavanadate. The site availability depends on the acid–base properties of the support oxide, on the size of the countercations for the decavanadate and on the presence of other adsorbates interfering with the formation of free MOH sites. The acid–base properties of the support must not tend to stabilize oxo-terminated surfaces (such as siloxane groups on silica) as this reaction opposes the availability of free coordination sites at the supporting structure. The complex reaction driven by the gradual acidification of the impregnated system during drying leads to tetrameric and dimeric metavanadate species existing as the most stable anions in not too extreme pH values.

$$\begin{aligned}\{[V_{10}O_{26}](MO)_2\}^{4-} + 4MOH + 4H^+ \rightarrow\, & 2\{[V_4(=O)_4(O)_4(OH)_2](MO)_2\} \\ & + \{[V_2(=O)_2(O)(OH)_2](MO)_2\} + H_2O\end{aligned} \quad (15.15)$$

Figure 15.5 illustrates that the mechanism of decomposition of the decavanadate may occur through the opening of the inner V–O bonds (colored in Figure 15.5) being elongated in the parent anions. Such an opening mechanism from the inside would lead to two tetrameric and two monomeric fragments or one dimer as indicated in eqn (15.15).

The driving force for this acid–base reaction (15.15) is the acidity of the MOH groups of the support with respect to the fixed pK value of the oxovanadium species. Under constant concentration conditions this driving force is given by the nature of the M–O bond polarizing the surface terminal O–H group. The net charge of the support M species (also referred to in the literature as electronegativity effect[16,172–175] or as acid–base effect[81,176–179]) thus controls

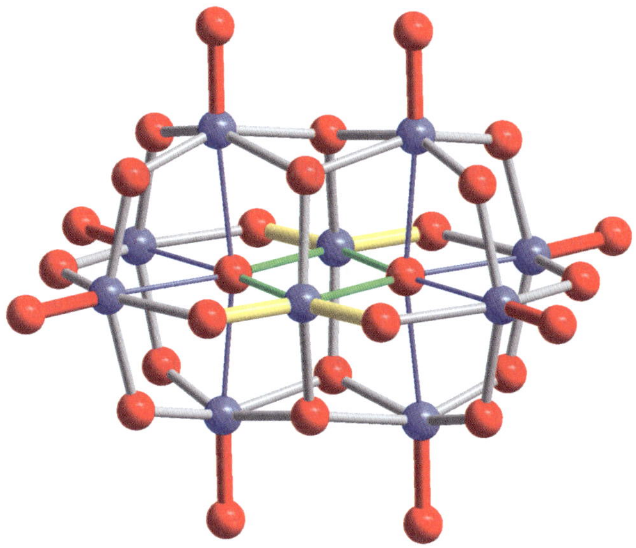

Figure 15.5 Experimental X-ray crystal structure of the decavanadate anion. The structure is a small section of a closed packing of oxo anions with V in octahedral voids. The inner bonds (green, yellow) are slightly elongated in comparison to the respective bonds in the peripheral octahedra.

the decomposition of the initially present isopolyvanadate species. Acid–base properties of the support exert multiple additional effects on the catalytic reaction process not discussed in this context. Besides the most stable dimeric ($[V_2O_7]^{4-}$) metavanadate species formulated in eqn (15.15), also the less-stable trimeric and monomeric species can form. Under conditions of highly dilute isopoly species and strongly polarizing M cations a high abundance of monomeric vanadium species can result, according to eqn (15.16).

$$\left\{[V_{10}O_{26}]\,(MO)_2\right\}^{4-} + 18MOH \rightarrow 10\left\{[V(=O)(OH)]^{2+}(MO)_2\right\} + 4OH^- + 2H_2O \quad (15.16)$$

It is obvious that a large excess of MOH is required and that the M–O–V bond has to have a substantial strength in order to overcome the driving force towards self-condensation of the decavanadate that also occurs upon heating dilute solutions. A case where this has been established[130,180] is the reaction of vanadium species with ceria where the most stable form is a mixed $CeVO_4$ compound. The dimeric metavanadate was found to exist as a dominating species[132] in a 10% vanadate-loaded SBA-15 sample using high-precision EXAFS. This result is in qualitative agreement with a surface spectroscopic study of the same system[101] using *in-situ* X-ray absorption spectroscopy at the oxygen edge as a site-sensitive detection method (see Chapter 12).

The total driving force for reaction (15.16) resulting from the formation of free water is only moderate given the large number of species involved. This means that the reverse reaction leading to hydrolysis and condensation of the supported vanadium species can easily occur in the presence of steam. The presence of steam is unavoidable during the use of these systems in ODH chemistry (see eqn (15.10)). Strongly acidic M–O systems are more susceptible to hydrolysis than less-polarizing M cations giving rise to more covalent bonding between support and vanadium species. Operation conditions determine the existence of the active phase as both a mixture of partly hydrated species as formulated in eqns (15.15) and (15.16) and a distribution of degrees of condensation of V_xO_y species. A distribution of species with various degrees of condensation[101] and even a dynamic steady state between them may well be present under reaction conditions with high load of substrates and water. The assumption that the electronic structure of the "support" determines (solely) the polarity of the M–O bond[136] and hence the reactivity of the vanadium–support interaction, is simplified as of course the local geometry of M–OH with all its structural variations ("roughness, porosity")[181] will also affect the interaction M–O–V.

Returning to the nature of active vanadium sites and their nuclearity it can be assumed that besides monomeric species as formulated in eqn (15.10) oligomeric species should also be present. In an excursion, the term "monomeric" is debated as follows. The nuclearity of the vanadium species can be clearly defined in terms of monomeric, dimeric, and higher oligomeric. This does, however, not define the catalytic function. Monomeric in the sense of "single site" only applies to a single vanadium species supported on a redox-inactive material. Such materials are silica and even alumina.[182] On all transition-metal supports this exclusion of inactivity of the support in redox chemistry cannot be maintained and a mononuclear V species on, for example, titania can well be a functional dimeric site, as a V(=O)(O)Ti(=O) species is likely[173,183–186] to exist. The case of vanadia on ceria[130,187–189] mentioned above is a documented case for such a heterodimeric structure. Another report[91] discusses such a structure for the vanadia on niobia system.

For the purpose of the following discussion we thus assume that there is at least the possibility that redox-active sites are functional dimers that can store two redox equivalents as symbolized in eqn (15.17).

$$\langle 2V{=}O\rangle \equiv \{[V_2({=}O)_2(O)(OH)_2](MO)_2\} \equiv O{=}V{-}O{-}V{=}O \qquad (15.17)$$

Here we assume for simplicity a homonuclear dimer $[V_2O_7]$ unit that is linked with a minimum of two bonds to the "support". The partial pressure of water decides[20,65,66,106] over the number of support interactions: in cases where the "support" forms stable hydroxyl terminations (such as in silica) the vanadate is only loosely fixed to the support and can aggregate back to isopolyvanadates, as described in eqns (15.14)–(15.16) (see Chapter 12). In the case of alumina as "support" the interaction seems to be stronger with multiple V–O–Al bonds and the characterization results[182] speak for a large abundance of a hetero-dimer as frequent species. We do not see why the characterization excludes a homonuclear dimer, whereas large oligomers are clearly not present.

It occurs that the praised "simplicity" of monolayer-supported redox catalysts[143,172,190,191] as opposed to bulk oxides with their complex structural chemistry[88,112,147,192–194] is only valid at first sight. Detailed examination of the factors controlling reactivity and structure of an active species in monolayer systems reveals that a distribution of structures and properties will prevail under operation conditions[181,183,195–198] thus limiting the ability to control the reactivity by obvious[45,199–202] synthetic measures such as the vanadium loading on a given support.

The regeneration of two reduced dimer sites will then occur by dioxygen according to eqn (15.18):

$$2(OH)-V-O-V-(OH) + O_2 \rightarrow 2\,O{=}V-O-V{=}O + 2H_2O \qquad (15.18)$$

At this stage, another boundary condition for the structure of an active site can be derived. According to eqn (15.18) it is necessary that both dimer sites are in electronic contact to supply the necessary 4 redox equivalents (electrons) for the activation of the dioxygen molecule. In the framework of site isolation it is expected that a dimer of dimers is supported on an electronically insulating (or semiconducting) support and that a chemical bond exists between the two dimers. Alternatively, activated molecular oxygen in the form of a peroxo-species[85,100,108] would have to move from one reduced dimer to a different reduced dimer species to "collect" the stored electrons. The tetrameric ring structure (Figure 15.6(A)) occurring as the most stable hydrolysis product of the decavandate (see eqn (15.15)) would be a perfect model for an isolated active site as it can undergo the full catalytic cycle for propane ODH without having to rely on transportation of atoms or redox equivalents from outside of its own molecular structure. The dimeric structure form Figure 15.6(B) is the next stable structure occurring in hydrolysis and is equivalent to the site described with eqn (15.17).

The structures of the smallest representatives of the metavanadate series are shown in Figure 15.6 as partly salified (bound to the support M) and hydroxylated species; in free form they occur as solvated ions. The forms shown in Figure 15.6 are still reactive with respect to further dehydration[66] of "structural water" in analogy to eqn (15.15) requiring additional bonds to the support. This fully dehydrated state can be prepared by activation[102] in dry oxygen (see Chapter 12). It is anticipated that this fully dehydrated state will partly rehydrate at the partial pressures of water present under reaction conditions. Complete rehydration transforming the coordination polyhedron from tetrahedral to octahedral, such as in dilute solutions, is not likely to occur at elevated temperatures.

Whereas the tetramer (Figure 15.6(A)) can regenerate the reduced form back to the pentavalent form without transport of activated oxygen species, this is not possible for the dimer, assuming the validity[108] of the limited reduction to

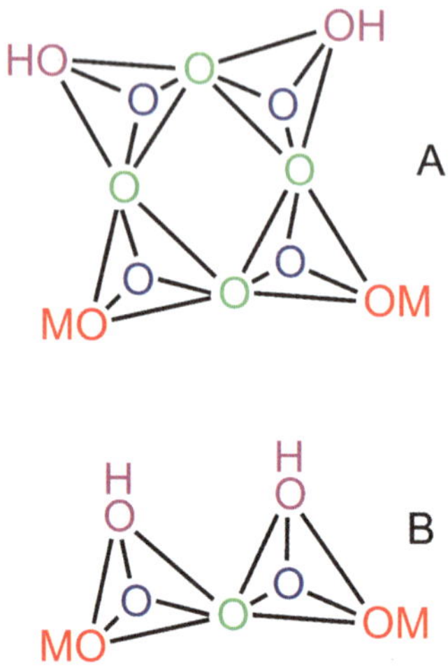

Figure 15.6 Schematic representation of possible active sites for supported V_xO_y species occurring from partial hydrolysis of decavanadate. For clarity, only the oxygen atoms are shown. The different colors of oxygen sites denote their different chemical structure: vanadyl (blue), bridging (green) and terminal (magenta). The anchoring to the support (M) occurs through the sites in red. The species are shown in their oxidized forms featuring V^{+5} (cation not shown) in the central location of each tetrahedron.

the tetravalent state. In this case, activated dioxygen would have to diffuse to the reduced site following the transformation of propane (eqn (15.6)). The products are then likely two oxidation equivalents in the form of "OH^-" species that may exhibit sufficient mobility on the surface. This scenario is probably meant when in the literature "surface lattice oxygen"[76] or "lattice oxygen"[36,61,90,93,97,98,203–207] are designated as the reactive species for catalytic oxidation. This mode of spatially separated sites for alkane oxidation and oxygen reduction is frequently termed[201,208] yet rarely clearly defined[3,61,209] as "Mars–van Krevelen" mode (MvK). In this mode the reactive surface contains active dimers, coexisting in different oxidation states. Some of them may be oxidized and thus form the active sites for propane (alkane) oxidation. Others will be present in the reduced form and be capable of reducing dioxygen. Two oxidation equivalents will remain as ("OH") that need reduction at a different reduced vanadia dimer. Motion of activated oxygen species to another independent reduced site would be necessary to fully activate the oxygen. This "molecular" traffic of activated oxygen[126] requires a careful choice of the kinetics of product desorption relative to the diffusion time ("spillover process")[96,210] in order to avoid overoxidation (see discussion of eqn (15.12)). Such a scenario may be feasible for the ODH process, although serious attempts to identify such processes by isotope labelling experiments were unsuccessful.[211]

A mode in which dimers and coadsorbed organic substrates cooperate to activate a dioxygen molecule is possible but will most likely not operate selectively as the peroxo-intermediate can easily overoxidize the organic adsorbate. This is also valid for the suggested "ozonide" active species[212] that should be present on VPO catalysts; such a species would be extremely labile and cause rapid deep oxidation if it were to survive hydrolysis under moisture at elevated temperature. A facile and selective operation is thus expected when tetrameric species such as shown in Figure 15.6A are operating. Even if they may be rare in abundance their effective catalytic cycling may render them as highly productive sites.

Deeper oxidation of the olefin to the desired product acrylic acid can occur at suitable conditions of oxygen chemical potential and availability of multiple redox sites. For such a process eqns (15.6) and (15.10) describe the formation of an intermediate propene that will readsorb according to eqn (15.19) rather than form a product. The conditions must allow facile rebonding of the olefin and must provide a limited flow of reactive oxygen species capable of inserting into C–H bonds. This form of oxygen is electronically different from active oxygen abstracting hydrogen from a C–H compound. The analysis of selective oxidation reaction pathways is hampered by the insufficient discrimination[61–64,207,213,214] of activated oxygen into nucleophilic (C–H activating) and electrophilic (oxygen-inserting) function.

$$\begin{aligned} &\mathrm{O{=}V{-}O{-}V{=}O + C_3H_6 \rightarrow O{=}V - O{-}V{-}OC_3H_5 + H^+ + e^-} \\ &\mathrm{O{=}V{-}O{-}V{-}OC_3H_5 + H^+ + e^- \rightarrow HO{-}V{-}O{-}V{-}OC_3H_5} \end{aligned} \qquad (15.19)$$

The first step is the oxidative adsorption of propene to the allylic alkoxide shown in eqn (15.19). This is a one-electron oxidation leaving behind on the organic moiety one electron to enable the allylic stabilization. The active site has absorbed one redox equivalent but contains one V^{5+} center, being responsible for holding the substrate in place.

The next combined step is a three-electron oxidation leading to the adsorbed unsaturated aldehyde still with allylic stabilization.

$$\begin{aligned} \mathrm{HO{-}V{-}O{-}V{-}OC_3H_5 + 3\,OH^-} &\rightarrow \mathrm{HO{-}V{-}O{-}V{-}OC_3H_4O} \\ &\quad + \mathrm{2H_2O} + 3e^- \end{aligned} \qquad (15.20)$$

This oxidation step in eqn (15.20) leads to electronically stabilized acrolein that is still held to the active site, as the electrons released from the organic substrate do not reduce the vanadium center. These electrons are transferred to neighboring sites. It is speculated that the large number of redox equivalents that need to be exchanged in this step presents a substantial hurdle to the overall reaction and hence may represent the kinetic bottleneck of the whole oxidation process. According to the assumed concept of one redox equivalent per metal sites and according to the assumption of dimer sites being the prevalent species available as active sites, the following complex overall reaction described in eqn (15.21) occurs. It is obvious that if other oligomers being capable of accepting the redox equivalents in a single ensemble site were available, then the overall reaction may be simplified and the kinetically complex cooperation of a minimum of three sites could be avoided to achieve one turnover of the substrate. At this point it becomes clear how the details of the oligomerization kinetics during catalyst synthesis may have a strong influence on the kinetics of the selective oxidation of propane.

$$\begin{aligned}&2\mathrm{HO{-}V{-}O{-}V{-}OC_3H_5} + 3\mathrm{O{=}V{-}O{-}V{=}O} + 2\mathrm{H_2O} \rightarrow \\ &2\mathrm{HO{-}V{-}O{-}V{-}OC_3H_4O} + 3\mathrm{HO{-}V{-}O{-}V{-}OH}\end{aligned} \tag{15.21}$$

Likewise, the concept of a functional heterodimer would help to simplify the reaction; the redox equivalents may be stored onto the redox-active "support". It becomes clear that silica or alumina are not suitable supports for partial oxidation catalysts but that molybdenum oxide or other reducible mixed oxides with a stable geometric structure[215–222] are excellent choices for reactions involving both C–H activation and oxygen transfer. The exchange of electrophilic oxygen for redox equivalents is a facile process when the site can easily change its oxidation state without having to break bonds to neighboring atoms. Handing over an oxygen atom to the adsorbate requires a change in local coordination such as from a corner-sharing dimeric form to an edge-sharing form. An adsorbate-type structure of the active site allowing dynamical response will be useful, whereas a rigid 3-dimsional crystalline state will hinder such a process with high activation barriers. Such "adsorbate-like" active phases can be identified with probe molecule reactions[17,186,213,223–227] and have been rationalized for covering some bulk catalysts[54,71,115,144,228–230] as a consequence of surface-energy minimization.

The regeneration of the active reduced dimers can occur according to eqn (15.18). This may happen in parallel with the continuation of the oxidation. It can be expected that the intermediate aldehyde can be easily oxidized[53,231] further to the more stable carboxylic acid. In a facile one-electron oxidation this can be achieved by a mobile oxidation equivalent as shown in eqn (15.22). The resulting redox equivalent can be stored in the active site that would lose its ability to hold the substrate.

$$\begin{aligned}&\mathrm{HO{-}V{-}O{-}V{-}OC_3H_4O} + \mathrm{OH^-} \rightarrow \mathrm{C_3H_3OOH} + \langle[\mathrm{OH{-}V{-}O{-}V^+{-}O}]^-\rangle \\ &+ e^- + \mathrm{H^+}\langle[\mathrm{OH{-}V{-}O{-}V^+{-}O}]^-\rangle + e^- + \mathrm{H^+} \rightarrow \mathrm{HO{-}V{-}O{-}V{-}OH}\end{aligned} \tag{15.22}$$

The species in brackets does not occur as product but is merely written to illustrate how the autoreduction of the binding site of the organic substrate gives rise to the liberation of the product.

The change of the redox state of the fixing metal site is characteristic of the concept of an adaptive site. In a sequence of reaction steps the bond between substrate and active site is adapted by the electronic-structure modification of the substrate: in the present case the allylic stabilization is only lost in the last step enabling the redox equivalent from the substrate to autoreduce its own binding site. Electron-accepting sites are required to be easily available to avoid premature reduction of the adaptive binding site. This means that the number of surface locations where several active sites are at the same time available in the correct oxidation state will be quite low, giving a hint to the low levels of productivity[232,233] observed for such catalysts.

The formal description did not make any distinction in the type of active oxygen that must fulfil two fundamentally different functions: one type acts as a deprotonating agent and is referred to as "nucleophilic", whereas another type inserts into C–H bonds and reacts thus "electrophilically". In the equations above both types are denoted with "OH^-". But the stoichiometry of these redox equivalents indicates their different function when ending either as water (ex-nucleophilic) or as oxygen in the organic molecule (ex-electrophilic). The formulation of redox equivalents as "OH^-" pays tribute to the necessary management of hydrogen species that is frequently neglected in other formulations of selective oxidation reactions. It further is inspired by the critical role that water plays in all selective oxidation reactions. Only in the strict absence of any moisture would it be possible to formulate "lattice oxygen" as "O^{2-}" in the role of a mobile oxidation equivalent, as exemplified with eqn (15.13). This notion is not realistic at finite conversion levels as the above-mentioned management of hydrogen leads inevitably in all selective oxidation and ODH processes to stoichiometric amounts of water, which would immediately hydrolyze with lattice oxygen forming the "OH^-" species.

The scenario thus predicts that oxidation equivalents are hydroxyl species moving between active sites. If no higher V_xO_y oligomers are assumed to be present, then also protonated peroxy species found to exist[108] by EPR on V–Ti systems would have to move between dimer sites to "harvest" electrons for their complete activation into redox-active hydroxyl species.

The coverage of the supporting surface with active vanadate must be below one monolayer in order to prevent unwanted condensation and eventual crystallization in reverse processes of eqns (15.15) and (15.16). How dense the active vanadia species can be adsorbed onto its support depends on the ratio of bond strength of vanadia–vanadia interactions against vanadia–substrate interactions. Theoretical studies for the vanadia–alumina system show that a high propensity exists for the vanadia–vanadia interaction,[114] thus limiting the density of stable surface-grafted species substantially below one monolayer. This is in line with an experimental study[124] reporting on redox-labile vanadia possibly due to extensive condensation of the vanadia species. It is difficult to generalize this, as the extent of hydroxylation of the substrate will be a critical factor controlling the loading. This property is a complex function of synthesis conditions and prehistory of a substrate and requires individual determination. We may estimate that the density of redox-active active sites is on the order of 20% or less of the geometric catalyst surface.

The extent of vanadia condensation should be limited to a few units, as never more than one formula unit of redox equivalents may be held on an active site during one turnover in order to avoid overoxidation. The oligomers must also not condense to stable oxide crystals, as these expose inactive terminations, removing the active material from functionality. This stability argument limits the internal size as well as the surface density of oligomers. It is argued that the chemical stability of metavanadate oligomers $(VO_3^-)_n$ with $n = 2,3,4$ limits the size to these values under operating conditions of selective oxidation.

Another direction of optimizing the performance of oxidation systems by chemical means is the choice of the "support". The formation of heteronuclear oligomers[173,234] (such as V–Ti, V–Ce, V–Mo) with a redox-active partner for the vanadia is one option. The successful application of heteropoly oxometallates in selective oxidation catalysis[29,43,71,155,157,235,236] is a worked example of the usefulness of this concept. On supported systems the same chemical concept is applied in a different structural realization. The technical V–Ti system for PSA synthesis[237–239] may be taken as an important example (see Chapter 16). It is noted that it was found that only oligomeric vanadia was active for conversion, whereas monomeric vanadium sites did only adsorb the organic educt. The often stressed direct effect of a support system on the local electronic structure of V_xO_y does exist,[67,80,94,133,173,175,191,204,240–245] is, however, not a variable of free choice as then the stability of the system is compromised. The observation of a systematic influence of the nature of the support cation on the reactivity of one type of V_xO_y species[137,191] may thus well find its explanation in a modified structure[108] (degree of condensation and degree of metal–support interaction) of active sites rather than in the claimed modified V=O local electronic structure of a supposedly unchanged geometric structure of V_xO_y (see Chapter 12).

15.3 Self-Supported Systems

We return to the case of a self-supported system where we assume that V_xO_y species exist also in monolayer form supported on the structures of multiphase–multielement oxide (MMO) systems. Typical cases are the M1 phase for C3 oxidation (see Chapter 4) and the VPO phase for C4 oxidation (see Chapter 6). In the case of the M1 phase a correlation of activity with the V+Te abundance at the surface was found[115] in good agreement with the unusual dynamics of V^{5+} on these systems[138] and in the VPO case irregularities of the phosphate abundance at the surface[144] indicate as well as direct spectroscopic[116,145,146] observations the possibility of a V_xO_y termination. A broad agreement exists about the structural dynamics of the VPO system[29,246–257] giving ample opportunity to restructure the surface[75,229,250,258,259] of even well-crystalline VPO phases. Statements about the particular activity of the (100) plane of the vanadyl pyrophosphate phase[249,260] are not in contradiction with the present notion of a metavanadate termination, as their molecular termination was not investigated with sufficient resolution. Other phosphate systems relevant for selective oxidation[72,157,261–272] were not studied in sufficient detail to know about their surface structure under working conditions, although reports about "extra-framework" vanadia[262,273] are not in contradiction to the metavanadate hypothesis.

A self-supported system exists when, driven by the chemical potential of the surrounding atmosphere, some components of the bulk crystal form an enriched surface layer[274,275] representing the monolayer of active phase on its support system. The situation is similar to metallic alloy systems where chemically induced surface segregation[276–278] leads to strong deviations from the bulk structure and composition. In both cases the gradients in chemical potential of constituents of the solid across its interface to the environment are responsible for the surface modifications. In oxide catalysts it is not likely that conclusions about the mode of operation are realistic when their active sites are constructed from bulk structures of the catalyst. This is observed, however, in two prominent cases namely of the VPO system for C4 oxidation[88,212,248,253,279,280] and of the M1 system[87,224,281–286] for C3 oxidation. Much earlier, the concept of an active surface layer dissimilar to its supporting bulk[287] was discovered for the vanadium-based catalysts. Another example for a supported active phase is the iron-oxide catalyst for styrene synthesis by dehydrogenation of ethylbenzene where an alkali ferrate surface phase[288–290] is relevant (see Chapter 11). A deliberate realization of the concept of a supported immobilized active phase with embedded active sites is the recent synthesis of SILP[291] catalysts. An inorganic support carries a monolayer of an ionic liquid in which homogeneous catalysts are dispersed.

The general situation at the active interface is shown in Figure 15.7. It occurs that a series of interfaces exists at the contact between catalyst and reactants. These interfaces comprise the boundary between gas phase and adsorbate layer, (1) the adsorbate/solid interface (2) the interface between the surface terminating layer and the end of the periodic bulk crystal (3) and the interface between the solid depletion or enrichment layer and the crystalline bulk phase (4).

In Figure 15.7 the potential of oxygen is indicated. It is customary to select the gas-phase potential (partial pressure) such that the catalyst has under operation conditions a slight tendency to lose oxygen, supporting the formation of oxygen defects[95,222,292–295] counted with respect to the highest possible oxidation state of the bulk oxide. It is assumed that the selection of the oxygen gradient supports the formation of the active phase layer. The resulting oxygen defects require a rearrangement of cations through which the active species such as V_xO_y can enrich at the surface and can get separated from each other through concomitant segregation of additional species such as phosphate, antimonate or tellurate made responsible in the literature for controlling the oxygen chemical potential in albeit other scenarios[126,296–299] and by the formation of interfacial nanostructures.[147,300,301] It is a hypothesis here that this process is addressed when bulk "lattice oxygen" is made responsible[60,302–308] for catalytic oxidation activity. The experimentally observed correlation of mobile oxygen[303,309–313] and of catalytic activity is based upon its involvement in the formation of the active phase.

At substantial chemical potentials of oxygen the beneficial effect of lattice oxygen is not due to its consumption in the substrate conversion. Under conditions where the chemical potential of oxygen is zero in the gas phase[96–98] (so-called "riser" operation conditions) a substantial flow of lattice oxygen through

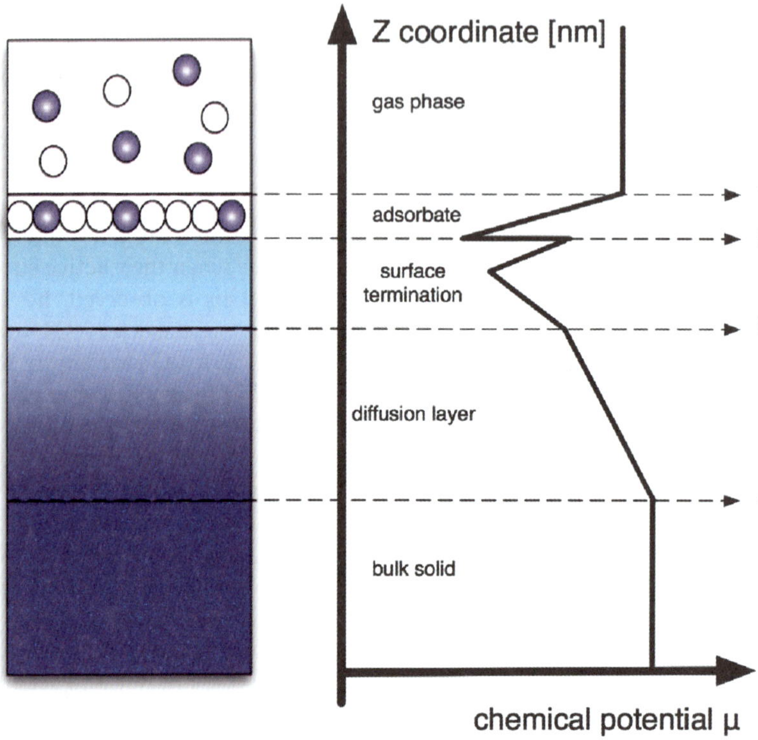

Figure 15.7 Schematic representation of the interface region of a bulk oxide catalyst in contact with a reactant atmosphere containing dioxygen (blue balls) and a strongly adsorbing organic species (white balls). The chemical potential gradient is shown for oxygen, other gradients of interest would be that of water, hydrogen, carbon and of the constituents of the solid. The numbers designate the interfaces for oxygen exchange (see text).

the gas/solid interface 2 (Figure 15.7) will occur according to eqns (15.11)–(15.13). If not carefully controlled[99] this reaction can lead to the formation of metallic solids if sufficient organic reactant is offered. On that scale the gradients in chemical potential shown in Figure 15.7 are small. Figure 15.7 assumes conventional operation conditions under which oxygen is present in the gas phase.

Under such conditions the catalyst is reduced to suboxides such as shear structures of Magnelli phases.[77,96,306,314–316] The formation of defects and segregation of components do occur out of the crystalline solid to stabilize the outer surface (interface 3 in Figure 15.7). These processes create a subsurface layer (between interfaces 3 and 4) constituting the suboxide. We can now discriminate two classes of catalysts. In highly stable systems such as the M1 phase for C3 oxidation the whole bulk represents a suboxide with respect to the oxygen content required for all cations being present in their highest oxidation state. In such systems the diffusion layer between interfaces 3 and 4 is absent, which may be one reason for their exceptional stability. The formation of the surface termination layer occurs not in contact with the reactants but during crystallization under low oxygen chemical potential[137,192,193] (second heat treatment step in inert gas atmosphere for the M1 phase). Such systems do not need an equilibration phase with the reactants and are active from the initial start-up phase. The oxygen chemical potential gradient in the surface termination layer is then steeper, as indicated in Figure 15.7. This can only be maintained if the chemical bonding of M–O within the bulk structure is strong. A high covalence enabled by an optimal orbital overlap indicated by nondense packing motifs of the bulk structure is a prerequisite for such systems. They will be rare among the many oxide structural types, as the structure still has to allow the formation of a catalytically reactive surface-terminating layer. In M1 this is guaranteed by the topological anisotropy of the channel structure.[286,317] Layered motifs with strong contributions of M=O bonds are poor catalysts as their surface termination layer is identical to its equilibrium structure and hence does not contain high-energy defect sites. Examples[318,319] are the stable allotropes of MoO_3 and V_2O_5.

The other class of catalysts contains solids that convert from their as-prepared state into the layer structure containing a suboxide phase as seen in Figure 15.7 (layer between interfaces 3 and 4). The Mo_5O_{14} phase-based systems[215,219,222,295,320] used for the technical two-step oxidation of propene to acrylic acid are of that type and show extensive equilibration effects. An extreme case where both bulk phases and the suboxide layer form in the chemical potential of a reactant atmosphere from a precursor system is the VPO catalyst exhibiting critical equilibration effects.[67,252,257] In this context it is understood that the choice and handling of organic alcohols in the VPO synthesis[321–323] plays a crucial role as it affects the internal redox chemistry and as such the requirement of the reactants to bring about the phase equilibration. It is not excluded that preparation procedures for VPO may be found that arrive at a stable solid with only a dynamical surface structure. The presently found complex influence of the activation period and bulk structural dynamics may then be not an intrinsic property of VPO but rather be related to the present synthesis strategies.

The surface-termination layer is the functional equivalent of the active phase supported on monolayer catalysts. Irrespective of its origin in the life cycle of the catalyst we expect to find there the active-site ensembles of structures similar as shown in Figure 15.6. The suggested conceptual unity between bulk and supported oxidation catalysts finds no clear confirmation in kinetic data; in many reactions bulk catalysts perform better per unit surface area than monolayer catalysts supported on nonreducible oxides. Two explanations may be offered without finding sufficient evidence to rule out any of the scenarios in a rigorous way: the complexity of the selective oxidation process may well allow for both answers to be valid in different examples of catalyst chemistry and reaction conditions.

In one scenario the diffusion layer supporting the active phase provides for transport of redox equivalents between active-site ensembles. This may enhance the activation of molecular oxygen in parallel with the substrate oxidation. Even if oxygen activation is often not kinetically limiting from its reaction order[65,94,191,214,240,315,324–326] it has many control function in the process network and may thus be a critical component.

Alternatively, the average electronic structure of the semiconducting bulk catalyst phase (all material below interface 2 in Figure 15.7) influences the local electronic structure of the active sites. This can either occur through direct modification of metal–oxygen bonding within the network of covalent interactions[115,327–331] in oxide systems. Alternatively, the energy of the HOMO–LUMO gap of the metavanadate ensemble can be influenced through electrostatic effects. Much more detailed information about the surface electronic structure of active catalysts is needed to further understand this conceptual question. The frequently used UV-Vis or EXAFS analyses are useful in that respect but are of insufficient surface sensitivity to deliver this information.

The chemical potential of oxygen is indicated in Figure 15.7 to follow a sharp kink at the interface between catalyst and adsorbate. This structure at interface 2 accounts for the prevalence of nonoxidic adsorbates in the presence of activated hydrocarbon species in the adsorbate layer. The surface is "blocked" for the adsorption of oxygen from the gas phase. The dip in the potential right above the solid is equivalent to a low sticking coefficient of oxygen at the working catalyst. Either high oxidation states of the oxide prevent activation of gas-phase oxygen or at lower oxidation states of the surface insufficient free sites are available for oxygen. This causes the need to operate at undesirable high partial pressures of oxygen that may account for unwanted consecutive reactions with the reactive products of the catalytic process. In practical catalytic testing the formation of CO or soot can occur from condensation processes of adsorbed organic species; the testing conditions were then chosen such that a complete conversion of oxygen was achieved within the catalyst bed, causing a strong gradient of oxygen chemical potential through the reactor. This phenomenon[124] is not discussed further here although it frequently obscures the interpretation of kinetic data.

In the catalyst interface the presence of oxygen-terminating features such as the common –yl termination[160,161] creates a local excess of oxygen. An oxide surface exposes mainly oxygen-derived electronic states to the adsorbate. The high stability of these terminating features presents a barrier against bulk oxygen diffusing out into the gas phase and equilibrating the chemical potential. These species provide the kinetic stability of the catalyst in a reducing atmosphere. Their abundance controls the stability but reduces the number of sites at which redox processes can occur. Defects in the oxygen-terminated surface are required for reactivity: they can either be formed through synthetic measures or during the activation phase of the catalyst in the feed or can occur as "cus sites"[332–335] already in the structural motif of the termination. Under conditions of selective oxidation and in the presence of steam it expected that such cus sites will not be available and that terminations with maximum oxygen and hydroxyl coverage will prevail. Consequently, it occurs that reactive oxidation catalysts are inherently unstable against their decomposition; in this light it is clear that high kinetic barriers for bulk structural rearrangements are protecting the catalyst. They can be created by a high crystallinity of the catalyst or by structural promotors inhibiting restructuring of the crystallite bulk or by a refractory support. These barriers ensure long-term stability of the catalyst also under modulated or fluctuating chemical potential as occurs in reactors or even within the granular structure of a catalyst particle.

From Figure 15.7 we learn that the understanding and the control of the oxygen chemical potential through the sequence of reactant/catalyst interfaces is a central challenge in selective oxidation catalysis. The material stability and catalytic performance are strongly coupled to each other. The diffuse interface depicted in Figure 15.7 is an indication of the multiple role of oxygen in the system. Oxygen is the key structural factor as a constituent of the anion lattice of the catalyst and oxygen is a reactant in the catalytic process. Oxygen is further present in several chemically different forms such as molecular and activated oxygen, bridging and terminal oxygen, electrophilic and nucleophilic oxygen and finally oxide and hydroxide anionic species. This adds further to the complexity of the picture. The dilemma of the dual role of oxygen as reactant and as material constituent is tried to heal conceptually through the introduction of the lattice-oxygen concept. In this concept the catalyst is considered as a sponge for active oxygen and as a dynamical structure regenerating itself in a catalytic cycle separated from the conversion of the organic substrate.

We do not see the conceptual necessity or the compelling experimental evidence having to follow this concept that exhibits some critical consequences. The phenomenological correlation of catalytic activity and oxygen mobility is well accounted for by the processes leading to the situation depicted in Figure 15.7: the stated observation from isotope tracer experiments that total oxidation is faster than selective oxidation can also be accounted for by considering the presence of adsorbed oxygen in either nucleophilic or electrophilic form. The latter is required most for combustion and thus depleted more readily than the former being strongly adsorbed.

In the MvK concept oxidation catalysts cannot exist without redox-active bulk phases. Introducing the concept of surface lattice oxygen, which, however, is not clearly discriminated from adsorbates, healed the clear inconsistency with experiment. If the driving forces for outward and inward oxygen diffusion are assumed to be the same, which is necessary for preserving the structure of the catalyst, then an equilibrium state would quickly stop the reaction. Differences in transport kinetics between the gas-phase oxygen delivery and the removal of oxidation products may perturb the equilibrium and so enable a continuous operation of the catalytic cycle. The resulting dynamical state is sensitive to destruction by the formation of stable oxide products[336–338] consisting of structures with oxygen species being immobile at reaction conditions. This deactivation through recrystallization is observed in steady-state operation only after long times on-stream[339] but may be a rapid process in the riser operation mode with cyclic feed of oxygen and organic reactant. The driving force to terminate the MvK operation would be the Madelung energy of the newly formed static crystal, which can but need not be the same phase as present in the fresh operating cataylst.

It is concluded that the MvK concept is valid for cyclic operation but not necessary for steady-state operation where the gradient of oxygen potential is limited to the subsurface region of the oxide (interface 3 in Figure 15.7). This statement understands that no distinction is made between adsorbed activated oxygen and "surface lattice oxygen" that clearly exists on working selective oxidation catalysts.

In contrast to many other functional materials the operation of a catalyst requires its chemical dynamics at least in the near-surface region. It is part of the function that constituents of the active species do change their location and electronic structure in the material. This opens many avenues for structural deactivation as discussed above. It is thus a target of catalyst design confining the chemical dynamics to the smallest possible units of the catalysts that are the active sites. Strategies for stable operation of the catalyst are sufficient interaction of the active-site ensembles (metavanadates) to their matrix or their facile recreation from deactivated species. The required chemical dynamics of the active surface layer either calls for a highly refined preparation procedure (possibly in the M1 and Mo_5O_{14} catalysts) or for an *in-situ* formation during the equilibration period (in the VPO system) of the catalyst. In any case, the thickness of the active phase must be strictly limited to a two-dimensional surface coverage to avoid the restructuring processes leading to stable 3-dimensional oxides with little or detrimental catalytic activity.

Resulting from this idea is the prediction of a correlation between increasing catalytic activity and increasing surface free energy. This correlation is of complex quantitative form as the above-mentioned multiple species of oxygen enter into the equation for the surface energy, being a function of the oxygen chemical potential. All other existing species, namely the cations of the solid, water and the adsorbed reacting species, also enter into this equation with their respective potentials. Thus it seems at present impossible to quantify this concept. In terms of a structure–function correlation[340] it cannot be expected to find linear correlations, as multiple influences are connected. The analysis of nonlinear relationships requires excellent data quality and qualitative information, suggesting the validity of such correlations.

One way of increasing the surface free energy in a given system is its nanostructuring.[248,341] The smaller the volume of a functional unit of an oxide material becomes, the more important is the surface contribution and hence the fraction of a supported active phase. The increasing structural complexity and concomitant electronic structure complications of surfaces of increasingly smaller particles kinetically stabilize smaller particles with larger surface free energy; smaller particles can carry more surface defects than larger ones before, at a critical size, recrystallization removes the defects, stabilizes the solid and hence reduces the possible catalytic activity. For semiconductors as frequently found in oxide catalysis, the nanosize has further implications on the surface electronic structure such as the position of the valence and conduction bands; the screening length of the surface charges may well be in the dimension of the particle and thus modify the energy levels by substantial amounts of the distance of the energy levels from the vacuum level.[342–346] This effect is rarely considered in discussions of the size effects in oxide catalysis.

15.4 Conclusions

The present analysis used a formal concept of stoichiometrically correct redox processes to establish that the nature of high-performing active sites in many selective oxidation catalysts may be that of metavanadates. From the synthesis procedure and from inorganic chemistry it was concluded that oligomers with 2, 3, 4 centers may be the most stable structures. Larger units tend to condense to crystallites of vanadium oxide, being less potent in providing active sites.

It is suggested that these oligomers are always the active sites within the vast collection of practical catalysts that were developed over time. These sites should be capable of performing all selective oxidations of small organic molecules. The reaction is in each case a series of consecutive and parallel reactions of C–H activations and oxygen insertion reactions. To perform these functions the active sites must provide nucleophilic oxygen for C–H activation and electrophilic oxygen for desired oxygen functionalization of the primary product olefin.

It is argued that these sites exist in similar form both on supported systems and on bulk catalysts. It was stated that a functional separation between "support" and "cocatalyst" may be difficult and may thus be discontinued. The picture of a heterodimer between vanadia and "support" was introduced. This can be a dimer with one metal species each but also designates the interaction of a vanadia oligomer with a matching number of heterometallic sites. On grafted catalyst systems the key synthesis parameters are the availability of coordination sites for the vanadia and the choice of the supporting material, such as to sufficiently strongly bind the vanadia without breaking it into single vanadium sites that are overall poor catalysts. In bulk catalysts the restructuring processes driven by the free-energy minimization under the chemical potential of the reaction atmosphere bring about the vanadia oligomers. The metavanadates are not constituents of the structural motif of the bulk phase and need being formed in enigmatic "activation" processes of the precursor oxide systems.

The need for crystalline bulk compounds and the complex solid-state chemistry found in this class of catalysts can be rationalized as being necessary to generate and maintain the active sites as vanadia oligomers with a maximum structural flexibility and a minimum tendency to agglomerate to a crystalline phase. On grafted systems the selection of the cation and the structure of the hetero-oxide are the control variables; a wide span exists between single crystals and rafts of large oligomers floating on an inert geometric support (titania on SBA-15 for example). In bulk systems, many more variables exist in the form of structural motifs, defect structure, lattice doping, optimization of particle size and morphology, formation of composite materials with mineral spaces and inert carriers or with main-group glass-forming elements. These variables work on the maintenance of the structural integrity of the ensembles between metavanadate and the heteroatoms under the widely varying chemical potentials of the different reaction environments. Besides maintaining the stability of the interaction between vanadia and the heteroelements it is speculated that one beneficial effect of the glass-forming elements P, Sb, Te consists in reducing the reactivity of the metavanadates against hydrolysis and condensation. This is achieved by binding to some of the peripheral bonding sites (see Figures 15.5 and 15.6) and replacing terminal hydroxyls through larger units of phosphates, antominates and tellurates.

Selective oxidation catalysts thus contain platform active sites (the vanadates) that are kept in a dynamical state through a wide variety of partly collaborating measures of choosing the composition and morphology of the solid catalyst. The dynamic response to the nature of the adsorbate and the chemical potential of the environment allows a cycle of consecutive reactions with changes of local coordination and redox state to be repeated many times without losing the metavanadate structure.

Such a general architecture is in principle not limited to vanadates but to all elements allowing the existence of stable oligometaoxoacids. This is true for vanadium but also for molybdenum, chromium and to a much lesser extent for titanium, niobium and tungsten. These other elements exhibit, however, a lower range of stability of their oligometaoxoacids and may only exist in smaller ranges of chemical potential defined between oxygen, water and organic reductants from educts and products of the selective oxidation.

The material properties of the matrix that bring about the metavanadates and stabilize them on the surface are also responsible for the acid–base properties of the catalysts. An independent assessment of the acidity and the redox activity from the catalyst performance is thus difficult and may explain the unclear situation. No attempt was made here to analyze the individual role of acidity that is important for the reaction mechanisms, as can be seen from the fact that water and OH groups play crucial roles in the redox chemistry as shown here. No attempt was made to discuss the frequently used term of Lewis acidity[81,93,201,251,347–352] as a measure for the nature of interaction of the adsorbate-catalyst complex in relation to the redox chemistry. A Lewis-acid–base complex occurs through a weaker interaction between electron pairs residing at the complexing partners rather than a redox reaction where electrons are completely transferred between atoms. In the absence of a clear description of the strong redox chemistry of the working catalysts and considering that many Lewis sites are also redox-active, it seems premature to come up with clear statements about this important part of the selective oxidation chemistry.

It may be instructive to follow the detailed descriptions of reaction systems throughout this book with the concept in mind that the enormous phenomenological complexity described may be the expression of the ingenuity of the inorganic chemists. They have to cope with the challenges of stabilizing an oligomeric active site in an environment hostile for its existence. It tends to drive the active site into condensation and continuous reduction with respect to the required highest possible oxidation state being only capable of activating the organic substrate. Static solutions for the active species may do the activation step but are either inactive for oxygen transfer (V_2O_5) or overoxidize (MoO_3) the desired products being in all cases thermodynamically more instable than the educt to be activated. It is little surprise that this delicate optimization challenge has not led to a commercially useful production[353] of activated C3 molecules through selective oxidation from propane. The enormous increase in experiments provided many new systems that, however, still do not solve the challenge of bringing about the optimal vanadate site and keep it in place. The use of other active centers such as molybdena follows much the same rules with the disadvantage that the recrystallization product MoO_3 is quite active for combustion catalysis. Meeting the challenge may thus not need to generate new compositions of catalysts[354–356] but to close the gap[117,357] between the beginning generalized understanding and the synthesis of optimized catalysts exhibiting many additional parameters for tuning than the elemental composition.

References

1. G. Centi, F. Cavani and F. Trifiro, *Selective Oxidation by Heterogeneous Catalysis*, Kluwer Academic/Plenum Publishers, New York, 2000.
2. H. H. Kung, *Transition-metal Oxides Surface Chemistry and Catalysis*, Stud. Surf. Sci. Catal., Vol. 45, Elsevier, Amsterdam, 1988.
3. A. Bielanski and J. Haber, *Oxygen in Catalysis*, Marcel Dekker, New York, 1991.
4. B. K. Hodnett, *Heterogeneous Catalytic Oxidation*, John Wiley & Sons Ltd, Chichester, 2000.
5. J. L. G. Fierro (ed.), *Metal Oxides*, CRC Press, Taylor & Francis Group, Boca Raton, 2006.
6. S. D. Jackson, J. S. J. Hargreaves (ed.), *Metal Oxide Catalysis*, Wiley-VCH, Weinheim, 2008.
7. N. Mizuno (ed.), *Modern Heterogeneous Oxidation Catalysis*, Wiley-VCH, Weinheim, 2009.
8. B. Gryzbowska-Swierkosz, *Top. Catal.*, 2000, **11/12**, 23.
9. R. Schlögl, A. Knop-Gericke, M. Hävecker, U. Wild, D. Frickel, T. Ressler, R. E. Jentoft, J. Wienold, G. Mestl, A. Blume, O. Timpe and Y. Uchida, *Top. Catal.*, 2001, **15**, 219.
10. R. Schlögl, in *Modern Heterogeneous Oxidation Catalysis*, N. Mizuno (ed.),Wiley-VCH, Weinheim, 2009, 1.
11. R. K. Grasselli, *Top. Catal.*, 2002, **21**, 79.
12. G. Emig and M. A. Liauw, *Top. Catal.*, 2002, **21**, 11.
13. J. M. L. López Nieto, *Top. Catal.*, 2006, **41**, 3.
14. N. Magg, B. Immaraporn, J. B. Giorgi, T. Schroeder, M. Bäumer, J. Döbler, Z. Wu, E. Kondratenko, M. Cherian, M. Baerns, P. C. Stair, J. Sauer and H.-J. Freund, *J. Catal.*, 2004, **226**, 88.

15. I. E. Wachs, J.-M. Jehng, G. Deo, B. M. Weckhuysen, V. V. Guliants, J. B. Benziger and S. Sundaresan, *J. Catal.*, 1997, **170**, 75.
16. A. Khodakov, B. Olthof, A. T. Bell and E. Iglesia, *J. Catal.*, 1999, **181**, 205.
17. T. V. Venkov, C. Hess and F. C. Jentoft, *Langmuir*, 2007, **23**, 1768.
18. K. Chen, E. Iglesia and A. T. Bell, *J. Catal.*, 2000, **192**, 197.
19. A. Dinse, B. Frank, C. Hess, D. Habel and R. Schomäcker, *J. Mol. Catal. A: Chemical*, 2008, **289**, 28.
20. A. Dinse, S. Khennache, B. Frank, C. Hess, R. Herbert, S. Wrabetz, R. Schlögl and R. Schomäcker, *J. Mol. Catal. A: Chemical*, 2009, **307**, 43.
21. X. Rozanska, R. Fortrie and J. Sauer, *J. Phys. Chem. C*, 2007, **111**, 6041.
22. X. Rozanska and J. Sauer, *J. Phys. Chem. A*, 2009, **113**, 11586.
23. B. Schiott, K. A. Jorgensen and R. Hoffmann, *J. Phys. Chem. B*, 2001, **95**, 2297.
24. Elke Goos, Institute of Combustion Technology, DLR, German Aerospace Center, private Communication, Elke.Goos@dlr.de.
25. L. Leveles, K. Seshan, J. A. Lercher and L. Lefferts, *J. Catal.*, 2003, **218**, 296.
26. S. Gaab, J. Find, R. K. Grasselli and J. A. Lercher, *Stud. Surf. Sci. Catal.*, 2004, **147**, 673.
27. V. Fattore, Z. A. Fuhrman, G. Manara and B. Notari, *J. Catal.*, 1975, **37**, 215.
28. J. A. Labinger, *J. Mol. Catal. A: Chemical*, 2004, **220**, 27.
29. F. Cavani and F. Trifiro, *Catal. Today*, 1999, **51**, 561.
30. A. B. Anderson, D. W. Ewing, Y. Kim, R. K. Grasselli, J. D. Burrington and J. F. Brazdil, *J. Catal.*, 1985, **96**, 222.
31. L. Chen, J. Liang, W. Weng, Y. Wang, H. Wan and J. C. Vedrine, *Catal. Commun.*, 2004, **5**, 697.
32. R. K. Grasselli, J. D. Burrington, in *Catalysis of Organic Reactions*, W. R. Moser (ed.), Marcel Dekker, New York, 1981, **5**, 17.
33. Y.-H. Han, W. Ueda and Y. Moro-Oka, *Appl. Catal. A: General*, 1999, **176**, 11.
34. C. J. Huang, W. Guo, X. D. Yi, W. Z. Weng and H. L. Wan, *Stud. Surf. Sci. Catal.*, 2004, **147**, 661.
35. O. V. Udalova, D. P. Shashkin, M. D. Shibanova and O. V. Krylov, *Kinet. Catal.*, 2005, **46**, 535.
36. C. Zhao and I. E. Wachs, *J. Phys. Chem. C*, 2008, **112**, 11363.
37. M. Ai, *J. Catal.*, 1986, **101**, 389.
38. J. N. Al-Saeedi, V. V. Guliants, *Abstracts of Papers, 224th ACS National Meeting*, Boston, MA, United States, August 18–22, 2002, PETR.
39. M. Baerns and O. V. Buyevskaya, *Erdoel, Erdgas, Kohle*, 2000, **116**, 25.
40. E. Balcells, F. Borgmeier, I. Grisstede, H. G. Lintz and F. Rosowski, *Appl. Catal. A: General*, 2004, **266**, 211.
41. U. Bentrup, A. Brueckner, S. Kant, S. Kolf, U. Dingerdissen, S. Jansen, D. Maschmeyer, H. Sieger and H. W. Zanthoff, *DGMK Tagungsbericht*, 2005, **2005–2**, 115.

42. P. Botella, P. Concepcion, J. M. L. Nieto and Y. Moreno, *Catal. Today*, 2005, **99**, 51.
43. P. Botella, B. Solsona, E. Garcia-Gonzalez, J. M. Gonzalez-Calbet and J. M. Lopez Nieto, *Chem. Commun.*, 2007, 5040.
44. A. Celaya Sanfiz, T. W. Hansen, A. Sakthivel, A. Trunschke, R. Schloegl, A. Knoester, H. H. Brongersma, M. H. Looi and S. B. A. Hamid, *J. Catal.*, 2008, **258**, 35.
45. L. Chen, C. Bouchy, J. Tabatabatei, N. Winterton, J. C. Vedrine and E. G. Derouane, *NATO Science Series, II: Mathematics, Physics and Chemistry*, 2002, **69**, 291.
46. A. Godefroy, G. S. Patience, T. Tzakova, D. Garrait and J. L. Dubois, *Chem. Eng. Technol.*, 2009, **32**, 373.
47. R. K. Grasselli, G. Centi, F. Trifiro, *Proc. Int. Symp. Uses Selenium Tellurium, 4th*, 1989, 609.
48. K. I. Grobova, V. A. Selyakova, E. M. Lukina, N. A. Korobova, V. A. Semenova, G. V. Fedchenko and A. A. Tupitsina, *Khimicheskaya Promyshlennost (Moscow, Russian Federation)*, 1971, **47**, 412.
49. V. V. Guliants, R. Bhandari, J. N. Al-Saeedi, V. K. Vasudevan, R. S. Soman, O. Guerrero-Perez and M. A. Banares, *Appl. Catal., A: General*, 2004, **274**, 123.
50. V. V. Guliants, R. Bhandari, H. H. Brongersma, A. Knoester, A. M. Gaffney, S. Han, in *JCAT04-634*, 2004.
51. Y.-F. Han, H.-M. Wang, H. Cheng and J.-F. Deng, *Chem. Commun.*, 1999, 521.
52. S. A. Holmes, J. Al-Saeedi, V. V. Guliants, P. Boolchand, D. Georgiev, U. Hackler and E. Sobkow, *Catal. Today*, 2001, **67**, 403.
53. P. Kampe, L. Giebeler, D. Samuelis, J. Kunert, A. Drochner, F. Haass, A. H. Adams, J. Ott, S. Endres, G. Schimanke, T. Buhrmester, M. Martin, H. Fuess and H. Vogel, *Phys. Chem. Chem. Phys.*, 2007, **9**, 3577.
54. G. Landi, L. Lisi and G. Russo, *J. Mol. Catal. A: Chemical*, 2005, **239**, 172.
55. M. Lin, M. Linsen, C. Long, F. Kaiser, P. Klugherz, J. McGregor, M. Clark, *Abstracts of Papers, 222nd ACS National Meeting*, Chicago, IL, United States, August 26–30, 2001, COLL.
56. J. M. Lopez Nieto, J. M. Oliver, P. Botella, *Abstracts of Papers, 226th ACS National Meeting*, New York, NY, United States, September 7–11, 2003, COLL.
57. K. Oshihara, T. Hisano, Y. Kayashima and W. Ueda, *Stud. Surf. Sci. Catal.*, 2001, **136**, 93.
58. W. Ueda, Y. Kobayashi, Y. Mangaya, (Nippon Kayaku Co., Ltd., Japan). Application: JP, 2008, p. 8.
59. T. Ushikubo, *Catalysts in Petroleum Refining and Petrochemicals, Proceedings of the Annual Saudi-Japanese Symposium, 11th*, Dhahran, Saudi Arabia, Nov. 11–12, 2001, paper 14/1.
60. R. Zhao, J. Wang, *Xinshiji De Cuihau Kexue Yu Jishu, Quanguo Cuihuaxue Jihuiyi Lunwenji, 10th*, Zhangjiajie, China, Oct. 15–19, 2000, 391.
61. J. Haber and W. Turek, *J. Catal.*, 2000, **190**, 320.

62. D. L. Stern and R. K. Grasselli, *Catal. Lett.*, 1996, **42**, 139.
63. W. Ueda, W. Li, N. Chen, M. Kida and K. Oshihara, *Res. Chem. Intermed.*, 2000, **26**, 137.
64. J. C. Vedrine, J. M. M. Millet and J.-C. Volta, *Catal. Today*, 1996, **32**, 115.
65. B. Frank, R. Fortrie, C. Hess, R. Schlögl and R. Schomäcker, *Appl. Catal. A: General*, 2009, **353**, 288.
66. C. Hess, G. Tzolova-Muller and R. Herbert, *J. Phys. Chem. C*, 2007, **111**, 9471.
67. G. J. Hutchings, *J. Mater. Chem.*, 2009, **19**, 1222.
68. D. E. Keller, T. Visser, F. Soulimani, D. C. Koningsberger and B. M. Weckhuysen, *Vibr. Spectrosc.*, 2007, **43**, 140.
69. G. Landi, L. Lisi and J. C. Volta, *Catal. Today*, 2004, **91–92**, 275.
70. G. Landi, L. Lisi and J. C. Volta, *Chem. Commun.*, 2003, 492.
71. G. Mestl, T. Ilkenhans, D. Spielbauer, M. Dieterle, O. Timpe, J. Krohnert, F. Jentoft, H. Knozinger and R. Schlogl, *Appl. Catal. A-General*, 2001, **210**, 13.
72. J. M. M. Millet, M. Forisier, D. Rouzies, P. Bonnet and J. C. Vedrine, *Stud. Surf. Sci. Catal.*, 1996, **101**, 1011.
73. E. K. Novakova, E. G. Derouane and J. C. Vedrine, *Catal. Lett.*, 2002, **83**, 177.
74. T. Okuhara, N. Mizuno and M. Misono, *Appl. Catal.: General*, 2001, **222**, 63.
75. N. Ryumon, H. Imai, Y. Kamiya and T. Okuhara, *Appl. Catal. A: General*, 2006, **297**, 73.
76. U. A. Schubert, J. Spengler, R. K. Grasselli, B. Pillep, P. Behrens and H. Knozinger, *Stud. Surf. Sci. Catal.*, 1997, **110**, 817.
77. J. C. Vedrine, E. K. Novakova and E. G. Derouane, *Catal. Today*, 2003, **81**, 247.
78. B. M. Weckhuysen and D. E. Keller, *Catal. Today*, 2003, **78**, 25.
79. S. Xie, E. Iglesia and A. T. Bell, *Langmuir*, 2000, **16**, 7162.
80. F. Arena, F. Frusteri, A. Parmaliana, G. Martra and S. Coluccia, *Stud. Surf. Sci. Catal.*, 1998, **119**, 665.
81. T. Blasco, J. M. Lopez Nieto, A. Dejoz and M. I. Vazquez, *J. Catal.*, 1995, **157**, 271.
82. A. N. Desikan, W. M. Zhang and S. T. Oyama, *J. Catal.*, 1995, **157**, 740.
83. B. Grzybowska-Swierkosz, *Top. Catal.*, 2000, **11**, 23.
84. E. Iglesia, D. G. Barton, J. A. Biscardi, M. J. L. Gines and S. L. Soled, *Catal. Today*, 1997, **38**, 339.
85. X. Rozanska, E. V. Kondratenko and J. Sauer, *J. Catal.*, 2008, **256**, 84.
86. R. K. Grasselli, *React. Kinet. Catal. Lett.*, 1987, **35**, 327.
87. R. K. Grasselli, *Top. Catal.*, 2002, **21**, 79.
88. R. K. Grasselli, J. D. Burrington, *Handbook of Heterogeneous Catalysis*, G. Ertl, H. Knözinger, F. Schüth, J. Weitkamp (Ed.), Wiley-VCH, Weinheim, 2008, **7**, 3479.
89. R. K. Grasselli, J. D. Burrington and J. F. Brazdil, *Faraday Disc. Chem. Soc.*, 1981, **72**, 203.

90. R. K. Grasselli, D. J. Buttrey, J. D. Burrington, A. Andersson, J. Holmberg, W. Ueda, J. Kubo, C. G. Lugmair and A. F. Volpe Jr., *Top. Catal.*, 2006, **38**, 7.
91. X.-K. Li, W.-J. Ji, J. Zhao, Z. Zhang and C.-T. Au, *Appl. Catal. A: General*, 2006, **306**, 8.
92. R. Schlögl, A. Knop-Gericke, M. Hävecker, U. Wild, D. Frickel, T. Ressler, R. E. Jentoft, J. Wienold, G. Mestl, A. Blume, O. Timpe and Y. Uchida, *Top. Catal.*, 2001, **15**, 219.
93. K. Chen, A. T. Bell and E. Iglesia, *J. Phys. Chem. B*, 2000, **104**, 1292.
94. K. Chen, A. Khodakov, J. Yang, A. T. Bell and E. Iglesia, *J. Catal.*, 1999, **186**, 325.
95. T. Ressler, O. Timpe, T. Neisius, J. Find, G. Mestl, M. Dieterle and R. Schlögl, *J. Catal.*, 2000, **191**, 75.
96. B. Delmon, *Catalyst Deactivation 1997*, 1997, **111**, 39.
97. J. Wang, R.-S. Zhao and Z.-D. Xu, *Yingyong Huaxue*, 2000, **17**, 313.
98. R. Zhao, J. Wang, Q. Dong and J. Liu, *J. Nat. Gas Chem.*, 2005, **14**, 88.
99. K. W. Hutchenson, C. La Marca, G. S. Patience, J. P. Laviolette and R. E. Bockrath, *Appl. Catal. A: General*, 2010, **376**, 91.
100. M. Abu Haija, S. Guimond, Y. Romanyshyn, A. Uhl, H. Kuhlenbeck, T. K. Todorova, M. V. Ganduglia-Pirovano, J. Dobler, J. Sauer and H. J. Freund, *Surf. Sci.*, 2006, **600**, 1497.
101. M. Cavalleri, K. Hermann, A. Knop-Gericke, M. Hävecker, R. Herbert, C. Hess, A. Oestereich, J. Döbler and R. Schlögl, *J. Catal.*, 2009, **262**, 215.
102. C. Hess, *ChemPhysChem*, 2009, **10**, 319.
103. C. Hess, I. J. Drake, J. D. Hoefelmeyer, T. D. Tilley and A. T. Bell, *Catal. Lett.*, 2005, **105**, 1.
104. C. Hess, *Surf. Sci.*, 2006, **600**, 3695.
105. J. P. Thielemann, J. Kröhnert and C. Hess, *J. Phys. Chem. C*, 2010, **114**, 17092.
106. C. Hess, *J. Catal.*, 2007, **248**, 120.
107. J.-M. Jehng, H. Hu, X. Gao and I. E. Wachs, *Catal. Today*, 1996, **28**, 335.
108. A. Dinse, A. Ozarowski, C. Hess, R. Schomäcker and K.-P. Dinse, *J. Phys. Chem. C*, 2008, **112**, 17664.
109. A. Goodrow, A. T. Bell and M. Head-Gordon, *J. Phys. Chem. C*, 2009, **113**, 19361.
110. J. M. López Nieto, *Top. Catal.*, 2006, **41**, 3.
111. M. M. Bettahar, G. Costentin, L. Savary and J. C. Lavalley, *Appl. Catal. A: General*, 1996, **145**, 1.
112. J. M. Jehng, I. E. Wachs, W. Ueda, *Abstracts of Papers, 228th ACS National Meeting*, Philadelphia, PA, United States, August 22–26, 2004, COLL.
113. M. D. Argyle, K. Chen, E. Iglesia and A. T. Bell, *J. Phys. Chem. B*, 2005, **109**, 2414.
114. V. Brazdova, M. V. Ganduglia-Pirovano and J. Sauer, *J. Phys. Chem. C*, 2010, **114**, 4983.

115. A. Celaya Sanfiz, T. W. Hansen, D. Teschner, P. Schnörch, F. Girgsdies, A. Trunschke, R. Schlögl, M. H. Looi and S. B. A. Hamid, *J. Phys. Chem. C*, 2010, **114**, 1912.
116. H. Bluhm, M. Havecker, E. Kleimenov, A. Knop-Gericke, A. Liskowski, R. Schlogl and D. S. Su, *Top. Catal.*, 2003, **23**, 99.
117. R. Schlögl and S. B. A. Hamid, *Angew. Chem. Int. Ed.*, 2004, **43**, 1628.
118. T. Blasco, A. Galli, J. M. Lopez Nieto and F. Trifiro, *J. Catal.*, 1997, **169**, 203.
119. R. Grabowski, *Catal. Rev. Sci. Eng.*, 2006, **48**, 199.
120. R. K. Grasselli, *Top. Catal.*, 2002, **21**, 79.
121. N. Ballarini, F. Cavani, M. Cimini, G. M. Mascanzoni, F. Trifiro, J.-M. M. Millet, M. Banares, M. O. Guerrero-Perez, U. Cornado and R. Catani, *Stud. Surf. Sci. Catal.*, 2007, **172**, 145.
122. E. Bordes and R. M. Contractor, *Top. Catal.*, 1996, **3**, 365.
123. M. M. Koranne, J. G. Goodwin and G. Marcelin, *J. Catal.*, 1994, **148**, 369.
124. F. Klose, T. Wolff, H. Lorenz, A. Seidel-Morgenstern, Y. Suchorski, M. Piurkowska and H. Weiss, *J. Catal.*, 2007, **247**, 176.
125. R. K. Grasselli, *Catal. Today*, 2005, **99**, 23.
126. L. Moens, P. Ruiz, B. Delmon and M. Devillers, *Appl. Catal. A: General*, 1998, **171**, 131.
127. A. Goodrow and A. T. Bell, *J. Phys. Chem. C*, 2008, **112**, 13204.
128. J.-M. Herrmann, *Catal. Today*, 2006, **112**, 73.
129. M. Baron, H. Abbott, O. Bondarchuk, D. Stacchiola, A. Uhl, S. Shaikhutdinov, H.-J. Freund, C. Popa, M. V. Ganduglia-Pirovano and J. Sauer, *Angew. Chem. Int. Ed.*, 2009, **48**, 8006.
130. M. V. Ganduglia-Pirovano, C. Popa, J. Sauer, H. Abbott, A. Uhl, M. Baron, D. Stacchiola, O. Bondarchuk, S. Shaikhutdinov and H.-J. Freund, *J. Am. Chem. Soc.*, 2010, **132**, 2345.
131. R. K. Grasselli, A. Andersson, D. J. Buttrey, J. D. Burrington, C. G. Lugmair, A. F. Volpe, *Abstracts of Papers, 228th ACS National Meeting*, Philadelphia, PA, United States, August 22–26, 2004, COLL.
132. A. Walter, R. Herbert, C. Hess and T. Ressler, *Chem. Central J.*, 2010, **4**, 3.
133. G. C. Bond and S. F. Tahir, *Appl. Catal.*, 1991, **71**, 1.
134. J. Dobler, M. Pritzsche and J. Sauer, *J. Am. Chem. Soc.*, 2005, **127**, 10861.
135. Y.-M. Liu, Y. Cao, N. Yi, W.-L. Feng, W.-L. Dai, S.-R. Yan, H.-Y. He and K.-N. Fan, *J. Catal.*, 2004, **224**, 417.
136. H. Tian, E. I. Ross and I. E. Wachs, *J. Phys. Chem. B*, 2006, **110**, 9593.
137. I. E. Wachs, *Catal. Today*, 2005, **100**, 79.
138. I. E. Wachs, J. M. Jehng and W. Ueda, *J. Phys. Chem. B*, 1994, **109**, 2275.
139. A. Goodrow and A. T. Bell, *J. Phys. Chem. C*, 2007, **111**, 14753.
140. C. Huang, W. Guo, X. Yi, W. Weng and H. Wan, *Catal. Commun.*, 2006, **8**, 162.

141. J. Lebars, J. C. Vedrine, A. Auroux, S. Trautmann and M. Baerns, *Appl. Catal. A-General*, 1992, **88**, 179.
142. G. Li and X. S. Zhao, *Ind. Eng. Chem. Res.*, 2006, **45**, 3569.
143. L. Tessier, E. Bordes and M. Gubelmannbonneau, *Catal. Today*, 1995, **24**, 335.
144. R. Schlogl, A. Knop-Gericke, M. Havecker, U. Wild, D. Frickel, T. Ressler, R. E. Jentoft, J. Wienold, G. Mestl, A. Blume, O. Timpe and I. Uchida, *Top. Catal.*, 2001, **15**, 219.
145. M. Havecker, R. W. Mayer, A. Knop-Gericke, H. Bluhm, E. Kleimenov, A. Liskowski, D. Su, R. Follath, F. G. Requejo, D. F. Ogletree, M. Salmeron, J. A. Lopez-Sanchez, J. K. Bartley, G. J. Hutchings and R. Schlogl, *J. Phys. Chem. B*, 2003, **107**, 4587.
146. E. Kleimenov, H. Bluhm, M. Havecker, A. Knop-Gericke, A. Pestryakov, D. Teschner, J. A. Lopez-Sanchez, J. K. Bartley, G. J. Hutchings and R. Schlogl, *Surf. Sci.*, 2005, **575**, 181.
147. J. B. Wagner, O. Timpe, F. A. Hamid, A. Trunschke, U. Wild, D. S. Su, R. K. Widi, S. B. A. Hamid and R. Schloegl, *Top. Catal.*, 2006, **38**, 51.
148. H. Brongersma, R. Veen, A. Knoester, A. Trunschke and R. Schloegl, *Preprints-American Chemical Society, Division of Petroleum Chemistry*, 2007, **52**, 67.
149. D. Goebke, Y. Romanyshyn, S. Guimond, J.-M. Sturm, H. Kuhlenbeck, J. Doebler, U. Reinhardt, M.-V. Ganduglia-Pirovano, J. Sauer and H.-J. Freund, *Angew. Chem. Int. Ed.*, 2009, **48**, 3695.
150. N. Magg, B. Immaraporn, J. B. Giorgi, T. Schroeder, M. Baumer, J. Dobler, Z. L. Wu, E. Kondratenko, M. Cherian, M. Baerns, P. C. Stair, J. Sauer and H. J. Freund, *J. Catal.*, 2004, **226**, 88.
151. Y. Romanyshyn, S. Guimond, H. Kuhlenbeck, S. Kaya, R. Blum, H. Niehus, S. Shaikhutdinov, V. Simic-Milosevic, N. Nilius, H. J. Freund, M. Ganduglia-Pirovano, R. Fortrie, J. Döbler and J. Sauer, *Top. Catal.*, 2008, **50**, 106.
152. N. Dimitratos and J. C. Vedrine, *J. Mol. Catal. A: Chemical*, 2006, **255**, 184.
153. N. Dimitratos, J. C. Vedrine, *Catalysis in Application, Papers presented at the International Symposium on Applied Catalysis.*, Glasgow, United Kingdom, July 16–18, 2003, 145.
154. H. Liu and E. Iglesia, *J. Catal.*, 2004, **223**, 161.
155. J. Arichi, M. Eternot and B. Louis, *Catal. Today*, 2008, **138**, 117.
156. X.-K. Li, J. Zhao, W.-J. Ji, Z.-B. Zhang, Y. Chen, C.-T. Au, S. Han and H. Hibst, *J. Catal.*, 2006, **237**, 58.
157. N. Mizuno, D.-J. Suh, W. Han and T. Kudo, *J. Mol. Catal. A: Chemical*, 1996, **114**, 309.
158. A. Stein, F. Fendorf and T. E. Mallouk, *Chem. Mater.*, 1995, **7**, 304.
159. T. Adschiri, *Chem. Lett.*, 2007, **36**, 1188.
160. A. Bandara, M. Abu-Haija, F. Hobel, H. Kuhlenbeck, G. Rupprechter and H. J. Freund, *Top. Catal.*, 2007, **46**, 223.
161. C. Kolczewski, K. Hermann, S. Gulmond, H. Kuhlenbeck and H. J. Freund, *Surf. Sci.*, 2007, **601**, 5394.

162. K. Hermann, M. Witko, R. Druzinic and R. Tokarz, *Top. Catal.*, 2000, **11–12**, 67.
163. D. K. Bohme and H. Schwarz, *Angew. Chem. Int. Ed.*, 2005, **44**, 2336.
164. J. Jarupatrakorn and T. D. Tilley, *J. Am. Chem. Soc.*, 2002, **124**, 8380.
165. H. Nair, M. J. Liszka, J. E. Gatt and C. D. Baertsch, *J. Phys. Chem. C*, 2008, **112**, 1612.
166. J. Thomas and R. Raja, *Top. Catal.*, 2006, **40**, 3.
167. O. Glemser and E. Preisler, *Z. anorg. allg. Chem.*, 1960, **303**, 303.
168. D. Hou, K. S. Hagen and C. L. Hill, *J. Am. Chem. Soc.*, 1992, **114**, 5864.
169. K. H. Tytko, *Chem. uns. Zeit*, 1979, **13**, 184.
170. K. H. Tytko and J. Mehmke, *Z. anorg. allg. Chem.*, 1983, **503**, 67.
171. O. Glemser, *Nachr. Akad. Wiss. Gottingen, Math. physik.-Kl. Abt.*, 1955, 121.
172. M. A. Banares, *Catal. Today*, 1999, **51**, 319.
173. B. Olthof, A. Khodakov, A. T. Bell and E. Iglesia, *J. Phys. Chem. B*, 2000, **104**, 1516.
174. C.-B. Wang, Y. Cai and I. E. Wachs, *Langmuir*, 1999, **15**, 1223.
175. L. J. Burcham, M. Badlani and I. E. Wachs, *J. Catal.*, 2001, **203**, 104.
176. M. Ai, *J. Catal.*, 1978, **54**, 426.
177. P. Concepcion, A. Galli, J. M. L. Nieto, A. Dejoz and M. I. Vazquez, *Top. Catal.*, 1996, **3**, 451.
178. J. M. L. Nieto, R. Coenraads, A. Dejoz and M. I. Vazquez, *Stud. Surf. Sci. Catal.*, 1997, **110**, 443.
179. J. C. Védrine, *Top. Catal.*, 2002, **21**, 97.
180. M. Taylor, A. Carley, T. Davies and S. Taylor, *Top. Catal.*, 2009, **52**, 1660.
181. R. Zukerman, L. Vradman, L. Titelman, C. Weidenthaler, M. V. Landau and M. Herskowitz, *Microp. Mesop. Mater.*, 2008, **116**, 237.
182. H.-S. Kim, S. A. Zygmunt, P. C. Stair, P. Zapol and L. A. Curtiss, *J. Phys. Chem. C*, 2009, **113**, 8836.
183. H.-M. Lin, S.-T. Kao, K.-M. Lin, J.-R. Chang and S.-G. Shyu, *J. Catal.*, 2004, **224**, 156.
184. U. Scharf, M. Schraml-Marth, A. Wokaun and A. Baiker, *J. Chem. Soc. Faraday Trans.*, 1991, **87**, 3299.
185. W. C. Vining, A. Goodrow, J. Strunk and A. T. Bell, *J. Catal.*, 2010, **270**, 163.
186. I. E. Wachs, *Catal. Today*, 1996, **27**, 437.
187. M. Baron, O. Bondarchuk, D. Stacchiola, S. Shaikhutdinov and H. J. Freund, *J. Phys. Chem. C*, 2009, **113**, 6042.
188. V. C. Corberán, *Catal. Today*, 2005, **99**, 33.
189. W. Daniell, A. Ponchel, S. Kuba, F. Anderle, T. Weingand, D. H. Gregory and H. Knözinger, *Top. Catal.*, 2002, **20**, 65.
190. T. V. Malleswara Rao, E. Vico-Ruiz, M. A. Bañares and G. Deo, *J. Catal.*, 2008, **258**, 324.
191. G. Deo and I. E. Wachs, *J. Catal.*, 1994, **146**, 323.

192. P. Beato, A. Blume, F. Girgsdies, R. E. Jentoft, R. Schloegl, O. Timpe, A. Trunschke, G. Weinberg, Q. Basher, F. A. Hamid, S. B. A. Hamid, E. Omar and L. Mohd Salim, *Appl. Catal. A: General*, 2006, **307**, 137.
193. A. Celaya Sanfiz, T. W. Hansen, F. Girgsdies, O. Timpe, E. Roedel, T. Ressler, A. Trunschke and R. Schloegl, *Top. Catal.*, 2008, **50**, 19.
194. R.-M. Feng, X.-J. Yang, W.-J. Ji, H.-Y. Zhu, X.-D. Gu, Y. Chen, S. Han and H. Hibst, *J. Mol. Catal. A: Chemical*, 2007, **267**, 245.
195. H. Eckert and I. E. Wachs, *J. Phys. Chem.*, 1989, **93**, 6796.
196. B. Grzybowska-Swierkosz, *Appl. Catal. A-General*, 1997, **157**, 263.
197. M. V. Landau, L. Vradman, X. Wang and L. Titelman, *Microp. Mesop. Mater.*, 2005, **78**, 117.
198. J. Nickl, D. Dutoit, A. Baiker, U. Scharf and A. Wokaun, *Appl. Catal. A: General*, 1993, **98**, 173.
199. M. Baltes, K. Cassiers, P. Van Der Voort, B. M. Weckhuysen, R. A. Schoonheydt and E. F. Vansant, *J. Catal.*, 2001, **197**, 160.
200. R. Schloegl, A. Trunschke, (Germany). Application: EP EP, 2008, 25.
201. J. C. Vedrine, G. Coudurier and J.-M. M. Millet, *Catal. Today*, 1997, **33**, 3.
202. Z. Yang, Y. Lu and Z. Yang, *Chem. Commun.*, 2009, 2270.
203. M. D. Argyle, K. Chen, C. Resini, C. Krebs, A. T. Bell and E. Iglesia, *J. Phys. Chem. B*, 2004, **108**, 2345.
204. A. Dinse, R. Schomacker and A. T. Bell, *Phys. Chem. Chem. Phys.*, 2009, **11**, 6119.
205. E. V. Kondratenko, N. Steinfeldt and M. Baerns, *Phys. Chem. Chem. Phys.*, 2006, **8**, 1624.
206. O. Ovsitser, M. Cherian and E. V. Kondratenko, *J. Phys. Chem. C*, 2007, **111**, 8594.
207. A. Pantazidis, S. A. Bucholz, H. W. Zanthoff, Y. Schuurman and C. Mirodatos, *Catal. Today*, 1998, **40**, 207.
208. P. Mars and D. W. Van Krevelen, *Chem. Eng. Sci.*, 1954, **3**, 41.
209. J. Haber, *Catal. Today*, 2009, **142**, 100.
210. J. L. Bronkema, D. C. Leo and A. T. Bell, *J. Phys. Chem. C*, 2007, **111**, 14530.
211. H. Knözinger, *Catal. Today*, 1996, **32**, 71.
212. P. A. Agaskar, L. DeCaul and R. K. Grasselli, *Catal. Lett.*, 1994, **23**, 339.
213. J. Haber and E. Lalik, *Catal. Today*, 1997, **33**, 119.
214. H. Launay, S. Loridant, D. L. Nguyen, A. M. Volodin, J. L. Dubois and J. M. M. Millet, *Catal. Today*, 2007, **128**, 176.
215. P. Afanasiev, *J. Phys. Chem.*, 2005, **109**, 18293.
216. P. Botella, E. Garcia-Gonzalez, J. M. Lopez Nieto and J. M. Gonzalez-Calbet, *Solid State Sci.*, 2005, **7**, 507.
217. N. Haddad, E. Bordes-Richard and A. Barama, *Catal. Today*, 2009, **142**, 215.
218. T. Kardash, L. Plyasova, V. Bondareva and A. Shmakov, *J. Struct. Chem.*, 2008, **49**, 701.

219. O. Ovsitser, Y. Uchida, G. Mestl, G. Weinberg, A. Blume, J. Jäger, M. Dieterle, H. Hibst and R. Schlögl, *J. Mol. Catal. A: Chemical*, 2002, **185**, 291.
220. W. Ueda, D. Vitry, T. Kato, N. Watanabe, *Abstracts of Papers, 226th ACS National Meeting*, New York, NY, United States, September 7–11, 2003, COLL.
221. J. B. Wagner, D. S. Su, S. A. Schunk, H. Hibst, J. Petzoldt and R. Schlögl, *J. Catal.*, 2004, **224**, 28.
222. G. A. Zenkovets, G. N. Kryukova, V. Y. Gavrilov, S. V. Tsybulya, V. A. Anufrienko, T. A. Larina, D. F. Khabibulin, O. B. Lapina, E. Rödel, A. Trunschke, T. Ressler and R. Schlogl, *Mater. Chem. Phys.*, 2007, **103**, 295.
223. F. Cavani and F. Trifiro, *Appl. Catal. A: General*, 1997, **157**, 195.
224. V. V. Guliants, R. Bhandari, A. R. Hughett, S. Bhatt, B. D. Schuler, H. H. Brongersma, A. Knoester, A. M. Gaffney and S. Han, *J. Phys. Chem. B*, 2006, **110**, 6129.
225. V. V. Guliants and S. A. Holmes, *J. Mol. Catal. A: Chemical*, 2001, **175**, 227.
226. A. Kudelski, *Surf. Sci.*, 2009, **603**, 1328.
227. Z. Luan and J. A. Fournier, *Microp. Mesop. Mater.*, 2005, **79**, 235.
228. H. Brongersma, R. ter Veen, A. Knoester, A. Trunschke, R. Schlogl, *Abstracts of Papers, 234th ACS National Meeting*, Boston, MA, United States, August 19–23, 2007, PETR.
229. G. J. Hutchings, J. A. Lopez-Sanchez, J. K. Bartley, J. M. Webster, A. Burrows, C. J. Kiely, A. F. Carley, C. Rhodes, M. Havecker, A. Knop-Gericke, R. W. Mayer, R. Schlogl, J. C. Volta and M. Poliakoff, *J. Catal.*, 2002, **208**, 197.
230. T. Ilkenhans, B. Herzog, T. Braun and R. Schlogl, *J. Catal.*, 1995, **153**, 275.
231. T. V. Andrushkevich, *Catal. Rev. Sci. Eng.*, 1993, **35**, 213.
232. H. Berndt, A. Martin, A. Bruckner, E. Schreier, D. Muller, H. Kosslick, G. U. Wolf and B. Lucke, *J. Catal.*, 2000, **191**, 384.
233. P. Botella, J. M. Lopez Nieto, B. Solsona, A. Mifsud and F. Marquez, *J. Catal.*, 2002, **2**, 445.
234. S. Yang, E. Iglesia and A. T. Bell, *J. Phys. Chem. B*, 2006, **110**, 2732.
235. U. Bentrup, A. Bruckner, M. Kant, S. Kolf, U. Dingerdissen, S. Jansen, D. Maschmeyer, H. Siegert and H. W. Zanthoff, *Oil, Gas (Hamburg, Germany)*, 2006, **32**, 145.
236. J. M. M. Millet, M. Langpape, *Book of Abstracts, 215th ACS National Meeting*, Dallas, March 29–April 2 1998, CATL.
237. S. Anniballi, F. Cavani, A. Guerrini, B. Panzacchi, F. Trifiro, C. Fumagalli, R. Leanza and G. Mazzoni, *Catal. Today*, 2003, **78**, 117.
238. S. Besselmann, C. Freitag, O. Hinrichsen and M. Muhler, *Phys. Chem. Chem. Phys.*, 2001, **3**, 4633.
239. C. Freitag, S. Besselmann, E. Loffler, W. Grunert, F. Rosowski and M. Muhler, *Catal. Today*, 2004, **91–2**, 143.

240. S. L. T. Andersson, *Appl. Catal. A: General*, 1994, **112**, 209.
241. M. D. Argyle, K. Chen, A. T. Bell, E. Iglesia, *Abstracts of Papers, 223rd ACS National Meeting*, Orlando, FL, United States, April 7–11, 2002, CATL.
242. H. Dai, A. T. Bell and E. Iglesia, *J. Catal.*, 2004, **221**, 491.
243. D. L. Hoang, S. Farage, A. Dittmar, A. Trunschke, H. Lieske and A. Martin, *Catal. Lett.*, 2006, **112**, 173.
244. A. A. Lemonidou, L. Nalbandian and I. A. Vasalos, *Catal. Today*, 2000, **61**, 333.
245. X. Li and E. Iglesia, *Appl. Catal. A: General*, 2008, **334**, 339.
246. F. Cavani, S. Ligi, T. Monti, F. Pierelli, F. Trifirò, S. Albonetti and G. Mazzoni, *Catal. Today*, 2000, **61**, 203.
247. M. Conte, G. Budroni, J. K. Bartley, S. H. Taylor, A. F. Carley, A. Schmidt, D. M. Murphy, F. Girgsdies, T. Ressler, R. Schlogl and G. J. Hutchings, *Science*, 2006, **313**, 1270.
248. N. Duvauchelle and E. Bordes, *Catal. Lett.*, 1999, **57**, 81.
249. V. V. Guliants, J. B. Benziger, S. Sundaresan and I. E. Wachs, *Stud. Surf. Sci. Catal.*, 2000, **130B**, 1721.
250. V. V. Guliants, J. B. Benziger, S. Sundaresan, N. Yao and I. E. Wachs, *Catal. Lett.*, 1995, **32**, 379.
251. J. Haber, J. Stoch, V. A. Zazhigalov, I. V. Bacherikova and E. V. Cheburakova, *Polish J. Chem.*, 2008, **82**, 1839.
252. G. J. Hutchings, *J. Mater. Chem.*, 2004, **14**, 3385.
253. I.-C. Marcu, I. Sandulescu, Y. Schuurman and J.-M. M. Millet, *Appl. Catal. A: General*, 2008, **334**, 207.
254. T. Okuhara and M. Misono, *Catal. Today*, 1993, **16**, 61.
255. Y. H. Taufiq-Yap, C. S. Saw and R. Irmawati, *Catal. Lett.*, 2005, **105**, 103.
256. V. A. Zazhigalov, *Preprints-American Chemical Society, Division of Petroleum Chemistry*, 1996, **41**, 207.
257. J.-C. Volta, *Catal. Today*, 1996, **32**, 29.
258. A. Brückner, *Top. Catal.*, 2006, **38**, 133.
259. C. J. Kiely and G. J. Hutchings, *Appl. Catal. A: General*, 2007, **325**, 194.
260. V. V. Guliants, J. B. Benziger, S. Sundaresan, I. E. Wachs, J. M. Jehng and J. E. Roberts, *Catal. Today*, 1996, **28**, 275.
261. A. Benarafa, M. Kacimi, S. Gharbage, J. M. M. Millet and M. Ziyad, *Mater. Res. Bull.*, 2001, **35**, 2047.
262. T. Blasco, P. Concepcion, P. Grotz, J. M. Lopez Nieto and A. Martinez-Arias, *J. Mol. Catal. A: Chemical*, 2000, **162**, 267.
263. P. Bonnet and J. M. M. Millet, *J. Catal.*, 1996, **161**, 198.
264. P. Concepcion, A. Corma, J. M. L. Nieto and J. Perez-Pariente, *Appl. Catal. A: General*, 1996, **143**, 17.
265. P. Concepcion, J. M. Lopez Nieto and J. Perez-Pariente, *J. Mol. Catal. A: Chemical*, 1995, **97**, 173.
266. P. Concepcion and J. M. L. Nieto, *Catal. Commun.*, 2001, **2**, 363.

267. I.-C. Marcu, J.-M. M. Millet and J.-M. Herrmann, *Catal. Lett.*, 2002, **78**, 273.
268. J.-M. M. Millet and J. C. Vedrine, *Top. Catal.*, 2001, **15**, 139.
269. J. M. L. Nieto, *Top. Catal.*, 2001, **15**, 189.
270. B. Solsona, J. M. Lopez-Nieto, M. Alcantara-Rodriguez, E. Rodriguez-Castellon and A. Jimenez-Lopez, *J. Mol. Catal. A: Chemical*, 2000, **153**, 199.
271. W. Ueda and Y. Suzuki, *Chem. Lett.*, 1995, **541**.
272. H.-L. Zubowa, H. Kosslick, D. Muller, M. Richter, M. Schneider, A. Trunschke and R. Fricke, *Revue Roumaine de Chimie*, 2000, **44**, 1025.
273. S. Shylesh, S. P. Mirajkar and A. P. Singh, *J. Mol. Catal. A: Chemical*, 2005, **239**, 57.
274. N. Ballarini, F. Cavani, M. Cimini, F. Trifiro, J. M. M. Millet, U. Cornaro and R. Catani, *J. Catal.*, 2006, **241**, 255.
275. B. Delmon, *J. Therm. Anal. Calor.*, 2007, **90**, 49.
276. O. A. Barias, A. Holmen and E. A. Blekkan, *J. Catal.*, 1996, **158**, 1.
277. A. Christensen, A. V. Ruban, P. Stoltze, K. W. Jacobsen, H. L. Skriver, J. K. Norskov and F. Besenbacher, *Phys. Rev. B*, 1997, **56**, 5822.
278. V. Stamenkovic, T. J. Schmidt, P. N. Ross and N. M. Markovic, *J. Phys. Chem. B*, 2002, **106**, 11970.
279. J. M. Lopez Nieto, P. Concepcion, A. Dejoz, H. Knozinger, F. Melo and M. I. Vazquez, *J. Catal.*, 2000, **189**, 147.
280. E. Bordes and P. Courtine, *Top. Catal.*, 2000, **11–12**, 61.
281. P. DeSanto, D. Buttrey, R. Grasselli, W. Pyrz, C. Lugmair, A. Volpe, T. Vogt and B. Toby, *Top. Catal.*, 2006, **38**, 31.
282. R. K. Grasselli, D. J. Buttrey, P. DeSanto, J. D. Burrington, C. G. Lugmair, A. F. Volpe Jr. and T. Weingand, *Catal. Today*, 2004, **91–92**, 251.
283. J. N. Al-Saeedi, V. K. Vasudevan and V. V. Guliants, *Catal. Commun.*, 2003, **4**, 537.
284. A. M. Gaffney, S. Han, V. V. Guliants, R. Bhandari, P. Gonzalez, H. H. Brongersma, *Abstracts of Papers, 227th ACS National Meeting*, Anaheim, CA, United States, March 28–April 1, 2004, IEC.
285. V. V. Guliants, H. H. Brongersma, A. Knoester, A. M. Gaffney and S. Han, *Top. Catal.*, 2006, **38**, 41.
286. W. D. Pyrz, D. A. Blom, N. R. Shiju, V. V. Guliants, T. Vogt and D. J. Buttrey, *J. Phys. Chem. C*, 2008, **112**, 10043.
287. S. Albonetti, F. Cavani and F. Trifiro, *Catal. Rev. Sci Eng.*, 1996, **38**, 413.
288. M. Muhler, R. Schlogl and G. Ertl, *J. Catal.*, 1992, **138**, 413.
289. M. Muhler, J. Schutze, M. Wesemann, T. Rayment, A. Dent, R. Schlogl and G. Ertl, *J. Catal.*, 1990, **126**, 339.
290. W. Weiss and R. Schlogl, *Top. Catal.*, 2000, **13**, 75.
291. H. P. Steinruck, *Surf. Sci.*, 2010, **604**, 481.
292. J. F. Brazdil, L. C. Glaeser and R. K. Grasselli, *J. Catal.*, 1983, **81**, 142.

293. M. Dieterle, G. Weinberg and G. Mestl, *Phys. Chem. Chem. Phys.*, 2002, **4**, 812.
294. J.-M. M. Millet, *Top. Catal.*, 2006, **38**, 83.
295. E. Roedel, O. Timpe, A. Trunschke, G. A. Zenkovets, G. N. Kryukova, R. Schloegl and T. Ressler, *Catal. Today*, 2007, **126**, 112.
296. A. Corma, J. M. Lupez-Nieto, N. Paredes, M. Perez, Y. Shen, H. Cao, S. L. Suib and P. R. a. B. Delmon, *Stud. Surf. Sci. Catal.*, 1992, **72**, 213.
297. M. Lin, *Abstracts of Papers, 226th ACS National Meeting*, New York, NY, United States, September 7–11, 2003, COLL.
298. W. Ueda, *Shokubai*, 2003, **45**, 23.
299. A. P. Vieira Soares, L. D. Dimitrov, M. C.-R. Andrede Oliveira, L. Hilaire, M. F. Portela and R. K. Grasselli, *Appl. Catal. A: General*, 2003, **253**, 191.
300. J. F. Brazdil, L. C. Glaeser and R. K. Grasselli, *J. Phys. Chem.*, 1983, **87**, 5485.
301. S. V. Yanina and R. L. Smith, *J. Catal.*, 2003, **213**, 151.
302. D. Creaser, B. Andersson, R. R. Hudgins and P. L. Silveston, *Appl. Catal.: General*, 1999, **187**, 147.
303. Z. Deng, H. Wang, W. Chu and W. Yang, *Cuihua Xuebao*, 2008, **29**, 1032.
304. S. Hamakawa, S. Yoshino, J. Nakamura, Y. Y. Liu, A. L. Tsyganok, K. Suzuki and K. Murata, *Electrochem. Solid State Lett.*, 2001, **4**, D9.
305. A. D. Mayernick and M. J. Janik, *J. Phys. Chem. C*, 2008, **112**, 14955.
306. T. Ressler, J. Wienold, R. E. Jentoft and T. Neisius, *J. Catal.*, 2002, **210**, 67.
307. Y. Uchida, G. Mestl, O. Ovsitser, J. Jager, A. Blume and R. Schlogl, *J. Mol. Catal. A: Chemical*, 2002, **187**, 247.
308. L. Xuebing and E. Iglesia, *J. Phys. Chem. C*, 2008, **112**, 15001.
309. Y. H. Han, W. Ueda and Y. Moro-oka, *J. Catal.*, 1999, **186**, 75.
310. A. Kaddouri, R. Del Rosso, C. Mazzocchia, P. Gronchi and D. Fumagalli, *J. Therm. Anal. Calor.*, 2001, **66**, 63.
311. G. Mestl, *Top. Catal.*, 2006, **38**, 69.
312. W. Ueda, Y. Moro-Oka and T. Ikawa, *J. Catal.*, 1984, **88**, 214.
313. D. Zhonghua, W. Hongxin, C. H. U. Wenling and Y. Weishen, *Chin. J. Catal.*, 2008, **29**, 1032.
314. E. V. Antipov, A. M. Abakumov and S. Y. Istomin, *Inorg. Chem.*, 2008, **47**, 8543.
315. J. F. Brazdil, D. D. Suresh and R. K. Grasselli, *J. Catal.*, 1980, **66**, 347.
316. H. Hibst, F. Rosowski and G. Cox, *Catal. Today*, 2006, **117**, 234.
317. W. D. Pyrz, D. A. Blom, M. Sadakane, K. Kodato, W. Ueda, T. Vogt and D. J. Buttrey, *Proceedings of the National Academy of Sciences of the United States of America*, **107**, 6152.
318. I. Czekaj, K. Hermann and M. Witko, *Surf. Sci.*, 2003, **545**, 85.
319. K. Hermann, A. Michalak and M. Witko, *Catal. Today*, 1996, **32**, 321.
320. M. Dieterle and G. Mestl, *Phys. Chem. Chem. Phys.*, 2002, **4**, 822.

321. M. A. Carreon and V. V. Guliants, *Stud. Surf. Sci. Catal.*, 2002, **141**, 301.
322. V. V. Guliants and M. A. Carreon, *Catalysis*, 2005, **18**, 1.
323. V. Martin, J. M. M. Millet and J. C. Volta, *J. Them. Anal. Calor.*, 1998, **53**, 111.
324. P. A. Batist, H. J. Prette and G. C. A. Schuit, *J. Catal.*, 1969, **15**, 267.
325. D. X. Wang and M. A. Barteau, *Appl. Catal. A: General*, 2002, **223**, 205.
326. C. Zhao, I. E. Wachs, *Abstracts of Papers, 232nd ACS National Meeting*, San Francisco, CA, United States, Sept. 10–14, 2006, COLL.
327. X. Gao and I. E. Wachs, *J. Phys. Chem. B*, 2000, **104**, 1261.
328. M. Lu, W. M. Nolte, T. He, D. A. Corley and J. M. Tour, *Chem. Mater.*, 2009, **21**, 442.
329. E. Stoyanov, F. Langenhorst and G. Steinle-Neumann, *Am. Mineral.*, 2007, **92**, 577.
330. J. C. Woicik, M. Yekutiel, E. J. Nelson, N. Jacobson, P. Pfalzer, M. Klemm, S. Horn and L. Kronik, *Phys. Rev. B*, 2007, **76**, 16510176.
331. C. W. Bauschlicher, C. J. Nelin and P. S. Bagus, *J. Chem. Phys.*, 1985, **82**, 3265.
332. B. Zheng, W. M. Hua, Y. H. Yue and Z. Gao, *J. Catal.*, 2005, **232**, 143.
333. X. Q. Gong, R. Raval and P. Hu, *Phys. Rev. Lett.*, 2004, **93**, 106104.
334. Y. Wang, A. Lafosse and K. Jacobi, *J. Phys. Chem. B*, 2002, **106**, 5476.
335. B. Z. Sun, W. K. Chen, X. Wang and C. H. Lu, *Appl. Surf. Sci.*, 2007, **253**, 7501.
336. D. S. Su and R. Schlogl, *Catal. Lett.*, 2002, **83**, 115.
337. J. N. Al-Saeedi, V. V. Guliants, O. Guerrero-Perez and M. A. Banares, *J. Catal.*, 2003, **215**, 108.
338. M. A. Carreon, V. V. Guliants, M. Olga Guerrero-Perez and M. A. Banares, *Microp. Mesop. Mater.*, 2004, **71**, 57.
339. J. Cejka and S. Mintova, *Catal. Rev. Sci. Eng.*, 2007, **49**, 457.
340. U. S. Ozkan and R. B. Watson, *Catal. Today*, 2005, **100**, 101.
341. J. Macht and E. Iglesia, *Phys. Chem. Chem. Phys.*, 2008, **10**, 5331.
342. M. Batzill, K. Katsiev, D. J. Gaspar and U. Diebold, *Phys. Rev. B*, 2002, 66.
343. C. T. Campbell, A. W. Grant, D. E. Starr, S. C. Parker and V. A. Bondzie, *Top. Catal.*, 2001, **14**, 43.
344. K. Jacobi, G. Zwicker and A. Gutmann, *Surf. Sci.*, 1984, **141**, 109.
345. T. Ogama, *J. Appl. Phys.*, 1988, **64**, 753.
346. M. W. Roberts and R. S. Smart, *Surf. Sci.*, 1985, **151**, 1.
347. M. Baca, A. Pigamo, J. L. Dubois and J. M. M. Millet, *Catal. Commun.*, 2005, **6**, 215.
348. P. Concepcion, P. Botella and J. M. L. Nieto, *Appl. Catal. A: General*, 2004, **278**, 45.
349. R. K. Grasselli and J. D. Burrington, *Ind. Eng. Chem. Prod. Res. Dev.*, 1984, **23**, 393.

350. S. T. Korhonen, M. A. Bañares, J. L. G. Fierro and A. O. I. Krause, *Catal. Today*, 2007, **126**, 235.
351. S. T. Oyama, G. T. Went, K. B. Lewis, A. T. Bell and G. A. Somorjai, *J. Phys. Chem.*, 1989, **93**, 6786.
352. I. E. Wachs, Y. Chen, J.-M. Jehng, L. E. Briand and T. Tanaka, *Catal. Today*, 2003, **78**, 13.
353. F. Cavani, N. Ballarini and A. Cericola, *Catal. Today*, 2007, **127**, 113.
354. J. N. Al-Saeedi and V. V. Guliants, *Appl. Catal. A: General*, 2002, **237**, 111.
355. L. Chen, E. G. Derouane and J. C. Vedrine, *Appl. Catal. A: General*, 2004, **270**, 157.
356. R. K. Grasselli, C. G. Lugmair and A. F. Volpe, Jr., *Top. Catal.*, 2008, **50**, 66.
357. G. Li, D. Hu, G. Xia and Z. Zhang, *Top. Catal.*, 2010, **53**, 40.

CHAPTER 16

Reaction Engineering of Oxidation Reactions

ARNE DINSE AND REINHARD SCHOMÄCKER

Department of Chemistry, Technical University of Berlin, Straße des 17. Juni 124, Sekr. TC 8, D-10623 Berlin, Germany

16.1 Introduction

Going from analytical characterization of solid-state catalysts to their application in selected reactor systems on a laboratory or industrial scale, the chemist is confronted with several challenges that are related to the field of chemical engineering. The most important engineering aspects are summarized in the understanding of reaction kinetics, mass- and heat-transfer effects, residence-time distribution, catalyst stability and reactor safety that shall therefore be addressed in this chapter. While the minimum requirements for the implementation of a chemical process are provided by the catalyst performance, the interplay of the mentioned parameters together with reactor design and operation mode are a powerful tool for further optimization of a chemical process. The goal of the chemical engineer is thereby to maximize product yields, minimize energy input and provide a safe and ecologically compatible process environment. For a better understanding, the influence of different engineering aspects on the selection of reactor type and operation mode is discussed in this chapter based on different industrially realized large-scale processes. They are namely, the oxidation of ethylene, sulfur dioxide and naphthalene to ethylene oxide (EO), sulfur trioxide and phthalic anhydride (PSA), respectively. Please note that the field of reaction engineering mostly

RSC Nanoscience & Nanotechnology No. 19
Nanostructured Catalysts: Selective Oxidations
Edited by Christian Hess and Robert Schlögl

Published by the Royal Society of Chemistry, www.rsc.org

deals with questions regarding solely the reactor unit, while an important attribute of an industrial-scale process is also the extensive product processing. The theory of process engineering exceeds the scale of this book and the interested reader is referred to the respective literature.[1,2]

16.2 Basic Reactor Models

For the understanding of the behavior of certain reactor types, different models can be implied. Two basic reactor models exist that can be used to describe real reactors applied for continuous oxidation reactions on an industrial or laboratory scale, namely the plug-flow tubular reactor (PFTR, Figure 16.1 left) and the continuous stirred-tank reactor (CSTR, Figure 16.1 right).

The PFTR model describes a reactor tube through which a continuous reactant flow is passed. The reactor can be divided into differential volume segments in which the reaction takes place causing concentration and temperature gradients over the length of the reactor. Characteristics of this model are the distinct temperature gradients along the reactor length as well as its plug-flow behavior with a low degree of backmixing. The PFTR model is mostly used for the design of tubular and multitubular fixed-bed reactors (Figure 16.2 left, middle). The fixed-bed reactor is one of the most established reactor systems providing simple design and implementation. One of its disadvantages is the difficult catalyst replacement that is a problem in the case of fast deactivating catalysts.

The CSTR model represents a continuous stirred-tank reactor, in which reactants and products are ideally mixed under continuous throughput. The recirculation and/or backmixing of reactor contents in the case of the CSTR model concurrently leads to a dilution of reactants and products. This can influence the reaction rates both favorably and unfavorably towards higher or lower product yields and will be discussed below using the model equations for the specific reactor type. The recirculation also goes along with a uniform temperature distribution making reactors of this type easier to handle.

The CSTR model can also be used to describe the fluidized-bed reactor (Figure 16.2 right) that is used for some oxidation reactions on an industrial scale. In this case, the stirring effect is achieved by a high-velocity reactant

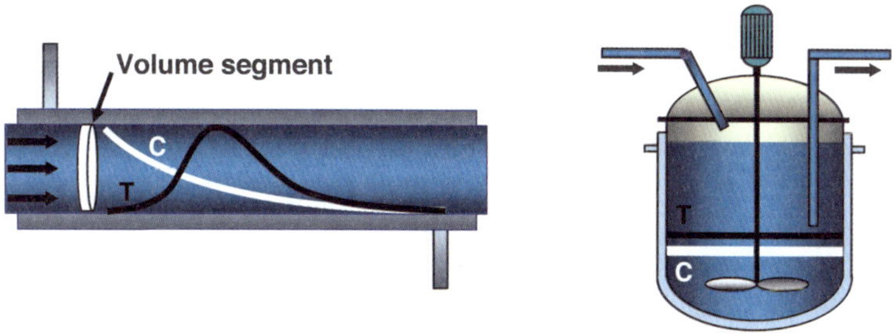

Figure 16.1 Schematic illustration of PFTR (left) and CSTR (right) models.

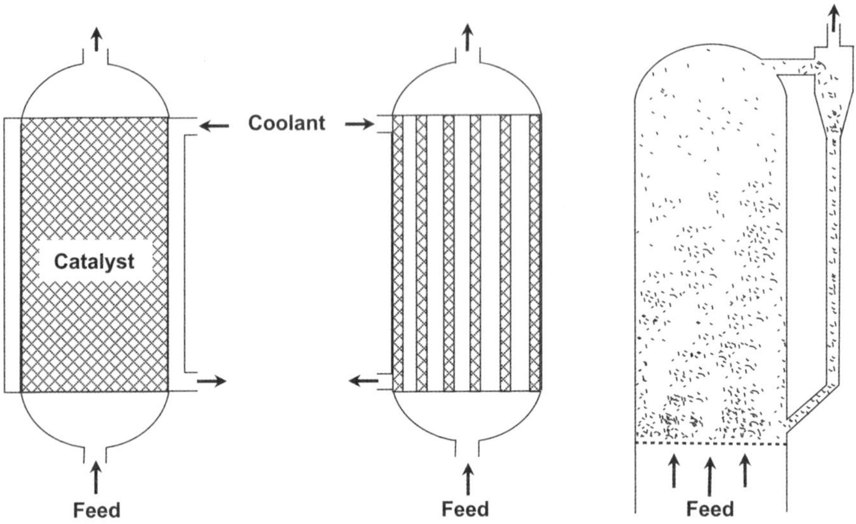

Figure 16.2 Schematic illustration of two different fixed bed and a fluidized-bed reactor including a cyclone for discharged catalyst.

stream onto a catalyst bed until the particles are held suspended that, as in case of a stirred-tank reactor, results in a nearly gradient-free distribution of reactants and products and a uniform temperature profile. Another positive property of the fluidized bed is the possible on-stream catalyst exchange and regeneration of strongly deactivating catalysts. The advantages of the fluidized-bed reactor are, however, often devaluated by the complex built-in components such as cyclones against catalyst discharge and heating coils, which lead to catalyst attrition and complicated scaleup procedures. In addition, large-scale fluidized-bed reactors also exhibit the problem of complicated fluid dynamics and are the subject of intensive ongoing research.[3] A summary of the properties of tubular and fluidized-bed reactors is given in Table 16.1.

Because of their isothermal heat distribution, fluidized sand beds are sometimes used instead of ovens as heating devices for fixed-bed laboratory reactors, commonly using quartz sand as heat carriers (Figure 16.3).

Table 16.1 Characteristics of tube and fluidized-bed reactors.

Tube reactor	*Fluidized-bed reactor*
Complex catalyst exchange	On-stream catalyst exchange (for deactivating catalysts)
Negligible particle attrition	Particle attrition
Distinct hotspots in case of exothermic reactions	Efficient heat exchange
Simple reactor design	Complicated scale-up
Low degree of backmixing	Pronounced backmixing (lowering reactant concentration)

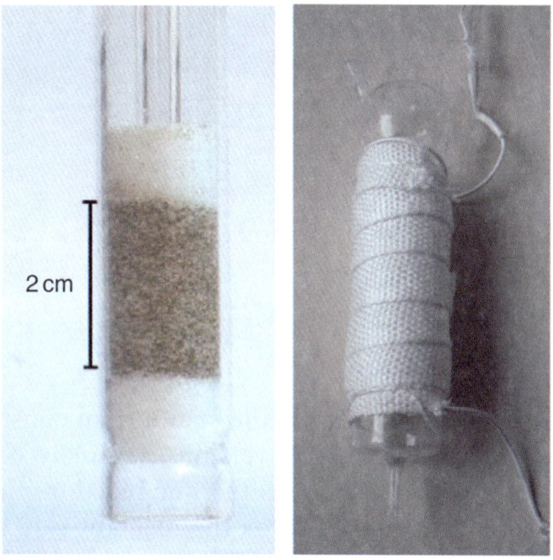

Figure 16.3 Laboratory-scale fixed-bed reactor (left) and fluidized-bed reactor (right).

16.3 Conversion, Selectivity and Yield

Considering a simple first-order irreversible reaction $A \xrightarrow{r_A} B$, with the reaction rate r_A, the conversion X_A with respect to a component A is defined by eqn (16.1), describing the ratio of reacted moles A per moles of A fed into a reactor. In the case of a constant reaction volume, the conversion can also be expressed as a function of the respective concentrations or partial pressures.

$$X_A = \frac{n_{A,0} - n_A}{n_{A,0}} = \frac{c_{A,0} - c_A}{c_{A,0}} = \frac{p_{A,0} - p_A}{p_{A,0}} \tag{16.1}$$

If a simple reversible reaction (equilibrium reaction) of the type $A \underset{r_A}{\overset{r_A}{\rightleftharpoons}} B$ is considered, the reaction rate with respect to component A can be described by eqn. (16.2)

$$-r_A = k_A c_A - k_{-A} c_B \tag{16.2}$$

At chemical equilibrium, the effective rates are zero and eqn (16.2) can be written as eqn (16.3)

$$k_A \left(c_A - \frac{k_{-A}}{k_A} c_B \right) = 0 \tag{16.3}$$

And with the definition of eqns (16.4)–(16.6)

$$c_A = c_{A,0}(1 - X) \tag{16.4}$$

$$c_B = c_{A,0}X \tag{16.5}$$

$$K_e = \frac{k_A}{k_{-A}} \tag{16.6}$$

one can rearrange eqn (16.3) to eqn (16.7)

$$X_A = \frac{K_e}{1 + K_e} \tag{16.7}$$

From thermodynamics it is known that the equilibrium constant depends only on the temperature and, therefore, conversion can solely be influenced by a change in temperature. In the case of exothermic reactions, high temperatures lead to a shift of the equilibrium towards the reactant side and a decrease in conversion, while low temperatures on the contrary favor high conversion. For both the simple irreversible as well as the reversible reaction, the overall yield of product B, Y_B, is equal to the reactant conversion X_A, because every molecule of A reacted is selectively transformed into a product molecule B.

In selective oxidation reactions, one is often confronted with the more sophisticated case of a reaction network. This describes the contribution of different simultaneous (parallel and/or consecutive) reactions, for each of which a rate law can be formulated. In selective oxidation reactions, *e.g.* the partial oxidation of ethylene to EO, the overall reaction network consists of partial oxidation and parallel combustion of reactant A to products B and C, respectively as well as consecutive combustion of product B to product C (usually carbon oxides and water) as described in Figure 16.4. The impact of an individual side reaction is often small compared to other participating side reactions, simplifying the reaction network to a simple consecutive (A to B to C) or parallel reaction (A to B and C).[4,5]

As opposed to the simple reaction types discussed before, reactant A is not selectively transformed to product B any longer. To account for competing side reactions it is necessary to introduce a second parameter for the calculation of the overall product yield, namely selectivity, which can be subdivided into differential and integral selectivity.

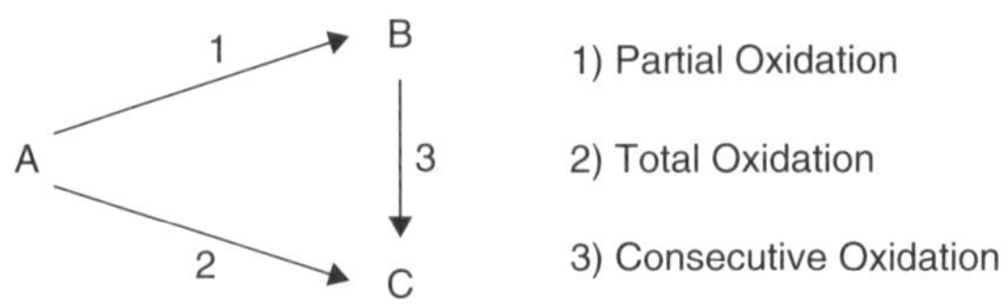

Figure 16.4 General reaction network of selective oxidation reactions including partial reactant oxidation, direct reactant combustion and consecutive product combustion.

The differential selectivity towards the desired product B is a direct function of the reaction rates, defined by eqn (16.8). It describes the interplay of product forming (1) as well as reactant (2) and product (3) consuming reactions and their reaction rates r_i.

$$\varphi_B = \frac{r_1 - r_3}{r_1 + r_2} \tag{16.8}$$

The integral selectivity, analytically measured at the reactor outlet, is obtained by integration of the differential selectivities at any location z between reactor inlet and outlet, as expressed in eqn. (16.9) and describes the ratio of desired product B and consumed reactant A.

$$S_B = \int_{i=\text{inlet}}^{\text{outlet}} \varphi_{i,B} dz = \frac{c_B}{c_{A,0} - c_A} \tag{16.9}$$

The substrate conversion (16.1) describes the concentration change between reactor inlet ($c_{A,0}$) and outlet (c_A) and is also a measure for the activity of the respective catalyst. The product of reactant conversion and integral selectivity towards the desired product B finally lead to the overall product yield Y_i, defined in eqn (16.10). This value is one of the most important aspects related to the feasibility of an industrial application, that is, efficient catalysts need to reveal both high catalytic activity and product selectivity.

$$Y_B = X_A S_B \tag{16.10}$$

In the following, the influence of different reactor types on the reaction parameters above shall be discussed with the example of a simple consecutive reaction scheme consisting of a desired reaction $A + B \rightarrow C$ and an undesired reaction $C + B \rightarrow D$. In this case the differential product selectivity is described by eqn (16.11) if the reaction orders of side and main reaction are the same, which is actually the case for many catalytic oxidation reactions.

$$\varphi_B = \frac{r_1 - r_3}{r_1} = \frac{k_1 c_A^n c_B^m - k_3 c_C^n c_B^m}{k_1 c_A^n c_B^m} = 1 - \frac{k_3 c_C^n}{k_1 c_A^n} \tag{16.11}$$

The differential (and integral) selectivity is now a function of reactant and product concentrations and therefore sensitive to the concentration distribution of reactants and products inside the reactor. This case of a simple consecutive reaction results in different selectivity curves with the side reaction either being slower or faster than the main reaction. The latter case leads to a fast decline of product selectivity as a function of reactant conversion. A slower side reaction, in contrast, results in high selectivities even at elevated conversions, depicted in Figure 16.5, the latter case usually representing applied industrial processes. It can also be seen that it is desirable to operate the reactor at limited reactant conversion in order to provide sufficient product selectivity.

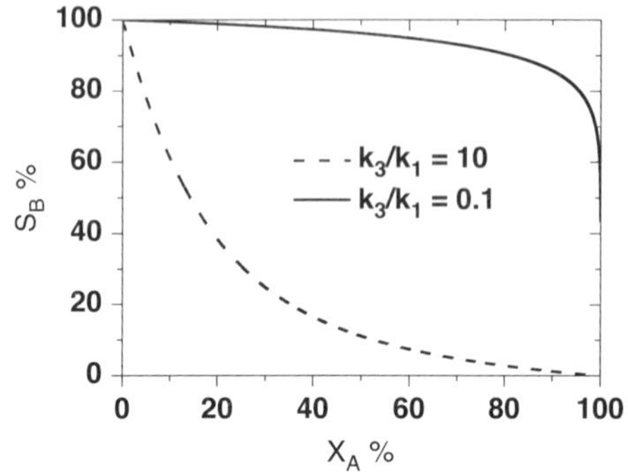

Figure 16.5 Product selectivity as a function of reactant conversion for a simple consecutive reaction with different ratios of rate constants. (solid) $k_3/k_1 = 10$, (dashed) $k_3/k_1 = 0.1$.

In addition, such operation modes commonly result in a better temperature control, because side reactions are often highly exothermic.

In the case of a simple parallel reaction network in which $A + B \rightarrow C$ and $A + B \rightarrow D$, again assuming equal reaction orders for main and side reaction, the differential selectivity can be described by eqn (16.12)

$$\varphi_B = \frac{r_1}{r_1 + r_2} = \frac{k_1 c_A^n c_B^m}{k_1 c_A^n c_B^m + k_2 c_A^n c_B^m} = \frac{k_1}{k_1 + k_2} \tag{16.12}$$

In this case, the differential (and integral) selectivity is only a function of the respective rate constants and can only be adjusted by temperature, being independent of the degree of reactant conversion. This is, for example, the case in the selective oxidation of ethylene to EO.[6] Because selectivity and yield are not a function of the reactor design in this case, the following chapter discussing the influence of reactor design on these parameters is based on a consecutive reaction network. It should, however, be mentioned that certain reaction mechanisms including different combinations of kinetic parameters (reaction order, rate) can lead to reverse conclusions as derived later in this chapter. In addition, maximum selectivity in the case of a parallel reaction network is a function of temperature, thus temperature gradients within tubular reactors (see below) can also have an undesirable influence on the product yield, making a different reactor design advantageous. For a more thorough understanding, the interested reader is referred to respective textbooks.[7,8]

16.4 Mass Balance

For a comparison of the performance of different reactor types it is necessary to solve the mass balance, which describes substrate accumulation *via* molecular flux and chemical reaction. For a given chemical reaction in an ideal reactor it is generally defined by eqn (16.13) assuming no axial reactant dispersion.

$$\frac{\partial c_i}{\partial t} = -\text{div}(uc_i) + \sum_j \nu_{ij} r_j \tag{16.13}$$

with c_i being the reactant concentration, u the linear flow rate and r_i the respective reaction rate. For a simple irreversible reaction $A \rightarrow B$ it simplifies to eqn (16.14)

$$\frac{\partial c_A}{\partial t} = \text{div}(uc_A) + r_A \tag{16.14}$$

For a dimensionless consideration of the mass balance, eqn (16.13) can be rewritten as eqn (16.15)

$$\tau \frac{\partial X_A}{\partial t} = \frac{\tau}{c_{A,0}} \text{div}(uc_A) - \frac{r_{A,0}\tau}{c_{A,0}} \Phi(X_A) \tag{16.15}$$

with τ being the residence time of reactants inside the reactor, X_A the conversion of component A, $c_{A,0}$ the initial concentration of component A at the reactor inlet and $\Phi(X)$ the conversion dependence of the reaction kinetics. In this case the characteristic, hydrodynamic residence time is defined as eqn (16.16).

$$\tau = \frac{V_R}{\dot{V}} \tag{16.16}$$

with $\dot{V}$ the volumetric flow and V_R the reactor volume.

A characteristic of the dimensionless description of mass balances and direct comparison of different reactor types is based on the Damköhler number, which is defined by eqn (16.17)

$$\text{Da} = \frac{r_0 \tau}{c_{A,0}} \tag{16.17}$$

It describes the ratio of the residence time τ and a characteristic reaction time. For a given chemical reaction Da is a measure for the conversion X_A at a certain residence time τ.

In the case of the PFTR the following assumptions are made:

- The concentration profile inside the reactor is stationary, that is there is no accumulation.
- Plug flow is assumed, no backmixing occurs.
- A one-dimensional consideration is sufficient, no radial concentration gradients are considered.

It can be shown that the relation of Damköhler number and reactant conversion can be expressed as:

$$\mathrm{Da} = \int_0^{X_e} \frac{dX}{\Phi(X)} \tag{16.18}$$

Derivation of the mass balance of a CSTR model is based upon the following assumptions:

- Ideal mixing, *i.e.* concentration at reactor exit equal to concentration inside the reactor at any given time.
- The concentration is time independent, *i.e.* steady state is assumed resulting in zero accumulation.

In the case of a CSTR the dimensionless mass balance can be expressed as eqn (16.19)

$$\mathrm{Da} = \frac{X_e}{\Phi(X_e)} \tag{16.19}$$

The reactant conversion for a given first-order chemical reaction as a function of Da is shown in Figure 16.6.

It can be seen that a higher conversion is attained in the case of a PFTR as compared to a CSTR at a given value of Da and residence time, respectively. The reason for this behavior can be found in the different mixing behavior of the reactors. While a PFTR shows ideal plug-flow behavior with no backmixing, a CSTR is an ideally mixed reactor with complete backmixing leading to lower reactant concentrations. For a reaction with positive reaction orders the CSTR, therefore, reveals lower reaction rates. For zero-order reactions that are independent of the reactant concentration the differences between the reactor models disappear and for reactions with negative orders inverse results can be found.

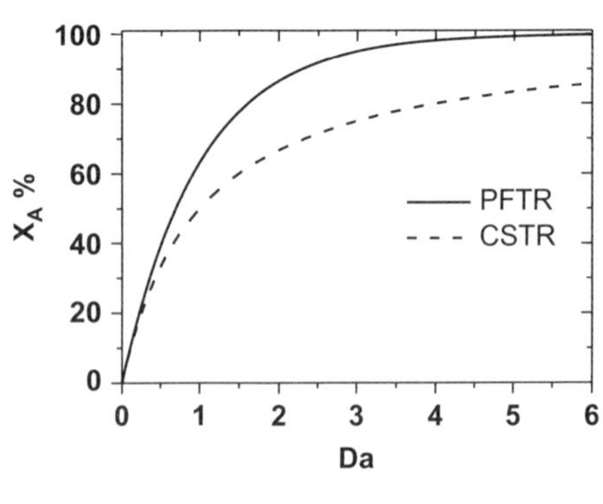

Figure 16.6 Simulated reactant conversion as a function of Da for PFTR and CSTR for a first-order reaction.

The different mixing behavior of the two reactor models also influences the integral selectivity of a given reaction. The lower product selectivities observed using a CSTR for a fast consecutive reaction, as illustrated in Figure 16.7, can be explained by the increased probability of formed product to undergo further reaction if it is mixed with unreacted substrate.

The difference in integral selectivities for a given reactor can also be understood by taking a closer look on the correlation between integral and differential product selectivities for PFTR and CSTR, as shown in Figure 16.8. For a

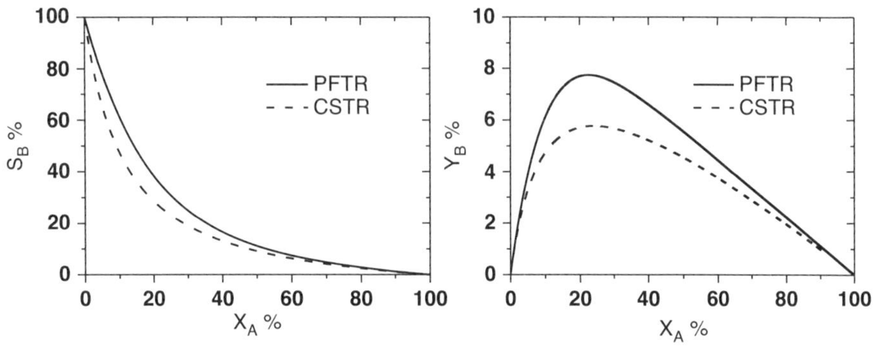

Figure 16.7 Simulated selectivity and yield as a function of reactant conversion for a first-order consecutive reaction with respect to reactant A and product B, a rate constant $k_1 = 1\ s^{-1}$ and a ratio of $k_3/k_1 = 10$ using PFTR and CSTR model approaches.

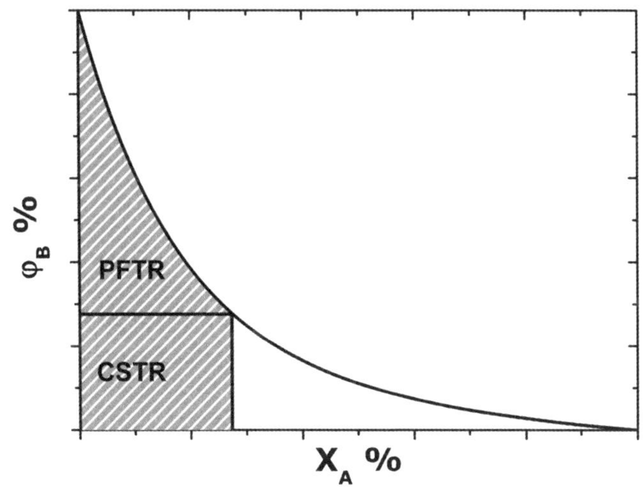

Figure 16.8 Differential (trace) and integral (area) selectivity for CSTR (rectangle) and PFTR (whole area) as a function of conversion for a simple consecutive reaction for a ratio of reaction rate constants $k_3/k_1 = 10$.

given reactant conversion, the integral selectivity of a PFTR is the integral of all differential selectivities over the length (or residence time) of the reactor. In contrast, for a CSTR, differential and integral selectivity are the same, because of negligible concentration gradients inside the reactor. For a fast consecutive reaction, the achievable integral selectivity inside a CSTR is consequently lower than in a PFTR, as illustrated in Figure 16.8. At low reactant conversion, the PFTR can also be interpreted as a gradient-free reactor, because in this case concentration gradients inside the reactor can be considered as linear. This is especially useful for kinetic experiments on a laboratory scale, in which a differential reactor system is often sufficient for evaluation of kinetic data.[7]

16.5 Heat-Balance and Heat-Transfer Effects

One of the most important aspects of chemical reaction engineering is the control of heat formation. In almost all chemical processes this has a direct influence on reactor type and operation mode. In analogy to the mass balance, the heat balance can be written as is done in eqn (16.20). It is noteworthy that this equation is based on the assumption that no axial heat diffusion takes place and heat transfer occurs solely through the reactor wall.

$$\frac{\partial(\rho c_{\mathrm{P}} T)}{\partial t} = -\mathrm{div}(\rho c_{\mathrm{p}} T u) + \sum_j r_j(-\Delta H_{\mathrm{R}j}) \tag{16.20}$$

In this equation, ρ is the density of the reaction mixture, c_{p} the heat capacity, T the temperature, and ΔH_{R} the reaction enthalpy. The heat balance consists of an accumulation term that is the sum of heat production and removal through convective flow, transfer through the reactor wall and chemical reaction.

With respect to temperature control a reactor can be operated in the following operation modes:

- isothermal;
- adiabatic (autotherm);
- polytropic.

Isothermal conditions are only achievable if the temperature of the coolant is controlled in a way that the wall temperature is the same at every point of the reactor. While this is realizable for CSTR-type reactors, it is not achievable for PFTR-type reactors. In the latter case a polytropic operation mode (*e.g.* selective oxidation of ethylene to EO) in which the coolant temperature changes along the reactor length is common for established industrial processes. The adiabatic mode, in which no heat exchange occurs with the environment, is applied for highly selective reactions such as sulfuric-acid production. An autotherm operation mode is present, if no additional external heating (heat exchanger) is necessary to start the reaction at the reactor inlet.

Especially in the case of highly exothermic chemical reactions using tubular reactors heat-transfer effects have to be considered in order to prevent thermal runaway phenomena. Heat formation can easily generate temperature gradients of 100 K and more, which can lead to undesirable side reactions and dangerous operating points. In the case of laboratory reactors significantly lower temperature profiles (less than 10 K) are usually observed because of smaller dimensions.[4] The temperature profiles inside a reactor can be calculated by solving the heat balances for a given reactor. The appearance of heat generation is especially challenging using fixed-bed reactors in polytropic operation mode. In this case, the heat balance is given by eqn (16.21).

$$\frac{dT}{dZ} = \frac{4}{d_R}\frac{U\tau}{\rho_0 c_p}(T_W - T) + \Delta T_{ad}\frac{dX}{dZ} \tag{16.21}$$

Heat and mass balance have to be solved simultaneously by numerical integration here in order to determine conversion and temperature profiles.

In this equation Z is the normalized reactor length, d_R the reactor diameter, U the heat-transfer coefficient, T_W the wall temperature and ΔT_{ad} the adiabatic temperature increase. The latter characteristic number is the maximum possible temperature increase in the case of zero heat exchange with the environment (*e.g.* perfect isolation, adiabatic operation).

The axial temperature gradients for polytropic and adiabatic tubular reactors are schematically illustrated in Figure 16.9 (left).

Radial and axial temperature gradients inside reactors lead to "hot spots" which can lead to a burst of reactor tubes. In addition, product selectivity and yield can be significantly influenced by distinct temperature profiles. In the case of a higher activation energy of the side reaction as compared to the main reaction a selectivity decrease will be observed. This is attributed to the stronger acceleration of the side reactions at high temperatures.[9] If the activation energy of the side reaction is lower than the main reaction (*e.g.* ODP), hot spots can

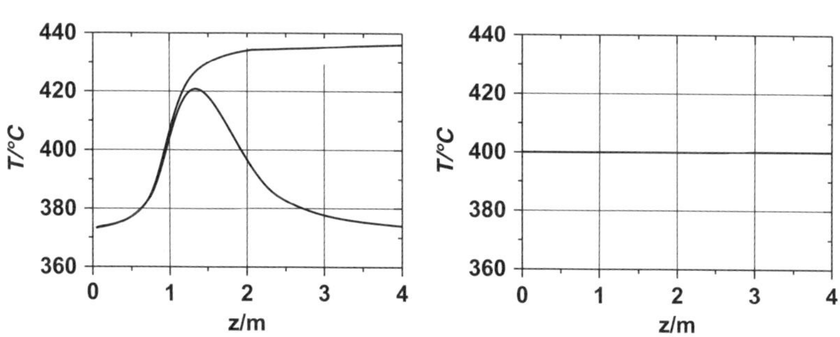

Figure 16.9 (Left) Schematic axial temperature profile of a polytropic and adiabatic tube reactors. (Right) Schematic temperature profile in a fluidized-bed reactor for the same reaction.

also favor a selectivity increase towards the desired product.[4] This favorable scenario, however, is devaluated by the fact that hot spots generally lead to a deactivation of catalysts. In fluidized beds with sufficient homogeneity, temperature gradients will usually not occur to a large extent (Figure 16.9, right). Apart from proper cooling, hot spots can also be avoided by an operation mode adjusted to lower conversions.

16.6 Effects of Residence Time and Residence-Time Distribution

The behavior of real chemical reactors often deviates from the ideal models of PFTR and CSTR. One reason for this can be found in the ideal assumption of either no backmixing (PFTR) or complete backmixing (CSTR). For the description of real reactors, which show properties in between the two idealized cases, the cascade as well as the dispersion model can be implied.

Using the latter as an exemplary model for the description of the residence behavior of a tube reactor, the molar flux of a nonreacting marking substance A is described by eqn (16.22)

$$\frac{dc_A}{dt} = -u\frac{\partial c_A}{\partial z} + D_{\text{ax}}\frac{\partial^2 c_A}{\partial z^2} \tag{16.22}$$

with u being the linear velocity and z the reactor length. This model describes backmixing effects caused by built-in components and gas-flow fluctuation using an axial dispersion coefficient D_{ax}. Equation (16.22) can be rewritten in dimensionless form to give eqn (16.23)

$$\frac{\partial C}{\partial \theta} = -\frac{\partial C}{\partial Z} + \frac{1}{\text{Bo}}\frac{\partial^2 C}{\partial Z^2} \tag{16.23}$$

with $\theta = \frac{t}{\tau}$, $Z = \frac{z}{L}$, $C = \frac{c_A}{c_{A,0}}$ and $\text{Bo} = \frac{uL}{D_{\text{ax}}}$.

In the case of a both-way open system the response function to a tracer impulse at the reactor inlet can be described by eqn (16.24), which is also known as the residence-time distribution function.

$$C(\theta) = \frac{1}{2}\sqrt{\frac{\text{Bo}}{\pi\theta}}\exp\left(\frac{-(1-\theta)^2\text{Bo}}{4\theta}\right) \tag{16.24}$$

The cumulative residence-time distribution $F(\theta)$ can be derived by integration of eqn (16.24).

On a laboratory scale, mixing behavior of reactor systems is usually studied by pulse- or step-marking experiments. In the first case, an inert substance is injected at the reactor inlet as a short pulse, while in the latter case a step marking is used (Figures 16.10(A) and (B)). The resulting concentration

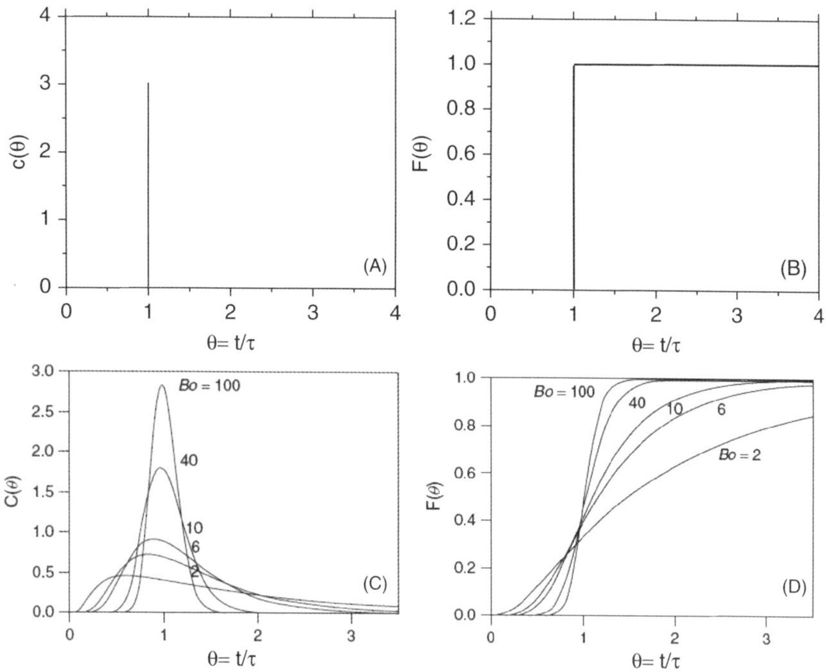

Figure 16.10 Pulse (A) and step marking (B) at reactor inlet and resulting concentration distribution as a function of normalized residence time for different values of Bo for (C) impulse marking and (D) step marking. Values predicted by dispersion model using Bo as parameter.

distribution at the reactor outlet as a function of a dimensionless time for different values of the Bodenstein number Bo is illustrated in Figures 16.10(C) and (D). A low degree of backmixing as is the case for most applied tubular reactors with turbulent flow profile leads to low axial dispersion coefficients and, therefore, high Bo numbers. This case is represented by a narrow residence-time distribution. With increasing degree of backmixing usually found in the laminar flow region, the residence-time distribution broadens remarkably. A disadvantageous residence-time distribution can significantly lower product selectivity and overall yield.

16.7 Mass-Transfer Limitations

As a result of the high pressure drop in laboratory- or industrial-scale reactors, the minimum catalyst particle sizes are limited. Together with the properties of the catalyst particles often consisting of highly porous materials used to support and disperse active sites, this may lead to the appearance of mass-transfer limitations. In such a case, the pore structure of the catalyst leads to the competition of diffusion processes with the actual reaction. This leads to

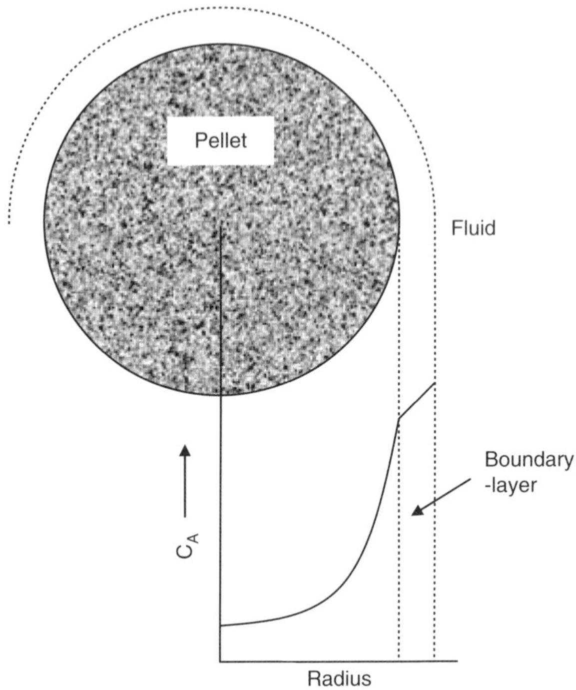

Figure 16.11 Schematic concentration and temperature gradients within catalyst particles.

concentration gradients of the substrate between the catalyst surface and particle interior (see Figure 16.11) which leads to a decrease of the local reaction rate, lowering the overall yield of a chemical reaction.

The ratio of the effective rate influenced by diffusion r_{eff} and the noninfluenced rate r_i is also known as the effectiveness factor eqn (16.25)

$$\eta = \frac{r_{\mathrm{eff}}}{r_i} \tag{16.25}$$

As η approaches the value 1, no diffusional limitations are present, while at lower values the reaction is more or less influenced by mass transport. For an empirical estimation of mass-transport phenomena the effectiveness factor can be plotted as a function of the Weisz modulus Ψ' illustrated in Figure 16.12.

The Weisz modulus describes the ratio of reaction rate to reactant diffusion rate as shown in eqn (16.26).

$$\Psi' = L^2 \frac{m+1}{2} \frac{r_{\mathrm{eff}} \rho_{\mathrm{cat}}}{D_{\mathrm{eff},i} c_i} \tag{16.26}$$

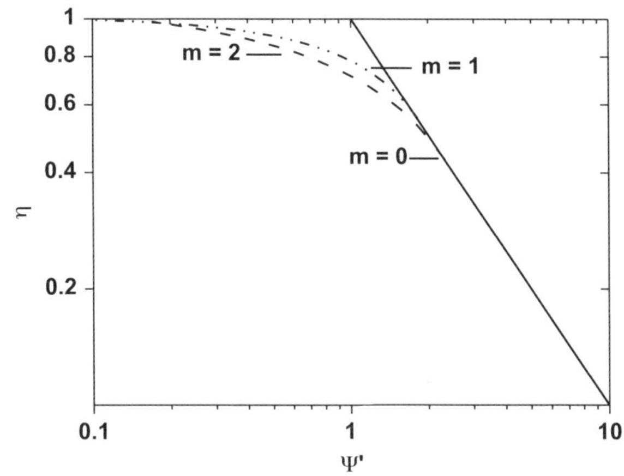

Figure 16.12 Effectiveness factor η as a function of the Weisz modulus Ψ' for different reaction orders m.

where L is the characteristic length of catalyst particle (defined as the ratio of particle volume and external surface area), m the reaction order of the limiting reactant, r_{eff} the measured effective reaction rate, ρ the catalyst density, D_{eff} the effective diffusivity and c_i the reactant concentration. With decreasing values of Ψ' and thus smaller particles, lower effective reaction rates and/or higher diffusion rates, the value of η increases, thus reflecting a decrease of the impact of pore diffusion on the effective rate. Generally, the influence of mass-transfer effects leads to an undesirable decrease of reaction rate and conversion. The influence of such effects on the product selectivity was investigated by Frank *et al.*[10] It was shown that in the case of large catalyst particles the desired product propene is trapped in the pore system of the catalyst, leading to a higher probability of its consecutive combustion, causing poor propene selectivities using large catalyst particles (>350 μm). Thus, laboratory investigation and industrial application of selective oxidation reactions require a compromise between the pressure drop inside the reactor and mass-transfer limitations within the catalyst particles. The study also revealed that especially for the laboratory application catalyst particle diameters can be usually decreased to an extent such that mass-transfer effects are negligible. To overcome mass-transfer limitations in industrial processes egg-shell catalysts can be used. In this case, only the outer shell of the support material is coated with catalytically active material.[11] This also leads to a better temperature control in the case of highly exothermic reactions.

16.8 Applied Industrial Reactor Designs

As discussed so far, reactor design and operation mode were shown to be important tools in order to maximize yield and, hence, cost efficiency of industrially applied chemical processes. But catalysts developed for industrial processes are often remarkably selective even at high reactant conversion and different criteria besides the optimization of product yields often play the key role for the final selection of a certain reactor type. A proper combination of reactor and operation mode is needed in order to compensate heat-generation effects that not only leads to undesirable side reactions but also to unsafe operation points. A further important aspect of process safety is, for example, the abidance of explosion limits. Additionally, material properties of catalyst particles can make a certain reactor type inapplicable for a chosen catalyst even though a sufficient yield can be attained in a laboratory- or pilot-scale experiment. Summarizing, the key criteria for the selection of certain reactor types for a chemical reaction are:

- yield (*e.g.* equilibrium limitations, mass transfer);
- safety/temperature control (*e.g.* explosion limits, hot spots);
- catalyst properties (*e.g.* deactivation, attrition).

In order to demonstrate the discussion and balancing the advantages and disadvantages of different reactors the influence of these factors on the reactor design of established industrial processes is discussed in the following. The examples will demonstrate the requirements for the catalysts catalytic, thermal and mechanical properties.

As a representative example, the reaction network of the selective oxidation of ethylene to EO is depicted in Figure 16.13. EO is an important raw material for the production of glycols (72% of world demand), mainly ethylene glycol (EG), which is subsequently used for the production of polyester fibers (53% of world demand) and polyethylene terephthalate (PET, 26% of world demand).[12] Today, EO is produced using air or oxygen as an oxidant. In larger plants oxygen is preferred due to higher yields and less downtime.[13]

The reaction kinetics can be described by a network of simultaneous reactions that are highly exothermal and irreversible. The consecutive EO combustion is considered to be negligible, while the occurrence of parallel total ethylene combustion leads to high degrees of heat formation.

The flow scheme of the industrial process for the production of EO is illustrated in Figure 16.14.[14] It shows how ethylene is converted into EO in a

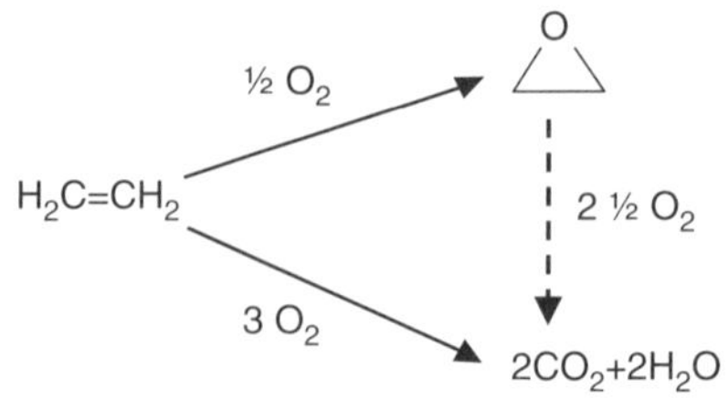

Figure 16.13 General reaction network of EO including partial ethylene oxidation, direct ethylene combustion and consecutive EO combustion.

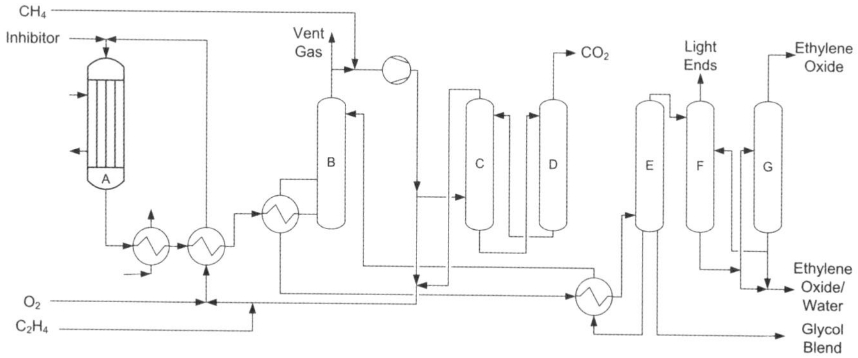

Figure 16.14 Flow scheme of ethylene oxidation with oxygen using a fixed-bed reactor (A) and product processing units (B-G) based on Ullmann's Encyclopedia of Industrial Chemistry.[14]

multitube fixed-bed reactor (A). Heated coolants are subsequently cooled *via* steam generation. Because the reaction enthalpies of the side reaction are considerably higher than for the main reaction, even operation modes revealing high EO selectivities lead to a remarkable heat formation. Catalyst deactivation effects leading to a decrease in product selectivities need to be considered here and cooling systems need to be adjusted to higher heat formation with ongoing time on-stream. Methane is used as a diluent to further improve temperature control. In the subsequent product processing steps, EO is first scrubbed using potassium carbonate solution (B), which is fed into a second scrubber for separation of low boiling components (E) and subsequently processed in a desorber column (F). In a final step, the head product of the desorber column is distilled to give the final product (G). Recycle gas processing (C, D) proceeds *via* CO_2 absorption in potassium carbonate solution under formation of potassium hydrogencarbonate. Processed recyle gas is returned into the reactor (A). The hydrogencarbonate solution is sent to a desorber column and regenerated CO_2 is released into the atmosphere while carbonate-containing solution is reused for CO_2 absorption. As a catalyst, alumina-supported silver catalysts are almost exclusively used today.[14] They can be prepared by impregnating low surface area α-alumina with silver oxalate.[15] For enhanced catalytic performance chlorine is often added to the reaction as an inhibitor of the catalyst in order to increase EO selectivity. It is assumed that it reduces the negative charge of adsorbed oxygen atoms, thus making them less reactive and, hence, inhibiting the parallel combustion of ethylene.[16] Additional selectivity enhancing effects can also be achieved by using metal promoters such as Cs.[17] It was found that caesium selectively binds and blocks the stepped part of the silver surface which reveals lower product selectivity than the silver (111) surface (see Chapter 9).

The main challenge of this process is the temperature control. Despite the advantages of fluidized-bed reactors for highly exothermic reactions, a multitube fixed-bed reactor is used for this reaction.[18,19] The reason for this can be found in the increased abrasion of the catalytically active Ag particles from the

Table 16.2 Reaction parameters of ethylene oxidation to EO.

Catalyst	*Temperature* °*C*	*Conversion* %	*Selectivity* %	*Pressure* *MPa*	*Capacity* $t\ a^{-1}$
Ag/Al_2O_3	200–300	7–15	80–90	1–3	5×10^5

carrier material because of the powerful oxygen flow inside the reactor. To increase heat dissipation, the numerous (up to 10 000) reactor tubes are embedded into a cooling system, in which either water or molten salts are the coolants of choice. The diameters of such tubes are held below a certain value called the critical tube diameter (usually <10 cm) in order to prevent distinct radial temperature gradients. The heat generation inside the reactor has to be controlled by an operation mode adjusted to low reactant conversions (see Table 16.2), resulting in a cost-intensive reprocessing of unreacted ethylene on the side of downstream processing. A further aspect of this operation mode is the abidance of explosion limits and the suppression of catalyst deactivation and a decrease in product selectivity due to product inhibition.[5,9]

It is noteworthy that other chemical reactions such as ODP could achieve similar product yields at comparable operating conditions (low conversion). However, the difficult separation of propane and propylene makes this process uneconomical. In addition, more efficient product routes such as steam cracking of naphtha or refinery off-gases and to a small extent the nonoxidative dehydrogenation of propane cover today's demand in propylene.[20] Such aspects have to be considered with the application of all industrial processes and prevented ODP from becoming industrially relevant to today despite decades of extensive research activities.[21–25]

Fixed-bed reactors are also used for the oxidation of methanol to formaldehyde using Ag or Cu (>600 °C) as well as MoO_3/Fe_2O_3 (400–425 °C due to Mo sublimation) catalysts.[26–28] Another important reaction using this reactor type is the production of vinyl acetate, which is used for the production of polyvinyl acetate. It is today almost exclusively produced from ethylene and acetic acid in a catalytic gas-phase oxidation process using palladium metal catalysts on SiO_2 or Al_2O_3.[11] For an early process by U.S. Industrial Chemicals reaction conditions of 175 °C and 1–7 bar were indicated.[29] Although it has now been applied for four decades and is still the most widespread process, it should be mentioned that a fluidized-bed process for this reaction has been developed in recent years supposedly lowering construction and operation costs by 30%.[30]

A special fixed-bed reactor is used for the oxidation of sulfur dioxide to sulfur trioxide as part of the production of sulfuric acid. The latter is one of the most important basic chemicals because it is utilized for the production of fertilizers (75% of world demand).[31] In this case the optimum adjustment of the equilibrium towards maximum product yield is the key factor of the reactor design. The oxidation of sulfur dioxide on V_2O_5, expressed in Figure 16.15, is an exothermic equilibrium reaction, thus low temperatures favor high conversion.[32] The minimum temperature, however, is limited by the ignition point of the catalyst, which is about 400 °C.

$$SO_2 + 1/2\ O_2 \rightleftharpoons SO_3 \quad \Delta H = -99\ \text{kJ/mol}$$

Figure 16.15 Exothermic oxidation of sulfur dioxide to sulfur trioxide.[14]

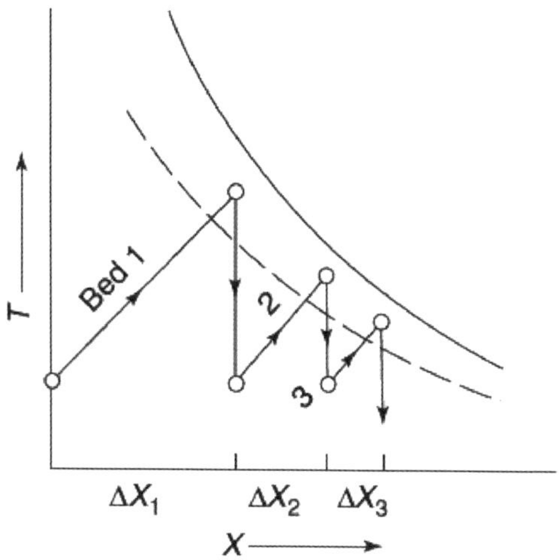

Figure 16.16 Temperature–conversion profile of sulfur dioxide oxidation with intermediate cooling.[3]

With increasing conversion, the high negative reaction enthalpy leads to an equilibrium shift towards the reactant side limiting the overall product yield. To overcome this problem, a multistage, nearly adiabatic reactor with intermediate cooling is used. In this case, intermediate cooling of the reaction participants will freeze the present conversion and subsequently lead to a higher overall conversion after the next catalytic stage. This intermediate cooling procedure leads to a so-called "saw-tooth" conversion profile as a function of temperature, as depicted in Figure 16.16.

In sulfuric acid production the multistage oxidation reactor is embedded in a complex pre- and postprocessing reaction system, which is presented in Figure 16.17.

Initially, atmospheric air is preconditioned using processed sulfuric acid (A). It is subsequently sent to the sulfur furnace (B) where it reaches a temperature above 1000 °C. The following heat exchanger discharges heat to a steam drum (C) that is fed by a feed-water tank (D). The generated steam produced in the steam drum, can be used for heating of further reaction units. After the mentioned preprocessing steps generated sulfur dioxide and oxygen are fed into the adiabatic four-stage converter (E). After the first and second stages, respectively, the reaction mixture is sent to heat exchangers for intermediate cooling. The heat exchangers, following the third stage, are connected with an intermediate absorption unit (F), in which sulfur trioxide reacts to sulfuric acid in order to increase overall acid yields. Nonabsorbed gases are fed back into the

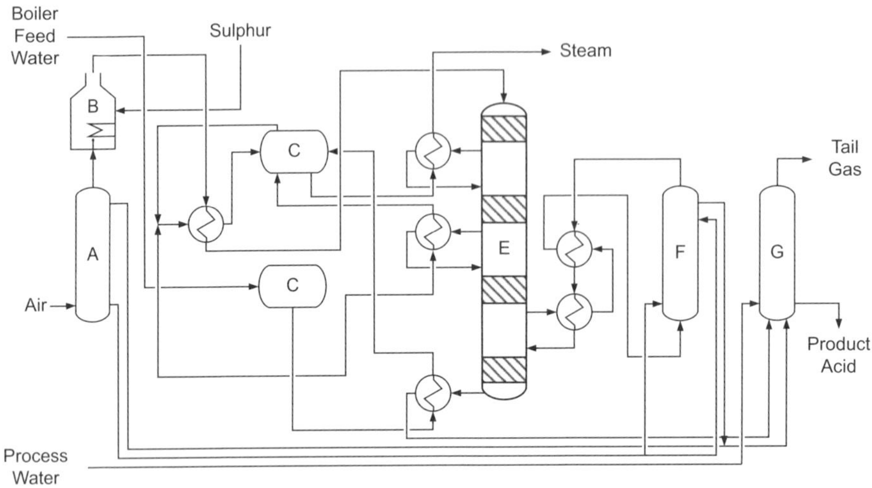

Figure 16.17 Flow scheme of sulfuric acid production using a multistaged fixed-bed reactor (p) with product preparation and processing units, respectively (a–q), based on Ullmann's Encyclopedia of Industrial Chemistry.[14]

Table 16.3 Reaction parameters of selective oxidation of sulfur dioxide to sulfur trioxide.

Catalyst	*Temperature* °*C*	*Conversion* %	*Selectivity* %	*Pressure* *MPa*	*Capacity* $t\,a^{-1}$
V_2O_5/SiO_2	380–430	99	100	–	6×10^4

fourth converter stage after passing the previous heat exchangers again. After final oxidation in the fourth converter stage, reaction gases are cooled using another heat exchanger (economizer) and sent to the final absorber unit (G). The most important process parameters are extracted in Table 16.3.

Today V_2O_5 catalysts are almost exclusively used for the industrial production of sulfuric acid. Generally, V_2O_5 is dispersed on SiO_2 support materials such as silica gels or zeolites. This can be achieved by procedures such as incipient wetness impregnation or grafting/ion exchange methods (see Chapter 12).[33,34]

Further examples of equilibrium-limited reactions using multistaged reactors or intermediate cold-gas feed are the synthesis of ammonia and methanol. Both reactions are known to be exothermic equilibrium limited reactions. For the synthesis of ammonia, the Haber–Bosch process, invented in the early 20th century still states the most common process. The reaction of nitrogen with hydrogen (produced by steam reforming of natural gas or other carbon sources) in the ratio of 1:3 is performed at 400–500 °C using promoted Fe_3O_4 catalysts. Because of high hydrogen diffusivity accompanied with high pressure (~ 300 bar) a special reactor design is used with a reactor jacket constructed from hard steel, while the inner parts are composed of hydrogen-withstanding soft iron.[14,35] Methanol is synthesized using synthesis gas (CO, H_2) produced from coal gasification, natural gas or partial oxidation of natural-gas hydrocarbons (CH_4).[36] The process is conducted at temperatures of 200 to 300 °C at elevated pressures (50–250 bar) using $Cu/ZnO/Al_2O_3$ and other copper-based catalysts catalysts.[26,37,38]

An oxidation reaction for which a fluidized-bed reactor is used can be found in the case of selective oxidation of naphthalene to PSA.[39] It is today mainly used for the production of plasticizers (57% of world demand) and resins (34% of world demand).[38]

Because of the great stability of the aromatic system a consecutive over-oxidation of PSA only occurs to a low extent leading to high product selectivities. As in the case of EO production this, again, leads to aspects other than yield optimization being key factors in the reactor design for this process. As can be seen in Figure 16.18 this reaction is also highly exothermic, hence, causing marked heat gradients using conventional fixed-bed reactors. As opposed to the EO production catalyst, however, particle erosion is manageable, making the advantages of the fluidized-bed technology accessible for this reaction.

For this process, illustrated in Figure 16.19, liquid naphthalene is fed into the fluidized bed (A) and subsequently evaporates on the hot catalyst particles. Simultaneously, air is added through distribution plates. Discharged catalyst

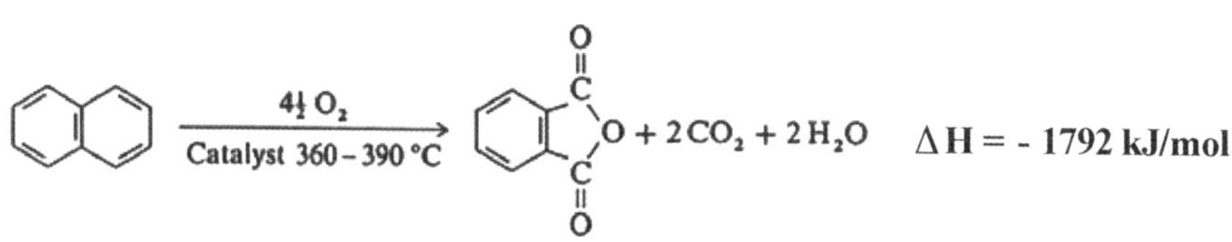

Figure 16.18 Exothermic selective oxidation of naphthalene to PSA.[14]

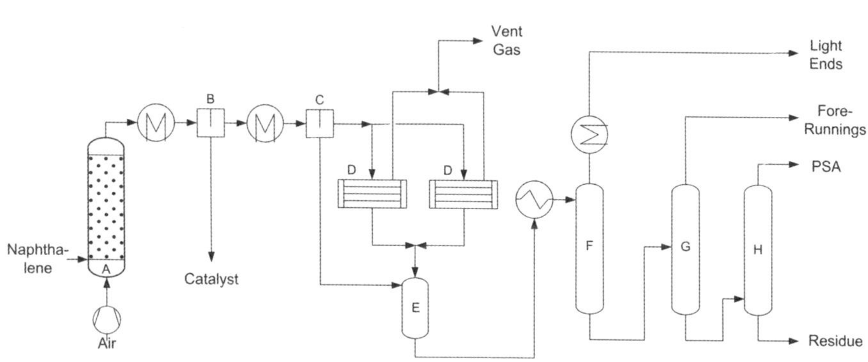

Figure 16.19 Flow scheme of PSA production using a fluidized-bed reactor (A) and product processing units, respectively (B–H), based on *Industrial Aromatic Chemistry*.[39]

Table 16.4 Reaction parameters for selective oxidation of naphthalene to phthalic acid.

Catalyst	*Temperature °C*	*Yield wt%*	*Pressure MPa*	*Capacity t a^{-1}*
V_2O_5/SiO_2 gel	340–380	90	0.4	6×10^4

particles can be separated using special filter plates or cyclones (B). Crude phthalic acid is held back *via* condensation (C) and gaseous remainders are transformed into solid *via* desublimation (D). Finally, liquid product is sent to a storage tank (E). Product distillation consists of three columns including pre-treatment (F), a light-ends column (G) and a phthalic-acid column (H). Process parameters are presented in Table 16.4. As a major advantage of PSA production it is possible to use o-xylene as an alternative feedstock with minimum process modifications in the same plant. Process parameters using naphthalene as a feedstock are depicted in Table 16.4.

In recent years, fixed-bed processes using o-xylene as feedstock have become more cost effective, decreasing the importance of fluidized-bed reactors for this reaction. Today's plants mostly use vaporized o-xylene as a feedstock, naphthalene providing only 16% of the feedstock requirements.[37,38]

Commercial catalysts usually consist of a support material such as silicate, silicon carbide, porcelain, alumina or quartz on which TiO_2 and V_2O_5 is deposited.[14] For a better catalyst performance potassium hydrogen carbonate is often used as a promoter since it is known to inhibit overoxidation of naphthalene or o-xylene.

Other processes using fluidized-bed reactors are the production of maleic anhydride using V–P–O catalysts with n-butane as a feedstock as well as cracking of crude oil with zeolite catalysts for olefine production. In the latter case, fluidized-bed processes offer fast on-stream exchange of catalyst, which is necessary because of its fast deactivation caused by coke deposition.[39] Maleic anhydride with n-butane as a low-cost feedstock is today produced in both fluidized-bed and fixed-bed reactors with the latter currently being the dominant process.[40]

16.9 New Developments in Reactor Design

In recent years, new conceptual developments in the field of reactor design have especially been represented by membrane and microstructured reactor systems. Both approaches have promised to overcome problems such as heat-transfer and yield limitations. In this chapter, advances and limitations of such systems are discussed.

Microstructured reactors offer an attractive alternative to conventional reactors because of their extremely efficient heat-transfer behavior along with minimum mass-transfer limitations, which is made possible by their high surface to volume ratio.[41,42] This reactor type also reveals the residence-time behavior of a PFTR-type reactor with minimum product backmixing, and thus is suitable for highly exothermic reactions with fast consecutive side reactions like oxidation of naphthalene to phthalic anhydride.[43] In the case of gas-phase reactions using solid-state catalysts, the reactor walls are used as support material for the catalyst itself. In this case, reactors are assembled in a three-step process consisting first of catalyst preparation, binding and spray coating of the catalyst to the reactor material, which is illustrated in Figure 16.20.

It has been shown that different binder materials have a distinct influence on selectivity and conversion of given chemical reactions.[44] Microchannels can be introduced into stainless steel platelets by wet chemical etching (Figure 16.21 left). To enhance surface areas and introduce support materials or catalysts into the microstructured system, various established methods such as anionic oxidation of aluminum, sol-gel deposition and wash coating can be implied.[45–47]

After application of the catalyst, the finalized reactor plates (Figure 16.21 right) are assembled as a distinct reactor unit, depicted in Figure 16.22, using standard soldering technology. Because of the difficult impregnation of the microchannels with solid-state catalysts, this reactor type is easier to use for homogeneously catalyzed reactions in liquid or gas phase, solely providing a reaction vessel with advantageous heat-transfer behavior and residence-time distribution.

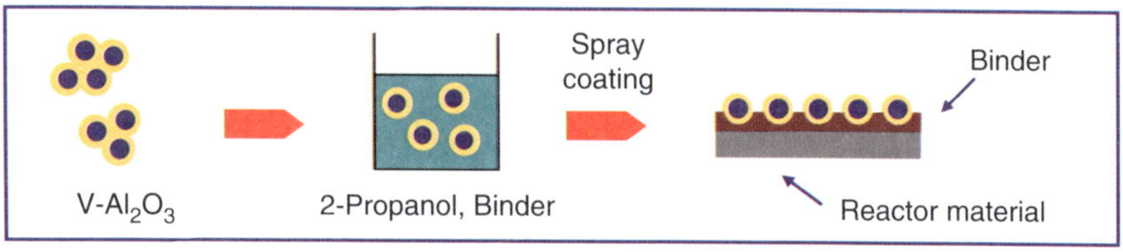

Figure 16.20 Three-way process for assembly of microstructured reactors.

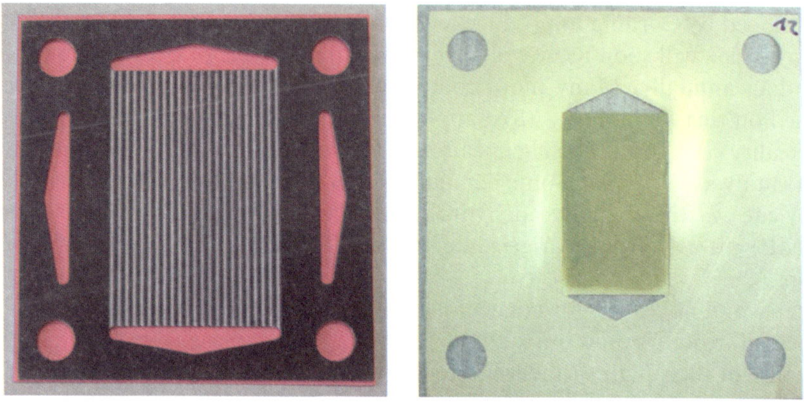

Figure 16.21 Bare reactor support plates (left) and finalized reactor plates (right).

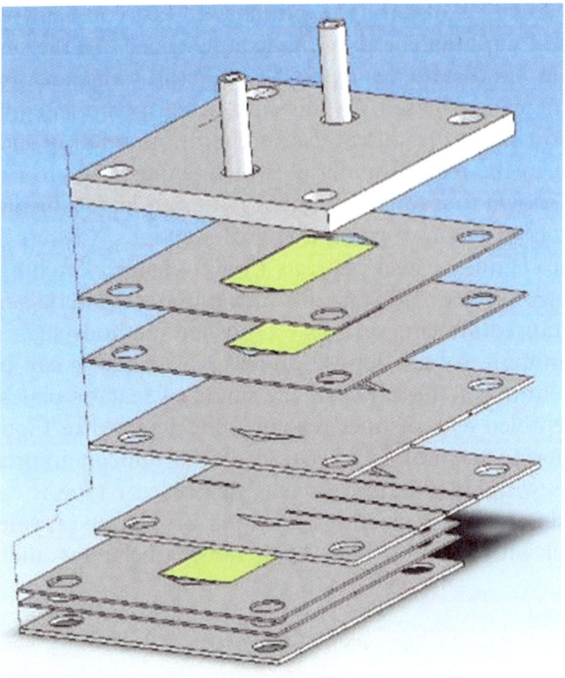

Figure 16.22 Assembly of microreactor plates.

The complicated and cost-intensive assembly techniques, however, are still the main drawback for a commercial implementation of a microstructured reactor, but despite many initial doubts, microreaction technology is on its way to maturity, leading the path to more efficient and environmentally advantageous chemical development and production processes. Today MRT is already applied by a growing number of manufacturers. Dutch fine chemical producer DSM, for example, which has used a microstructured reactor (MSR) developed by Corning to produce more than 25 tonnes of a nitration product in four weeks, will soon receive a Corning reactor capable of over 100 tons of the product annually. Many more companies are currently developing processes based on this technology. Most applications being developed for MSR involve speciality chemicals, fine chemicals and pharmaceuticals in liquid phase without catalysts. However, some companies also focus on bulk chemicals, such as ethylene *via* the oxidative dehydrogenation of ethane (ODE); dimethylether (DME) directly from syn-gas; the Fischer–Tropsch (FT) process and many more.[37]

A second novel reactor concept is represented by membrane reactors. These types of reactors not only allow for overcoming mass-transfer limitations because of their high surface area to active site ratio, but they are also used for yield increase of equilibrium-limited reactions such as the nonoxidative dehydrogenation of propane. In the latter case selective diffusion of hydrogen through a membrane can lead to an equilibrium shift towards the product accompanied with an increase in reactant conversion.[48] They can also be used to challenge the problem of cost-intensive catalyst separation in three-phase slurry processes, which are commonly used for hydrogenation reactions in liquid phase.[49] Another possible application is the on-stream separation of air into its components by using oxygen permeable membranes and, thus, economize on the energy and cost-intensive conventional air separation.[50]

Major drawbacks from a wide-scale application of this reactor type are mainly poor pressure resistance and oxygen permeation flux as a result of the thick walls of conventional tubular or disk-shaped membranes.[51,52] Another problem occurring with membrane type reactors is gas-tight sealing at the high temperatures. In a recent study by Caro *et al.*[50] these limitations could be overcome by using thin-walled perovskite-type membranes produced in a multiple step spinning procedure shown in Figure 16.23. The high temperatures

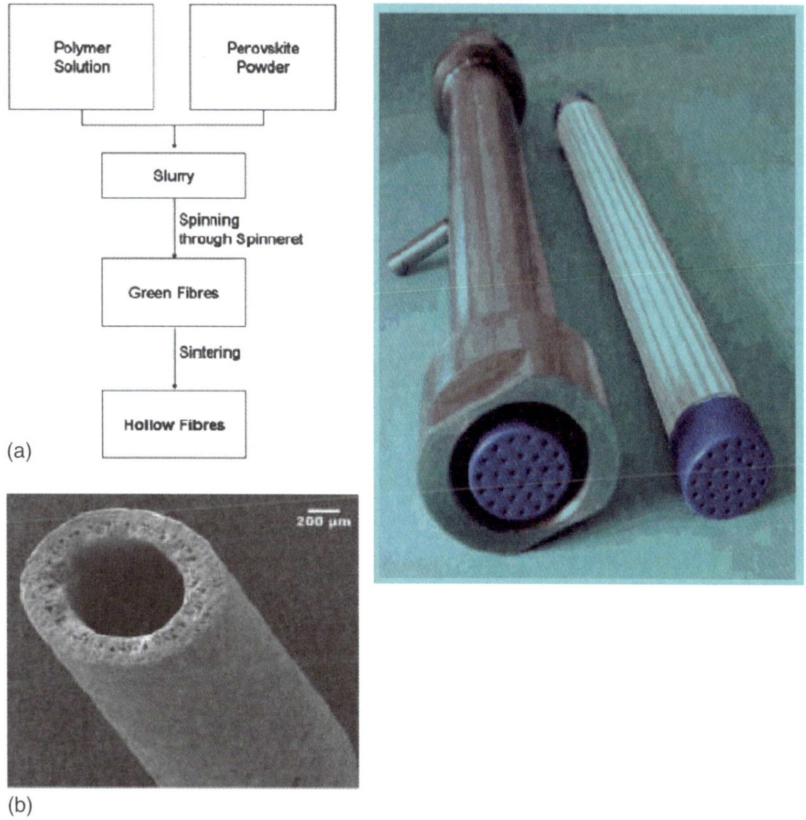

Figure 16.23 (Left) Processing scheme for production of (left, a) hollow fibers (left, b). (Right) Pilot-scale membrane reactor consisting of perovskite membrane and cooling bracket.

of methane oxidation investigated in this study were challenged by applying gold paste on the ends of the reactor and using gas-tight alumina tubes as jacket material. The gold paste then diffused into the slits at elevated temperatures providing sufficient sealing.

The membrane reactor can be applied as pore-through-flow units in which the membrane is utilized as a catalytically functionalized support material with a defined pore structure providing distinct residence times and minimum mass-transport limitations. In a different operation mode, used for the partial oxidation of methane to synthesis gas, the membrane acts as a permselective reactor, selectively separating oxygen from air.[50] Even though perovskite hollow-fiber membranes used for POM already meet the economic requirements of membrane area per cubic meter of permeator volume, industrial application of this reactor type is still prevented by a mismatch of thermal expansion of perovskite membrane, potting material and reactor housing, as well as solid-state diffusion leading to destabilization effects of the membrane structure.

References

1. J. R. Backhurst and J. H. Harker, *Process Plant Design*, Heinemann, London, 1973.
2. L. T. Biegler, I. E. Grossmann and A. W. Westerberg, *Systematic Methods of Chemical Process Design*, Prentice-Hall, Upper Saddle River, NJ, 1997.
3. G. Ertl, *Handbook of Heterogeneous Catalysis*, Wiley-VCH, Weinheim, 2008, Vol. 2.
4. A. Dinse, B. Frank, C. Hess, D. Habel and R. Schomäcker, *J. Mol. Catal. A: Chem.*, 2008, **289**, 28.
5. E. P. S. Schouten, P. C. Borman and K. R. Westerterp, *Chem. Eng. Process.*, 1996, **35**, 107.
6. W. M. H. Sachtler, C. Backx and R. A. Vansanten, *Cat. Rev. – Sci. Eng.*, 1981, **23**, 127.
7. H. S. Fogler, *Elements of Chemical Reaction Engineering*, Pearson, Upper SaddleRiver, NJ, 2006.
8. O. Levenspiel, *Chemical Reaction Engineering*, 3rd edn, Wiley, New York, 1999.
9. P. C. Borman and K. R. Westerterp, *Ind. Eng. Chem. Res.*, 1995, **34**, 49.
10. B. Frank, A. Dinse, O. Ovsitser, E. V. Kondratenko and R. Schomäcker, *Appl. Catal. A*, 2007, **323**, 66.
11. M. F. Lemanski, C. Paparizos, P. R. Blum, M. L. Cirjak, P. M.A., Process for the preparation of vinyl acetate catalyst, The Standard Oil Company (Cleveland, O., ed. USA, 1995, **5**, 665, 667).
12. EO-EG in *Chem. Week*, 2008, **170**, 1.
13. Ethylene Oxide in *ICIS Chemical Business*, 2005, **273**, 2.

14. *Ullmann's Encyclopedia of Industrial Chemistry*, 7th edn, Wiley-VCH, Weinheim, 2005.
15. S. Linic and M. A. Barteau, *J. Am. Chem. Soc.*, 2003, **125**, 4034.
16. S. A. Tan, R. B. Grant and R. M. Lambert, *J. Catal.*, 1986, **100**, 383.
17. M. Atkins, J. Couves, M. Hague, B. H. Sakakini and K. C. Waugh, *J. Catal.*, 2005, **235**, 103.
18. I. Kiguchi, T. Kumazawa and T. Nakai, *Hydrocarbon Process.*, 1976, **55**, 69.
19. E. Burridge, *Ethylene Oxide in Eur. Chem. News*, 2005, **83**, 1.
20. E. Burridge, *Propylene in Eur. Chem. News*, 2004, **80**, 1.
21. N. Boisdron, A. Monnier, L. Jalowieckiduhamel and Y. Barbaux, *J. Chem. Soc. Faraday Trans.*, 1995, **91**, 2899.
22. F. Cavani, N. Ballarini and A. Cericola, *Catal. Today*, 2007, **127**, 113.
23. M. A. Chaar, D. Patel and H. H. Kung, *J. Catal.*, 1988, **109**, 463.
24. R. Grabowski, S. Pietrzyk, J. Sloczynski, F. Genser, K. Wcislo and B. Grzybowska-Swierkosz, *Appl. Catal. A*, 2002, **232**, 277.
25. B. M. Weckhuysen and D. E. Keller, *Catal. Today*, 2003, **78**, 25.
26. H. J. Arpe, *Industrial Organic Chemistry*, 6th edn, Wiley-VCH, Weinheim, 2007.
27. E. Burridge, *Formaldehyde in Eur. Chem. News*, 2005, **82**, 1.
28. A. R. Chauvel, P. R. Courty, R. Maux and C. Petitpas, *Hydrocarbon Process*, 1973, **52**, 179.
29. K. R. Bedell and H. A. Rainbird, *Hydrocarbon Process*, 1972, **51**, 141.
30. VAM in *Chem. Week*, 2007, **169**, 1.
31. M. Kirschner, Sulfuric Acid (Virgin). In *Chemical Market Reporter*, 2005, **268**, 1.
32. A. Urbanek and M. Trela, *Cat. Rev.-Sci. Eng.*, 1980, **21**, 73.
33. C. Hess, U. Wild and R. Schlogl, *Microporous Mesoporous Mater.*, 2006, **95**, 339.
34. E. V. Kondratenko, M. Cherian, M. Baerns, D. S. Su, R. Schlögl, X. Wang and I. E. Wachs, *J. Catal.*, 2005, **234**, 131.
35. Ammonia, *Chem. Week*, 2007, **168**, 1.
36. Methanol, *Chem. Week*, 2007, **169**, 1.
37. www.icis.com. ICIS Chemical Business, 2009.
38. Phthalic Anhydride, *Chem. Week*, 2005, **167**, 1.
39. H. G. Franck, J. W. Stadelhofer, *Industrial Aromatic Chemistry*, Springer, Heidelberg, 1988.
40. Maleic Anhydride, *Chem. Week*, 2009, **171**, 1.
41. J. M. Commenge, L. Falk, J. P. Corriou and M. Matlosz, *Chem. Eng. Technol.*, 2005, **28**, 446.
42. W. Ehrfeld, V. Hessel and H. Löwe, *Microreactors*, Wiley-VCH, Weinheim, 2000.
43. O. Schwarz, P. Q. Duong, G. Schaefer and R. Schomacker, *Chem. Eng. J.*, 2009, **145**, 429.
44. O. Schwarz, P. Q. Duong, G. Schaefer and R. Schomacker, *Chem. Eng. J.*, 2009, **145**, 420.

45. O. Muraza, E. V. Rebrov, T. Khirnyak, B. F. G. Johnson, P. J. Kooyman, U. Lafont, M.H.J.M. De Croon and J. C. Schouten, *Chem. Eng. J.*, 2008, **135**, S99.
46. P. Pfeifer, K. Schubert, M. A. Liauw and G. Emig, *Appl. Catal. A*, 2004, **270**, 165.
47. G. Wiessmeier and D. Honicke, *J. Micromech. Microeng.*, 1996, **6**, 285.
48. T. Waku, J. A. Biscardi and E. Iglesia, *J. Catal.*, 2004, **222**, 481.
49. A. Schmidt, A. Wolf, R. Warsitz, R. Dittmeyer, D. Urbanczyk, I. Voigt, G. Fischer and R. Schomacker, *AlChE J.*, 2008, **54**, 258.
50. J. Caro, K. J. Caspary, C. Hamel, B. Hoting, P. Kolsch, B. Langanke, K. Nassauer, T. Schiestel, A. Schmidt, R. Schomacker, A. Seidel-Morgenstern, E. Tsotsas, I. Voigt, H. H. Wang, R. Warsitz, S. Werth and A. Wolf, *Ind. Eng. Chem. Res.*, 2007, **46**, 2286.
51. P. N. Dyer, R. E. Richards, S. L. Russek and D. M. Taylor, *Solid State Ion.*, 2000, **134**, 21.
52. H. H. Wang, Y. Cong and W. S. Yang, *J. Membr. Sci.*, 2002, **210**, 259.

CHAPTER 17
Outlook

CHRISTIAN HESS

Technische Universität Darmstadt, Eduard-Zintl-Institut für Anorganische und Physikalische Chemie, Petersenstr. 20, 64287 Darmstadt, Germany

Based on the previous discussion a selective oxidation catalyst can be pictured as a highly dynamical material that, starting from the precursor material, develops into its active state under the influence of the gas-phase reactants. This has important consequences for future strategies of rational catalyst development. As the precatalyst structure does not necessarily have a relation to the surface active phase structure meaningful structure–reactivity relationships can only be obtained under the conditions of catalyst operation by the use of suitable *in-situ* analysis. Such an approach certainly represents the ultimate goal in catalyst design as it allows for the direct identification of active-site structures. Table 17.1 lists a selection of *in-situ* methods commonly used or with future potential for characterization of real selective oxidation catalysts together with each method's most important information contents, conditions of use and typical time resolution.

In general, supported systems do not require a discrimination of the surface signal from the bulk contribution, in contrast to bulk systems. Inspection of Table 17.1 reveals that based on the obtained information *e.g.* synchrotron-based X-ray photoelectron spectroscopy (XPS) and X-ray absorption spectroscopy (XAS) are powerful methods for the characterization of selective oxidation catalysts. XAS is surface sensitive when applied in electron yield mode in the low-energy range, XPS can be used for depth profiling when the

RSC Nanoscience & Nanotechnology No. 19
Nanostructured Catalysts: Selective Oxidations
Edited by Christian Hess and Robert Schlögl

Published by the Royal Society of Chemistry, www.rsc.org

Table 17.1 Selection of *in-situ* methods for characterization of real selective oxidation catalysts. Realistic conditions: atmospheric pressure or higher, elevated temperatures (500 °C).

Method	*Information content*	*Conditions of use/time resolution*
Diffuse reflectance UV-Vis spectroscopy	Electronic (e.g. charge-transfer) transitions	Realistic conditions/ ~1 s
Electron paramagnetic resonance (EPR)	Paramagnetic transition-metal ions	Realistic conditions/ ~5 min
Fourier-transform infrared spectroscopy (FTIR)	Vibrational spectra of metal oxides, adsorbates and gas phase	Realistic conditions/ ~1 s
Nuclear magnetic resonance (NMR)	Identification of organic molecules (chemical shift)	Realistic conditions/ ~2 h
(Resonant) Raman spectroscopy, surface-enhanced Raman scattering (SERS)	Vibrational spectra of metal oxides, adsorbates and deposits (coke), surface specific (SERS)	Realistic conditions/ ~5 min
Sum-frequency generation (SFG)	Vibrational spectra of adsorbates, surface specific	Realistic conditions, real catalysts feasible/ ~5 min
Transmission electron microscopy (TEM)	Microstructure, particle size and shape, lattice structure, chemical composition	Realistic conditions feasible, so far typical pressure ~1 mbar/ ~1 min
X-ray absorption spectroscopy (XAS)	Short-range structure, coordination and oxidation state	Realistic conditions/ ~1 s
X-ray diffraction (XRD)	Long-range structure	Realistic conditions/ ~1 min
X-ray photoelectron spectroscopy (XPS), electron spectroscopy for chemical analysis (ESCA)	Composition, chemical state, oxidation state(s), surface sensitive	High-pressure XPS, synchrotron: ~1 mbar/ ~1 s

exciting photon energy is varied. However, their use is limited to reduced pressures (~mbar), which for selective oxidation catalysts may have a significant influence on phase diagrams. On the other hand, optical methods with inherent surface specificity (*e.g.* SFG) have only recently been applied to powder samples. In principle, surface-enhanced Raman scattering (SERS) is a surface-specific technique that can be used to probe real catalysts under realistic conditions. However, its use has so far been limited to studies on coin metals (Cu, Ag, Au).

This short discussion already shows that there is a desperate need for the more extensive application and development of *in-situ* methods, which can selectively probe the surface region of bulk systems and at the same time allow for experiments at realistic conditions. A further important aspect regarding future development will include the combination of *in-situ* spectroscopy and microscopy to monitor the catalyst dynamics spatially resolved in real time.

The controlled manufacture of selective oxidation catalysts will be a formidable future task. This is due to the fact that the active sites are not prepared during synthesis of the catalyst precursor but are formed by reaction with the gas-phase reactants. The chemical potential of the gas phase crucially depends on parameters such as the partial pressures of reactants or the presence of water and is difficult to determine with catalytic processes. To this end, important clues may be expected from high-throughput experiments designed to systematically elucidate the influence of the gas-phase reactants parameter space on catalyst performance. It is of particular importance to clarify the role of water, which is omnipresent in selective oxidation reactions. On the other hand, very little is known of its adsorption behavior and/or reaction with the catalyst surface under realistic conditions.

Besides the identification of active-site structures using *in-situ* analysis, predictions on candidates for active sites may also be provided by *ab initio* theory. Finally, it will be the tremendous task of synthetic chemists to find strategies to assemble these sites into well-defined (single phase) nanostructured inorganic polymers with high surface area. *In-situ* analysis as well as high-throughput instrumentation will be indispensable tools for systematic investigations on catalyst preparation parameters. The improvement of selective oxidation catalysts will rely on a highly interdisciplinary approach based on a combination of physical chemistry, chemical engineering, inorganic chemistry and material science.

Subject Index

References to figures are given in *italic* type. References to tables are given in **bold** type.